Progress in Mathematics
Volume 184

New Horizons in pro-*p* Groups

Marcus du Sautoy
Dan Segal
Aner Shalev
Editors

Springer Science+Business Media, LLC

Marcus du Sautoy
DPMMS
Centre for Mathematical Sciences
Cambridge CB3 0WB
United Kingdom

Dan Segal
All Souls College
Oxford OX1 4AL
United Kingdom

Aner Shalev
Institute of Mathematics
Hebrew University
Jerusalem 91904
Israel

Library of Congress Cataloging-in-Publication Data

New horizons in pro-p groups / Marcus du Sautoy, Dan Segal, Aner Shalev, editors.
p. cm. — (Progress in mathematics ; v. 184)
Includes bibliographical references and index.
ISBN 978-1-4612-7122-2 ISBN 978-1-4612-1380-2 (eBook)
DOI 10.1007/978-1-4612-1380-2
1. Nilpotent groups. 2. p-adic groups. I. du Sautoy, Marcus. II. Segal, Daniel, Ph.D. III. Shalev, Aner, 1958- IV. Progress in mathematics (Boston, Mass.) ; vol. 184.
QA177.N48 2000
00-023027
CIP

AMS Subject Classifications: 11R23, 11S15, 11S20, 20D15, 20E6, 20E7, 20E8, 20E18, 20F05, 20F10, 20F40, 20F50, 20J05, 20M35, 22E20

Printed on acid-free paper.

Originally published by Birkhäuser Boston in 2000
Softcover reprint of the hardcover 1st edition 2000

ISBN 978-1-4612-7122-2 SPIN 10757748

Reformatted from editors' files by TEXniques, Inc., Cambridge, MA.

9 8 7 6 5 4 3 2 1

Contents

Introduction

A pro-p group is the inverse limit of some system of finite p-groups, that is, of groups of prime-power order where the prime — conventionally denoted p — is fixed. Thus from one point of view, to study a pro-p group is the same as studying an infinite family of finite groups; but a pro-p group is also a compact topological group, and the compactness works its usual magic to bring 'infinite' problems down to manageable proportions.

The p-adic integers appeared about a century ago, but the systematic study of pro-p groups in general is a fairly recent development. Although much has been discovered, many avenues remain to be explored; the purpose of this book is to present a coherent account of the considerable achievements of the last several years, and to point the way forward. Thus our aim is both to stimulate research and to provide the comprehensive background on which that research must be based. The chapters cover a wide range. In order to ensure the most authoritative account, we have arranged for each chapter to be written by a leading contributor (or contributors) to the topic in question.

Pro-p groups appear in several different, though sometimes overlapping, contexts. The first main impetus towards their study came from number theory, specifically in the context of Galois cohomology, and the first discussion in book form of pro-p groups was in Serre's influential *Cohomologie galoisienne* [6]. A burst of activity in the 1960s produced substantial results on the cohomology of pro-p groups, and the classification of Demushkin groups. A self-contained account of some of this theory is given in Wilson's book [7]. The Demushkin groups are 'large', being rather close to free pro-p groups. The other major achievement of the 1960s was Lazard's theory of p-adic analytic groups [3]; these are rather 'small' pro-p groups, having *finite rank* in the sense of profinite group theory. The group theoretic aspect of Lazard's work was taken up and extended in the 1980s by Lubotzky and Mann [5], who brought into prominence the concept of *powerful* pro-p groups; this approach and developments arising out of it are expounded in detail in the book *Analytic pro-p groups* [1]. In a sense, the present book is a sequel to that one: it aims to cover those aspects of pro-p group theory that lay outside the scope of [1].

The impetus for current research on pro-p groups could be broadly described as coming from four directions. The first is *number theory*, which continues to be a source of deep and challenging problems. Chapter 10 gives an account of the higher ramification theory of local fields and the corresponding pro-p Galois groups of totally ramified extensions. Chapter 11, on the other hand, explores exciting developments in the theory of p-adic Galois representations in the context of algebraic number fields, recently brought into prominence by Wiles's proof of Fermat's Last Theorem.

The second source is the problem of classifying *finite p-groups*. Traditionally considered a hopeless undertaking, this problem acquired an entirely different aspect in 1980 thanks to the brilliant insight of Leedham-Green and Newman [4], who proposed classifying p-groups by *coclass*, and conjectured (1) that almost all finite p-groups of a given coclass r can be obtained in a natural way from infinite pro-p groups of coclass r, and (2) that infinite pro-p groups of finite coclass r are *p-adic space groups*, and moreover, that there are only *finitely many* of them for each r. With the subsequent successful proof of the coclass conjectures, the pro-p method has come to dominate this part of finite group theory. Attention has now turned to the problem of classifying p-groups by *width*, a broader concept than coclass; this project is still in its early stages and promises to be extremely challenging. It is explored in Chapter 2; some special aspects are also discussed in Section 6 of Chapter 1. New results on groups of fixed coclass are discussed in Section 7 of Chapter 9 (for a self-contained account of the 'pro-p' part of the coclass theory, see Chapter 10 of [1]).

The third source is the theory of infinite groups: the formation of profinite, and in particular pro-p, completions is a natural procedure that casts certain kinds of group-theoretic problem in a new light, and can lead to significant insights — a paradigm is 'Lubotzky's linearity criterion' [3] (see [1], Interlude B). Sections 5–7 of Chapter 3 apply this method to the study of subgroups in free groups, while Section 1 of Chapter 9 illustrates the approach in the context of subgroup growth.

Finally, there is the 'umbrella' subject of 'profinite group theory': the whole range of questions that arise when one begins to investigate profinite groups as groups of interest in their own right. The theory of pro-p groups is only a part of this, but a particularly well developed part. Of course, all the above-mentioned topics also fit under this umbrella, in the sense that problems originating elsewhere are tackled by turning them into problems about pro-p groups; and whatever the original motivation, most of the chapters focus most of the time on the structure, properties and classification of pro-p groups.

The crucial requirement for the healthy development of a subject still in its infancy is to have lots of examples: only by studying these can we hope to eventually build up a reliable picture of the universe of pro-p groups. A large part of this book is devoted to the construction of interesting families of groups, mainly on the 'large' side; the 'small', that is the p-adic analytic, pro-p groups being already well served by existing literature: [1] for their general properties, [2] for the construction of many examples. In particular, Chapter 3 develops the theory of *groups acting on pro-p trees*, analogous to the Bass–Serre theory of groups acting on trees; here we find the free pro-p groups and more generally pro-p groups obtained by 'free constructions', such as free pro-p products. Going 'one step down in size', Chapter 4 describes the construction of *branch groups*, commonly known as groups of 'Grigorchuk type'; these are also groups of tree automorphisms, with many interesting properties. They tend to be fairly 'small from above', some of them for example having finite width, but 'large from below', containing all countably based pro-p groups as subgroups. Another group having these two properties is the *Nottingham group* $\mathcal{N}(\mathbb{F}_p)$: the group of normalised automorphisms of the power series ring $\mathbb{F}_p[[t]]$. From the point of view of classification, $\mathcal{N}(\mathbb{F}_p)$ appears in many ways to be the 'smallest' of the non-soluble, non-p-adic analytic pro-p groups; it also arises naturally as a sort of universal Galois group in the theory of totally ramified extensions of local fields. Chapter 6 gives the construction and elementary properties of this group, while Chapter 10 explores its role in higher ramification theory of local fields, as well as giving two proofs of Camina's theorem that $\mathcal{N}(\mathbb{F}_p)$ contains all countably based pro-p groups.

Having found, or constructed, a healthy spread of types of pro-p group, what we also need are some general principles of classification. This issue is discussed in Chapter 2; but many results in the other chapters may also be construed as 'classification principles'. Let us mention here two specific ones. The Nottingham group, and many of the branch pro-p groups, have the property of being *just infinite*: they are infinite but have no proper infinite quotients. In an obvious sense, the just infinite pro-p groups are the basic building blocks for all pro-p groups, and an understanding of these must be a prerequisite for understanding the rest. Chapter 5 examines just infinite groups in general, and establishes a fundamental dichotomy: the profinite just infinite groups fall into two disjoint classes, the *branch groups* and the *hereditarily just infinite* groups. This is also discussed in Section 6 of Chapter 4. Next, a pro-p group is said to have a 'small set of relations' if it has a presentation by generators and relations that violates the *Golod–Shafarevich inequality*. The deep theorem of Zelmanov asserts that every finitely generated pro-p group having a small set of relations contains *a non-abelian free pro-p subgroup*. This is proved for the first time in Chapter 7.

Several chapters concentrate on particular methodological approaches to pro-p groups. Chapter 1 is a comprehensive survey of the role of Lie algebras, and the related concept of analytic groups, in the theory. Here both structural and classification issues are discussed; indeed, this chapter also serves as an introduction to many of the problem areas in pro-p group theory currently receiving attention. As mentioned above, Chapter 3 explains the use of *pro-p trees*, in what might be called the geometrical approach to combinatorial pro-p group theory. In Chapter 9 the concept of *integration* on p-adic analytic pro-p groups is explained, and applied to the study of certain Poincaré series associated to these groups; this has applications both to questions of subgroup growth and to the classification of finite p-groups.

The specific question of subgroup growth in pro-p groups is discussed in Chapter 8, where among others probabilistic methods are introduced.

Two chapters are devoted primarily to number-theoretic matters. Chapter 10 is concerned with totally ramified extensions of local fields, both in characteristic zero and in characteristic p; a key issue is the recognition in group-theoretic terms of the arithmetically defined higher ramification groups inside a pro-p Galois group. A self-contained proof of Sen's theorem on p-adic analytic Galois extensions illustrates the subtle interplay between arithmetic and group-theoretic information; and an application of the field-theoretic construction of the 'field of norms' is made to give a proof of Camina's theorem on the Nottingham group, mentioned above.

Chapter 11, on the other hand, gives a survey of applications of pro-p group theory in algebraic number fields; here, number-theoretic problems, such as the Fontaine–Mazur conjecture, give rise to interesting new questions about pro-p groups. These are mostly still wide open, and present a lively challenge to the algebraic community.

Finally, Chapter 12 gives a unified treatment of Lazard's theorems on the cohomology of p-adic analytic pro-p groups. This chapter is *not* intended as an introduction to the topic (such as may be found in Chapters 9–11 of Wilson's book [7]). It presents a new approach based on the introduction of a suitable module category and the systematic use of spectral sequences, and should serve as the standard reference for further research in this area.

A large number of *open problems* are presented throughout the text; a selection of these, with additional problems suggested by many of the authors, are collected together in the Appendix.

Acknowledgements

This book would not exist but for the dedication and effort of the individual chapter authors, who all responded magnificently to our request for a contribution. Several of the authors also contributed significantly during the process of mutual reviewing of chapters, as did a small number of conscientious referees who remain anonymous.

The arduous task of translating chapters written in various dialects of TEX into a uniform format was skillfully undertaken by Aaron Wilson. Chapter 4 was translated from the original Russian by Olga Tabachnikova and Geoff Smith, who also created the numerous pictures in that chapter. We thank all three for their indispensable contribution.

We have greatly appreciated the encouragement and practical help of Ann Kostant of Birkhäuser Boston, and the patience and skill of Elizabeth Loew of TEXniques, Inc., who made an excellent job of the final layout and production while accommodating the last-minute whims of a gang of authors and editors.

Prerequisites and Notation

As well as notation defined within the individual chapters, the following will be used as standard:

$\mathbb{N}$, $\mathbb{N}_0$, $\mathbb{Z}$, $\mathbb{Q}$, $\mathbb{Z}_p$, $\mathbb{Q}_p$, $\mathbb{F}_q$ for the positive integers, non-negative integers, integers, rationals, p-adic integers, p-adic numbers, and the field with q elements, respectively.

$[x]$, $\lceil x \rceil$: greatest integer $\leq x$, least integer $\geq x$.

$H \leq G$, $H \leq_c G$, $H \leq_o G$: H is a subgroup, respectively closed subgroup, open subgroup of G.

$H \lhd G$, $H \lhd_c G$, $H \lhd_o G$: H is a normal subgroup, respectively closed normal subgroup, open normal subgroup of G.

$\overline{H}$: the closure of H.

$\Phi(G)$: Frattini subgroup of G.

G', $\gamma_n(G)$, $Z(G)$: derived group, nth term of lower central series, centre of G, respectively.

$x^y = y^{-1}xy$, $[x, y] = x^{-1}x^y$

$\langle S \rangle$: group generated by the set S

$[S, T] = \langle \{[x, y] \mid x \in S,\ y \in T\} \rangle$

$P_n(G) = \overline{P_{n-1}(G)^p[P_{n-1}(G), G]}$ is the nth term of the lower p-series (or 'lower p-central series') of the pro-p group $G = P_1(G)$. In particular $P_2(G) = \Phi(G)$ if G is a pro-p group.

$\mathrm{d}(G)$: the minimal cardinality of a topological generating set for G.

A profinite group G has a property $\mathcal{P}$ *virtually* if G has an open normal subgroup with property $\mathcal{P}$.

The *Nottingham group* (see Chapter 6): for a finite field $F = \mathbb{F}_q$, this is the group of normalised automorphisms of the local field $F((t))$ and is denoted $\mathcal{N}(F)$.

A profinite group is said to be *countably based* if there is a countable base for the neighbourhoods of the identity (such groups are sometimes called *separable.*) It is easy to see that every finitely generated profinite group is countably based, and that a countably based group is topologically generated by a countable set. (However, the *converse* of this last statement is not true: see [6] §1.3, Exercise 2 (where 'dual' is a misprint for 'double dual')).

A profinite group is *just infinite* if every proper (continuous) quotient of the group is finite, or equivalently if *every closed normal subgroup is open*. It is *hereditarily just infinite* if every open subgroup (or equivalently, every open normal subgroup) is just infinite.

For basic properties of profinite groups and pro-p groups see [7], Chapters 1–4 and [1], Chapter 1. See also the forthcoming book *Profinite Groups* by Ribes and Zalesskii.

For the definition and properties of free pro-p groups and free profinite groups, see [7], Chapter 5 and Ribes & Zalesskii, *op. cit.*

For pro-p groups of finite rank and p-adic analytic groups, see [1]. We recall here some definitions and results from there that will crop up several times.

1. A profinite group G has *rank* $r = \mathrm{rk}(G)$ if every closed subgroup of G can be generated (topologically) by r elements, and r is minimal with this property (so r is either a positive integer or else ∞). (Here 'closed' can be replaced by 'open'.) *Warning*: the word 'rank' is also used in a different sense in the context of free profinite (or pro-p) groups; tradition insists that a 'free group of rank r' means a group having a free generating set (or *basis*) of cardinality r. Since a non-abelian free profinite (or pro-p) group necessarily has infinite rank in the first sense, it will be clear from the context which usage is in force.

2. A pro-p group G is *powerful* if $G/\overline{G'G^p}$ is abelian ($p \neq 2$), or if $G/\overline{G'G^4}$ is abelian ($p = 2$).

3. A pro-p group G is *uniform* (= 'uniformly powerful') if G is finitely generated and powerful and $|P_n(G) : P_{n+1}(G)| = p^{\mathrm{d}(G)}$ for all $n \in \mathbb{N}$. This holds if and only if G is finitely generated, powerful and torsion-free.

4. **Theorem** (Lubotzky & Mann): *A pro-p group has finite rank if and only if it is finitely generated and virtually powerful.*

5. *A finitely generated powerful pro-p group is virtually uniform.*

6. **Theorem** (Lazard): *A pro-p group is p-adic analytic if and only if it is finitely generated and virtually powerful. Every compact p-adic analytic group is profinite and virtually pro-p.*

Combining (4), (5) and (6), we see that the following are equivalent for a pro-p group G: (a) G is p-adic analytic, (b) G has finite rank, (c) G is finitely generated and virtually powerful, (d) G is finitely generated and virtually uniform. Interlude A of [1] lists 14 further conditions equivalent to these, and six equivalent definitions for the *dimension* $\dim(G)$ of such a group G. The simplest is

$$\dim(G) = \mathrm{d}(H)$$

where H is any uniform open subgroup of G.

References

[1] J. D. Dixon, M. P. F. du Sautoy, A. Mann and D. Segal, *Analytic pro-p Groups, 2nd edition*, Cambridge Studies in Advanced Math. 61, Cambridge Univ. Press, Cambridge, 1999.

[2] G. Klaas, C. R. Leedham-Green and W. Plesken, *Pro-p groups of finite width*, Lecture Notes in Math., **1674**, Springer-Verlag, Berlin, 1997.

[3] M. Lazard, Groupes analytiques p-adiques, *Publ. Math. I.H.E.S.* **26** (1965), 389–603.

[4] C. R. Leedham-Green and M. F. Newman, Space groups and groups of prime-power order I, *Arch Math. (Basel)* **35** (1980), 193–202.

[5] A. Lubotzky and A. Mann, Powerful p-groups, I and II, *J. Algebra* **105** (1987), 484–505 and 506–515.

[6] J-P. Serre, *Galois Cohomology*, Springer-Verlag, Berlin–Heidelberg, 1997.

[7] J. S. Wilson: *Profinite Groups*, OUP, 1998.

1

Lie Methods in the Theory of pro-p Groups

Aner Shalev

1 Introduction

The purpose of this chapter is to describe the use of Lie-theoretic methods in the study of pro-p groups. I shall also discuss briefly some related objects, such as finite p-groups and residually finite groups. Aspects of this topic feature in several books and survey papers; see for instance [62], [45] Chapter VIII, [42], [137], [27], [57], [58], [148], [121]. In this survey I will try to focus on the most recent developments and applications, which are mostly not covered in the sources mentioned above.

The association of Lie algebras with groups originated in the 19th century in the context of Lie groups. The idea of constructing Lie rings from abstract groups was raised by Magnus [85], in the context of the restricted Burnside problem. Since then this construction has proved useful in a variety of applications, including the restricted Burnside problem, the study of some group identities, the theory of fixed points of automorphisms, the coclass theory for p-groups and pro-p groups, the investigation of just-infinite pro-p groups, and the recent study of Hausdorff dimension and the spectrum of pro-p groups. The Lie ring method provides a recipe for translating some group-theoretic questions to Lie-theoretic ones. This translation is straightforward in some cases, and quite sophisticated and non-trivial in others; once this is done, the resulting Lie-theoretic questions have to be tackled directly, with varying degrees of difficulty.

Let me briefly describe the classical Lie ring construction. Let G be a residually nilpotent group. For example, G can be a pro-p group, or a finite nilpotent group, or an infinite torsion-free nilpotent group, etc. Let $\{G_i\}_{i\geq 1}$ be a series of normal subgroups of G such that $G_1 = G$, $\bigcap_{i\geq 1} G_i = 1$, and $[G_i, G_j] \leq G_{i+j}$. Here, if H, K are subgroups of G, we denote by $[H, K]$ the subgroup generated by all commutators $[x, y] = x^{-1}y^{-1}xy$ where $x \in H$ and $y \in K$. Define

$$L(G) = \bigoplus_{i\geq 1} G_i/G_{i+1}.$$

Then multiplication in G induces on $L(G)$ the structure of an abelian group (with additive notation). A Lie product in $L(G)$ is induced by taking commutators in G. We first define it for homogeneous elements by

$$[xG_{i+1}, yG_{j+1}] = x^{-1}y^{-1}xyG_{i+j+1} \quad (x \in G_i, y \in G_j),$$

and then extend it to arbitrary elements using bi-additivity. The resulting object $L(G)$ has the structure of a Lie ring. Clearly, if G is finite, then $|G| = |L(G)|$.

Classically $\{G_i\}$ is taken to be the lower central series of G, and G is assumed to be nilpotent. In this case G and $L(G)$ have the same nilpotency class and the same number of generators. However, if G is a pro-p group we sometimes require that $L(G)$ be a Lie algebra over the field $\mathbb{F}_p$, or even a restricted Lie algebra (which is a Lie algebra with a formal pth power operation; see [48], V.7). We then have to find a filtration $\{G_i\}$ satisfying $G_i^p \leq G_{i+1}$, or $G_i^p \leq G_{pi}$, respectively. One takes $\{G_i\}$ to be the lower central p-series of G (defined by $G_i = [G_{i-1}, G]G_{i-1}^p$) in the first case, and the dimension subgroup series (defined by $G_i = [G_{i-1}, G]G_{\lceil i/p \rceil}^p$) in the second case. Here H^n denotes the (closed, if H is a topological group) subgroup generated by all nth powers in H.

The first major applications of this Lie ring construction occur in Kostrikin's solution of the restricted Burnside problem for groups of prime exponent [61], [62] (following work by Magnus and others), and in Higman's work on nilpotent groups with a regular automorphism of prime order [44]. These two results were extended in various directions, as described in Sections 2 and 3.

While the main object of this chapter is to describe applications of Lie algebras to pro-p groups, it is impossible to ignore another important tool, which can be considered Lie-theoretic as well. By this I mean the theory of p-adic Lie groups and the related theory of powerful p-groups [27]. This theory, developed by Lazard, Lubotzky and Mann, and others, has been instrumental in the solution of several problems concerning pro-p groups (as well as arbitrary finite, and residually finite groups). One can find its traces in the study of subgroup growth, of group identities, of coclass, and even in the modern approach to more classical topics such as fixed points of automorphisms and Burnside-type problems. A typical scheme is to reduce a certain group-theoretic problem to one regarding powerful groups. The latter problem can often be solved using some "linear" features of powerful groups, or by passing to associated Lie rings (classical, or of Lazard type). Indeed, in recent years it has been shown in many examples that the Lie ring method is significantly more powerful when applied to a powerful group.

Recall that a pro-p group G is called powerful if G/G^p (G/G^4 if $p = 2$) is abelian. A pro-p group G has the structure of a p-adic Lie group if and only if it is finitely generated (topologically) and has an open (hence of finite index) powerful subgroup. This is equivalent to G having finite rank, where the rank of G is the supremum over the minimal numbers of generators of closed subgroups. The Lie ring of 'Lazard type' associated to G is then essentially the p-adic Lie algebra of the the p-adic Lie group G (so it is really more classical than the 'classical' Lie ring!). For an extensive account of powerful pro-p groups, p-adic Lie groups and their applications, see [27].

In 1994 I wrote a long survey paper [121] on finite p-groups, to which the reader is referred. This chapter gives me an opportunity to outline developments occurring mainly between 1994 and 1999. Of course, some references to older results are unavoidable, but I will try to cut these to a minimum. Usually older results will be stated in a brief form, but ideas behind proofs of more recent results will be discussed in some detail (hopefully without getting into too many technicalities). Putting emphasis on ideas and methods may make the uniformity of the subject apparent, while the various applications demonstrate its diversity. Though this chapter is mainly a survey, some new results are scattered here and there (Theorem 4.10, and the results of Sections 5.1 and 5.2. Section 5.1 is joint work with Dan Segal.)

2 Burnside-type problems and Zelmanov's work

The deepest applications of Lie methods to pro-p groups are undoubtedly those provided in the past ten years by the work of Efim Zelmanov. Their importance is twofold: first, solving an outstanding classical problem, the restricted Burnside problem; and second, providing machinery for solving a wide range of other, more recent, questions.

2.1 PI Lie algebras and torsion groups

Let me begin this section by quoting the most general Lie-theoretic result of Zelmanov proved in the context of Burnside-type problems. This result was announced in [148], where its proof is sketched. The proofs of major particular cases (which yield the solution to the restricted Burnside problem) appeared in [145], [146], [147]. An element a of a Lie ring L is said to be *ad-nilpotent* if $\mathrm{ad}(a)^n = 0$ for some n. L is said to satisfy a *polynomial identity* (is PI, for short) if there is a non-zero element $f(x_1, \dots, x_k)$ in the free Lie algebra on $x_1, \dots, x_k$ such that $f(a_1, \dots, a_k) = 0$ for all $a_1, \dots, a_k \in L$.

Theorem 2.1. (Zelmanov, 1993) *Let L be a Lie algebra generated by finitely many elements $a_1, \dots, a_d$. Suppose that*

(i) every commutator in the generators a_i is ad-nilpotent, and

(ii) L satisfies a polynomial identity.

Then L is nilpotent.

A similar result follows for restricted Lie algebras, by applying the theorem to the ordinary subalgebra generated by $a_1, \dots, a_d$. This version of the theorem will be used freely below.

This result has many interesting consequences for pro-p groups, which I list below. Many of these consequences have been proved without relying on the full power of Theorem 2.1, but on some particular cases of it. For example, the case where the Lie algebra L satisfies the linearized Engel identity $\sum_{\sigma \in S_n} [y, x_{\sigma(1)}, \dots, x_{\sigma(n)}] = 0$, which is proved in [147], already yields results 2.4, 2.5, and 2.6 below.

Given a pro-p group G, let G_i be its dimension subgroup filtration (see [27], Chapters 11,12), and let $L_p(G) = \bigoplus_{i\geq 1} G_i/G_{i+1}$ be the associated Lie ring. Then $L_p(G)$ is generated as a restricted Lie algebra by G_1/G_2. Note that $L_p(G)$ is sometimes defined as the ordinary subalgebra of $\bigoplus_{i\geq 1} G_i/G_{i+1}$ generated by G_1/G_2 (see [148]), but this does not make a real difference.

We say that G is *infinitesimally PI* if the Lie algebra $L_p(G)$ is PI. Note that many natural pro-p groups have this property. These include p-adic analytic groups (in which $L_p(G)$ is nilpotent), analytic groups over $\mathbb{F}_p[[t]]$ and other commutative pro-p rings (in which $L_p(G)$ is finite-dimensional over some commutative ring; see [84], [27], Chapter 13), as well as the Nottingham group $\mathcal{N}(\mathbb{F}_p)$ (in which $L_p(G)$ is a loop algebra associated with the Witt algebra).

A leading question in the family of Burnside-type problems is as follows: *which residually finite p-groups are locally finite?* Examples due to Golod [39], Grigorchuk, and others, show that finitely generated residually finite p-groups need not be finite. See Chapter 4 in this volume for more details.

The following finiteness criterion will play a major role below.

Corollary 2.2. (Zelmanov, 1993) *Let G be a finitely generated residually finite p-group. If G is infinitesimally PI, then G is finite.*

To prove the corollary, consider the Lie algebra $L_p(G)$. Let $g_1, \dots, g_d$ be generators for G. Then $L_p(G)$ is a restricted Lie algebra over $\mathbb{F}_p$, generated by the images $a_i = g_i G_2$ of the generators. The fact that every element of G has finite order implies that every commutator in the generators a_i is ad-nilpotent. Since $L_p(G)$ is PI by our assumption, Theorem 2.1 (in its restricted version) shows that $L_p(G)$ is nilpotent. By a result of Lazard [68] (see [27], Chapter 11) this implies that the pro-p completion of G is p-adic analytic, hence linear. Thus G is a finitely generated periodic linear group, hence is finite.

Corollary 2.3. (Zelmanov, 1993) *Let G be a finitely generated residually finite p-group. If G satisfies some group identity, then G is finite.*

Indeed, a group identity in G implies that $L_p(G)$ is PI.

We can weaken the assumption of the above corollary, by requiring only that the group G satisfies some pro-p identity. This means that some non-trivial word $w(x_1, \dots, x_k)$ in the free pro-p group on $x_1, \dots, x_k$ always takes the value 1 when we substitute elements of the pro-p completion of G for $x_1, \dots, x_k$. Indeed, it is shown ([148], p. 571) that pro-p groups satisfying a pro-p identity are infinitesimally PI.

Yet another consequence is the main result of [147]:

Corollary 2.4. (Zelmanov, 1992) *A torsion pro-p group is locally finite.*

To outline the proof, let me first introduce the notion of a *coset identity*. We say that a pro-p group G satisfies a coset identity if there is an open subgroup H of G and elements $g_1, \dots, g_k \in G$ and a non-identity element $w(x_1, \dots, x_k)$ in the free group on $x_1, \dots, x_k$ such that $w(g_1h_1, \dots, g_kh_k) = 1$ for all $h_1, \dots, h_k \in H$. Now suppose that G is a finitely generated torsion pro-p group. Then a standard Baire category argument shows that some coset gH of an open subgroup H of G satisfies the identity $x^{p^k} = 1$ for some k. By a theorem of Wilson and Zelmanov [141], Theorem 1, a coset identity in G, implies a polynomial identity in the Lie algebra $L_p(G)$. We see that our group G is infinitesimally PI. Applying Corollary 2.2 we deduce that G is finite.

A particular case of Corollary 2.4 yields a positive solution to the restricted Burnside problem.

Corollary 2.5. (Zelmanov, 1991-2) *A finitely generated pro-p group of finite exponent is finite. In other words, there is a finite upper bound to the orders of finite d-generated groups of exponent p^k.*

The existence of a bound follows by applying the first assertion to the relatively free pro-p group on d generators of exponent p^k. Assuming the classification of finite simple groups, and relying on the Hall–Higman reduction theorems [41], a similar result follows for every exponent n.

2.2 Other identities

Let us now turn to problems which do not involve groups of bounded exponent. Recall that a group G is said to be *Engel* if for every $x, y \in G$ there exists a positive integer n (possibly depending on x, y) such that $1 = [x,_n y] := [x, y, \dots, y]$ where y occurs n times. If this holds with an integer n which is independent of x, y we say that G is n-Engel.

Clearly, nilpotent groups have this property; it is still unknown whether finitely generated n-Engel groups are always nilpotent. Zelmanov [144] and Wilson [140] showed that this is the case if the group G is residually finite. However, for pro-p groups a stronger result holds.

Corollary 2.6. (Wilson–Zelmanov, 1992) *Engel pro-p groups are locally nilpotent.*

The idea behind the proof is as follows. Let G be a finitely generated Engel pro-p group. First one shows, using a Baire category argument, that G satisfies some coset identity (of the type $[x,_n y] = 1$ for all $x \in aH$ and $y \in bH$, where H is some open subgroup). It follows that the Lie algebra $L_p(G)$ is PI. Moreover, the fact that G is Engel can be used to show that all commutators in the generators of $L_p(G)$ (in fact all homogeneous elements of $L_p(G)$) are ad-nilpotent. Thus $L_p(G)$ is nilpotent, and so G is p-adic analytic, hence linear. By a result of Gruenberg [40], G must be nilpotent.

Corollary 2.6 is valid for larger classes of topological groups. It holds for all profinite groups, as shown in [141], and for all compact groups, as shown by Medvedev in [98].

More precise information is now available on the structure of n-Engel groups, which is particularly interesting in the non finitely generated case.

Theorem 2.7. (Burns–Medvedev, 1998) *Let G be a residually finite n-Engel group. Then G has a nilpotent normal subgroup N of class at most $c(n)$ such that G/N has finite exponent dividing $e(n)$, where c, e are suitable functions of n only.*

See [20]. It is also shown there that the groups above are an extension of a group of bounded exponent by a nilpotent group of bounded class.

Curiously, the conclusion of Theorem 2.7 also holds when one replaces the Engel identity by other types of group identities. Recall that a positive law (or a semigroup law) is a law of the form $u = v$, where u, v are distinct words in some variables $x_1, \ldots, x_k$ which do not involve inverses. For example, $x^n = 1$ and $xy^2x = yx^2y$ are positive laws. In fact any nilpotent group, and hence any extension of a nilpotent group by a group of finite exponent, satisfies a positive law. The following partial converse is proved in [19].

Theorem 2.8. (Burns-Macedońska-Medvedev, 1997) *Let G be a residually finite group satisfying a positive law. Then G has a nilpotent normal subgroup N such that G/N has finite exponent.*

Again, the class of N, and the exponent of G/N, are bounded in terms of the positive law satisfied by G. In fact results 2.7 and 2.8 hold not only for residually finite groups, but for larger classes of groups. Theorem 2.8 extends a result from [116], where the normal subgroup N is only shown to be locally nilpotent.

The proof of results 2.7 and 2.8 combine tools from the theory of Lie algebras, as well as from the theory of powerful p-groups. The main new ingredient (which is not used in [116]) is a theorem of Zelmanov on the global nilpotency of n-Engel Lie algebras in characteristic zero [143]. This theorem implies that any n-Engel Lie ring L satisfies $eL^c = 0$ where e, c depend only on n (here L^c is the cth term in the lower central series of L). This can be regarded as the Lie ring form of Theorem 2.7. The deduction of the group-theoretic form (as well as of 2.8) uses a reduction to powerful p-groups and an exploitation of the 'linear' properties of such groups. As we shall see in several more examples, powerful p-groups seem to make the translation of Lie-theoretic results to group-theoretic ones somewhat smoother than it used to be.

There are several variations on and extensions of Theorem 2.8. A group G is termed *n-collapsing* if for any subset $S \subseteq G$ of size n, the set S^n of all products of length n of elements of S has cardinality less than n^n. Macedońska has recently shown that every n-collapsing residually finite group is an extension of a nilpotent group of class at most $c(n)$ by a group of exponent dividing $e(n)$. This strengthens results from [116].

2.3 Probabilistic identities

Various other Burnside-type results have been proved recently using Lie methods. For example, Shumyatsky has shown that if G is a residually finite group satisfying the identity $[x, y]^{p^k} = 1$, then the derived subgroup G' is locally finite [132]. There is also some current interest in probabilistic identities, namely, in laws which hold with high probability. Mann and Martinez showed that for some $\epsilon(d, n) > 0$ a d-generated finite group in which $x^n = 1$ holds with probability $1 - \epsilon(d, n)$ has exponent dividing n (and hence has d, n-bounded order) [88]. They also obtained results on Lie algebras which are nearly Engel, in the sense that $[x,_n y] = 0$ with high probability [89]. Many problems on probabilistic identities, or on laws satisfied by particular types of elements, are still very much open, and some arise naturally in various contexts. Let me make a few (very risky) conjectures, in decreasing order of strength.

Conjectures:

1. Let G be a finite p-group satisfying some identity $w(x_1, \dots, x_k) = 1$ with probability $\epsilon > 0$. Then G satisfies some identity depending only on w, p and ϵ.

2. Let G be a finitely generated pro-p group satisfying some identity $w(x_1, \dots, x_k) = 1$ with positive probability. Then G satisfies some identity.

3. Let G be a finitely generated pro-p group satisfying some coset-identity. Then G satisfies some identity.

4. Let G be a k-generated pro-p group satisfying a law $w(x_1, \dots, x_k) = 1$ on all generating k-tuples. Then G satisfies some identity.

The evidence supporting these conjectures is so far rather weak. Part 1 (and hence the others too) holds for the identities $[x_1, x_2] = 1$ and $x_1^2 = 1$, as follows from [103] and [87]. Part 2 holds for arbitrary identities if the group G is assumed to be Λ-standard (see [84], [27], Chapter 13), as shown by Levai and Pyber. Indeed, in this case the original identity w must hold in G [78]. Conjecture 3 holds for the coset-identity $x_1^p = 1$, as shown by Khukhro [53]. In this case the resulting pro-p groups G turn out to be virtually nilpotent. It was suggested that finitely generated pro-p groups G satisfying the identity $x^{p^k} = 1$ in some coset gH of an open subgroup H might be soluble [121], Problem 18, but this is still open. The groups G satisfying the conditions of Conjecture 4 with the identity $x_1^p = 1$ turn out to be nilpotent. Indeed, let G be a finite p-group and suppose all elements in $G \setminus \Phi(G)$ have order p. Then the Hughes subgroup $H_p(G)$ generated by the elements g satisfying $g^p \neq 1$ is contained in the Frattini subgroup $\Phi(G)$. Thus $|G/H_p(G)| > p$ and by a theorem of Khukhro [54] the nilpotency class of G is bounded above in terms of p and $d(G)$. This implies the nilpotency of the pro-p groups in **4**.

The above conjectures are related to interesting questions on the existence of free subgroups in some pro-p groups, and on residual properties of free groups. Some of these connections were recently discovered by Levai and Pyber [78].

Proposition 2.9. (Levai–Pyber, 1998) *Let G be a finitely generated pro-p group. Then G satisfies a coset-identity if and only if G does not contain a (nonabelian) dense free subgroup.*

The proof uses a Baire-category argument and additional ideas. With this result, we see that Conjecture 3 implies a positive answer to the following:

Problem 5. Let G be a finitely generated pro-p group. Does it follow that either G has a nonabelian free subgroup, or G satisfies an identity?

The conjectures above also imply a positive solution to the following:

Problems:

6. Let C be a collection of d-generated finite p-groups which is not contained in any proper variety of groups. Does it follow that the free group of rank d is residually C?

7. Let G be a finitely generated pro-p group which has a free subgroup of rank $d > 1$. Does it follow that almost all d-tuples in G generate a free subgroup of rank d?

For a probabilistic approach to residual properties of free groups, see [26]. As for Problem 7, note that it is still unknown whether almost all pairs of elements of the Nottingham group generate an (abstract) free subgroup.

3 Fixed point theorems and invariant theory for Lie algebras

3.1 Invariant theory and grading

Invariant theory often deals with the following situation. Suppose we have an algebraic object A acted on by some finite group H. Let

$$A^H = \{a \in A : a^h = a \text{ for all } h \in H\}$$

be the common fixed points of H on A, namely the sub-object of invariants. What are the relations between the properties of A and that of A^H? For example, under what condition is A^H finitely generated? Or, assuming A^H satisfies some nice properties, what can we deduce regarding the structure of A?

In some cases the action of H on A induces on A an H-grading. This is the case when H is abelian, and A is a (not necessarily associative) algebra over a field F whose characteristic does not divide $|H|$. Let E be a field extension of F containing a primitive root of unity of order $|H|$, and let $H^* = \mathrm{Hom}(H, E^*)$ be the dual group of H. We then have

$$A \otimes E = \sum_{\chi \in H^*} A_\chi,$$

where $A_\chi = \{a \in A : a^h = \chi(h)a \text{ for all } h \in H\}$. Note that $A_\chi A_\psi \leq A_{\chi\psi}$, and $A_1 = A^H$. We are therefore led to the problem of studying graded rings A under suitable assumptions on their unital (or zero, in additive notation) component A_1. In fact reductions to graded rings can sometimes be carried out in more general situations.

3.2 Lie algebras

Our main interest here is in the case where the object A acted on by H is a group (or even a p-group). However, in order to understand this situation it is necessary to start with actions on rings, and on Lie rings in particular. Of special importance is the case when the subring of fixed points is trivial, namely $A^H = 0$ (this applies for Lie rings, and associative rings without unity). In this case we say that the group H acts fixed point freely on A.

In his seminal paper [44], G. Higman proves the following.

Theorem 3.1. (Higman, 1957) *Let L be a Lie ring acted on fixed point freely by a cyclic group H of prime order p. Then L is nilpotent, and its nilpotency class is bounded above by some function $h(p)$ of p alone.*

The core of the proof deals with a C_p-graded Lie ring $L = \oplus_{i=0}^{p-1} L_i$, where $L_0 = 0$. Combinatorial arguments are employed to show that for such a ring we have $L^{h(p)} = 0$.

Higman gave a polynomial lower bound on the so-called Higman function $h(p)$, and a doubly-exponential upper bound was subsequently provided by Kostrikin and Kreknin [66]. More specifically, we have $h(2) = 1$ and for odd p

$$\frac{p^2 - 1}{4} \leq h(p) \leq \frac{(p-1)^{2^{p-1}-1} - 1}{p-2},$$

with equality on the left for $p = 3, 5, 7$. Recently P. Shumyatsky [128] slightly reduced the upper bound above, showing that for $p \geq 11$

$$h(p) < \frac{(p-2)^{2^{p-3}+1}}{(p-3)^2} - 2^{p-3}.$$

It is still an open problem to determine $h(p)$, or even to find reasonable estimates for it.

The next Lie-theoretic result dealt with the case where the acting group G is still cyclic, but not of prime order [65]. In this case examples show that the Lie ring need not be nilpotent. However, we do have

Theorem 3.2. (Kreknin, 1963) *Let L be a Lie ring acted on fixed point freely by a cyclic group H of order n. Then L is soluble, and its derived length is at most $2^n - 2$.*

Again, the core of the proof deals with a C_n-graded Lie ring L satisfying $L_0 = 0$. It is likely that the bound on the derived length is far too big.

What can be said about fixed point free actions of non-cyclic finite groups? We claim that the group $C_2 \times C_2$ acts fixed point freely on the simple Lie algebra $sl_2(F)$, where F is any field of characteristic $\neq 2$. Indeed, consider the maps $i, j : sl_2(F) \to sl_2(F)$ such that i sends a matrix B to $-B^t$, and j sends B to $C^{-1}BC$, where C is the 2-by-2 matrix $e_{11} - e_{22}$. Then i, j are commuting involutory automorphisms of $sl_2(F)$. The fixed points of i are the anti-symmetric matrices with zero diagonal, and the fixed points of j are the diagonal matrices. Therefore the only common fixed point of i and j is the zero matrix.

This example shows that Lie rings admitting fixed point free actions of non-cyclic finite groups need not be soluble. Can we still say something about such Lie rings? For example, do they satisfy some other Lie identity, if not solubility?

In associative algebras A there is a classical answer to this question, even if we weaken our assumption on A^H.

Theorem 3.3. (Kharchenko, 1974) *Let A be an associative algebra over a field F acted on by a finite group H. Suppose the characteristic of F does not divide $|H|$. If the subring of invariants A^H is PI, then so is A.*

See [52], and the book by S. Montgomery [101].

Results in associative algebras can sometimes be extended to Lie algebras. For example, Zelmanov's Theorem 2.1 can be regarded as an analogue of a similar result for associative algebras, which was obtained much earlier by Shirshov and others in the context of the Kurosh–Levitzki problem. However, proving such Lie-theoretic analogues often requires considerable efforts and a whole range of new techniques. It is quite striking that Theorem 3.3 above has just been obtained for Lie algebras, with a rather short proof (based on growth considerations). The case where the acting group G is soluble was established by Bahturin and Zaicev [7] Theorem 2. The extension for arbitrary finite groups was given by Linchenko [80], so we have

Theorem 3.4. (Bahturin–Zaicev–Linchenko, 1998) *Let L be a Lie algebra over a field F acted on by a finite group H, and suppose the characteristic of F does not divide $|H|$. If the subalgebra of invariants L^H is PI, then so is L.*

As we shall see below, this result, when combined with the powerful tools of Zelmanov described in Section 2, opens the way to a variety of new group-theoretic applications.

3.3 Groups: Higman's method

The study of fixed point theorems in (abstract) groups goes back to Burnside at the beginning of the 20th century. Lie methods were applied in this context since the mid-1950s, in several phases. The first phase is based on Higman's Lie ring method, and uses Theorem 3.1 and related results as main tools. The second phase, which starts in the 1990s, applies ideas from the theory of p-adic analytic groups (such as powerful groups and their associated Lie rings) in addition to older tools. In the third and current phase, yet more tools come into play, and they include Zelmanov's Lie-theoretic results, as well as Theorem 3.4 above. For other aspects of this topic, such as automorphisms of free pro-p groups, see Chapter 3.

Let me give some more details. The basic question we now examine is: given a group G acted on by a finite group H, and given some properties of the subgroup $G^H = C_G(H)$ of fixed points, what can we deduce regarding the structure of G?

The basic example of the free group on x, y which admits a fixed point free automorphism of order two switching x, y shows that some extra assumptions on G are needed. Traditionally, one assumes that G is a torsion group. Often, more assumtions are required; for example, that G is finite, or residually finite.

Burnside showed that finite groups admitting fixed point free automorphisms of order p are abelian when $p = 2$ and nilpotent of class at most two if $p = 3$. B.H. Neumann extended these results for torsion groups [102].

Higman applied the graded Lie ring construction described in the introduction to prove the following:

Theorem 3.5. (Higman, 1957) *Let G be a nilpotent group acted on a fixed point freely by a cyclic group H of prime order p. Then the nilpotency class of G is at most $h(p)$.*

To show this, he considered the Lie ring $L(G)$ obtained using the lower central series of G. The group H can be made to act on $L(G)$ in a natural manner, and this action is fixed point free. By Higman's Lie-theoretic result (3.1) we see that the class of $L(G)$ is at most $h(p)$. But G and $L(G)$ have the same class. Hence G has class at most $h(p)$.

There are various extensions of Higman's theorem. First, by a celebrated result of Thompson, finite groups admitting a fixed point free automorphism of prime order are nilpotent [135]. Therefore Theorem 3.5 holds for finite groups. Secondly, the case where the subgroup $C_G(H)$ of fixed points is not trivial but of bounded order, was studied by several people. The most general result obtained is as follows.

Theorem 3.6. (Khukhro–Medvedev, 1994) *Let G be a nilpotent group acted on by a cyclic group H of prime order p. Suppose the subgroup $C_G(H)$ of fixed points is finite of order m. Then G has a nilpotent subgroup G_1 of index at most $f(p,m)$ and nilpotency class at most $g(p)$, where f, g are suitable functions.*

The case where G is a torsion group was obtained by Khukhro in [55], and the general case by Medvedev in [95].

3.4 Groups: p-adic methods

While the machinery of Higman and its variants were powerful enough to obtain strong results on actions of groups of prime orders, the case of automorphisms of composite order remained much less understood. In the 1960s Kovács proved (without using Lie rings) that finite groups admitting a fixed point free automorphism of order 4 are central-by-metabelian. However, the case of automorphisms of larger composite orders remained unapproachable for many years (and in some sense is still unapproachable; see Conjecture 8 below).

Tools and ideas from the theory of p-adic analytic groups changed the situation, and gave rise to new methods and results. The first such result is a fixed point theorem concerning p-groups with an automorphism of p-power order [117].

Theorem 3.7. (Shalev, 1993) *Let G be a finite p-group acted on by a cyclic group H of order p^k, and suppose $|C_G(H)| = p^m$. Then the derived length of G is at most $f(p^k, p^m)$, where f is a suitable function.*

The case where H has order p was established by Alperin [3] in the 1960s, using Higman's methods. The general case implies that the derived length of a finite p-group is bounded in terms of the minimal order of a centralizer $C_G(x)$ where $x \in G$, a result which may not look very intuitive at first sight.

The idea behind the proof of Theorem 3.7 is as follows. First, one observes that the rank of the group G is bounded above in terms of p^k, p^m. Then one concludes, using a result of Lubotzky and Mann [81] (see also [27], Chapter 2), that G has a powerful characteristic subgroup whose index is bounded above in terms of p^k, p^m. This reduces the theorem to the case of powerful p-groups. A further, more technical, reduction allows us to assume that G is a *uniform* p-group; that is, G resembles a uniform pro-p group 'truncated' at some term of its lower p-series. At this stage a p-adic Lie ring construction, analogous to the Lie algebra of a uniform pro-p group [27], Chapter 4, reduces the theorem to the case of Lie rings, which is then solved using Kreknin's theorem.

We point out that one of the difficulties in proving results like Theorem 3.7 using classical methods stems from the fact that the derived length of G need not be bounded in terms of the derived length of the classical graded Lie ring $L(G)$ associated with G. However, if G is assumed to be uniform, then its derived length is controlled by that of suitable Lie rings associated with it. This crucial point enables Lie-theoretic reductions which were not possible before. Using this approach, Khukhro obtained a more detailed structure theorem [56], as follows.

Theorem 3.8. (Khukhro, 1993) *Let G be a finite p-group acted on by a cyclic group H of order p^k, and suppose $|C_G(H)| = p^m$. Then G has a subgroup G_1 of p^k, p^m-bounded index, whose derived length is p^k-bounded.*

A related structure theorem was conjectured by Khukhro and proved by Medvedev [96, 97].

Theorem 3.9. (Medvedev, 1998) *Let G be a finite p-group acted on by a cyclic group H of order p^k, and suppose $|C_G(H)| = p^m$. Then*

(i) G has a subgroup G_1 of p^k, p^m-bounded index, whose derived length is m-bounded.

(ii) if $k = 1$ then G has a subgroup G_1 of p^m-bounded index, whose nilpotency class is m-bounded.

Medvedev's proof starts with reductions to powerful and uniform p-groups, and then uses the p-adic Lie ring construction, as well as new ideas such as p-divided multiplication. Part (ii) of his theorem extends a phenomenon discovered in the theory of p-groups of maximal class. Such groups have an element of order p whose centralizer has order p^2; it was proved that they have a subgroup of class two and p-bounded index. See Section 6.2 for extensions and related results.

In spite of the progress described above, the following plausible conjecture, posed by Khukhro in the Kourovka Notebook, is still open.

Conjecture 8. Let G be a finite p-group acted on fixed point freely by a cyclic group of order n. Then the derived length of G is n-bounded.

Note that in this situation we may assume that n is prime to p. In my paper [118] a positive answer is given provided the p-group in question has bounded rank. Some progress was recently obtained by Y. Brenner and by P. Shumyatsky in the case $n = 8$, but even this special case is still very much open.

3.5 Groups: applying PI theory

The third phase of the Lie-theoretic applications to fixed point theorems in groups involves Zelmanov's Theorem 2.1, as well as the Bahturin–Linchenko–Zaicev Theorem 3.4. For quite some time the area of Lie methods in group theory branched into two separate main streams: Burnside type problems, and problems regarding fixed points of automorphisms. It is only in the past two years or so that these two main branches have begun to interact.

While the results described above concern groups that are assumed to be finite, or nilpotent, let us now turn to possibly infinite residually finite torsion groups. A major question in the field is to provide criteria which force a residually finite torsion group

G to be locally finite. This is the case if G is soluble, or if G has finite exponent (by the solution of the restricted Burnside problem). It has been recently shown that if these conditions are imposed on certain subgroups of fixed points in G then the local finiteness of G still follows [123].

Theorem 3.10. (Shalev, 1998) *Let G be a residually finite torsion group with no 2-torsion. Let H be a finite 2-group acting on G, and suppose that $C_G(H)$ is soluble, or of finite exponent. Then G is locally finite.*

Before sketching the proof of this result, and its relation to p-groups and pro-p groups, let me make some remarks. First, the condition that H is a 2-group is essential here, even if one assumes that $C_G(H)$ is trivial. Indeed, Miller and Obraztsov [99] have just constructed, for each pair of odd primes p, q, an infinite finitely generated residually finite p-group admitting a fixed point free automorphism of order q. Second, the special behaviour of the prime 2 is already seen in a classical result of Shunkov [133], showing that a torsion group which admits an automorphism of order 2 with finitely many fixed points is soluble (hence locally finite). Shumyatsky [129] showed that a residually finite torsion group which admits an automorphism of order 2 whose centralizer is abelian is locally finite; this can be regarded as a particular case of Theorem 3.10.

Let me now describe the ideas behind the proof of Theorem 3.10. Let G, H be as above.

Stage 1. One first observes that if $N \lhd G$ is H-invariant, then $C_{G/N}(H) = C_G(H)N/N$. This result, obtained in [110], is crucial to the rest of the proof. The way to show it is to reduce to the case $|H| = 2$, say $H = \langle\alpha\rangle$, and then to notice that any element $x \in G$ has a unique square root $x^{1/2}$ in the subgroup it generates. Now, if $xN \in C_{G/N}(\alpha)$ then $x^{-1}x^{\alpha} \in N$, and $x = yz$, where $y = x(x^{-1}x^{\alpha})^{1/2} \in C_G(\alpha)$ and $z = (x^{-1}x^{\alpha})^{-1/2} \in N$.

Stage 2. Reduce to the case where G is residually nilpotent.

This is done as follows. First, note that by the Odd Order Theorem every finite quotient of G is soluble; in particular G is residually soluble. Now let $C = C_G(H)$, which is assumed to be soluble (say of derived length l) or of finite exponent (say n). Let $N \lhd G$ be an H-invariant subgroup of finite index. Then $C_{G/N}(H) = CN/N$, which is a finite group of bounded derived length (at most l) or bounded exponent (dividing n). Either way (using the Hall-Higman theory in the latter case) it follows that CN/N has Fitting Height bounded by some number h. A theorem of Thompson [136] on coprime actions of finite soluble groups then shows that the Fitting height of G/N is bounded in terms of $|H|$ and h alone (namely, independently of N). Therefore G is residually-(finite soluble of bounded Fitting height), and it follows that G is poly-(residually nilpotent). One now easily sees that the theorem for residually nilpotent groups G implies the theorem in general.

Stage 3. We may assume that G is residually p for some odd prime p. Hence G is a p-group. Assuming G is finitely generated, we have to show it is finite. By Corollary 2.2 of Zelmanov it suffices to show that G is infinitesimally PI.

Stage 4. Let $L_p(G) = \oplus_{i\geq 1} G_i/G_{i+1}$, and note that the group H acts on $L_p(G)$ by Lie algebra automorphisms. Moreover, it follows from stage 1 that the subalgebra of fixed points is $L_p(G)^H = \oplus(C \cap G_i)G_{i+1}/G_{i+1}$. Since the group C is soluble, or of finite exponent, it follows that $L_p(G)^H$ is PI. Applying Theorem 3.4 of Bahturin–Linchenko–Zaicev, we conclude that $L_p(G)$ is PI, so G is infinitesimally PI.

This completes the proof.

For residually finite p-groups the same arguments yield a somewhat stronger result.

Theorem 3.11. (Shalev, 1998) *Let G be a residually finite p-group ($p > 2$) acted on by a finite 2-group H. Suppose that $C_G(H)$ satisfies some group identity. Then G is locally finite.*

The same holds if $C_G(H)$ (that is, its pro-p completion) is assumed to satisfy a pro-p identity. Theorem 3.11 can be viewed as an extension of Corollary 2.3: instead of requiring that the whole group G satisfies an identity, it suffices to require this for some prescribed subgroup (namely $C_G(H)$).

It would be nice to extend Theorem 3.11 for groups G without 2-torsion. If the identity satisfied by $C_G(H)$ has the property that it bounds the Fitting height of finite soluble groups, then the same argument goes through. This leads to the following question, which has some independent interest.

Problem 9. Find the identities w such that if G is a finite soluble group satisfying $w = 1$ then $h(G) < f(w)$, where $f(w)$ depends on w alone. Does every identity have this property? (Here $h(G)$ denotes the Fitting height of G.)

This calls for an extension of the Hall–Higman theory to families of identities other than $x^n = 1$.

Theorem 3.10 has several consequences. For example, it yields

Corollary 3.12. (Shalev, 1998) *Let G be a residually finite torsion group, and suppose G has a finite 2-subgroup H whose centralizer $C_G(H)$ is finite. Then G is locally finite.*

This extends recent results obtained in [130] and [110] for the cases $H = C_2 \times C_2$ and $H = C_{2^n}$. To prove Corollary 3.12 one first finds an H-invariant finite index subgroup G_1 of G intersecting trivially with $C_G(H)$. Hence H acts on G_1 and $C_{G_1}(H) = 1$. One then deduces that G_1 has no 2-torsion. Theorem 3.10 can now be applied (using the fact that the trivial group is soluble, or of finite exponent!). We conclude that G_1, and therefore G as well, is locally finite.

Shumyatsky has extended Theorem 3.10 to groups H which are not 2-groups, at the price of some strong finiteness conditions on the group G [131]. In particular he shows the following.

Theorem 3.13. (Shumyatsky, 1998) *Let q be a prime, and let G be a residually finite group with no q-torsion such that every set of $q - 1$ elements of G generates a finite subgroup. Suppose that G is acted on by a finite q-group H, and that $C_G(H)$ is soluble or has finite exponent. Then G is locally finite.*

The idea is to show that, under the above assumptions, the property $C_{G/N}(H) = C_G(H)N/N$ still holds, and therefore the same scheme of proof applies. Another result of a similar spirit from [131] is as follows.

Proposition 3.14. (Shumyatsky, 1998) *Let G be a residually finite group in which every pair of elements generates a finite subgroup. Let q be a prime and suppose G has a finite q-subgroup H whose centralizer $C_G(H)$ is finite. Then G is locally finite.*

It is likely that the combination of Zelmanov's results and the invariant theory of Bahturin, Linchenko and Zaicev will have yet more applications to residually finite groups and p-groups in particular. An interesting challenge would be to solve the following.

Problem 10. Let G be a residually finite torsion group with no 2-torsion. Let H be a finite 2-group acting on G, and suppose that $C_G(H)$ is locally finite. Does it follow that G is locally finite?

This would constitute a far-reaching generalization of existing results. It suffices to show that this is the case if $|H| = 2$, as the general case would then follow by induction on $|H|$. This leads naturally to the following Lie-theoretic question.

Problem 11. Let $L = L_0 \oplus L_1$ be a C_2-graded Lie algebra. Suppose that L is generated by finitely many elements $a_1, \ldots, a_d$ such that every commutator in the a_i is ad-nilpotent. Suppose further that L_0 is locally nilpotent. Does it follow that L is nilpotent?

4 Hausdorff dimension and Kac–Moody algebras

4.1 The Hausdorff spectrum of profinite groups

It seems useful, when dealing with pro-p groups, to have a concept of the size of a group, or of the relative size of a subgroup. Of course, in finite groups, the cardinality is a natural measure. And in some types of infinite pro-p groups, such as the p-adic analytic ones, there is a concept of dimension which is commonly used. What we are looking for is some general dimension function, which applies not only to well-behaved groups; it would be nice, though, if, when restricted to well-behaved pro-p groups, this function coincides with some classical dimension function.

Pro-p groups, profinite groups, and compact groups in general, admit an invariant Haar measure, which can be used to measure the size of subsets. However, this measure is not very informative when applied to subgroups. Indeed, all closed subgroups of infinite index will have measure 0, and the (normalized) measure of an open subgroup is simply the inverse of its index. Our main interest in this section, as well as in the preceding one, is in closed subgroups which are not open, and for this purpose another concept of size should be adopted.

Such a concept emerges naturally once we regard profinite groups as metric spaces, rather than measure spaces. Indeed, in any metric space there is a well defined *Hausdorff dimension* function, which attaches to every subset of the space a non-negative real number (or possibly infinity). This notion, which arose at the beginning of this century, has recently assumed importance in the study of fractals (see Falconer [35]). The definition, for a general metric space, is rather intricate. However, just as for the Haar measure, it assumes a more transparent form when applied to a profinite group.

In order to apply the notion of Hausdorff dimension to (infinite) profinite groups G, we need to regard them as metric spaces. This can be done once we fix a filtration $\{G_i\}_{i\geq 1}$ of G, namely a descending chain of open normal subgroups (with $G_1 = G$) which form a base for the neighbourhoods of the identity.

Definition. Given a filtration $\{G_i\}$, define an invariant metric d on G by

$$d(x, y) = \inf\left\{ |G : G_i|^{-1} \ : \ xy^{-1} \in G_i \right\}.$$

The balls in G with respect to this metric are the cosets of G_i, and the diameter of such a ball is $|G : G_i|^{-1}$. In particular, G has diameter 1. It turns out that, in the metric space (G, d), the Hausdorff dimension of a subgroup H is easily determined from the orders of the projections of H into the finite quotients G/G_n. For the present purpose, we can take the following theorem as *defining* the Hausdorff dimension $\mathrm{Dim}(H)$ for a closed subgroup H of G:

Theorem 4.1. (Abercrombie, 1994) *Let G be a profinite group with a filtration $\{G_i\}_{i\geq 1}$, and let $H \leq G$ be a (closed) subgroup. Then*

$$\mathrm{Dim}(H) = \liminf_{i\to\infty} \frac{\log |HG_i/G_i|}{\log |G/G_i|},$$

where the Hausdorff dimension is computed with respect to the metric associated with the filtration $\{G_i\}$.

In fact in [1], 2.6 Abercrombie proved the main part of the theorem, namely the inequality $\geq$, while the easier inequality $\leq$ was obtained by Barnea and myself in [12], 2.1. As far as I know Abercrombie was the first to apply the Hausdorff dimension concept to profinite groups.

It follows from the theorem that $\mathrm{Dim}(H) = 1$ for all open subgroups H of G (recall that G is infinite). Similarly, finite subgroups of G have Hausdorff dimension zero, and commensurable subgroups have the same Hausdorff dimension. The Hausdorff dimension may be the right tool in studying intermediate (i.e., not finite or of finite index) subgroups of profinite groups. One setback though is that $\mathrm{Dim}(H)$ may depend on the filtration $\{G_i\}$ which defines the underlying metric; indeed, examples illustrating this dependence are given in [12], 2.5. So strictly speaking, the Hausdorff dimension is a concept which is attributed to a filtered profinite group. The question of finding a canonical (or at least a natural) filtration arises. While I do not have good candidates for a canonical filtration in arbitrary profinite groups, there seem to be such candidates for pro-p groups G. One (which we use below, unless stated otherwise) is the filtration $\{G_i\}$ where $G_i = G^{p^{i-1}}$, the subgroup generated by all p^{i-1}th powers in G. There are other natural choices, such as the lower p-series, the dimension subgroup series, congruence subgroups when available, etc. Experience shows that these various natural filtrations often give rise to the same Hausdorff dimension function. But we do not yet have a general result guaranteeing this.

Let me now comment on the Lie-theoretic nature of the Hausdorff dimension concept in pro-p groups. Suppose G is a pro-p group equipped with a filtration $\{G_i\}$

satisfying $[G_i, G_j] \leq G_{i+j}$ and $G_i^p \leq G_{i+1}$. Then we can construct our standard graded Lie ring $L(G) = \bigoplus_{i\geq 1} G_i/G_{i+1}$, which will be a Lie algebra over $\mathbb{F}_p$. Furthermore, with a closed subgroup H of G we can associate a graded subalgebra of $L(G)$ which by abuse of notation we shall denote by $L(H)$. It is defined by

$$L(H) = \bigoplus_{i\geq 1}(H \cap G_i)G_{i+1}/G_{i+1}.$$

In order to measure the relative size of graded subalgebras $K = \sum_{i\geq 1} K_i$ inside $L = L(G) = \sum_{i\geq 1} L_i$, consider the following.

Definition. The density of K in L is defined by

$$D(K) = \liminf_{n\to\infty} \frac{\sum_{i\leq n} \dim K_i}{\sum_{i\leq n} \dim L_i}.$$

Theorem 4.1 shows that the density function on graded Lie subalgebras of $L(G)$ is closely related to the Hausdorff dimension function on the subgroups of G. Indeed we have

$$\mathrm{Dim}(H) = D(L(H)).$$

This observation will be useful in what follows.

We now come to the main definition of this section. The idea is to attach to every profinite group a certain "picture" (a subset of the unit interval) reflecting its subgroup structure.

Definition. Let G, d be as above. Define the *spectrum* of G by

$$\mathrm{Spec}(G) = \{\mathrm{Dim}(H) : H \text{ is a closed subgroup of } G\}.$$

Then we have

$$\{0, 1\} \subseteq \mathrm{Spec}(G) \subseteq [0, 1].$$

The rest of this section is devoted to the study of the spectrum of various types of pro-p groups. We start with p-adic analytic groups, then move on to some $\mathbb{F}_p[[t]]$-analytic ones, and finally consider the Nottingham group and free pro-p groups. Most of the results are taken from my work [12] with Barnea, but some more recent results will also be described. In many cases we only obtain partial information on the spectrum, leaving some curious questions unresolved.

4.2 p-adic analytic groups

If G is p-adic analytic then its dimension $\dim G$ (as a p-adic manifold) is well defined ([27] Chapters 4, 8); moreover, if $H \leq G$ then $\dim H$ is also well defined, since H too is p-adic analytic. The following result shows that the Hausdorff dimension of H is proportional to its dimension as a manifold.

Theorem 4.2. (Barnea–Shalev, 1998) *Let G be a p-adic analytic pro-p group. Then*

$$\mathrm{Dim}(H) = \frac{\dim H}{\dim G}$$

for all closed subgroups H of G.

This result shows that the notion of Hausdorff dimension extends the notion of dimension in Lie groups; of course it is also applicable when no Lie group structure is available.

Corollary 4.3. *The spectrum of a p-adic analytic pro-p group is finite and consists of rational numbers.*

It is natural to ask whether the converse holds.

Problem 12. Let G be a finitely generated pro-p group such that $\mathrm{Spec}(G)$ is finite. Does it follow that G is p-adic analytic?

This problem is still open, even in the case $\mathrm{Spec}(G) = \{0, 1\}$. Conjecturably, pro-$p$ groups with this spectrum are virtually procyclic (it suffices to show that they are p-adic analytic).

Some characterization of p-adic analytic groups in terms of Hausdorff dimension can nevertheless be obtained. Note that it follows from Theorem 4.2 that infinite subgroups of a p-adic analytic pro-p group have positive Hausdorff dimension (since they have positive dimension as Lie groups). It is shown in [12] that this property characterizes p-adic analytic groups.

Theorem 4.4. (Barnea–Shalev, 1997) *The following are equivalent for a finitely generated pro-p group G.*

(i) The Hausdorff dimension of every infinite closed subgroup of G is positive.

(ii) G is p-adic analytic.

The proofs of Theorems 4.2 and 4.4 rely on the theory of p-adic analytic groups; see [27] Chapter 4, Exercises 14–16. It is interesting that the proof of Theorem 4.4 also relies on the local finiteness of torsion pro-p groups (see Corollary 2.4 above). This result shows that, if G is not p-adic analytic, then it has an infinite procyclic subgroup H; assuming $\mathrm{Dim}(H) > 0$ and using Theorem 4.1 we then deduce that $|G : G^{p^n}| \leq p^{nc}$ for some c and for infinitely many n. A result of Lazard [68] (see the proof of [27], Theorem 3.16 and the improvement in [115]) now implies that G is p-adic analytic, so (i) implies (ii).

While p-adic analytic pro-p groups have discrete spectrum, a completely different behaviour is manifested by the wreath product $C_p \wr \mathbb{Z}_p$, which is in some sense the 'minimal' finitely generated non p-adic analytic pro-p group. Indeed we have

Proposition 4.5. (Levai, 1999) *The spectrum of $C_p \wr \mathbb{Z}_p$ is the whole unit interval* $[0, 1]$.

It would be interesting to find other finitely generated pro-p groups whose spectrum is the whole unit interval.

4.3 The spectrum of $\mathrm{SL}_d(\mathbb{F}_p[[t]])$

Having studied the spectrum of p-adic analytic groups, we now turn to a group with analytic structure over $\mathbb{F}_p[[t]]$. Let $G = SL_d^1(\mathbb{F}_p[[t]])$ be the first congruence subgroup of $\mathrm{SL}_d(\mathbb{F}_p[[t]])$. It turns out that the group G presents a mixed behvaiour: its spectrum has a continuous part and a discrete part. For simplicity assume $p > 2$. The filtration we use to define the metric on G is the congruence filtration $\{G_i\}_{i\geq 1}$, so that

$$G_i = \ker(G \to SL_d(\mathbb{F}_p[[t]]/(t^i))).$$

We note that this series coincides with both the lower central series and the lower central p-series of G (see [84], [27], Chapter 13). The results below will also be valid if we use the filtration $G^{p^{i-1}}$ instead (this can be seen using the equality $G^{p^{i-1}} = G_{p^{i-1}}$).

It is easy to see that the spectrum of G is not discrete. For example, let H be a standard root subgroup (so H is isomorphic to the additive group of $\mathbb{F}_p[[t]]$). Then one easily verifies that $\mathrm{Dim}(H) = \frac{1}{d^2-1}$, and that for each $\alpha \in [0, \frac{1}{d^2-1}]$ there is a closed subgroup H_α of H such that $\mathrm{Dim}(H_\alpha) = \alpha$. Therefore

$$\mathrm{Spec}(G) \supseteq [0, \frac{1}{d^2-1}],$$

so the spectrum of S contains intervals. By constructing subgroups of the Borel subgroup B of upper triangular matrices (with arbitrary diagonal) one can also deduce that

$$\mathrm{Spec}(G) \supseteq [0, \frac{d(d+1)-2}{2(d^2-1)}] \supset [0, \frac{1}{2}].$$

It will turn out that the spectrum of G contains isolated points; in fact 1 is such a point. To study this phenomenon we need the following.

Definition. Let G be a finitely generated pro-p group (equipped with a filtration defining the metric). Define

$$\lambda(G) = \sup\{\mathrm{Dim}(H) : H < G \text{ is not open}\}.$$

Thus $\lambda(G)$ can be regarded as the supremal size of a closed subgroup of infinite index in G. Since the open subgroups of G have Hausdorff dimension 1, we see that

$$\mathrm{Spec}(G) \cap (\lambda(G), 1) = \emptyset.$$

Therefore, if $\lambda(G) < 1$, then 1 is an isolated point in $\mathrm{Spec}(G)$.

Let us return to $G = \mathrm{SL}_d^1(\mathbb{F}_p[[t]])$ and let $H < G$ be the stabilizer in G of a line, or of a subspace of codimension 1. Then it is easy to verify that

$$\mathrm{Dim}(H) = \frac{d^2-d}{d^2-1} = 1 - (d+1)^{-1}.$$

This yields

$$\lambda(G) \geq 1 - (d+1)^{-1}.$$

Do we have equality?

Theorem 4.6. (Barnea–Shalev, 1997) *Let p be an odd prime and let d ≥ 2. Then*

$$\lambda(SL_d^1(\mathbb{F}_p[[t]])) = 1 - (d+1)^{-1}.$$

The proof, which is essentially Lie-theoretic, will now be outlined in some detail. Let H be a non-open subgroup of G. We have to show that $\mathrm{Dim}(H) \leq 1 - (d+1)^{-1}$. Let $L = L(G)$ be the Lie algebra associated with G with respect to the filtration $\{G_n\}$. Let $L(H)$ be the graded subalgebra corresponding to H. Then $L(H)$ has infinite codimension in L, and by previous remarks we have $\mathrm{Dim}(H) = D(L(H))$, the density of $L(H)$. It therefore suffices to show that the density of any graded subalgebra of infinite codimension in $L(G)$ is at most $1 - (d+1)^{-1}$.

Let us say that a subalgebra of L is *weakly maximal* if it is graded of infinite codimension, and it is maximal with respect to these properties. It can be shown that every graded subalgebra of infinite codimension in L can be extended to a weakly maximal one. By a slight abuse of notation, set

$$\lambda(L) = \sup\{D(K) : K < L \text{ is weakly maximal}\}.$$

Then it suffices to show that $\lambda(L) \leq 1 - (d+1)^{-1}$.

So far the Lie-theoretic reduction has been fairly general, and can actually be applied for other types of virtually pro-p groups. However, in our case the Lie algebra L can be easily identified. Let $\mathcal{G}$ be the finite Lie algebra $sl_d(\mathbb{F}_p)$ consisting of the $d \times d$ matrices of trace zero over $\mathbb{F}_p$. Then by identifying L_i with $\mathcal{G} \otimes t^i$ we obtain a Lie algebra isomorphism

$$L \cong \bigoplus_{i \geq 1} \mathcal{G} \otimes t^i = \mathcal{G} \otimes t\mathbb{F}_p[t].$$

We are therefore interested in $\lambda(sl_d(\mathbb{F}_p) \otimes t\mathbb{F}_p[t])$. A description of the weakly maximal subalgebras of $sl_d(\mathbb{F}_p) \otimes t\mathbb{F}_p[t]$ will be rather useful at this point.

Recall that p is odd. Hence the finite Lie algebra $sl_d(\mathbb{F}_p)$ is simple when p does not divide d, and simple modulo its 1-dimensional center otherwise. For simplicity I will assume below that p does not divide d, but the arguments can be adapted to the general case (see [12]). Since $sl_d(\mathbb{F}_p)$ is simple, the infinite-dimensional Lie algebra $sl_d(\mathbb{F}_p) \otimes t\mathbb{F}_p[t]$ can be regarded as the positive part of an affine Kac–Moody algebra, namely $sl_d(\mathbb{F}_p) \otimes \mathbb{F}_p[t, t^{-1}]$. Let me now discuss the subalgebra structure of affine Kac–Moody algebras in general, and then study the implications to the spectrum of G.

4.4 Subalgebras of Kac–Moody algebras

In the joint work [13] with Barnea and Zelmanov we initiate the study of the maximal subalgebras of affine Kac–Moody algebras and related objects, such as loop algebras. Our results are fairly general, in that they hold over arbitrary fields F, though they take a particularly simple form in the important case $F = \mathbf{C}$. The case of finite fields serves as a tool in the proof of Theorem 4.6 above.

Given a finite-dimensional Lie algebra $\mathcal{G}$ over a field F, consider the infinite-dimensional F-algebra

$$L(\mathcal{G}) = \mathcal{G} \otimes_F F[t, t^{-1}],$$

where $F[t, t^{-1}]$ is the ring of Laurent polynomials. Then (derived) affine Kac–Moody algebras (corresponding to an indecomposable extended Cartan matrix) $\tilde{L}$ can be realized as central extensions of $L(\mathcal{G})$ by a 1-dimensional center Z, where $\mathcal{G}$ is a simple finite-dimensional Lie algebra over F. The study of the maximal subalgebras of $\tilde{L}$ is therefore reduced to the study of the maximal subalgebras of $L(\mathcal{G})$. The latter algebra is $\mathbb{Z}$-graded, and so it is natural to focus first on its maximal graded subalgebras.

Recall that a simple Lie algebra $\mathcal{G}$ over F is said to be *central simple* if F coincides with the centroid of $\mathcal{G}$, which consists of all elements $T \in \mathrm{End}_F\mathcal{G}$ satisfying $[T(x), y] = T([x, y])$ $(x, y \in \mathcal{G})$. Clearly, if p does not divide d, then $sl_d(\mathbb{F}_p)$ is central simple over $\mathbb{F}_p$.

We need to define the important notion of a loop algebra. It will play in this section, as well as in Section 6 below.

Definition. Let k be a positive integer, and let $\mathcal{G} = \oplus_{i=0}^{k-1}\mathcal{G}_i$ be a C_k-grading of $\mathcal{G}$. Let α be the k-tuple $(\mathcal{G}_0, \dots, \mathcal{G}_{k-1})$. Define

$$L(\mathcal{G}, k, \alpha) = \oplus_{n\in\mathbb{Z}}\mathcal{G}_{n \bmod k} \otimes t^n.$$

Then $L(\mathcal{G}, k, \alpha)$ is said to be a *loop algebra* on $\mathcal{G}$ (corresponding to α). Loop algebras are often regarded as twisted Kac–Moody algebras.

We allow the trivial C_k-grading $\alpha = (\mathcal{G}, 0, \dots, 0)$, which gives rise to the loop algebra $L(\mathcal{G}, k, \alpha) = \mathcal{G} \otimes F[t^k, t^{-k}]$.

We can now state the main result of [13].

Theorem 4.7. (Barnea–Shalev–Zelmanov, 1998) *Let F be any field and let $\mathcal{G}$ be a central simple finite-dimensional Lie algebra over F. Let M be a maximal graded F-subalgebra of $L(\mathcal{G})$. Then one of the following holds:*

(i) $M = L(\mathcal{G}, r, \alpha)$ for some prime r and a C_r-grading α of $\mathcal{G}$.

(ii) $M = L(\mathcal{H}) = \mathcal{H} \otimes F[t, t^{-1}]$, where $\mathcal{H}$ is a maximal subalgebra of $\mathcal{G}$.

(iii) M is commensurable with $\mathcal{G} \otimes F[t]$ or with $\mathcal{G} \otimes F[t^{-1}]$.

Let me make some comments. First, note that the subalgebras of types (ii) and (iii) (as well as $\mathcal{G} \otimes F[t]$ and $\mathcal{G} \otimes F[t^{-1}]$) are all maximal. Secondly, the main result of [13] is slightly more general, in that simple algebras that are not central simple are also dealt with; this is done by allowing a certain twisted version of case (ii). Thirdly, let us examine the situation in the case $F = \mathbf{C}$. In this case $\mathcal{G}$ is one of the known (classical or exceptional) simple complex Lie algebras. According to the theorem, in order to determine all the possibilities for M (up to commensurability), one needs to know

(a) the maximal subalgebras of $\mathcal{G}$, and

(b) the elements of prime order in $\mathrm{Aut}\mathcal{G}$ (these elements determine all cyclic gradings of prime period of $\mathcal{G}$).

These data are available in the literature (see for instance Dynkin [33] and Kac [50] p. 96), and so the description in Theorem 4.7 can be made explicit.

For the group-theoretic applications we need to define the positive part of Kac–Moody algebras and of loop algebras. For $\mathcal{G}$ and F as above, set

$$L^+(\mathcal{G}) = \mathcal{G} \otimes_F tF[t].$$

Define the positive part of the loop algebra $L(\mathcal{G}, k, \alpha)$ by

$$L^+(\mathcal{G}, k, \alpha) = L(\mathcal{G}, k, \alpha) \cap L^+(\mathcal{G}).$$

We aim to describe the weakly maximal subalgebras of Lie algebras of type $L^+(\mathcal{G})$. This can be done using our methods, and the description is even simpler than the one obtained in Theorem 4.7 (as subalgebras of type (iii) do not occur).

Theorem 4.8. *Let $\mathcal{G}$ be a central simple finite-dimensional Lie algebra over a field F, and let M be a weakly maximal subalgebra of $L^+(\mathcal{G})$. Then one of the following holds.*

(i) $M = L^+(\mathcal{G}, r, \alpha)$ for some prime r and a C_r-grading α of $\mathcal{G}$.

(ii) $M = L^+(\mathcal{H}) = \mathcal{H} \otimes tF[t]$, where $\mathcal{H}$ is a maximal subalgebra of $\mathcal{G}$.

Recall that we are interested in bounding the density $D(M)$ of weakly maximal subalgebras M of $L^+(\mathcal{G})$, where $\mathcal{G} = sl_d(\mathbb{F}_p)$. Apply Theorem 4.8. If M is as in part (i) (i.e., the positive part of a loop algebra of period r), then we readily obtain

$$D(M) = \frac{1}{r},$$

while if $M = L^+(\mathcal{H})$ as in part (ii), then

$$D(M) = \frac{\dim \mathcal{H}}{\dim \mathcal{G}}.$$

Defining

$$m(\mathcal{G}) = \max\{\dim \mathcal{H} : \mathcal{H} \text{ is a proper subalgebra of } \mathcal{G}\},$$

we see that

$$\lambda(L^+(\mathcal{G})) = \max\{\frac{1}{2}, \frac{m(\mathcal{G})}{\dim \mathcal{G}}\}.$$

Note that, in our case $\mathcal{G} = sl_d(\mathbb{F}_p)$, and $m(\mathcal{G}) \geq \frac{1}{2} \dim \mathcal{G}$. It follows that

$$\lambda(L^+(sl_d(\mathbb{F}_p))) = \frac{m(sl_d(\mathbb{F}_p))}{d^2 - 1}.$$

It remains to compute $m(sl_d(\mathbb{F}_p))$.

If F has characteristic zero, then it follows from results of Dynkin that $m(sl_d(F)) = d^2 - d$ (the maximum being attained by the obvious parabolic subalgebra). Though Dynkin's methods do not seem to apply in characteristic p, we show directly in [12] that $m(sl_d(\mathbb{F}_p)) = d^2 - d$. This yields

$$\lambda(sl_d(\mathbb{F}_p) \otimes t\mathbb{F}_p[t]) = \frac{d^2 - d}{d^2 - 1} = 1 - (d+1)^{-1}.$$

Putting everything together we conclude that.

$$\lambda(SL_d(\mathbb{F}_p[[t]])) \leq 1 - (d+1)^{-1},$$

so equality holds. This is how Theorem 4.6 is proved.

Rephrasing this result, we can say that among all non-open subgroups of $SL_d(\mathbb{F}_p[[t]])$, the parabolic subgroups stabilizing a line or a subspace of codimension 1 are of largest size. This can be viewed as a local field analogue of the result of Patton [106] from 1972, determining the maximal size of a proper subgroup of $\mathrm{SL}_d(q)$ (it is usually attained by similar parabolic subgroups).

It would be interesting to obtain more information on the spectrum of $SL_d(\mathbb{F}_p[[t]])$, apart from its 'top' and its 'bottom'. In the case $d = 2$ we know that the spectrum is $[0, 2/3] \cup \{1\}$. In the general case let me state the following

Conjecture 13. The spectrum of $\mathrm{SL}_d^1(\mathbb{F}_p[[t]])$ is a finite union of closed intervals (including points).

Barnea has verified the conjecture for $d \leq 4$ (in which case $\mathrm{Spec}(G) = [0, 1 - (d+1)^{-1}] \cup \{1\}$), and so the first open case is $d = 5$. An even broader question is as follows.

Problem 14. Study the spectrum of classical groups over $\mathbb{F}_p[[t]]$, and over other pro-p rings such as $\mathbb{F}_p[[t_1, \dots, t_k]]$.

This is related to

Problem 15. Study graded subalgebras of $\mathcal{G} \otimes F[t_1, \dots, t_n]$ where $\mathcal{G}$ is a finite-dimensional simple Lie algebra over the field F.

4.5 The spectrum of the Nottingham group

Let me now turn to the Nottingham group $G = \mathcal{N}(\mathbb{F}_p)$, consisting of all normalized power series $t + a_2t^2 + a_3t^3 + \dots$ $(a_i \in \mathbb{F}_p)$ via substitution. It can be shown that, for $p \geq 5$, 1 is an isolated point in $\mathrm{Spec}(G)$. Moreover, for large p, additional isolated points can be found in the spectrum, since it is shown in [12] that

$$\{d^{-1} : d = 1, 2, 3, \dots\} \subseteq \mathrm{Spec}(G) \subseteq \{d^{-1} : d = 1, 2, 3, \dots\} \cup [0, 3p^{-1}].$$

While the left inclusion stems from an easy construction, the proof of the inclusion on the right applies Lie methods, and is related to results on the subalgebra structure of the Witt algebra.

More progress has recently been made in [9] concerning the spectrum of the Nottingham group, by considering the so called index subgroups, namely subgroups of the form $\{\sum_{n \in S} a_n t^n\}$ for suitable subsets $S \subseteq \mathbb{N}$ (containing 1). In particular it was shown that the spectrum of the Nottingham group also starts with an interval.

Theorem 4.9. (Barnea–Klopsch, 1998) *Let $G = \mathcal{N}(\mathbb{F}_p)$ be the Nottingham group. Then*

$$Spec(G) \supseteq [0, p^{-1}] \cup \{p^{-1} + p^{-j}\}_{j=1}^{\infty} \cup \{d^{-1}\}_{d=1}^{\infty}.$$

We see that $\mathrm{Spec}(G)$ is completely determined below p^{-1} and above $3p^{-1}$. Its precise shape in the interval $(p^{-1}, 3p^{-1}]$ is still a mystery.

4.6 Free pro-p groups

For a prime p and an integer $d \geq 2$ let $F(p,d)$ denote the free pro-p group on d generators. For some time it was unclear whether $F(p,d)$ has closed subgroups of Hausdorff dimension strictly between 0 and 1. Let me discuss some Lie-theoretic ideas which lead to the construction of such subgroups. First, the filtration I will use to define the metric in $G = F(p,d)$ will be the dimension subgroup filtration $\{G_i\}$. The reason is that the indices of these subgroups are known, whereas the indices of $G^{p^{i-1}}$ are largely unknown (this is the difficult problem of computing the Burnside numbers). Let $L = L(G)$ be the graded Lie algebra associated with G with respect to the filtration $\{G_i\}$. Then $L = \oplus_i L_i$ is the free restricted Lie algebra on d generators over $\mathbb{F}_p$. The dimensions $\dim L_i$ can be found in [112].

Let α be an automorphism of $G = F(p,d)$ of finite p'-order e. As in Section 3, α acts on the Lie algebra L. Let $H = C_G(\alpha)$ and $K = L^\alpha$ be the centralizers of α in the group and the Lie algebra respectively. Since the order of α is prime to p, it follows that the subalgebra $L(H)$ of L corresponding to H is precisely K. This shows that the Hausdorff dimension of H with respect to the filtration $\{G_i\}$ equals the density $D(K)$ of K in L.

At this point we can apply results from invariant theory of Lie algebras again. Since $K_n = L_n^\alpha$, what we need is a computation of the dimensions of the fixed spaces L_n^α. It can be shown that, if the action of α on L_1 is non-scalar, $\dim L_i^\alpha \sim e^{-1} \dim L_i$, where e is the order of α. This implies that $D(K) = e^{-1}$, and, in view of the above, that $\mathrm{Dim}(H) = e^{-1}$. We therefore obtain

Theorem 4.10. *Let α be an automorphism of $F(p,d)$ of finite order e which is prime to p. Suppose the action of α on the Frattini quotient $F(p,d)/\Phi(F(p,d))$ is non-scalar. Then the centralizer $C_{F(p,d)}(\alpha)$ of α has Haudorff dimension e^{-1}. Consequently, Spec$(F(p,d))$ contains e^{-1} whenever e is the order of a non-scalar p'-element of the general linear group $GL_d(\mathbb{F}_p)$.*

In fact more points in the spectrum can be realized using extensions of this method, but it is still unclear how to construct subgroups of $F(p,d)$ whose Hausdorff dimension is irrational.

4.7 Other spectra

Let me draw attention to some natural subsets of Spec(G). Consider the normal spectrum of a pro-p group G, namely, the set of Hausdorff dimensions of closed normal subgroups of G. The normal spectrum is often much smaller than the spectrum. For example, if G is just-infinite then its normal spectrum is $\{0,1\}$. The wreath product $C_p \wr \mathbb{Z}_p$ is an example of a group whose spectrum is $[0,1]$ and whose normal spectrum is $\{0,1\}$.

Abercrombie showed that the normal spectrum the iterated wreath product of countably many groups of order 2 is also $\{0,1\}$. Note that this group, which is the automorphism group of a rooted binary tree, is not even finitely generated; still there is a natural congruence filtration there, which is used to define the metric.

It turns out that a similar phenomenon holds for free pro-p groups $F(p, d)$: their normal spectrum is $\{0, 1\}$. More precisely, if H is any non-trivial normal subgroup of $F(p, d)$, then $\mathrm{Dim}(H) = 1$. As usual, the proof is Lie-theoretic, and consists of showing that any non-zero graded ideal in a free Lie algebra has density 1. This is a consequence of some more detailed results from [105] on the entropy of graded algebras.

Of course the normal spectrum of some finitely generated pro-p groups can be larger than $\{0, 1\}$ (consider abelian pro-p groups, for example). Still, it seems that the most familiar pro-p groups have finite normal spectrum. I therefore pose the following.

Problem 16. Construct finitely generated pro-p groups with infinite normal spectrum. Can the normal spectrum of such a group contain intervals?

In a similar manner one can consider the finitely generated spectrum of a pro-p group G, namely, the set of Hausdorff dimensions of finitely generated closed subgroups of G. Again, it seems that this subset is usually considerably smaller than the whole spectrum (of course, it coincides with the spectrum in p-adic analytic groups). The finitely generated spectrum of $\mathrm{SL}_d^1(\mathbb{F}_p[[t]])$ contains the harmonic series n^{-1} (since the subgroups $\mathrm{SL}_d^1(\mathbb{F}_p[[t^n]])$ are finitely generated). A similar remark applies for the Nottingham group. It seems likely that the finitely generated spectrum of free pro-p groups is $\{0, 1\}$ (the centralizers constructed in the previous subsection are are never finitely generated). It is natural to ask how large the finitely generated spectrum can be.

Problem 17. Are there finitely generated pro-p groups with uncountable finitely generated spectrum? Can the finitely generated spectrum of such a group contain irrational numbers? Intervals?

The work of Pink, described in Section 5.2 below, suggests a promising approach to this question in some $\mathbb{F}_p[[t]]$-analytic pro-p groups. For example, it is likely to imply that the Hausdorff dimension of a finitely generated closed subgroup of $\mathrm{SL}_d^1(\mathbb{F}_p[[t]])$ is always rational.

5 Analytic groups, algebraic groups and subgroup structure

5.1 Analytic groups and the adjoint representation

The groups $\mathrm{SL}_d(\mathbb{F}_p[[t]])$ are familiar examples of groups having an analytic structure, in this case over the pro-p ring $\mathbb{F}_p[[t]]$. Here, the analytic structure comes from the natural linear representation of the group. However, the following is open:

Problem 18. Is every $\mathbb{F}_p[[t]]$-analytic pro-p group linear over $\mathbb{F}_p((t))$, or at least linear over some field?

In view of Pink's theorem, discussed in the following subsection, a positive answer would imply that $\mathbb{F}_p[[t]]$-analytic pro-p groups are in a sense not far removed from algebraic groups, reflecting the analogous situation in complex semi-simple Lie groups.

If we replace $\mathbb{F}_p[[t]]$ (and $\mathbb{F}_p((t))$) by $\mathbb{Z}_p$ in Problem 18, the answer is yes ([27], §7.3). This is deduced from Ado's Theorem on the linearity of finite-dimensional Lie algebras, using the tight correspondence between representations of G and those of $L(G)$, where G is a p-adic analytic group and $L(G)$ is its Lie algebra. While Ado's Theorem still holds in characteristic p (it is then Iwasawa's Theorem), the links between an $\mathbb{F}_p[[t]]$-analytic group and its Lie algebra seem to be weaker (in particular, the Campbell–Hausdorff formula is no longer available to translate back and forth between the group and the Lie algebra). However, every analytic group does have a natural linear representation, namely the adjoint representation. As this is not usually faithful it does not solve Problem 18 in general, but it does lead to an answer in some interesting cases. The purpose of this subsection is to present such a partial answer, and to use it in order to show that the Nottingham group is not analytic over any pro-p ring.

Throughout this subsection, we work with an infinite, commutative Noetherian local ring $(R, \mathfrak{m})$ whose residue field $R/\mathfrak{m}$ is finite, of characteristic p; i.e., R is a *pro-p ring*. For technical reasons, it is assumed further that the associated graded ring $\mathrm{gr}(R) = \bigoplus_{n=0}^{\infty} \mathfrak{m}^n/\mathfrak{m}^{n+1}$ is an integral domain (which implies that R is itself a domain). Thus R could be $\mathbf{Z}_p[[t_1, \dots, t_m]]$ with $m \geq 0$ or $\mathbb{F}_p[[t_1, \dots, t_m]]$ with $m \geq 1$. The concept of *R-analytic group* is defined in [27], Chapter 13, where it is shown that every R-analytic group has an open subgroup H which is a so-called *standard group*: that is, the underlying set of H may be identified with $(\mathfrak{m}^h)^{(d)}$, the group operation being given by a *formal group law* (see [43]); here, $h \geq 1$ is called the *level* and d is the *dimension* of H. The formal group law is a d-tuple $\mathbf{F} = (F_1, \dots, F_d)$ of power series over R in $2d$ variables, and satisfies

$$\mathbf{x} \cdot \mathbf{y} = (F_1(\mathbf{x}, \mathbf{y}), \dots, F_d(\mathbf{x}, \mathbf{y}))$$

for $\mathbf{x}, \mathbf{y} \in H = (\mathfrak{m}^h)^{(d)}$. Let $B_i(\mathbf{X}, \mathbf{Y})$ be the sum of all terms of $F_i(\mathbf{X}, \mathbf{Y})$ having total degree 1 in the X's and total degree 1 in the Y's. The *Lie algebra* $L(H)$ of H is the R-module $R^{(d)}$ with Lie product

$$[\mathbf{x}, \mathbf{y}] = \mathbf{B}(\mathbf{x}, \mathbf{y}) - \mathbf{B}(\mathbf{y}, \mathbf{x})$$

for $\mathbf{x}, \mathbf{y} \in R^{(d)}$. To each morphism of standard groups $f : H \to K$ corresponds a Lie algebra homomorphism $Df : L(H) \to L(K)$: if f is represented by a d-tuple of power series $\Phi = (\Phi_1, \dots, \Phi_d)$, then Df is simply given by the linear parts of $\Phi_1, \dots, \Phi_d$. The construction is functorial, so $D(f \circ g) = Df \circ Dg$ (the chain rule!). For details, see [27], Chapter 13. It follows that if a^* denotes the inner automorphism of H effected by an element $a \in H$, then $\mathrm{Ad}(a) := Da^*$ is an automorphism of $L(H)$ and $\mathrm{Ad} : H \to \mathrm{Aut}(L(H)) \leq \mathrm{GL}_d(R)$ is a group homomorphism.

As we are interested in seeing how close H is to being a linear group, it is useful to know how small the kernel of Ad has to be. Write $\iota : H \to L(H)$ for the inclusion mapping of $H = (\mathfrak{m}^h)^{(d)}$ into $R^{(d)} = L(H)$, and for $n \geq h$ let H_n denote the subset $(\mathfrak{m}^n)^{(d)}$ of H; in fact H_n is a normal subgroup of H.

Lemma 5.1. *Let $a \in H_m$. If $\mathrm{Ad}(a) = 1$ then for each $n \geq 0$,*

$$[\mathfrak{m}^n L(H), \iota(a)] \subseteq \mathfrak{m}^{2m+n} L(H).$$

To prove this, one proceeds as follows (recall that we are identifying H with $(\mathrm{m}^h)^{(d)} \subseteq R^{(d)}$). For $\mathbf{y} = (y_1, \ldots, y_d) \in H_m$ and $\mathbf{x} = (x_1, \ldots, x_d) \in \mathrm{m}^n L(H)$, we have

$$\mathbf{x}\mathrm{Ad}(\mathbf{y}) = \mathbf{x} + \mathbf{B}(\mathbf{x}, \mathbf{y}) - \mathbf{B}(\mathbf{y}, \mathbf{x}) + \sum_{j=1}^{d} \mathbf{V}_j(\mathbf{y}) x_j, \qquad (*)$$

where each $\mathbf{V}_j$ is a d-tuple of power series in d variables in which all terms have total degree at least 2; this is a straightforward consequence of the definition and of properties of the group law $\mathbf{F}$, summarized in [27], Proposition 13.16. It follows that

$$\begin{aligned} \mathbf{x}\mathrm{Ad}(\mathbf{y}) - \mathbf{x} &\equiv \mathbf{B}(\mathbf{x}, \mathbf{y}) - \mathbf{B}(\mathbf{y}, \mathbf{x}) \pmod{\mathrm{m}^{2m+n} R^{(d)}} \\ &= [\mathbf{x}, \iota(\mathbf{y})], \end{aligned}$$

which implies the lemma.

Properties of the formal group law (given in [27], Proposition 13.16) imply that if $a \in H_m$ and $b \in H_n$ where $m \leq n$, then

$$\iota([b, a]) \equiv [\iota(b), \iota(a)] \pmod{\mathrm{m}^{2m+n} L(H)}.$$

So if in addition we have $\mathrm{Ad}(a) = 1$ then Lemma 5.1 shows that

$$\iota([b, a]) \in \mathrm{m}^{2m+n} L(H),$$

which is equivalent to $[b, a] \in H_{2m+n}$. Thus we have

Lemma 5.2. *Let* $K = \ker(\mathrm{Ad})$. *Then*

$$[H_n, K \cap H_m] \leq H_{2m+n}$$

whenever $n \geq m \geq h$.

As a first application, we consider the so-called *R-perfect* groups. The standard group H as above is called R-perfect if $H = H_1$ and $[H, H] = H_2$; this is equivalent to the stipulation that the graded Lie algebra associated to $L(H)$ is perfect (equal to its derived algebra); see §13.5 of [27]. It is shown there that if H is R-perfect then $[H_n, H_m] = H_{m+n}$ for all $m, n \geq 1$. It follows therefore from Lemma 5.2 that $K \cap H_m \neq H_m$ for every m, hence that K is not open in H. We may thus derive the (rather weak) conclusion that *every R-perfect group has an infinite quotient which is linear over R*. In particular, then, if an R-perfect group is also *just-infinite*, then it is itself linear over R.

A similar approach provides the answer to a question which has been posed in several places (see [31], p. 127):

Theorem 5.3. *Let F be a finite field with $|F|$ not divisible by 4. Then the Nottingham group $\mathcal{N}(F)$ is not analytic over any pro-p ring (satisfying the conditions stated above).*

The result is undoubtedly true for every finite field F; at present our knowledge of the Nottingham groups over large fields of characteristic 2 is insufficient to confirm this. Let $G = \mathcal{N}(F)$ be the Nottingham group over the finite field $F = \mathbb{F}_q$ with $q = p^e$ elements, where $e = 1$ if $p = 2$, and let $\{G^{(n)}\}_{n\geq 0}$ be its derived series. It is known (see Chapter 6) that G is hereditarily just-infinite, that it is not linear over any field, and that $|G : G^{(n)}| = q^{2^{n+1}-2}$ $(q \neq 2)$, $|G : G^{(n)}| \leq 2^{4c^n-1}$ $(q = 2)$, where $c < 3$. Therefore Theorem 5.3 follows from

Proposition 5.4. *Let G be an R-analytic pro-p group. If G is hereditarily just-infinite then one of the following holds: (a) G is linear over either $\mathbb{Z}_p$ or $\mathbb{F}_p[[t]]$; (b) there exists $s \geq 1$ such that*

$$|G : G^{(n)}| \geq p^{3^{n-s}d} \text{ for all large } n,$$

where $d = \dim(G)$.

Here, $G^{(n)}$ denotes the nth term of the derived series of G. To prove the proposition, we start by reducing to the case where $R = \mathbf{F}_p[[t]]$. Let $H = (\mathbf{m}^h)^{(d)}$ be a standard open subgroup of G, and let P be a prime ideal of height 1 in R. Then $H(P) = H \cap P^{(d)}$ is a nontrivial closed normal subgroup of H ([27], Lemma 13.21); since G is hereditarily just infinite it follows that $H(P)$ has finite index in H, hence that P has finite index in R. Thus $\mathrm{Dim}(R) = 1$. It follows that R is finite over either $\mathbf{Z}_p$ or $\mathbf{F}_p[[t]]$, and hence that G is analytic over one of these rings ([27], Example 13.9(iv)). If G is p-adic analytic then it is linear over $\mathbb{Z}_p$; otherwise, G is $\mathbf{F}_p[[t]]$-analytic. Thus we may assume that $R = \mathbf{F}_p[[t]]$. As above, we shall write $H_n = (\mathbf{m}^n)^{(d)}$ for each $n \geq h$, where $\mathbf{m} = tR$.

Let K be the kernel of Ad on H. Then H/K is linear over $\mathbf{F}_p[[t]]$; as $|G : H|$ is finite it follows that either G is linear over $\mathbf{F}_p[[t]]$ or $K \neq 1$. Assuming the latter, we shall see that (b) holds.

Evidently K is a closed normal subgroup of H, so as H is just-infinite it follows that K is open in H, hence $K \geq H_m$ for some m. We take $m \geq 2$. Now Lemma 5.2 shows that

$$[H_n, H_n] \leq H_{3n}$$

for each $n \geq m$. There exists s such that $G^{(s)} \leq H_m$. Then $G^{(n)} \leq H_{3^{n-s}m}$ for each $n \geq s$, and as $|H_i : H_{i+1}| = p^d$ for each i it follows that

$$|G : G^{(n)}| \geq |H : H_{3^{n-s}m}| = p^{d(3^{n-s}m-1)} \geq p^{3^{n-s}d}$$

for all $n > s$. Thus we have (b).

A Lie algebra version of the last result can be phrased as follows:

Proposition 5.5. *Let G be either (a) a hereditarily just-infinite R-analytic pro-p group, or (b) a just-infinite standard group over R. Then either G is linear over R or the Lie algebra of G is abelian.*

Recall that the Lie algebra of an R-analytic group G is defined to be $E \otimes_R L(H)$ where E is the field of fractions of R and H is some standard open subgroup of G. In case (b) of the proposition, we take $H = G$. Let $K \lhd H$ be the kernel of Ad on H. If $K = 1$

then H is linear over R, and then so is G since $|G : H|$ is finite. Otherwise, $K \geq H_n$ for some n. Using the notation of Lemma 5.1, we see from $(*)$ that

$$\mathbf{B}(\mathbf{x},\mathbf{y}) - \mathbf{B}(\mathbf{y},\mathbf{x}) + \sum_{j=1}^{d} \mathbf{V}_j(\mathbf{y})x_j = \mathbf{0}$$

for each $\mathbf{x} \in R^{(d)}$ and all $\mathbf{y} \in (\mathrm{m}^n)^{(d)}$. Now fix $\mathbf{x} \in R^{(d)}$. The expression on the left is the result of evaluating a d-tuple of power series at the point $\mathbf{y}$; since it takes the value $\mathbf{0}$ for all $\mathbf{y}$ in a neighbourhood of $\mathbf{0}$, it follows that each of the power series is identically zero ([27], Proposition 6.49). Evaluating the linear parts of these power series at $\mathbf{z} \in R^{(d)}$ gives $\mathbf{B}(\mathbf{x},\mathbf{z}) - \mathbf{B}(\mathbf{z},\mathbf{x}) = \mathbf{0}$. Thus the Lie product $[\mathbf{x},\mathbf{z}]$ vanishes for all $\mathbf{x},\mathbf{z} \in R^{(d)} = L(H)$ and $L(H)$ is abelian.

Problem 19. What can be said of an R-analytic group whose Lie algebra is abelian?

Note that such a group need not be abelian if R has characteristic $p > 0$; the 2-dimensional R-standard group G with group law

$$(X_1 + Y_1 + X_1Y_1, X_2 + Y_2 + X_1^pY_2)$$

has abelian Lie algebra, but its centre $Z(G) = 0 \times \mathbf{m}$ is 1-dimensional. This example is due to Chevalley ([24], Chapter II, §10, Example V).

5.2 Pink's work

In 1998 Richard Pink published the paper [108] on the structure of compact subgroups of algebraic groups over a local field of positive characteristic. His work provides a powerful tool for handling various problems in pro-p groups, mostly concerning subgroup structure and generation properties. Applications of this type were provided by Barnea and Larsen [10], [11], who immediately realized the relevance of Pink's methods. It is likely that additional problems will be solved using these methods, including ones which before seemed quite unapproachable.

The results of Section 4 provide some information on sizes (i.e. Hausdorff dimension) of closed subgroups of groups such as $\mathrm{SL}_d(\mathbb{F}_p[[t]])$. But the subgroup structure of such groups remained unclear. For example, does $\mathrm{SL}_d(\mathbb{F}_p[[t]])$ have a closed subgroup which is a free nonabelian pro-p group? Can a non-soluble p-adic analytic pro-p group be embedded in $\mathrm{SL}_d(\mathbb{F}_p[[t]])$? These questions can now be answered.

Pink's work analyses compact subgroups Γ of $G(F)$, where G is a simple algebraic group defined over a local field F of characteristic $p > 0$. Exact formulations of his results require many definitions and technical assumptions, so here we will be content with a rough description of them. In the case where Γ is Zariski-dense in G the conclusion is that Γ must be open in an algebraic subgroup which is roughly of type $G(E)$, where E is a closed subfield of F. See [108], 0.7 for the precise formulation. An example of such a subgroup is $\mathrm{SL}_d(\mathbb{F}_p[[t^m]])$ inside $\mathrm{SL}_d(\mathbb{F}_p[[t]])$. If Γ is not Zariski-dense then we form its Zariski-closure, say H. Factoring out the soluble radical of H, we obtain an embedding of some quotient of Γ with respect to a soluble normal subgroup into a certain semisimple algebraic group. Analysis of this situation eventually shows that Γ has a filtration of normal subgroups $\Gamma_3 \leq \Gamma_2 \leq \Gamma_1 \leq \Gamma$ such that

1. Γ/Γ_1 is finite.
2. Γ_1/Γ_2 is abelian of finite exponent.
3. Γ_2/Γ_3 is isomorphic to some open subgroup of (possibly a cover of) $H(E)$, where E is a local field of characteristic p and H is a (connected adjoint) algebraic group over E.
4. Γ_3 is soluble.

Let me start with an interesting consequence of the result above.

Proposition 5.6. *Let Γ be a p-adic analytic pro-p subgroup of $SL_d(\mathbb{F}_p[[t]])$. Then Γ is virtually abelian.*

To prove this, let Γ_i be the filtration as above. Since Γ has finite rank, we see that Γ_1/Γ_2 is finite. Now, Γ_2/Γ_3 is analytic both over the p-adic field and over a local field of characteristic p, hence it is finite (see [84] or [27], Theorem 13.23). This shows that Γ/Γ_3 is finite. By the Lie–Kolchin theorem there is a finite index subgroup Γ_4 of Γ_3 which is represented by upper triangular matrices (with respect to a suitable basis). Let Γ_5 be the intersection of Γ_4 with the uni-triangular matrices (in that basis). Then Γ_4/Γ_5 is abelian, and the unipotent subgroup Γ_5 is of finite exponent. Hence Γ_5 is finite. It follows that Γ is finite-by-abelian-by-finite, hence abelian-by-finite (by standard arguments). Using the fact that Γ is p-adic analytic it follows that it has a finite index subgroup which is a free p-adic module of finite rank.

Note that any free p-adic module of finite rank can be embedded in $SL_d(\mathbb{F}_p[[t]])$ using diagonal matrices, since it can be embedded in the multiplicative group of units in $\mathbb{F}_p[[t]]$.

Applying Proposition 5.6 we see, in particular, that open pro-p subgroups of $SL_m(\mathbb{Z}_p)$ ($m \geq 2$) cannot be embedded in $SL_d(\mathbb{F}_p[[t]])$, a problem which has been around for some time.

Another nice consequence of [108] is the following, a sort of converse to Proposition 5.4:

Proposition 5.7. *Let Γ be a just-infinite pro-p group which is linear over $\mathbb{F}_p((t))$. Then either Γ is $\mathbb{F}_p[[t]]$-analytic, or it is a p-adic space group (hence $\mathbb{Z}_p$-analytic).*

To show this we embed Γ in $GL_n(\mathbb{F}_p((t)))$, and we let Γ_i be as in Pink's result. If Γ_3 is non-trivial then (being normal) it is of finite index in Γ. This implies that Γ is soluble, and it easily follows that Γ is a p-adic space group. So suppose $\Gamma_3 = 1$. Note that, since Γ is finitely generated, Γ/Γ_2 is finite. Now Γ_2 is isomorphic to an open subgroup of an algebraic group $H(E)$ over a local field of characteristic p. Since algebraic groups are analytic (see Margulis [91], Proposition 2.5.2 on page 55) this implies that Γ_2 is $\mathbb{F}_p[[t]]$-analytic, and so is Γ.

On the other hand, a pro-p group that is linear over $\mathbb{Z}_p$ is p-adic analytic [27] Interlude A. Therefore we obtain:

Corollary 5.8. *A just-infinite pro-p group which is linear over a local field is analytic over a local field.*

5.3 More applications

Let me now discuss the work of Barnea and Larsen, which solves some interesting problems in pro-p groups. The first problem concerns the linearity of nonabelian free pro-p groups. Zubkov [149] showed that, if $p > 2$, then such groups cannot be embedded in $GL_2(R)$ where R is any commutative pro-p ring. This implies (using a standard generic matrices argument) that the pro-p part of $GL_2(R)$ satisfies some non-trivial pro-p identity. In [84] it is conjectured that nonabelian free pro-p groups cannot be embedded in $GL_n(R)$ for any p and n. While this conjecture is still unresolved, the case of local fields is settled in [10] as follows.

Theorem 5.9. (Barnea–Larsen, 1998) *Nonabelian free pro-p groups are not linear over non-archimedean local fields.*

Let me outline the proof. Let $\Gamma = F(p, d)$ be the free pro-p group on $d \geq 2$ generators, and suppose Γ is embedded in $GL_n(F)$, where F is a non-archimedean local field. Then F has characteristic p (since Γ is not p-adic analytic). Let Γ_i be as in Pink's result. Then we immediately see that $\Gamma_3 = 1$ and that Γ_2 is a free pro-p group of finite index in Γ. Replacing Γ by Γ_2 we can assume that Γ is open in some nice algebraic group $H(E)$. It is easy to see that $H(E)$ has an abelian pro-p subgroup of infinite rank. It follows that the free pro-p group Γ has such a subgroup, a contradiction.

Note that Theorem 5.9 does not imply that pro-p groups which are linear over non-archimedean local fields satisfy a pro-p identity. To show this one has to prove that free pro-p groups are not linear over $\mathbb{Z}_p[[t_1, \ldots, t_d]]$ for all $d \geq 1$, and Pink's work is no longer applicable in that context.

The next application is of a probabilistic nature. A profinite group G can be regarded as a probability space with respect to a normalized Haar measure. Given a profinite group G and a positive integer k, let $Q(G, k)$ denote the probability that k randomly chosen elements of G generate (topologically) an open subgroup of G. Mann [86] showed that profinite groups G with polynomial subgroup growth satisfy $Q(G, k) = 1$ for some k. In particular, this is the case for p-adic analytic pro-p groups. For some time it was unclear whether the property $Q(G, k) = 1$ also implies polynomial subgroup growth. It is now known that this is not the case: counterexamples are constructed in [90]. The pro-p case of this question was more persistent: are pro-p groups G satisfying $Q(G, k) = 1$ necessarily p-adic analytic? This was posed by Mann in [86]. I conjectured that this is not the case, and that the group $SL_d^1(\mathbb{F}_p[[t]])$ might be a counterexample. This was verified in [11] in a rather general form, which also settles the conjecture posed in [124], §4, provided the characteristic is not too small.

Theorem 5.10. (Barnea–Larsen, 1998) *Let G be a simply connected simple algebraic group over a non-archimedean local field F of characteristic p. Suppose $p \geq 5$, and let Γ be a compact open subgroup of $G(F)$. Then the probability that two randomly chosen elements x, y of Γ generate an open subgroup of Γ is* 1.

In other words, the groups above satisfy $Q(G, 2) = 1$. This result can be regarded as the local field analogue of Dixon's conjecture (now a theorem), that two randomly chosen elements of a finite simple group G generate G with probability $\to 1$ as $|G| \to \infty$; see [25], [51], [79]. The proof of Theorem 5.10 is rather technical, and I will not outline it here; again, it depends crucially on Pink's theorem. Instead, let me pose some related questions.

Problems.

20. Let G be the Nottingham group. Show that $Q(G, 2) = 1$.
21. Find just-infinite pro-p groups G with $Q(G, k) < 1$ for all k.

It is likely that Grigorchuk's groups have the latter propery.

Next, let us turn to some applications of Theorem 5.10. Recall that the *lower rank* of a pro-p group G is the minimal integer r (possibly infinity) such that G has a base for the neighbourhoods of the identity consisting of r-generated subgroups. Clearly, p-adic analytic pro-p groups have finite lower rank. In fact the lower rank of such groups coincides with the number of generators of their associated p-adic Lie algebra [81]. For sometime it was unclear whether non p-adic analytic pro-p groups can have finite lower rank. This was resolved in [84], where it is shown that $SL_d^1(\mathbb{F}_p[[t]])$ (and R-perfect groups in general) have finite lower rank. The bounds given in [84] on the lower rank of such groups were recently improved by Barnea [8]. In particular he shows that $SL_2^1(\mathbb{F}_p[[t]])$ has lower rank 2, thus improving the upper bound 3 given in [84]. The relation between the lower rank of an $\mathbb{F}_p[[t]]$-simple group and the number of generators of its Lie algebra is also studied in [8] (an $\mathbb{F}_p[[t]]$-standard group H is $\mathbb{F}_p[[t]]$-*simple* if the finite $\mathbb{F}_p$-Lie algebra $L(H)/tL(H)$ is simple). Let me now indicate the relevance of the probabilistic question discussed above to the study of lower rank.

We claim that, if G is a pro-p group satisfying $Q(G, k) = 1$, then the lower rank of G is at most k. Indeed, if H is an open subgroup of G, then we must have $Q(H, k) = 1$. In particular, any open subgroup H of G contains a k-generated open subgroup, which proves the claim.

Corollary 5.11. (Barnea–Larsen, 1998) *Let Γ be as in Theorem 5.10. Then Γ has lower rank 2.*

This applies, for example, for $\Gamma = SL_d(\mathbb{F}_p[[t]])$ when $p \geq 5$. It would be interesting to construct explicitly a series of 2-generated open subgroups of these groups which converges to the identity.

Some more general questions can also be considered. For example

Problems.

22. Let G be a pro-p group of finite lower rank. Does it follow that $Q(G, k) = 1$ for some k?
23. Find just-infinite pro-p groups (of finite width?) whose lower rank is infinite.

Again, the groups constructed by Grigorchuk seem natural candidates.

While this section contains various recent results on the subgroup structure of pro-p groups, it does not discuss the subgroup structure of finite p-groups. For some exciting new results in this area, which were obtained using Lie methods, see [4] and [38].

6 Growth and classification

There are several ways to measure the growth a pro-p group G. The first is by fixing a filtration $\{G_i\}$ of G (usually a central series) and considering the orders (or ranks) of the respective factors G_i/G_{i+1}. The second is by considering the word growth, in the sense of Milnor, Gromov, and others. And the third is by focusing on the subgroup growth of G, namely the function attaching to each pth power n the number of index n subgroups of G.

In this section we focus on the first interpretation of growth, though we shall also discuss subgroup growth very briefly. For detailed discussion of subgroup growth and of word growth, see Chapters 4 and 8.

An ambitious goal in the context of growth is: classify the groups with prescribed growth behaviour. In most cases this is far too hard, so one aims at structure theorems, or characterisations, instead of detailed classification. Alternatively, one regards classification in a weak sense of division into natural and well behaved classes. But in many cases even this is too hard.

I shall describe the few cases where some kind of classification, or at least general structure theorems, can be obtained under suitable growth assumptions. I will also comment (and sometimes speculate) on possible future challenges and ways of attack.

6.1 Connection to Lie algebras

The first interpretation of growth leads naturally to questions on Lie algebras. This is because the natural filtrations $\{G_i\}$ can often be used for the graded Lie ring construction $L(G) = \sum_{n\geq 1} G_i/G_{i+1}$; assumptions on the growth of G then translate to conditions on the growth of $L(G)$ (in terms of the dimensions of the homogeneous components, or similar invariants).

Thus, to each group-theoretic growth related question, there corresponds a Lie-theoretic question. For example, if we are interested in groups G for which G_i/G_{i+1} have bounded order, we can first examine graded Lie algebras $L = \bigoplus_{i\geq 1} L_i$ such that the $|L_i|$ (or $\dim L_i$) are bounded.

It should be emphasized that this procedure does not usually lead to reductions in the strict sense (like some we encountered in Section 3). For suppose we solve the Lie-theoretic question and understand (or even classify) the Lie algebras L of given growth type; can we then understand (or classify) the respective groups? Not much is known about the pro-p groups with a given associated graded Lie algebra $L(G)$. For example, we don't even know if the Lie algebra of the Nottingham group (say, with respect to the lower central series) characterizes the Nottingham group. It is also unclear whether pro-p groups whose graded Lie algebra is isomorphic to $\mathcal{G} \otimes t\mathbb{F}_p[t]$ for some simple classical Lie algebra $\mathcal{G}$ over $\mathbb{F}_p$ are always analytic, or linear. In general $L(G)$ need not determine G, for example the groups $\mathrm{SL}_d^1(\mathbb{F}_p[[t]])$ and $\mathrm{SL}_d^1(\mathbb{Z}_p)$ $(p > 2)$ have the same graded Lie algebra $sl_d(\mathbb{F}_p) \otimes t\mathbb{F}_p[t]$ with respect to their lower central series. Thus, it may well be that, in spite of various nice results for Lie algebras, applications to pro-p groups would still be far away.

There is yet another difficulty which is quite disturbing. For the questions we have in mind, we are naturally interested in Lie algebras which come from groups. But this may be a special class of Lie algebras. For example, we now know that only a small fraction of the graded Lie algebras of maximal class over $\mathbb{F}_p$ come from p-groups or pro-p groups of maximal class. Unless we have a way to identify those Lie algebras which come from groups, the Lie-theoretic questions we end up considering are far too general (and possibly too difficult) for our original purpose. Indeed, going back to the previous example, there is only one Lie algebra coming from pro-p groups of maximal class, and it is metabelian and very easily described. The recent study of graded Lie algebras of maximal class over $\mathbb{F}_p$ (see [22] and [23]) required considerable effort, but it deals with algebras which seem to have very little to do with groups.

It seems to me that the two sources of difficulty described above are the main obstacles in understanding pro-p groups of given growth type. However, one should not be overly pessimistic. In some cases other Lie ring constructions (reflecting the group structure more closely) can be (and have been) helpful. Note also that in the context of pro-unipotent groups in zero characteristic, the relation to Lie algebras is tighter, and it gives rise to a wider spectrum of results.

6.2 Coclass theory for groups

The main area where growth assumptions have led to non-trivial structure and classification theorems in pro-p groups is the so-called coclass theory. I assume that the readers of this book are at least vaguely familiar with some of the results and the basic concepts of this theory, and so the background I give below is fairly brief. More details can be found in Chapter 2 and in [27], Chapter 10.

The coclass theory originated in the study of the so-called p-groups of maximal class, conducted in the 1950s through the 1970s. These groups have order p^n and nilpotency class $n-1$, so all their non-trivial central factors G_i/G_{i+1} have order p, except the first one which has order p^2. If $p = 2$ there are exactly three such groups for each $n \geq 5$, namely the dihedral group, the semi-dihedral group, and the generalized quaternions. Note that all these groups, while being very far from abelian on the one hand, have an abelian subgroup of index 2. Blackburn's seminal paper [17] provides a lot of information on the structure of such groups for odd primes p, obtaining complete classification for $p = 3$. Additional results were obtained in the 1970s by Shepherd (in his unpublished Ph.D. thesis) and by Leedham-Green and S. McKay [71], [72], [73]. Their results show that the case $p = 2$ is in a way typical: a p-group of maximal class always has a subgroup of class at most two whose index is bounded in terms of p alone.

In their influential 1984 paper [76], C.R. Leedham-Green and M.F. Newman proposed a far reaching extension of the theory of p-groups of maximal class. They defined the *coclass* of a group of order p^n and class c to be $n - c$. Thus the groups of maximal class are precisely those of coclass 1. The approach of [76] was to try to classify p-groups using the coclass (and not the class) as the primary invariant. The basic intuition was that the family of all p-groups of a given coclass is well behaved in some sense. The exact sense was expressed in a series of conjectures, which attracted significant attention in the following decade.

Some of these conjectures are really about pro-p groups; this in itself was perhaps the most seminal idea introduced in [76]. The coclass of a pro-p group G is the supremum of the coclasses of its finite quotients (possibly infinity). Thus an infinite pro-p

group G has finite coclass if and only if its lower central factors G_n/G_{n+1} are all finite and have order p for all large n.

We now state the coclass conjecture from the weakest to the strongest.

Conjecture E. *Given p and r, there are only finitely many isomorphism types of infinite soluble pro-p groups of coclass r.*

Conjecture D. *Given p and r, there are only finitely many isomorphism types of infinite pro-p groups of coclass r.*

Conjecture C. *Pro-p groups of finite coclass are soluble.*

Conjecture B. *For some function g, every finite p-group of coclass r has derived length at most $g(p, r)$.*

Conjecture A. *For some function f, every finite p-group of coclass r has a normal subgroup N of class at most* 2 (1, *if* $p = 2$) *and index at most $f(p, r)$.*

Some of these conjectures can be reformulated as combinatorial statements on the structure of certain graphs. Thus, let Γ be a (directed) graph whose vertices are all p-groups of coclass r, where (H, G) is an edge if there is an epimorphism from H to G whose kernel has order p. Then Conjecture D is tantamount to saying that there are only finitely many infinite chains in the graph Γ.

The coclass conjectures turned out to be extremely accurate, and they have all now become theorems. Conjecture E was proved by Leedham-Green, McKay and Plesken in [74], [75]. Focusing on Conjecture C, S. Donkin [28] showed that p-adic analytic pro-p groups of finite coclass are soluble, provided $p > 3$. The complicated proof applies the classification of simple p-adic Lie algebras, as well as Iwahori and Matsumoto's arithmetic theory of p-adic Chevalley groups. Soon afterwards Leedham-Green showed that pro-p groups of finite coclass are always p-adic analytic [69], and thus confirmed Conjecture C for $p > 3$. In [126] we used a different method and obtained a short proof of Conjecture C for all primes p. As was observed in [76], a soluble pro-p group G of finite coclass has the structure of a p-adic pre space group; this means that G is an extension of a free finitely generated p-adic module by a finite p-group (which may not act faithfully), so in particular it is abelian-by-finite. Assuming further that G is just-infinite, it follows that G is a p-adic space group. We therefore obtain

Theorem 6.1. (Donkin–Leedham-Green, Shalev–Zelmanov) *Let G be a just-infinite pro-p group of finite coclass. Then G has the structure of a p-adic space group.*

Let us now briefly discuss the situation for finite p-groups. In [70] Leedham-Green used an inverse limit argument to deduce Conjecture A from Conjecture C. This approach, being non-constructive, did not yield bounds on the function $f(p, r)$ appearing in Conjecture A. An effective and self-contained proof of Conjecture A is given in [119], where the following result is obtained.

Theorem 6.2. (Shalev, 1994) *Let G be a p-group of coclass r. If $p > 2$ then $G_{2(p^r-p^{r-1}-1)}$ has class at most* 2. *If $p = 2$ and $|G| \geq 2^{2^{2r+5}}$ then $G_{7\cdot 2^r-2}$ is abelian.*

This yields explicit bounds on the function $f(p, r)$.

The main tools in the proof are the theory of powerful p-groups and some Lie ring methods. Let us make some remarks on the role of Lie rings in the proofs in [126] and [119]. Let G be a pro-p group of finite coclass. Then $L(G) = \bigoplus_{i\geq 1} G_i/G_{i+1}$ is an $\mathbb{N}$-graded Lie ring, and $L = L(G)/pL(G)$ is an $\mathbb{N}$-graded Lie algebra of finite coclass over the field $\mathbb{F}_p$. Thus, for large n, the homogeneous components of L^n are all 1-dimensional. It turns out that, using the power map $g \mapsto g^p$, one can turn L^n into a finite-dimensional Lie algebra over the polynomial ring $\mathbb{F}_p[t]$. Assuming G is non-soluble, and passing to the fraction field $\mathbb{F}_p(t)$ one obtains a perfect Lie algebra K, graded by some finite cyclic group C_d, whose homogeneous components are all 1-dimensional. Furthermore, the uniserial action of G on G_n enables one (possibly by extending scalars further) to construct a non-singular derivation D of K, satisfying $D^d = 1$, which moves the homogeneous components of K cyclically. At this point a crucial piece of information is invoked: the dimension d of K, which coincides with the dimension of G as a p-adic analytic group, has the form $(p-1)p^s$ for some integer $s \geq 0$. Hence $E = D^{p^s}$ is a non-singular derivation of K satisfying $E^{p-1} = 1$. It follows that E is semisimple and its eigenvalues all lie in $\mathbb{F}_p^*$. Decomposing $K = \oplus_{a\in\mathbb{F}_p} K_a$ as a sum of eigenspaces for E, we find that K is C_p-graded with $K_0 = 0$. Now, for non-zero elements $a, b \in \mathbb{F}_p$ there is $i \leq p-1$ such that $a + ib = 0$, hence $L_a \mathrm{ad}(L_b)^i \subseteq L_{a+ib} = 0$. This implies that K satisfies the $(p-1)$-Engel condition on homogeneous elements. By Jacobson's graded version of Engel's theorem (see [48]) it follows that K is nilpotent, hence not perfect, a contradiction (for a slightly different argument, spelt out in detail, see Chapter 10 of [27]).

We see that the non-existence of a finite-dimensional perfect Lie algebra L with a non-singular derivation of order dividing $(p-1)p^s$ plays a role in some proofs of the coclass conjectures. In fact we can strengthen this as follows.

Proposition 6.3. (Shalev, 1999) *Let L be a non-nilpotent finite dimensional Lie algebra over a field of characteristic $p > 0$. Suppose L admits a non-singular derivation D satisfying $D^d = 1$, for some positive integer d. Then the p'-part of d is at least $p^2 - 1$.*

This result, proved in [125], is the best possible, in the sense that some simple Lie algebras of characteristic p admit non-singular derivations of multiplicative order $p^2 - 1$. Moreover, as we shall see below, these Lie algebras and derivations play a key role in the theory of graded modular Lie algebras of maximal class.

Let me close this subsection by referring the reader to a new set of coclass conjectures, recently posed by M.F. Newman and E.A. O'Brien in [104]. These detailed conjectures concern mostly the structure of the graph Γ of 2-groups of given coclass described earlier, and are stated in full in Chapter 2.

The Newman–O'Brien conjectures are supported by strong computer evidence. Moreover, Marcus du Sautoy [32] has just confirmed the peridocity statement in Conjecture P of [104]; the proof is given in Chapter 9. This proof uses model theory to prove rationality of certain integrals associated with zeta functions encoding the number of p-groups of given order and coclass. The fact that model-theoretic methods are applicable in this context is a pleasant surprise.

6.3 Coclass for Lie algebras

Let L be a residually nilpotent Lie algebra and let L^i be its lower central series. If $\dim(L^i/L^{i+1})$ are all finite, and $\dim(L^i/L^{i+1}) \leq 1$ for all sufficiently large i, we say that L has finite coclass. The coclass of such a Lie algebra is defined by $cc(L) = \sum_{i\geq 1, L^i \neq 0}(\dim(L^i/L^{i+1}) - 1)$. Lie algebras of coclass 1 are also referred to as Lie algebras of *maximal class* or as *algèbres filiformes*. In group-theoretic contexts it is natural to consider $\mathbb{N}$-graded Lie algebras $L = \bigoplus_{i\geq 1} L_i$ generated by L_1, since Lie rings associated with groups have these properties. Such a Lie algebra has finite coclass if $\dim L_i$ are all finite and are at most 1 for all large i.

We shall first examine Lie algebras of finite coclass over fields of characteristic zero, and then turn to the situation in characteristic $p > 0$.

Lie algebras of maximal class over fields of characteristic zero were considered by M. Vergne in the 1960s [138],[139]. More recently they have been used in [15] and [18] in the context of a conjecture of Milnor [100]. Some examples of Lie algebras of maximal class are given below.

Example 1. Let $M(\infty)$ be a linear space with a basis x, y_i $(i \geq 1)$. Define Lie products on M by $[y_i, x] = y_{i+1}$ and $[y_i, y_j] = 0$ $(i, j \geq 1)$. Then $M(\infty)$ is a Lie algebra of maximal class. By forming quotients of $M(\infty)$ we obtain finite-dimensional Lie algebras of maximal class, namely $M(d) = M(\infty)/M(\infty)^d$ $(d \geq 2)$. Clearly, these Lie algebras are all soluble (indeed, metabelian).

Example 2. Consider the Witt algebra $W_1 = \mathrm{Der}(F[x])$ where F is a field of characteristic zero. Then W_1 is $\mathbb{Z}$-graded, and its homogeneous component of degree n $(n \geq -1)$ is spanned by the derivation $e_n = x^{n+1}d/dx$. These basis elements satisfy the relation $[e_n, e_m] = (n - m)e_{n+m}$. Let W_1^+ be the positive part of W_1, namely the subalgebra spanned by e_n for $n \geq 1$. Then W_1^+ is an $\mathbb{N}$-graded Lie algebra of maximal class. It is generated by e_1 and e_2, and it is clearly non-soluble.

In the work [127] by Zelmanov and myself we study $\mathbb{N}$-graded Lie algebras $L = \bigoplus_{i\geq 1} L_i$ of finite coclass in zero characteristic. It turns out that if we also assume that L is generated by L_1, then quite a lot can be said about its structure. First, it was shown by Vergne [139] that if L is a finite-dimensional graded Lie algebra of maximal class which is generated in degree 1, then $L/Z(L) \cong M(d)$ for some d. In [127] we extend this as follows.

Theorem 6.4. (Shalev–Zelmanov, 1997) *There exists a function f such that, if L is an $\mathbb{N}$-graded Lie algebra of finite coclass r in zero characteristic which is generated in degree 1, then*

$$L/Z_{f(r)}(L) \cong M(d)$$

for some $d \leq \infty$.

Here $Z_i(L)$ denotes the ith term of the upper central series of L.

Corollary 6.5. *Let L, r be as above. Then*

(i) L has an ideal I of codimension 1 *satisfying* $\dim I^2 \leq f_1(r)$, *for a suitable function f_1.*

(ii) L has a nilpotent ideal J of class at most two satisfying $\dim L/J \leq f_2(r)$, *for a suitable function* f_2.

(iii) L is soluble of derived length at most $f_3(r)$, for a suitable function f_3.

(iv) If L is just-infinite, then $L \cong M(\infty)$.

This shows that $\mathbb{N}$-graded Lie algebras in zero characteristic generated in degree 1 satisfy the natural analogues of the coclass conjectures for p-groups and pro-p groups in a somewhat stronger form (for example, parts (i) and (iv) above do not hold for pro-p groups). It turns out that generation in degree 1 is an essential requirement: the positive part W_1^+ of the Witt algebra is a non-soluble $\mathbb{N}$-graded Lie algebra of maximal class (generated in degrees 1 and 2). However, Zelmanov and I conjecture that the Witt algebra is essentially the only non-soluble Lie algebra of maximal class in zero characteristic. More precisely

Conjecture 24. Let L be a Lie algebra of maximal class over an algebraically closed field of characteristic zero. Then the pro-nilpotent completion of L is either soluble or commensurable to the completion of W_1^+.

This conjecture can be reduced to the case of $\mathbb{N}$-graded Lie algebras, but we can no longer restrict our attention to algebras generated in degree 1. The following preliminary result from [127] provides some positive evidence.

Theorem 6.6. *Let $L = \bigoplus_{i\geq 1} L_i$ be a non-soluble infinite-dimensional $\mathbb{N}$-graded Lie algebra of maximal class, and suppose that $L = \langle L_1, L_2\rangle$. Then L is isomorphic to W_1^+.*

This result already follows from the work of A. Fialowski which is outlined in [36], of which we were unaware at the time our paper was published. We thank her for bringing her work to our attention. Additional evidence supporting the conjecture was recently provided by M.R. Vaughan-Lee. He used computer calculations to obtain a similar result under the assumption $L = \langle L_1, L_n\rangle$ where n is 3 or 4. Other partial results have been obtained by Zelmanov and myself, but the general case is still open.

Let me now turn to Lie algebras of finite coclass in positive characteristic. The case of restricted Lie algebras turns out to be very special. Riley and Semple showed in [109] that $\mathbb{N}$-graded restricted Lie algebras of coclass r in characteristic p are finite-dimensional, and their dimension is bounded above in terms of p and r. Non-restricted modular Lie algebras of finite coclass behave rather differently. Not only can their dimension be infinite, they do not satisfy the analogues of the coclass conjectures. Indeed, a construction from [120] gives rise to the following:

Proposition 6.7. (Shalev, 1994) *For each prime p there exists a non-soluble $\mathbb{N}$-graded Lie algebra $L = \bigoplus_{i\geq 1} L_i$ of maximal class over the field $\mathbb{F}_p$ which is generated by L_1.*

The idea behind the proof is to let L be the positive part of a certain loop algebra extended by a certain non-singular derivation of degree 1. More specifically, the construction starts with finite-dimensional simple Lie algebras $\mathcal{G}$ constructed by Albert and Frank [2]. These algebras have dimensions $p^k - 1$ $(k \geq 2)$ and they are $(C_p)^k$-graded. However, it can be shown that they also admit a cyclic grading $\mathcal{G} = \bigoplus_{0\leq i<p^k-1} \mathcal{G}_i$ with 1-dimensional homogeneous components, and a non-singular

derivation D satisfying $D(\mathcal{G}_i) = \mathcal{G}_{i+1}$ (cyclically). The required infinite-dimensional Lie algebra L is constructed by adjoining Dt to the positive part of the twisted loop algebra $\oplus_{n \in \mathbb{Z}} \mathcal{G}_{n \bmod p^k - 1} \otimes t^n$. Lie algebras of this type are termed AFS-algebras in [22].

Note that the above construction is in a way the inverse of the one described in our sketch of the proof of Conjecture C. There we passed from an infinite-dimensional Lie algebra of finite coclass to a finite-dimensional perfect Lie algebra admitting a certain non-singular derivation which cannot exist. Here we pass from a finite-dimensional perfect Lie algebra with a certain non-singular derivation to an infinite-dimensional Lie algebra of finite coclass.

For a systematic study of non-singular derivations of finite-dimensional simple modular Lie algebras of Cartan type, see the work [14] by Benkart, Kostrikin and Kuznetsov.

For some time it was hoped that modular graded Lie algebras of maximal class are still well behaved in some sense, e.g., being either soluble, or commensurable to the positive part of loop algebras (see [121] Problem 28). This was disproved in [22].

Theorem 6.8. (Caranti–Mattarei–Newman, 1997) *For each prime p there are $2^{\aleph_0}$ isomorphism types of $\mathbb{N}$-graded Lie algebras of maximal class over $\mathbb{F}_p$, generated in degree 1.*

The Caranti–Mattarei–Newman algebras were constructed using the algebras in [120] via a technical process which they call inflation, in which, roughly speaking, new homogeneous components are inserted between existing ones. When inflation steps are carried out countably many times in all possible ways, $2^{\aleph_0}$ pairwise non-isomorphic algebras of maximal class arise.

Does this mean that modular Lie algebras of maximal class cannot be classified? Not necessarily. Indeed, a brand-new surprising result of Caranti and Newman [23] does provide some sort of classification.

Theorem 6.9. (Caranti–Newman, 1999) *All $\mathbb{N}$-graded Lie algebras of maximal class in characteristic $p > 2$ generated in degree 1 are obtained via (finitely many or infinitely many) inflation steps from some AFS-algebra.*

Let me end this subsection with some

Problems.

25. Study graded modular Lie algebras of maximal class which are not generated in degree 1.

26. Study graded modular Lie algebras of finite coclass.

The first problem may be difficult, in view of the fact that even the zero characteristic case is still not completely resolved (see Conjeture 24 above). As for the second problem, it is natural to start with algebras generated in degree 1, and to check whether they can all be constructed from AFS-algebras using some extension of the Caranti–Newman method.

6.4 Pro-p groups of finite width

The (relatively) happy end of the coclass story for pro-p groups led people to examine pro-p groups of finite width in some detail. In fact there are several non-equivalent definitions of finite width. One requires that the lower central factors have bounded orders, the other only requires that these factors have bounded rank. One can also pose similar conditions on all central sections $N/[N, G]$ of G (where $N \lhd G$). Or one can use other filtrations (the lower p-series, the dimension subgroups series) instead of the lower central series. The right definition often emerges in the presence of a convincing theorem. In the absence of such a theorem the choice of precise definition is more a matter of personal taste.

Can one classify just-infinite pro-p groups of finite width (say up to commensurability)? or at least develop a good structure theory for them? It is natural to split the groups in question into several families, as follows:

1. p-adic analytic groups.
2. $\mathbb{F}_p[[t]]$-analytic groups.
3. other groups.

The recent book [60] by Klaas, Leedham-Green and Plesken focuses on the first family above, namely that of p-adic analytic pro-p groups of finite width. Some groups from the second family have been discussed in previous sections; among these for example are the so-called $\mathbb{F}_p[[t]]$-perfect groups (see also [84], [27] Chapter 13). As we saw in §5.1, the Nottingham group belongs to the third family; in spite of its narrowness, the Nottingham group has a very wild subgroup structure [21].

Theorem 6.10. (Camina, 1997) *Every finitely generated pro-p group is isomorphic to some closed subgroup of the Nottingham group.*

A proof of this result is given in Chapter 10. In particular it follows that the Nottingham group is not linear over any field, since it contains soluble subgroups of unbounded derived length. This fact was first discovered by C. Leedham-Green and A. Weiss.

The Nottingham group has a multi-dimensional version, which is the group of all normalized automorphisms of the power series ring $\mathbb{F}_p[[t_1, \ldots, t_d]]$. This group in turn has some distinguished subgroups, namely the centralizers of the various differential forms (Special, Contact and Hamiltonian). These groups, which we call *pro-p groups of Cartan type*, are described in more detail in [121]. They seem to be the pro-p analogues of the restricted modular Lie algebras of Cartan type. Since they have infinite width (though they still have slow growth in the sense that their Lie algebra has finite GK-dimension) I will not elaborate here on their structure.

It has recently been realized that the class of just-infinite pro-p groups of finite width contains certain groups which in a way are even wilder than the Nottingham group. These groups are the pro-p completions of certain finitely generated infinite p-groups constructed by Grigorchuk and others. Grigorchuk's groups, which are residually p, can be defined recursively as automorphisms of certain trees (or intervals). It is now known that these groups have finite width. For more details see Chapter 4.

Let L be the graded Lie ring associated with a Grigorchuk group G (or its completion). Then L has some interesting properties. Its homogeneous components have dimension at most 3, and their elements are ad-nilpotent, but L is not PI (by Theorem 2.1). It can be shown that Lie algebras as above do not exist in characteristic zero [93].

Leedham-Green suggested that, in studying pro-p groups of finite width, one should focus first on the groups which are hereditarily just-infinite. Since Grigorchuk's groups are not hereditarily just-infinite (some of their finite index subgroups decompose as direct products) they will not play a role in this part of the programme.

As this subsection illustrates, our understanding of pro-p groups of finite width is currently rather limited. The main achievements so far have been in the study of specific examples, and not in general structure theorems. In the next subsection I briefly describe the current knowledge of Lie algebras of finite width.

6.5 Lie algebras of GK-dimension 1

Let me discuss the notion of GK-dimension for algebras, which, in a way, generalizes the concept of width. In [37] Gelfand and Kirillov suggested a way to measure the growth of an infinite-dimensional algebra. Let A be an infinite-dimensional (not necessarily associative) algebra over a ground field F. Suppose that the algebra A is generated by a finite-dimensional subspace V. For $n \geq 1$ let V^n denote the F-linear span of all products in elements of V of length at most n with arbitrary arrangements of brackets. Clearly $V = V^1 \subseteq V^2 \subseteq \ldots$ is an ascending chain of finite-dimensional subspaces of A and $\bigcup_{n\geq 1} V^n = A$. The Gelfand–Kirillov dimension of A is

$$\operatorname{GKdim}(A) = \limsup_{n\to\infty} \frac{\log \dim V^n}{\log n}.$$

This turns out to be independent of the choice of V.

An algebra A has finite Gelfand–Kirillov dimension if $\dim V^n$ is polynomially bounded; it has Gelfand–Kirillov dimension 1 if $\dim V^n$ grows linearly with n. Note that graded Lie algebras of finite width – in the sense that their homogeneous components have bounded dimension – are of GK-dimension 1.

In the special case of associative algebras of GK-dimension 1 there is a nice structure theory developed in [134].

Theorem 6.11. (Small–Stafford–Warfield, 1985) *Let A be a finitely generated associative algebra of GK-dimension 1. Then*

(i) The Jacobson radical Rad(A) is nilpotent.

(ii) If Rad(A) $= 0$ then there is an element c in the center of A such that A is a finite module over the subalgebra $F[c]$ generated by c.

(iii) A is PI.

Note that part (iii) follows from (i) and (ii). A celebrated result of Bergman [16] shows that there are no associative algebras of GK-dimension strictly between 1 and 2. Some progress has recently been made in the study of associative algebras of GK-dimension 2 in the context of the so called non-commutative algebraic geometry [5].

Let us now turn to Lie algebras. Recently, V.M. Petrogradsky [107] has shown that for every $\alpha \geq 1$ there exists a Lie algebra whose GK-dimension is precisely α. In particular, Bergman's theorem does not hold for Lie algebras. Similarly, there is no analogue of Theorem 6.11 for Lie algebras; as we shall see below, results on Lie algebras of GK-dimension 1 are obtained under some additional assumptions.

In the mid-1960s Victor Kac initiated the study of $\mathbb{Z}$-graded simple Lie algebras $L = \bigoplus_{i \in \mathbb{Z}} L_i$ of *polynomial growth*, in the sense that $\dim L_i \leq |i|^n + m$ for some fixed $n, m \geq 0$ and for all $i \in \mathbb{Z}$. The ground field F is assumed to be algebraically closed of characteristic 0. Let us consider examples of such algebras.

Example 1. Let $\mathcal{G}$ be a finite-dimensional simple Lie algebra over F. Then the algebra $L(\mathcal{G}) = \mathcal{G} \otimes F[t, t^{-1}]$ defined in Section 4, and its associated loop algebras $L(\mathcal{G}, k, \alpha)$ with respect to C_k-gradings α, are graded simple, with homogeneous components of bounded dimension.

Example 2. Let $W_m = \mathrm{Der}(F[t_1, \ldots, t_m])$ be the full algebra of derivations of the polynomial ring with m variables. Then W_m, referred to as the mth Witt algebra, is simple, $\mathbb{Z}$-graded, and has GK-dimension m.

The algebras W_m contain three important series of Special (S), Contact (K) and Hamiltonian (H) subalgebras which annihilate the corresponding differential forms. The resulting algebras W_m, S_m, K_m, H_m are referred to as algebras of Cartan type (see Kostrikin and Shafarevich [63]).

Example 3. The (centerless) Virasoro algebra $L = \mathrm{Der}(F[t, t^{-1}])$ is $\mathbb{Z}$-graded with 1-dimensional homogeneous components.

It turns out that the examples above exhaust the $\mathbb{Z}$-graded simple Lie algebras of polynomial growth.

Theorem 6.12. (Mathieu, 1990) *Let F be an algebraically closed field of zero characteristic. A $\mathbb{Z}$-graded simple Lie algebra $L = \bigoplus_{i \in \mathbb{Z}} L_i$ of polynomial growth is either finite-dimensional or else isomorphic to one of the algebras of Examples 1–3.*

This result, proved in [94], confirms a conjecture of Kac (see also [49] for partial results). In the algebras of Examples 1 and 3 the dimensions $\dim L_i$ are uniformly bounded, which corresponds to the case of GK-dimension 1. For algebras of Cartan type this is true for the Witt algebra W_1.

The group-theoretic context motivates the consideration of problems in Lie algebras which are somewhat similar to those discussed above. The main difference is that we are no longer dealing with simple $\mathbb{Z}$-graded algebras. The algebras of type $L(G)$ are $\mathbb{N}$-graded and cannot be simple. A natural substitute for simplicity is the requirement that L be *just-infinite* in the sense that it is infinite-dimensional, but all its proper quotients are finite-dimensional.

Dropping the simplicity assumption seems to make the situation much more involved, and standard approaches (e.g., using representation theory) are no longer applicable. What we can offer at this stage is a conjecture, and some partial results supporting it. The conjecture, suggested by Zelmanov and myself, deals not just with $\mathbb{N}$-graded algebras, but with filtered algebras as well. Recall that a descending chain of ideals $L = L_1 \supset L_2 \supset \ldots$ is called a *filtration* if $[L_i, L_j] \subseteq L_{i+j}$ for all $i, j \geq 1$. A *filtered* algebra is an algebra equipped with a filtration. The *completion* of a filtered algebra L is the completion of L with respect to the topology generated by the L_i.

Note that results on filtered Lie algebras in zero characteristic can be reformulated in the language of pro-unipotent groups.

We propose the following

Conjecture 27. Let $L = L_1 \supset L_2 \supset \dots$ be a just-infinite filtered Lie algebra over an algebraically closed field of zero characteristic. Suppose that the dimensions $\dim(L_i/L_{i+1})$ $(i \geq 1)$ are uniformly bounded. Then one of the following holds.

1. L is soluble.

2. The completion of L is commensurable to the completion of the positive part of a loop algebra.

3. The completion of L is commensurable to the completion of W_1^+.

Furthermore, if L above is $\mathbb{N}$-graded, then either L itself is soluble or else L is commensurable to one of W_1^+ or the positive part of a loop algebra.

The following result from [92] confirms the conjecture for filtered algebras with sufficiently many derivations.

In a filtered algebra $L = L_1 \supset L_2 \supset \dots$ we let $L_0 = \{d \in \mathrm{Der}(L) | d(L_i) \subseteq L_i \text{ for all } i \geq 1\}$.

Theorem 6.13. (Martinez–Zelmanov, 1999) *Under the assumptions of the above conjecture, if L_0 is not soluble then the completion of L is commensurable to the completion of the positive part of some loop algebra.*

Another special case of the conjecture has just been proved by Zelmanov and myself; it concerns Lie algebras with so-called "polynomial multiplication," and the proof uses Mathieu's Theorem and ideas from vertex algebras.

In contrast with the zero characteristic case, we are not in a position to formulate a reasonable conjecture concerning the structure of just-infinite modular Lie algebras of finite width.

6.6 Subgroup growth

As indicated earlier, another way to measure the growth of pro-p groups is by studying their subgroup growth. While this study is not necessarily of Lie-theoretic flavour, it often applies Lie theory in the broad sense which includes p-adic Lie groups as well as Lie algebras. Here I will outline some of these applications, while focusing on recent results. The reader is referred to Chapter 8.

Given a finitely generated pro-p group G and a pth power n, let $a_n(G)$ be the number of subgroups of index n in G. Note that these subgroups are all open, by a result of Serre ([27], Chapter 1). It is known that if G is a nonabelian free pro-p group then $a_n(G)$ grows exponentially with n [46]. We are interested in the structure and possible classification of pro-p groups whose subgroup growth is slow in some sense.

We say that G has polynomial subgroup growth if $a_n(G) \leq n^\alpha$ for some $\alpha \geq 0$ and for all n. If G has this property, then its degree is defined by $\deg(G) = \limsup_{n\to\infty} \log a_n(G)/\log n$. An important discovery of Lubotzky and Mann is that pro-p groups of polynomial subgroup growth are p-adic analytic [82]. In fact, there are several characterizations of p-adic analytic pro-p groups in terms of subgroup growth (see [27], Interlude A):

Theorem 6.14. *The following are equivalent for a finitely generated pro-p group G:*
(i) G has polynomial subgroup growth.
(ii) $a_n(G) \leq n^{c \log n}$ for some $c < 1/8$ and for all large n.
(iii) The zeta function $\sum_{k \geq 0} a_{p^k}(G)(p^k)^{-s}$ is a rational function of p^{-s}.
(iv) G is p-adic analytic.

The equivalence of (ii) and (iv) was obtained in [115]. The equivalence of (iii) and (iv) is a deep result of du Sautoy [30], proved using model-theoretic methods. The equivalence of (i) and (iv) played a role in the characterization of finitely generated abstract groups of polynomial subgroup growth [83]. However, recently Dan Segal has shown that the use of p-adic analytic groups (and pro-p groups in general) in this characterization can be avoided [113]. As we shall see below, in some more recent results on the subgroup growth of discrete groups, pro-p groups and p-adic analytic groups still seem to play a vital role.

Theorem 6.14 is a fairly general result on the subgroup growth of pro-p groups; the results which followed it were usually of a more detailed nature. These include results concerning pro-p groups whose subgroup growth is slightly faster than polynomial, as well as results on pro-p groups of slow polynomial growth.

Pro-p groups whose subgroup growth is super-polynomial have growth at least $n^{c \log n}$. It is therefore tempting to find groups of growth type $n^{c \log n}$, and to ponder whether such groups could be characterized. The first task is handled in [84], where it is shown that the first congruence subgroup of $SL_d(\mathbb{F}_p[[t]])$ – and R-perfect groups in general – have subgroup growth of type $n^{c \log n}$. I obtained a similar result for the Nottingham group – as well as for its high dimensional generalizations (the normalized automorphisms of $\mathbb{F}_p[[t_1, \ldots, t_d]]$). The proofs of these estimates often have a Lie-theoretic flavour: using Lie algebras one can often obtain a level-index theorem, showing that a subgroup of small index (say n) must contain a congruence subgroup (say C) of small index (at most n^c); this reduces counting index n subgroups of G to counting index n subgroup of G/C, which is a lot easier.

The second task, that of characterizing pro-p groups of subgroup growth type $n^{c \log n}$, seems to be very difficult, and may be related to the challenge of understanding pro-p groups of finite width.

Problem 28. Let G be a just-infinite pro-p group of finite width. Does it follow that $a_n(G) \leq n^{c \log n}$ for some c and for all n?

In this context it would be interesting to determine the subgroup growth of the Grigorchuk groups.

In a different direction, it is possible that pro-p groups of growth type $n^{c \log n}$ have associated Lie algebras of finite GK-dimenion (with, say, respect to the lower central p-series). In other words, the question is about possible relations between our two ways of measuring growth.

Let me now return to pro-p groups of polynomial subgroup growth, and consider those whose growth is particularly slow.

We say that the subgroup growth of a group G is *sublinear* if $a_n(G)/n \to 0$ as $n \to \infty$. Residually finite groups with this property were characterized in [122] using a reduction to pro-p groups. In this case we have

Theorem 6.15. *A pro-p group has sublinear subgroup growth if and only if it has a central procyclic subgroup of finite index.*

The proof relies on a rather delicate property of powerful pro-p groups, which is of some independent interest (see [122] for the proof).

Proposition 6.16. *Let G be a finitely generated powerful pro-p group, and let $G_i = G^{p^{i-1}}$ ($i \geq 1$). Let H be a closed subgroup of G satisfying $N_G(H) = H$. Then there exists a constant c such that $|N_{G/G_i}(HG_i/G_i) : HG_i/G_i| \leq c$ for all i.*

In other words, if H is self-normalizing in G, then in the finite quotients G/G_i the images of H are almost self-normalizing.

To see how this is used in proving the theorem, let G be a pro-p group of sublinear subgroup growth. We claim that G is soluble. Suppose otherwise. Then since G is Noetherian, it has a maximal closed soluble subgroup H, which is necessarily of infinite index. It is clear that H is self-normalizing. Passing to the finite quotient G/G_i and applying the proposition above, we conclude the image of H and its conjugates give rise to at least $c^{-1}|G : HG_i|$ distinct subgroups of index $|G : HG_i|$. This yields $a_n(G) \geq c^{-1}n$ for infinitely many n, a contradiction. Therefore G is soluble, as claimed. At this stage it is not hard to complete the proof.

The next natural step is the following.

Problem 29. Characterize pro-p groups of linear subgroup growth.

Note that, while finitely generated residually finite groups of linear subgroup growth are virtually cyclic, the analogous result does not hold for pro-p groups; indeed $\mathbb{Z}_p \times \mathbb{Z}_p$ has linear subgroup growth. On the other hand, it follows from results of Ilani that the congruence subgroups of $SL_2(\mathbb{Z}_p)$ ($p > 2$) have super linear subgroup growth. This led me to speculate that pro-p groups of linear subgroup growth might always be soluble. However, this has just been refuted by Klopsch, whose work might eventually lead to a complete solution of Problem 29.

References

[1] A.G. Abercrombie, Subgroups and subrings of profinite rings, *Math. Proc. Cambr. Phil. Soc.*, **116** (1994), 209–222.

[2] A.A. Albert and M.S. Frank, Simple Lie algebras of characteristic p, *Rend. Torino*, **14** (1954/5), 117–139.

[3] J.L. Alperin, Automorphisms of solvable groups, *Proc. Amer. Math. Soc.*, **13** (1962), 175–180.

[4] J.L. Alperin and G. Glauberman, Limits of abelian subgroups of finite p-groups, *J. Alg.*, **203** (1998), 533–566.

[5] M. Artin and J.T. Stafford, Noncommutative graded domains with quadratic growth, *Invent. Math.*, **122** (1995), 231–276.

[6] M. Aschbacher, On the maximal subgroups of the finite classical groups, *Invent. Math.*, **76** (1984), 469–514.

[7] Y.A. Bahturin and M.V. Zaicev, Identities of graded algebras, *J. Alg.*, **205** (1998), 1–12.

[8] Y. Barnea, The lower rank of some pro-p groups and the number of generators of simple Lie algebras. Preprint, 1998.

[9] Y. Barnea and B. Klopsch, Index subgroups of the Nottingham group. Preprint, 1999.

[10] Y. Barnea and M. Larsen, A non-abelian free pro-p group is not linear over a local field, *J. Alg.*, **214** (1999), 338–341.

[11] Y. Barnea and M. Larsen, Random generation in simple algebraic groups over local fields. Preprint, 1998.

[12] Y. Barnea and A. Shalev, Hausdorff dimension, Pro-p groups, and Kac-Moody algebras, *Trans. Amer. Math. Soc.*, **349** (1997), 5073–5091.

[13] Y. Barnea, A. Shalev and E.I. Zelmanov, Graded subalgebras of affine Kac–Moody algebras, *Israel J. Math.*, **104** (1998), 321–334.

[14] G. Benkart, A.I. Kostrikin and M.I. Kuznetsov, Finite-dimensional Lie algebras with a nonsingular derivation, *J. Alg.*, **171** (1995), 894–916.

[15] Y. Benoist, Une nilvariété non affine, *C.R. Acad. Sci. Paris*, **315** (1992), 983–986.

[16] G.M. Bergman, A note on growth functions of algebras and semigroups, mimeographed notes, University of California, Berkeley, 1978.

[17] N. Blackburn, On a special class of p-groups, *Acta Math.*, **100** (1958), 49–92.

[18] D. Burde and F. Grunewald, Modules for certain Lie algebras of maximal class, *J. Pure Appl. Alg.*, **99** (1995), 239–254.

[19] R.G. Burns, O. Macedońska and Yu. Medvedev, Groups satisfying semigroup laws, and nilpotent-by-Burnside varieties, *J. Alg.*, **195** (1997), 510–525.

[20] R.G. Burns and Yu. Medvedev, A note on Engel groups and local nilpotence, *J. Austral. Math. Soc.* (Ser. A), **64** (1998), 92–100.

[21] R. Camina, Subgroups of the Nottingham group, *J. Alg.*, **196** (1997), 101–113.

[22] A. Caranti, S. Mattarei and M.F. Newman, Graded Lie algebras of maximal class, *Trans. Amer. Math. Soc.*, **349** (1997), 4021–4051.

[23] A. Caranti and M.F. Newman, Graded Lie algebras of maximal class. II, to appear.

[24] C. Chevalley, *Théorie des groupes de Lie: groupes algébriques, Théorèmes généraux sur les algèbres de Lie*, Hermann, Paris.

[25] J.D. Dixon, The probability of generating the symmetric group, *Math. Z.*, **110** (1969), 199–205.

[26] J.D. Dixon, L. Pyber, Á. Seress and A. Shalev, Residual properties of free groups and probabilistic methods. Preprint, 1999.

[27] J.D. Dixon, M.P.F. du Sautoy, A. Mann and D. Segal, *Analytic Pro-p Groups, 2nd Edition*, Cambridge University Press, Cambridge, U.K., 1999.

[28] S. Donkin, Space groups and groups of prime power order. VIII. Pro-p groups of finite coclass and p-adic Lie algebras, *J. Alg.*, **111** (1987), 316–342.

[29] E.B. Dynkin, Semisimple subalgebras of semisimple Lie algebras, *Amer. Math. Soc. Transl.*, **6** (2) (1957), 111–244.

[30] M.P.F. du Sautoy, Finitely generated groups, p-adic analytic groups, and Poincaré series, *Ann. Math.*, **137** (1993), 639–670.

[31] M.P.F. du Sautoy, Pro-p groups, in *CRM Proc. and Lecture Notes* **17** (1999) (Banff Summer School Proceedings), pp. 99–130.

[32] M.P.F. du Sautoy, p-Groups, coclass and model theory: a proof of conjecture P. Preprint, 1999.

[33] E.B. Dynkin, Semisimple subalgebras of semisimple Lie algebras, *Amer. Math. Soc. Transl.*, **6** (2) (1957), 111–244.

[34] E.B. Dynkin, Maximal subgroups of the classical groups, *Amer. Math. Soc. Transl.*, **6** (2) (1957), 245–378.

[35] K. Falconer, *Fractal Geometry: Mathematical Foundations and Applications*, John Wiley & Sons, New York, 1990.

[36] A. Fialowski, Classification of graded Lie algebras with two generators, *Moscow Univ. Math. Bull.*, **38** (1983), 76–79.

[37] I. M. Gelfand and A. A. Kirillov, Sur les corps liés aux algèbres enveloppantes des algèbres de Lie, *Publ. Math. IHES*, **31** (1966), 509–523.

[38] G. Glauberman, Large abelian subgroups of finite p-groups, *J. Alg.*, **196** (1997), 301–338.

[39] E.S. Golod, On nil algebras and finitely approximable groups, *Izv. AN SSSR*, **28** (1964), 273–276. (in Russian)

[40] K.W. Gruenberg, Two theorems on Engel groups, *Proc. Cambr. Phil. Soc.*, **49** (1953), 377–380.

[41] P. Hall and G. Higman, On the p-length of p-soluble groups and reduction theorems for Burnside's problem, *Proc. London Math. Soc.*, **6** (1956), 1–42.

[42] B. Hartley, *Topics in the Theory of Nilpotent Groups*, in: *Group Theory — Essays for Philip Hall*, eds: K.W. Gruenberg and J.E. Roseblade, Academic Press, London, 1984.

[43] M. Hazewinkel, *Formal Groups and Applications*, Academic Press, New York, 1978.

[44] G. Higman, Groups and Lie rings having automorphisms without non-trivial fixed points, *J. London Math. Soc.*, **32** (1957), 321–334.

[45] B. Huppert and N. Blackburn, *Finite Groups, II*, Springer, Berlin, 1982.

[46] I. Ilani, Counting finite index subgroups and the P. Hall enumeration principle, *Israel J. Math.*, **68** (1989), 18–26.

[47] I. Ilani, Analytic pro-p groups and their Lie algebras, *J. Alg.*, **176** (1995), 34–58.

[48] N. Jacobson, *Lie Algebras*, Wiley-Interscience, New York, 1962.

[49] V.G. Kac, Simple graded Lie algebras of finite growth, *Math. USSR Izv.*, **2** (1968), 1271–1311.

[50] V.G. Kac, *Infinite Dimensional Lie Algebras*, Progress in Mathematics **44**, Birkhäuser, Boston, 1983.

[51] W.M. Kantor and A. Lubotzky, The probability of generating a finite classical group, *Geom. Ded.*, **36** (1990), 67–87.

[52] V. Kharchenko, Galois extensions and rings of quotients, *Algebra i Logika*, **13** (4) (1974), 460–484.

[53] E.I. Khukhro, Locally nilpotent groups admitting a splitting automorphism of prime order, *Mat. Sbornik*, **130** (1986), 120–127. English Transl.: *Math. USSR Sbornik*, **58** (1987), 119–126.

[54] E.I. Khukhro, On the Hughes problem for finite p-groups, *Algebra i Logika*, **26** (1987), 642–646. English Transl.: *Algebra and Logic*, **26** (1988), 398–401.

[55] E.I. Khukhro, Groups and Lie rings admitting an almost regular automorphism of prime order, *Mat. Sbornik*, **181** (1990), 1207–1219. English transl.: *Math. USSR Sbornik*, **71** (1992), 51–63.

[56] E.I. Khukhro, Finite p-groups admitting p-automorphisms with few fixed points, *Mat. Sbornik*, **184** (1993), 53–64. English transl.: *Russian Acad. Sci. Sbornik Math.*, **80** (1995), 435–444.

[57] E.I. Khukhro, *Nilpotent Groups and their Automorphisms*, de Gruyter, Berlin, 1993.

[58] E.I. Khukhro, *p-Automorphisms of Finite p-Groups*, London Math. Soc. Lecture Note Series **246**, Cambridge University Press, Cambridge, 1998.

[59] I. Kiming, Structure and derived length of finite p-groups possessing an automorphism of p-power order having exactly p fixed points, *Math. Scand.*, **62** (1998), 153–172.

[60] G. Klaas, C.R. Leedham-Green and W. Plesken, *Linear Pro-p Groups of Finite Width*, Springer Lecture Note Series **1674**, Berlin, 1997.

[61] A.I. Kostrikin, On the Burnside problem, *Izv. AN SSSR*, **23** (1959), 3–34. (in Russian)

[62] A.I. Kostrikin, *Around Burnside*, Nauka, Moscow, 1986. English Transl.: Springer, Berlin, 1990.

[63] A.I. Kostrikin and I.R. Shafarevich, Graded Lie algebras of finite characteristic, *Math. USSR-Izv.*, **3** (1969), 237–304.

[64] G.R. Krause and T.H. Lenagan, *Growth of Algebras and Gelfand–Kirillov Dimension*, Pitman, London, 1985.

[65] V.A. Kreknin, Solvability of Lie algebras with a regular automorphism of finite period, *Soviet. Math. Dokl.*, **4** (1963), 683–685.

[66] V.A. Kreknin and A.I. Kostrikin, Lie algebras with a regular automorphism, *Soviet. Math. Dokl.*, **4** (1963), 355–358.

[67] M. Lazard, Sur les groupes nilpotents et les anneaux de Lie, *Ann. Sci. École Norm. Sup.*, **71** (1954), 101–190.

[68] M. Lazard, Groupes analytiques p-adiques, *Publ. Math. I.H.E.S.*, **26** (1965), 389–603.

[69] C.R. Leedham-Green, Pro-p groups of finite coclass, *J. London Math. Soc.*, **50** (1994), 43–48.

[70] C.R. Leedham-Green, The structure of finite p-groups, *J. London Math. Soc.*, **50** (1994), 49–67.

[71] C.R. Leedham-Green and S. McKay, On p-groups of maximal class I, *Quart. J. Math. Oxford*, **27** (1976), 297–311.

[72] C.R. Leedham-Green and S. McKay, On p-groups of maximal class II, *Quart. J. Math. Oxford*, **29** (1978), 175–186.

[73] C.R. Leedham-Green and S. McKay, On p-groups of maximal class III, *Quart. J. Math. Oxford*, **29** (1978), 281–299.

[74] C.R. Leedham-Green, S. McKay and W. Plesken, Space groups and groups of prime power order. V. A bound to the dimension of space groups with fixed coclass, *Proc. London Math. Soc.*, **52** (1986), 73–94.

[75] C.R. Leedham-Green, S. McKay and W. Plesken, Space groups and groups of prime power order. VI. A bound to the dimension of a 2-adic group with fixed coclass, *J. London Math. Soc.*, **34** (1986), 417–425.

[76] C.R. Leedham-Green and M.F. Newman, Space groups and groups of prime power order I, *Arch. Math.*, **35** (1980), 193–202.

[77] L. Levai, On identities of Λ-analytic groups. Preprint, 1999.

[78] L. Levai and L. Pyber, Coset-identities and dense free subgroups of profinite group. Preprint.

[79] M.W. Liebeck and A. Shalev, The probability of generating a finite simple group, *Geom. Ded.*, **56** (1995), 103–113.

[80] V. Linchenko, Identities of Lie algebras with actions of Hopf algebras, *Comm. Alg.*, **25** (1997), 3179–3187.

[81] A. Lubotzky and A. Mann, Powerful p-groups. I, II, *J. Alg.*, **105** (1987), 484–515.

[82] A. Lubotzky and A. Mann, Groups of polynomial subgroup growth, *Invent. Math.*, **104** (1991), 521–533.

[83] A. Lubotzky, A. Mann and D. Segal, Finitely generated groups of polynomial subgroup growth, *Israel. J. Math.*, **82** (1993), 363–371.

[84] A. Lubotzky and A. Shalev, On some Λ-analytic pro-p groups, *Israel J. Math.*, **85** (1994), 307–337.

[85] W. Magnus, Über Gruppen und zugeordnete Liesche Ringe, *Reine Angew. Math.*, **182** (1940), 142–159.

[86] A. Mann, Positively finitely generated groups, *Forum Math.*, **8** (1996), 429–459.

[87] A. Mann, Finite groups containing many involutions, *Proc. Amer. Math. Soc.*, **122** (1994), 383–385.

[88] A. Mann and C. Martinez, The exponent of finite groups, *Arch. Math.*, **67** (1996), 8–10.

[89] A. Mann and C. Martinez, Groups nearly of prime exponent and nearly Engel Lie algebras, *Arch. Math.*, **71** (1998), 5–11.

[90] A. Mann and A. Shalev, Simple groups, maximal subgroups, and probabilistic aspects of profinite groups, *Israel J. Math.*, **96** (1997) (Amitsur memorial issue), 449–468.

[91] G.A. Margulis, *Discrete Subgroups of Semisimple Lie groups*, Springer, Berlin, 1991.

[92] C. Martinez and E.I. Zelmanov, On pro-unipotent groups of finite width, to appear.

[93] C. Martinez and E.I. Zelmanov, Nil algebras and unipotent groups of finite width, to appear.

[94] O. Mathieu, Classification of simple graded Lie algebras of finite growth, *Invent. Math.*, **108** (1990), 455–519.

[95] Yu. Medvedev, Groups and Lie rings with almost regular automorphisms, *J. Alg.*, **164** (1994), 877–885.

[96] Yu. Medvedev, p-Groups, Lie p-rings and p-automorphisms, *J. London Math. Soc.*, **58** (1998), 27–37.

[97] Yu. Medvedev, p-Divided Lie rings and p-groups, *J. London Math. Soc.*, **59** (1999), 787–798.

[98] Yu. Medvedev, Compact Engel groups, *Israel. J. Math.*, to appear.

[99] C.F. Miller III and V.N. Obraztsov, Infinite periodic residually finite groups with all finite subgroups cyclic, in preparation.

[100] J. Milnor, On fundamental groups of complete affinely flat manifolds, *Adv. in Math.*, **25** (1977), 178–187.

[101] S. Montgomery, *Fixed Rings of Finite Automorphism Groups of Associative Rings*, Lect. Notes Math. **818**, Springer-Verlag, Berlin, 1980.

[102] B.H. Neumann, Groups with automorphisms that leave only the identity element fixed, *Arch. Math.*, **7** (1956), 1–5.

[103] P.M. Neumann, Two combinatorial problems in group theory, *Bull. London Math. Soc.*, **21** (1989), 456–458.

[104] M.F. Newman and E.A. O'Brien, Classifying 2-groups by coclass, *Trans. Amer. Math. Soc.*, **351** (1999), 131–169.

[105] M.F. Newman, C. Schneider and A. Shalev, The entropy of graded algebras, *J. Alg.*, **223** (2000), 85–100.

[106] W.H. Patton, The Minimum Index for Subgroups in Some Classical Groups: A Generalization of a Theorem of Galois, Ph.D. Thesis, University of Illinois at Chicago Circle, Chicago, 1972.

[107] V.M. Petrogradsky, On Lie algebras with non-integral q-dimensions, *Proc. Amer. Math. Soc.*, **123** (1997), 649–656.

[108] R. Pink, Compact subgroups of linear algebraic groups, *J. Alg.*, **206** (1998), 438–504.

[109] D.M. Riley and J.F. Semple, The coclass conjectures for restricted Lie algebras, *Bull. London Math. Soc.*, **26** (1994), 431–437.

[110] N.R. Rocco and P. Shumyatsky, On periodic groups having almost regular 2-elements, *Proc. Edinburgh Math. Soc.*, **41** (1998), 385–391.

[111] A. Rozhkov, The lower central series of one group of automorphisms of the trees, *Mat. Zametki*, **60** (1996), 225–237.

[112] C.M. Scoppola, Groups of prime power order as Frobenius–Wielandt complements, *Trans. Amer. Math. Soc.*, **325** (1991), 855–874.

[113] D. Segal, A footnote on residually finite groups, *Isr. J. Math.*, **94** (1996), 1–5.

[114] J.-P. Serre, *Lie groups and Lie algebras* (new edition), Lecture Notes in Math. **1500**, Springer, Berlin, 1991.

[115] A. Shalev, Growth functions, p-adic analytic groups, and groups of finite coclass, *J. London Math. Soc.*, **46** (1992), 111–122.

[116] A. Shalev, Combinatorial conditions in residually finite groups, II, *J. Alg.*, **157** (1993), 51–62.

[117] A. Shalev, On almost fixed-point-free automorphisms, *J. Alg.*, **157** (1993), 271–282.

[118] A. Shalev, Automorphisms of finite groups of bounded rank, *Israel J. Math.*, **82** (1993), (Thompson issue), 395–404.

[119] A. Shalev, The structure of finite p-groups: effective proof of the coclass conjectures, *Invent. Math.*, **115** (1994), 315–345.

[120] A. Shalev, Simple Lie algebras and Lie algebras of maximal class, *Arch. Math.*, **63** (1994), 297–301.

[121] A. Shalev, Finite p-groups, in *Finite and Locally Finite Groups*, eds: B. Hartley et al., NATO ASI Series, Kluwer, 1995, pp. 401–450.

[122] A. Shalev, Groups whose subgroup growth is less than linear, *International Journal of Algebra and Computation*, **7** (1997), 77–91.

[123] A. Shalev, Centralizers in residually finite torsion groups, *Proc. Amer. Math. Soc.*, **126** (1998), 3495–3499.

[124] A. Shalev, Probabilistic group theory, *Groups St Andrews 1997 in Bath, II*, London Math. Soc. Lecture Note Series **261**, Cambridge University Press, Cambridge, 1999, pp. 648–678.

[125] A. Shalev, The orders of nonsingular derivations, *J. Austral. Math. Soc.*, **67** (1999), 254–260.

[126] A. Shalev and E.I. Zelmanov, Pro-p groups of finite coclass, *Math. Proc. Cambr. Phil Soc.*, **111** (1992), 417–421.

[127] A. Shalev and E.I. Zelmanov, Narrow Lie algebras: a coclass theory and a characterization of the Witt algebra, *J. Alg.*, **189** (1997), 294–331.

[128] P. Shumyatsky, On finite nilpotent groups having fixed point free automorphisms. Preprint, Brasilia, 1996.

[129] P. Shumyatsky, Nilpotency of some Lie algebras associated with p-groups, *Canad. J. Math.*, **51** (1999), 658–672.

[130] P. Shumyatsky, On groups having a four-subgroup with finite centralizer, *Quart. J. Math.*, **49** (1998), 491–499.

[131] P. Shumyatsky, Centralizers in groups with finiteness conditions, *J. Group Th.*, **1** (1998), 275–282.

[132] P. Shumyatsky, On groups with commutators of bounded order, *Proc. Amer. Math. Soc.* **127** (1999), 2583–2586.

[133] V.P. Shunkov, On periodic groups with an almost regular involution, *Algebra and Logic*, **11** (1972), 260–272.

[134] L.W. Small, J.T. Stafford and R.B. Warfield, Affine algebras of Gelfand-Kirillov dimension one are PI, *Math. Proc. Cambr. Phil. Soc.*, **97** (1985), 407–414.

[135] J.G. Thompson, Finite groups with fixed-point-free automorphisms of prime order, *Proc. Nat. Acad. Sci. U.S.A.*, **45** (1959), 578–581.

[136] J.G. Thompson, Automorphisms of solvable groups, *J. Alg.*, **1** (1964), 259–267.

[137] M.R. Vaughan-Lee, *The restricted Burnside problem*, 2nd edition, Oxford Univ. Press, Oxford 1993.

[138] M. Vergne, Réductibilité de la Variété des algèbres de Lie nilpotentes, *C.R. Acad. Sc. Paris*, **263** (1966), A4–A6.

[139] M. Vergne, Cohomologie des algèbres de Lie nilpotentes. Application à l'étude de la variété des algèbres de Lie nilpotentes, *C.R. Acad. Sc. Paris*, **267** (1968), A867–A870.

[140] J.S. Wilson, Two-generator conditions for residually finite groups, *Bull. London Math. Soc.*, **23** (1991), 239–248.

[141] J.S. Wilson and E.I. Zelmanov, Identities for Lie algebras of pro-p groups, *J. Pure and Appl. Alg.*, **81** (1992), 103–109.

[142] I.O. York, The Group of Formal Power Series under Substitution, Ph.D. Thesis, Nottingham, 1990.

[143] E.I. Zelmanov, Engel Lie algebras, *Dokl. Akad. Nauk SSSR*, **292** (1987), 265–268.

[144] E.I. Zelmanov, On some problems in the theory of groups and Lie algebras, *Mat. Sb.*, **180** 159–167. (in Russian)

[145] E.I. Zelmanov, The solution of the restricted Burnside problem for groups of odd exponent, *Math. USSR-Izv.*, **36** (1991), 41–60.

[146] E.I. Zelmanov, The solution of the restricted Burnside problem for 2-groups, *Mat. Sb.*, **182** (1991), 568–592.

[147] E.I. Zelmanov, On periodic compact groups, *Israel J. Math.*, **77** (1992), 83–95.

[148] E.I. Zelmanov, Lie ring methods in the theory of nilpotent groups, in *Groups '93*, Galway-St Andrews, London Math. Soc. Lecture Note Series **212**, 1995, 567–586.

[149] A. Zubkov, Non-abelian free pro-p groups cannot be represented by 2-by-2 matrices, *Sib. Math. J.*, **28** (1987), 742–747.

Institute of Mathematics
The Hebrew University
Jerusalem 91904
Israel
e-mail:shalev@math.huji.ac.il

2

On the Classification of p-groups and pro-p Groups

C. R. Leedham-Green and S. McKay

1 Introduction

There is no possibility of finding a finite set of invariants that will define every p-group up to an isomorphism in a useful way. The coclass conjectures can be regarded as offering a way around this difficulty. The solution offered in these conjectures is to give a classification that is weaker than an isomorphism by taking a suitable invariant of a p-group; in this instance the *coclass* defined as $n - c$ where the group has order p^n and nilpotency class c; and seeking to classify p-groups by coclass. By this we mean that we prove a theorem that states that if G is any p-group of coclass r then G has a normal subgroup N, of order bounded by a function of p and r, such that G/N has a certain well-defined structure. This falls short of a classification up to isomorphism first by ignoring the subgroup N, and secondly by failing to classify precisely the groups G/N. We do however have a recipe for constructing the groups G/N which describes their structure closely and which can be followed through in detail in simple cases.

This recipe makes use of quotients of infinite pro-p groups, and so a clear theme of the coclass project has been that the primary source of finite p-groups is the set of finite quotients of just infinite pro-p groups; that is, pro-p groups with no proper infinite homomorphic images. Such pro-p groups play an analogous role in the theory of pro-p groups to the role that simple groups play in finite group theory.

Another theme is that we study p-groups 'top down'. That is, we regard the lower central quotients of a p-group as giving successive approximations to the group. So if $\gamma_i(G)$ denotes the ith term of the lower central series of G, we ask what information we can glean about G from knowing that $G/\gamma_i(G)$ is isomorphic to some finite p-group

P, or has some particular property such as small coclass.

Philip Hall developed many of the basic concepts, and considered the general issues involved in classifying p-groups, and in 1958 Norman Blackburn, as a student of Philip Hall, laid the foundations for the theory of p-groups of maximal class in [1]; this work, as described in Huppert [8], was the inspiration behind further developments in the theory of p-groups of maximal class in the 1970's. The coclass conjectures themselves were formulated by Leedham-Green and Newman as a result of these developments and were published in 1980. We state them here.

Conjecture A. For some function $f(p, r)$, every finite p-group of coclass r has a normal subgroup K of nilpotency class at most 2 and index at most $f(p, r)$. If $p = 2$, one can require K to be abelian.

Conjecture B. For some function $g(p, r)$, every finite p-group of coclass r has derived length at most $g(p, r)$.

Conjecture C. Every pro-p group of finite coclass is soluble.

Conjecture D. For fixed p and r, there are only finitely many isomorphism classes of infinite pro-p groups of coclass r.

Conjecture E. There are only finitely many isomorphism classes of infinite soluble pro-p groups of coclass r.

These conjectures, published in [18], are in order of decreasing strength in the sense that each conjecture may be deduced from the conjecture above it. They have all now been proved ([12] and [26]); moreover we can give far more detailed descriptions of the groups in question than is given by the strongest of the conjectures. A complete account of the whole coclass project [17] is in an advanced state of preparation.

In this chapter we will outline the proof of Conjecture A, describe its impact on our understanding of certain pro-p groups, and describe a recipe for constructing the groups G/N mentioned above. The aim for all of these is to find a subgroup H with 'small' index (that is, bounded by a function of p and r) such that 'commutation in H moves a long way down the lower central series of G'; the subgroup K of Conjecture A will then be a subgroup of small index in H and the normal subgroup N required in the construction will be a small subgroup of H. The bound on the derived length of G required by Conjecture B will also follow, and this implies that pro-p groups of finite coclass are soluble.

We then go on to outline ways in which these ideas might be extended to further classification of pro-p groups and finite p-groups. While a classification of just infinite pro-p groups seems for the time being out of reach, it may be that the class of all hereditarily just infinite pro-p groups is not much more extensive than those that we know. We include a conjecture on the structure of finite p-groups by width, rank and obliquity. By including rank among the parameters we ensure that all pro-p groups in question are p-adic analytic.

2 Maximal class groups and Conjecture A

A group of order p^n and nilpotency class $n - 1$, that is, of coclass 1, is said to be of maximal class if $n \geq 4$. A brief description of the structure of these groups is instructive as the basic ideas in the proof of Conjecture A are already to be seen in this special case.

The structure of a p-group of maximal class is determined by the structure of a certain characteristic subgroup P_1 of index p in G. This subgroup has small nilpotency class, in fact its nilpotency class is bounded by a function of p independent of n since $\gamma_3(P_1)$ has order at most p^{2p-8}, and P_1 contains a subgroup of nilpotency class 2 of index bounded by a function of p. Stronger results still hold for $p = 2$, where P_1 is in fact cyclic, and for $p = 3$, where P_1' has order at most p.

In establishing these results three crucial ideas are employed; all three were already present in Blackburn's paper [1] in 1958.

(i) Let $P_i = \gamma_i(G)$ for $i \geq 2$. The subgroup P_1 is defined to be the centraliser in P of P_2/P_4, and it is shown that the normal subgroups of G contained in P_1 are precisely those in the series $P_1 > \cdots > P_{n-1} > P_n = \langle 1 \rangle$, and each section P_i/P_{i+1} has order p. This is described by saying that the group G acts *uniserially* on P_1.

(ii) The *degree of commutativity* l of G is defined; this gives a measure of how fast commutation inside P_1 moves down the series $\{P_i\}$. By definition, l is the greatest integer such that $[P_i, P_j] \leq P_{i+j+l}$ for all $i, j \geq 1$ (and $l = n - 3$ if P_1 is abelian). Clearly P_1 will have small nilpotency class if l is large.

(iii) The behaviour of pth powers in P_1 is shown to be 'regular' in the sense that $P_i^p = P_{i+p-1}$, so that pth powers go down the series in P_1 by a fixed distance $p - 1$, the *dimension* of G.

An outline of the proof for $p \geq 5$ follows, the cases $p = 2, 3$ being much easier.

Choose $s \in P \setminus P_1$ and $s_1 \in P_1 \setminus P_2$ and define inductively, $s_{i+1} = [s_i, s]$. Then s_i is a generator for P_i/P_{i+1} for $i \geq 1$ as long as $l > 0$, which is shown to be the case if $|P| > p^{p+1}$, and the pth powers behave as follows: if $l > 0$ then $s_i^p s_{i+p-1} \in P_{i+p}$ for all $i > 1$ and $s_1^p s_p \in P_{1+p}P_{n-1}$.

Let $[s_i, s_j] \equiv s_{i+j+l}^{a(i,j)} \bmod P_{i+j+l+1}$ where $0 \leq a(i, j) < p$, and regard $a(i, j)$ as an element of $\mathbb{F}_p$. Define $b(i, j) = a(i_0, j_0)$ where $i \equiv i_0$ and $j \equiv j_0 \bmod (p - 1)$ and $1 \leq i_0, j_0 \leq p - 1$. The choice of $p - 1$ here comes from the behaviour of the pth powers which ensures that $a(i, j) = a(i_0, j_0)$ as long as i, j are not too big relative to n.

The commutator relations, including the Hall–Witt identity and results about the relationship between pth powers and commutators, now give a system of equations that must be satisfied by the numbers $b(i, j)$ if $2l \leq n - 2p + 3$. Their solution can be shown to be trivial using rather detailed calculations as in [14]. Since some $a(i, j)$ must be non-trivial from the definition of l, this gives a contradiction. Hence the assumption that $2l \leq n - 2p + 3$ is false.

Generalising these calculations looks impossible, but they can be replaced by an argument involving Lie algebras as follows. This argument is due to Shalev [26], and was shown by him to generalise to apply to groups of coclass r. We now outline a slightly modified version of his original proof.

Let A denote the free $\mathbb{F}_p[t]$-module generated by elements $y_1, \ldots, y_{p-1}$. Define y_i for $i \geq p$ by $y_i = t^{i_1} y_{i_0}$ where $i = i_1(p - 1) + i_0$ with $1 \leq i_0 \leq p - 1$. Define a product on A by defining $[y_i, y_j] = b(i, j) y_{i+j+l}$ for $i, j \geq 1$ and extending this to an

$\mathbb{F}_p$-bilinear product. Then A is a Lie algebra over $\mathbb{F}_p[t]$ of dimension $p-1$. Now let $L = A \otimes_{\mathbb{F}_p[t]} \mathbb{F}_p(t)$.

Define $\delta : L \to L$ by putting $y_i\delta = y_{i+1}$ and extending to give an $\mathbb{F}_p(t)$-linear map. (This corresponds to commutation with s in the group.) Then δ is an $\mathbb{F}_p(t)$-derivation of L and $\delta^{p-1} = t1$. The critical result is now the following:

Theorem. *Let L be a Lie algebra over a field of characteristic p. If L has a derivation δ such that δ^{p-1} is a non-zero scalar then L is nilpotent.*

This is proved in [26] using an ingenious argument based on Jacobson's version of Engel's theorem: if W is a Lie closed subset of the algebra of linear transformations on a finite-dimensional vector space over a field, and if every element of W is nilpotent, then the enveloping associative algebra of W is nilpotent.

But the definitions of δ and l imply that $y_i \in A'$ for all $i \geq u$ for some u and so L is perfect and non-trivial, giving a contradiction as before.

The condition $2l \leq n - 2p + 3$ comes from the Witt identity; we need to be able to apply the Witt identity to the triple commutators $[s_i, s_j, s_k]$ for $1 \leq i < j < k \leq p - 1$ to give one set of equations satisfied by the numbers $b(i, j)$. One would expect this to require $2l \leq n - 3p + 6$, but Fernández–Alcober in [6] shows that it is sufficient to take $i = 1$, thus giving the better bound. This is part of a program by Vera–López, Fernandez–Alcober, and others, to find better bounds for l in terms of various parameters.

The other equations satisfied by the numbers $b(i, j)$ only require one operation of commutation in P_1 and hold if $l < n - 2p + 3$. For $p = 3$ the Witt identity is not required, thus giving the much stronger bound below.

The precise results are as follows.

Theorem. *The 2-groups of maximal class are precisely the dihedral, quaternion and semi-dihedral groups. (Here P_1 is cyclic.)*

Theorem. *Let G be a 3-group of maximal class. Then the degree of commutativity l of G is at least $n - 4$. Hence P_2 is abelian, and P_1' has order at most p.*

Theorem. *Let G be a p-group of maximal class for $p \geq 5$ and let G have degree of commutativity l. Then $2l \geq n - 2p + 4$.*

The corollary that follows gives the information relevant to Conjecture A, the construction method and Conjecture B.

Corollary. *Let G be a p-group of maximal class for $p \geq 5$:*
(i) let $K = \gamma_i(G)$ where $i = \lceil (2p-5)/3 \rceil$; then K has nilpotency class at most 2 and index p^i;
(ii) let $N = \gamma_3(P_1)$; then $|N| \leq p^{2p-8}$;
(iii) G has derived length at most $\log_2(2p-2)$.

From the point of view of our 'top down' theme we have the following result.

Theorem. *Let p be an odd prime and H a group of order p^m with $m \geq p + 2$ such that $H/\gamma_{p+1}(H)$ is of maximal class. Then H is of maximal class.*

Similarly, if H is a non-abelian finite 2-group such that $H/\gamma_2(H)$ has order 4, then H is dihedral or quaternion or semi-dihedral.

3 Conjecture A; the general case

To generalise the arguments of the last section to the case $r > 1$ requires three steps. First an analogue G_1 of P_1 must be found, with a degree of commutativity l. Secondly, a perfect Lie algebra L over $\mathbb{F}_p(t)$ needs to be constructed from G_1 on the assumption that l is not too big. Thirdly, a suitable idempotent derivation of L needs to be constructed. This leads to a contradiction to the assumption on l, and the result follows. For $p = 2$ we need to find a bigger lower bound for l, but in this case we need only an algebra rather than a Lie algebra to give the contradiction, and so the Witt identity is not needed and the bigger bound follows.

The most fundamental property of P_1 is that G acts uniserially on it. In general, let N be any group of order p^n on which G acts and define $N_1 = N$ and $N_{i+1} = [N_i, G]$ for $i > 0$. If the series $N_1 > N_2 > \cdots > N_{n+1} = \langle 1 \rangle$ has all its factors of order p, then G is said to act *uniserially* on N. This is equivalent to the condition that $[H, G]$ has index p in H for every non-trivial G-invariant subroup H of N. It is also equivalent to the condition that the G-invariant subgroups of N are linearly ordered if G is a p-group and N is not cyclic, dihedral or quaternion. In the general case $r \geq 1$, one proves that G has a normal subgroup G_1 of index bounded by a function of p and r on which G acts uniserially.

We can now formulate the additional conditions that G_1 will be required to satisfy. Let G be a finite p-group of nilpotency class c, and let G_1 be a normal subgroup of G of order p^f. Then G is *settled of dimension d with respect to G_1* if the following conditions are satisfied.

(i) G acts uniserially on G_1.

Let $G_1 > G_2 > \cdots > G_{f+1} = \langle 1 \rangle$ be the G-invariant subgroups of G_1.

(ii) $G_i^p = G_{i+d}$ for all $i \leq f - d + 1$.

(iii) $G_f = \gamma_c(G)$.

(iv) $f > pd/(p-1)$.

Note that if G is of maximal class, then G is settled with respect to P_1 provided that P_1 has order greater than p^p. The term 'settled' comes from the following result which basically says that once a lower central quotient of the group is settled, then all further lower central factors have order p.

Theorem. *If H is a p-group of nilpotency class greater than c such that $H/\gamma_{c+1}(H)$ is settled with respect $H_1/\gamma_{c+1}(H)$ where $\gamma_{c+1}(H) < H_1 < H$, then H is settled with respect to H_1. In particular, the coclass of H is the coclass of $H/\gamma_{c+1}(H)$.*

If G is settled with respect to G_1, with the above notation we can now define the *degree of commutativity* of G with respect to G_1 as the biggest integer l such that $[G_i, G_j] \leq G_{i+j+l}$ for all i, j if G_1 is not abelian, and as $f - 2$ if G_1 is abelian.

The aim then is to prove that if G is a p-group of coclass r, then G is settled with respect to a subgroup of small index, the degree of commutativity being big.

It should come as no surprise that the dimension d referred to above will in general be of the form $(p-1)p^s$ for some s. This has been seen in the case of p-groups of maximal class when $s = 1$, and echoes the fact that a p-group can only act irreducibly over $\mathbb{Q}_p$ in dimension d if d is of this form (or $d = 1$). To prove this the following result is used; here, $\Pi_i(G)$ for $i \geq 0$ is defined inductively by: $\Pi_0(G) = G$ and $\Pi_i(G) = \Pi_{i-1}(G)^p = \langle g^p \mid g \in \Pi_{i-1}(G)\rangle$ for $i > 0$.

Theorem. ([13]) *Let G be a finite p-group acting uniserially on an abelian homocyclic p-group N of exponent p^k and rank d where d is not of the form $(p-1)p^s$ for $s \geq 0$. Let $f_d(0) = 1$ and $f_d(i+1) = \min(pf_d(i), f_d(i)+d)$ for all $i \geq 0$. Then with the above notation $[N_j, \Pi_i(G)] = N_{j+f_d(i)}$ for all $j \leq kd - f_d(i) + 1$. In particular, if $\Pi_i(G) = \langle 1\rangle$ then $kd \leq f_d(i)$.*

We now show how to find the required subgroup G_1, following the ingenious method of Shalev [26], and concentrating on the case of odd primes.

The central tool is the theory of powerful p-groups and powerfully embedded subgroups, published by Lubotzky and Mann in 1987 in [20] and [21]. The subgroup H of G is *powerfully embedded in* G if $[H, G] \leq H^p$ for odd p and $[H, G] \leq H^4$ for $p = 2$. The group G is *powerful* if it is powerfully embedded in itself.

We shall concentrate on the case of odd p; similar but not identical results hold for $p = 2$.

Define H to be *strongly hereditarily powerful* in G (in symbols H shp G) if $H \triangleleft G$ and every normal subgroup of G contained in H is powerfully embedded in H.

Note that if G is of maximal class, then P_1 is not in general strongly hereditarily powerful in G; this would require the degree of commutativity to be at least $p-3$. However, P_i shp G if $i \geq (p-1)/2$. Thus the group G_1 that we construct will not be P_1 in the case $r = 1$.

Being powerful is of course a generalisation of being abelian, and powerful p-groups have many of the simple properties of abelian groups. For example, If G is a powerful p-group and $H < G$, then $d(H) \leq d(G)$. This underlines the interaction between the theory of powerful groups and the theory of p-adic analytic groups. The search for G_1 relies on a result in the opposite direction.

The heart of the matter lying in the interaction between powers and commutators, it is perhaps not too surprising that the groups $X_i = X_i(G)$ defined by $X_i(G) = \Pi_i(G)\gamma_{p^i}(G)$ play a central role. Note that $X_i(G)$ is a verbal subgroup of G. A striking general result about verbal subgroups in [20] applied to this situation gives the following result:

Theorem. *Let p be odd. If $H \triangleleft G$, and $H \leq X_i$, and $d(H) \leq p^i$ for some $i \geq 0$, then H* shp *G.*

Using this result, one can prove the following:

Theorem. *For p odd, if $G/\Phi(X_r)\gamma_{2p^r}(G)$ has coclass at most r, then X_r shp G and $d(X_r) \leq p^r$.*

Note that this result already gives the case $p > 2$ of the following 'top down' result:

Theorem. *Let G be a finite p-group and let $n_1 = 2p^r$ if p is odd, and $n_1 = 2^{r+3}$ if $p = 2$. If $G/\gamma_{n_1}(G)$ has coclass at most r, then G has the same coclass as $G/\gamma_{n_1}(G)$.*

It is an immediate consequence of this result that if G has coclass r then G acts uniserially on $\gamma_{2p^r-1}(G)$.

Using the theory of powerful p-groups, one can now deduce that if G has coclass r and nilpotency class $c \geq 2p^r$, then G is settled of dimension d with respect to X_r, where $d = (p-1)p^s$ for some $s < r$. In fact, since G acts uniserially on $\gamma_{2p^{r-1}}(G)$, it turns out that G is settled of dimension d with respect to $G_1 = \gamma_m(G)$ where $m = (p-1)p^{r-1}$.

The next step is to construct the Lie algebra L. This follows very closely the lines of the maximal class case. Define l to be the greatest integer such that $[\gamma_i(G), \gamma_j(G)] \leq \gamma_{i+j+l}(G)$ for all $i, j \geq m$ if $\gamma_m(G)$ is not abelian, and to be $c - 2m$ otherwise, where c is the nilpotency class of G, so that G has degree of commutativity $l + m - 1$ with respect to G_1. To construct the Lie algebra L as in the maximal class case now requires $2l \leq c - 6(p-1)p^{r-1} + 6$. The crucial point is that the inequality bounds $2l - c$ as a function of r; this is what is needed for Conjecture A.

The next step is to construct the necessary derivation of L. Commutation by an element of G induces a derivation of L, and in this way we get a map from $G/\Phi(G)$ into the space of derivations of L. This gives rise to a derivation γ such that $\gamma^{(p-1)p^s} = \lambda 1$ for some non-zero λ in $\mathbb{F}_p(t)$, and hence to a derivation δ such that $\delta^{p-1} = \lambda 1$. A contradiction now arises as in the maximal class case.

For $p = 2$ the Lie structure for A and L is not needed as the existence of γ as a derivation of the $\mathbb{F}_2(t)$-algebra L is already impossible. In this case, if G has nilpotency class $c \geq 2^{r+3}$, and $m = 2^{r+2}$, then G is settled with respect to $\gamma_m(G)$ and this forces $l > c - 3.2^{r+2} + 2$.

The relevant corollaries of the main theorems can now be stated.

Corollary. *Let G be a finite 2-group of coclass r; then*

(i) G has an abelian normal subgroup K of index at most $2^{3.2^{r+1}+r-2}$;
(ii) let $N = \big(\gamma_{2^{r+2}}(G)\big)'$; then $\gamma_{2^{r+2}}(G)/N$ is abelian and $|N| \leq 2^{2^{r+2}-2}$;
(iii) G has derived length at most $r + 4$.

Corollary. *For p odd let G be a finite p-group of coclass r; then*

(i) G has a normal subgroup K of nilpotency class at most 2 and index at most $p^{2(p-1)p^{r-1}+r-3}$;
(ii) let $N = \gamma_3\big(\gamma_{(p-1)p^{r-1}}(G)\big)$; then $\gamma_{(p-1)p^{r-1}}(G)/N$ has nilpotency class at most 3 and $|N| \leq p^{3(p-1)p^{r-1}-6}$;
(iii) G has derived length at most $4 + \log_2\big((p-1)p^{r-1} - 1\big)$.

4 Connections with pro-p groups

Let us return to the idea of studying p-groups 'top down' by regarding the lower central quotients of a p-group as giving successive approximations to the group, and consider in more detail what information we can glean about G from knowing that $G/\gamma_i(G)$ is isomorphic to some finite p-group P. This depends on the exact structure of P. There are various possibilities. For example (assuming always that G is a finite p-group), it may be that we can deduce that $\gamma_i(G) = \langle 1 \rangle$. This is a very common but uninteresting situation. If this is the case, P is said to be *incapable*. For example, the quaternion and semi-dihedral groups are incapable.

We are mainly interested in the case when there is no bound to the possible nilpotency class of G. In this case, P is a lower central quotient of an infinite pro-p group; let us say that in this case P is *infinitely capable*. For example, this is the case if P is a dihedral 2-group. Given an infinitely capable group P, what can we say about the p-groups and pro-p groups G with $G/\gamma_i(G) \cong P$? If G is required to be an infinite pro-p group, there may be only one possibility for G (up to isomorphism), or a finite number greater than one, or even uncountably many. But even when there are uncountably many, the general structure of G may be very restricted. Typically we are interested in proving at least that the orders of the lower central factors of G are determined. In this case, we regard P as being 'settled'. We have above a technical definition for 'settled' in the finite coclass context; but in the above rather vague and more general context, the question of identifying settled p-groups is an important part of our philosophy.

Now consider the p-groups of maximal class and positive degree of commutativity. These can be visualised as the vertices of a graph as follows. (Recall that if $|G| > p^{p+1}$ then G has positive degree of commutativity, so that only a few small groups of maximal class are left out.) Let $\mathcal{G}(p)$ be the directed graph defined as follows:

The vertices of $\mathcal{G}(p)$ are p-groups of maximal class and positive degree of commutativity with one of each isomorphism type. If G and H are two such groups there is an edge $G \to H$ if H is a homomorphic image of G and the nilpotency class of G is one more than the nilpotency class of H. In this case G is called a *descendant* of H.

Clearly, $|G| = p|H|$, and if G has order p^n, then H is isomorphic to $G/\gamma_{n-1}(G)$. The graph $\mathcal{G}(p)$ is therefore a union of trees. It may be pictured as a set of points in the plane, with those corresponding to groups of order p^n for $n \geq 4$ having distance n below some horizontal axis of reference. The edges then join groups at one level to groups in the level above them. Note that we defined groups of maximal class to have order at least p^4, so the groups of order p^3 and nilpotency class 2, and those of order p^2 and abelian, have not been included in the graph or in the results that follow.

There is a natural relationship between infinite chains of descendants in $\mathcal{G}(p)$, and pro-p groups of coclass 1: the inverse limit of an infinite chain of descendants in $\mathcal{G}(p)$ is a pro-p group of coclass 1; and the lower central factors of any such pro-p group form an infinite chain of descendants in $\mathcal{G}(p)$. In fact $\mathcal{G}(p)$ has only one such chain and so for every prime p there is a unique pro-p group of maximal class.

It is also easy to see that only the groups with $s^p = (ss_1)^p = 1$ can have descendants in $\mathcal{G}(p)$, and any chain of descendants in $\mathcal{G}(p)$ going down from a group with P_1 non-abelian must be of finite length. We use the term *finite chain* for a finite chain going from a vertex on the infinite chain to a *leaf*, that is a vertex of valency 1. The *twig length* of the graph is the maximal length of these finite chains.

If $p = 2$, the groups on the infinite chain are the dihedral groups, and at each level there are 2 other groups which are the semi-dihedral and quaternion group of that order, and these have no descendants. Hence there are 3 groups at each level, and the finite chains have length 1. For $p = 3$, the situation is still under control. Here there are 6 or 7 groups alternately at each level after the level $n = 4$ and again the finite chains of descendants have length 1. These groups were classified by Blackburn in [1].

For $p \geq 5$, the situation is more complicated. The unique pro-p group S of maximal class is the split extension of $\mathcal{O}_p$, the ring of integers in the pth local cyclotomic field, by a cyclic group $\langle x \rangle$ of order p, with x acting as multiplication by a primitive pth root of unity θ. If G is a group of maximal class of order p^n, then any abelian normal section of P_1 of order $p^k \leq p^{n-2}$ is isomorphic to $\mathcal{O}_p/(\theta - 1)^k$. If P_1 is abelian, then $G/N \cong S/\gamma_i(S)$, for some i and a normal subgroup $N \leq P_1$ of order at most p. If P_1 is not abelian, then G/N is isomorphic to a 'twisted' version of $S/\gamma_i(S)$ for some i and some small normal subgroup $N \leq P_1$. In fact, modulo $\gamma_3(P_1)$, commutation corresponds in a natural way to a homomorphism from $\wedge^2\mathcal{O}_p/(\theta-1)^a$ to $\mathcal{O}_p/(\theta-1)^b$ for some a and b. This will be used to do the 'twisting' of S described in the next section, and makes it possible to produce an explicit construction for all p-groups of maximal class modulo a normal subgroup of order bounded as a function of p, as described in [14] and [15].

The basic ideas here extend to p-groups of coclass $r > 1$, although the situation is much more complicated. As before the groups under consideration can be visualised in a directed graph $\mathcal{G}(r, p)$ in which the vertices are finite p-groups of coclass r, and *descendants* are defined as in $\mathcal{G}(p)$. Now there is a natural relationship between infinite chains of descendants in $\mathcal{G}(r, p)$ and pro-p groups of coclass r.

Any chain of descendants in the graph $\mathcal{G}(r, p)$ going down from a group P with $\gamma_m(P)$ non-abelian must be of finite length, where $m = m(p, r) = (p-1)p^{r-1}$ for p odd, and $m(2, r) = 2^{r+2}$. There is however no bound to the length of such chains for odd p and $r > 1$ or for $p > 3$ and $r = 1$.

A pro-p group G has *coclass* r if there exists $u \geq 2$ such that $G/\gamma_i(G)$ is a finite p-group of coclass r for all $i \geq u$. The bound on the derived length of a finite p-group of coclass r given in the proof of Conjecture A clearly implies that every pro-p group of coclass r is soluble. It follows that such a pro-p group has finite centre, and modulo the centre is a *uniserial p-adic space group of dimension* $d = (p-1)p^s$ for some s with $0 \leq s \leq r-1$, see [22]; that is, it is an extension of a $\mathbb{Z}_pP$-lattice T (its *translation subgroup*) of rank d by P (its *point group*) where P is a finite p-group acting faithfully and uniserially on T. It can be shown that there are only finitely many such groups for given p and r; hence while we no longer have a unique infinite chain in $\mathcal{G}(p, r)$, we do know that there is only a finite number of infinite chains.

Now to construct all uniserial p-adic space groups of dimension $d = (p-1)p^s$ one would proceed as follows. First observe that all maximal p-subgroups of $GL((p-1)p^s, \mathbb{Q}_p)$ are conjugate and isomorphic to the iterated wreath product W_{s+1} of $s+1$ copies of C_p, with the one exception that a second conjugacy class arises for $p = 2$, see [19]. So find all conjugacy classes of subgroups of W_{s+1} that act irreducibly. For each such class, take a representative P and find the (finite) number of non-isomorphic P-lattices of dimension $(p-1)p^s$; that is, find how the conjugacy class of P in $GL((p-1)p^s, \mathbb{Q}_p)$ splits into $GL((p-1)p^s, \mathbb{Z}_p)$-classes. For each such lattice L, find the finitely many isomorphism classes of extensions of L by P. At this point we have constructed all uniserial p-adic space groups of dimension $d = (p-1)p^s$. The next step in the construction of finite p-groups of coclass r involves the twisting process; this is described in the next section.

5 Twisting

Let G be a finite p-group where p is odd, and let B be an abelian normal subgroup of G with a complement P. Twisting is a technique for constructing split extensions of a class 2 group D by P, the group D being constructed from the P-module B.

Our motivation for introducing this construction is to apply it to the case where G is the quotient of a split p-adic space group by some full sub-lattice of the translation group, so P will be the point group. If G has finite coclass, let us call the groups constructed in this way *constructible groups*. An unfortunate complication now arises, in that not all p-adic space groups are split. However, if G is a p-adic space group whose point group P acts irreducibly on the translation group T, then G lies in the split extension of $p^{-1}T$ by P, so we can reduce to the split case. Allowing this reduction, every p-group of coclass r has a normal subgroup N of order bounded in terms of p and r such that P/N is constructible.

We shall see that twisting can be made more explicit by slightly reducing its generality; this leads to a more explicit form of constructibility, at the cost of increasing the possible order of N as above.

The general construction is quite simple. Take a P-submodule A of B, and a P-homomorphism $\alpha : \wedge^2 C \to A$, where $C = B/A$. The aim is to construct D as an extension of A by C in which A is central and contains D', and in which α defines commutation.

As a first step, consider the group E with underlying set $C \times A$, and with multiplication defined by

$$(c_1, a_1)(c_2, a_2) = \big(c_1c_2, a_1 + a_2 + \tfrac{1}{2}(c_1 \wedge c_2)\alpha\big).$$

Note that E is naturally an extension of A by C with the required properties, but is not the group we are aiming to construct; note also the factor $1/2$, which requires p to be odd. Since α is a P-homomorphism, there is a natural action of P on E.

We now define D to be the Baer sum of E with B where B is regarded as an extension of A by C. That is, first form the subgroup H of $E \times B$ consisting of pairs (x, b) that each map to the same element of C. Now H contains a naturally embedded copy of $A \times A$; define D to be the quotient of H by the subgroup consisting of all pairs $(a, -a)$ with $a \in A$.

It is easy to see that D is a class 2 extension of A by C with commutation induced by α, and that P acts naturally on D. One might regard D as inheriting its commutator structure from E and its power structure from B.

The split extension of D by P is now a *twisting* of G.

In order to make this construction explicit, we need to compute $\mathrm{Hom}_P(\wedge^2 C, A)$ in cases of interest.

The most general case that concerns us is when P is a p-group acting irreducibly on the $\mathbb{Z}_p$-lattice T, so A and C are sections of T. If U and V are the inverse images of A and C in T, it is easy to see that a good approximation to $\mathrm{Hom}_P(\wedge^2 A, C)$ can be computed from $\mathrm{Hom}_P(\wedge^2 U, V)$. Also, any element of $\mathrm{Hom}_P(\wedge^2 U, V)$ extends uniquely to an element of $\mathrm{Hom}_P(\wedge^2 p^{-1}U, p^{-2}V)$, so by extending, restricting and multiplying by powers of p, it is sufficient to compute $\mathrm{Hom}_P(\wedge^2 T, T)$. For precise information one needs to be able to parametrise the homomorphisms, and compute the image of any sub-lattice of T under any of them, which leads to rather fussy and complicated details. Note that $\wedge^2 T$ is essentially the same whether the exterior square is taken over $\mathbb{Z}$ or $\mathbb{Z}_p$, since any $\mathbb{Z}$-bilinear map of $T \times T$ to T is continuous.

The most special case of interest arises with the p-groups of maximal class described above. In this case we may regard T as being the ring $\mathcal{O}_p$ of integers in the pth local cyclotomic number field; that is, the ring $\mathbb{Z}_p[\theta]/\big(f_p(\theta)\big)$, where $f_p(x) = (x^p - 1)/(x - 1)$. Then P is the cyclic group generated by x where x acts as multiplication by θ.

Let K be the field of fractions of $\mathcal{O}_p$. Then $V = \mathrm{Hom}_{\mathbb{Q}_p}(K \otimes K, K)$ (where the tensor product is over $\mathbb{Q}_p$) is a K-space, the action of K coming from the action of K on the codomain. Using the Jacobson independence theorem, and the fact that K is a Galois extension of $\mathbb{Q}_p$, one sees that V has a K-basis $\{\sigma_a \otimes \sigma_b | 1 \leq a, b \leq p-1\}$, where σ_c is defined by $\theta \mapsto \theta^c$. One then sees that $\{\sigma_a \otimes \sigma_b - \sigma_b \otimes \sigma_a | 1 \leq a < b \leq p-1\}$, is a K-basis for $\mathrm{Hom}_{\mathbb{Q}_p}(K \wedge K, K)$ (where the exterior square is over $\mathbb{Q}_p$).

It then follows that a K-basis for $\mathrm{Hom}_{\mathbb{Q}_p P}(K \wedge K, K)$ is

$$\{\sigma_a \otimes \sigma_{p+1-a} - \sigma_{p+1-a} \otimes \sigma_a \mid 1 < a < (p+1)/2\}.$$

This explains the particularly simple structure of p-groups of maximal class when $p = 3$, and the relatively simple structure when $p = 5$.

The real object of interest is $L = \mathrm{Hom}_{\mathbb{Q}_p P}(\wedge^2 \mathcal{O}_p, \mathcal{O}_p)$. The $(p-3)/2$ homomorphisms constructed as above generate a free $\mathbb{Z}_p$-module that is a full sub-lattice of L. Unfortunately this sub-lattice is proper for $p > 5$, and finding precise generators for $\mathrm{Hom}_P(\wedge^2 \mathcal{O}_p, \mathcal{O}_p)$ as $\mathbb{Q}_p$-linear combinations of these 'natural' homomorphisms is a rather messy business.

This case extends easily to P being a cyclic group of order p^c acting uniserially in dimension $p^c - p^{c-1}$. Then T can be taken to be the ring $\mathcal{O}_c$ of integers in the p^cth cyclotomic number field. The analysis proceeds as above, giving rise to $(p^c - p^{c-1})/2 - 1$ independent homomorphisms freely generating a full sub-lattice of $\mathrm{Hom}_P(\wedge^2 \mathcal{O}_c, \mathcal{O}_c)$. Note that in this case we get non-zero homomorphisms even when $p = 3$, so the trivial nature of 3-groups of maximal class does not extend to coclass 2.

Another extension is to the case when P is the wreath product of c copies of C_p acting in dimension $p^c - p^{c-1}$. In this case $\mathrm{Hom}_P(\wedge^2 T, T)$ is again of rank $(p - 3)/2$; but a new complication arises, in that the full sub-lattices of T are not pairwise isomorphic.

The most general case of interest to us is when P acts irreducibly on T, and hence is a subgroup of the full wreath product as above.

For further details see [16], [12], and [17].

6 Classification up to isomorphism

There is no hope of a useful classification of p-groups up to isomorphism. This universally believed maxim is hard to make precise. Obviously there is no theoretical difficulty in classifying the groups of order p^n for any value of p and n. The problem is that for fixed p as n tends to infinity, it seems that the complexity of the answer also, in some vague sense, tends to infinity. It seems unlikely that any classification up to isomorphism could avoid computing the automorphism group of every p-group G in question, and the action of the automorphism group of G on the Frattini quotient of G; and this would involve a complete understanding of the representations of all finite groups, see [2].

If on the other hand n is fixed and p goes to infinity, the general nature of the difficulty remains unknown. It is best encapsulated in G. Higman's conjecture in [7] that the number of isomorphism classes of p-groups of order p^n, for any fixed n, is a PORC function of p; that is, is a polynomial on residue classes. In other words for some integer k, there is a set of k polynomial functions $f_1, f_2, \ldots, f_k$ of one variable such that if $p \equiv i \bmod k$, then the number of isomorphism classes of groups of order p^n is $f_i(p)$. Of course k and the f_i would depend on n. This conjecture seems intractible. Even if valid, it is not clear that k could be computed, except for small values of n.

The situation becomes more hopeful if we include the coclass as a parameter, and try to classify up to isomorphism all groups of order p^n and coclass r. Here p plays the opposite role. It seems impossible to count in a useful way even the number of isomorphism classes of p-groups of maximal class as a function of p, and a relatively crude estimate shows that rather obscure properties of the prime p will play a role (see [16]). However, if we fix p and r and vary n there is some hope. That is, it seems likely that for any p and r a reasonably straightforward description can be given of the set of isomorphism classes of all p-groups of coclass r. This description should consist of an explicit list of the groups of order less than some bound, and a simple parametrisation of all bigger groups. That the bigger groups should have such a parametrisation might not be surprising to anyone who has glanced at the structure theorems we have given. The fact that the small groups require individual attention, where 'small' means 'bounded in terms of p and r', explains the irrelevance of such claims to Higman's PORC conjecture.

The best evidence for such a parametrisation arises in the case $p = 2$; it is also in this case that the conjectured answer is simplest, see the statement of Conjecture A.

To state these conjectures we need some terminology. A *root* is a p-group of coclass r that is isomorphic to $G/\gamma_i(G)$ for a unique pro-p group G (and for a unique value of i). So almost all finite quotients of a pro-p group of coclass r are roots. If P is a root, and Q is a root that is a descendant of P, and the tree of descendants of Q is isomorphic to the tree of descendants of P, then P is a *periodic root*. The existence of periodic roots if $p = 2$ has recently been proved by du Sautoy in [5]; for $p > 2$ a pro-p group of finite coclass will not in general give rise to periodic roots. If P is a periodic root, define its *period* to be the least value of $\log_p(|Q|/|P|)$ as above. This will correspond to a unique Q, given P. Define a periodic root to be *minimal* if it is not a descendant of a periodic root. Define a *family* to be the set of descendants of a fixed minimal periodic root, and define a group to be *sporadic* if it is not a member of a family. It is clear that there can be only finitely many families.

The conjectures of Newman and O'Brien in [24] are as follows:

Conjecture P. All but finitely many 2-groups of coclass r lie in some family; the period of every family divides 2^{r-1}.

Conjecture Q. All 2-groups of coclass r and order $2^{2^{r+1}}$ are settled.

Conjecture R. The minimal periodic roots of the families of 2-groups of coclass r have order at most $2^{2^{r+1}}$.

Conjecture S. All sporadic 2-groups of coclass r have order at most $2^{2^{r+2}}$.

Conjecture P as stated in [24] has a misprint; the period is stated there to be at most $r - 1$. We are grateful to the authors for pointing this out. The reason for the conjectured period is partly experimental and partly McKay's theorem about a 2-adic space group of coclass r having dimension at most 2^{r-1}, see [22].

If these conjectures are valid then clearly constructing all 2-groups of coclass r and order at most $2^{2^{r+2}}$, and noting the relation of being a descendant suffices to determine the complete tree of 2-groups of coclass r; the number of isomorphism classes of groups of order 2^n and coclass r being determined, for $n > 2^{r+1}$, by the residue of n modulo 2^{r-1}.

The conjectures above are known to hold for $r = 1$ (the essentially trivial maximal class case), and for $r = 2$ see [9] and [10]. The case $r = 3$ is extensively examined in [24], where the theoretical relationship between coclass and p-coclass is considered, where G is defined to have *p-coclass r* if G has order p^n, and lower p-central series of length $n - r$. They prove that the relationship between these concepts is very close, but the reader should be warned that they use the term 'coclass' to mean what we have defined as 'p-coclass'. While we feel that coclass, as we have defined it, may be the more useful definition for theoretical arguments, there is no doubt that p-coclass is more convenient for computation. These computations for $r \leq 3$ have been carried out, and support the above conjectures, as reported in [24].

The programs necessary to perform these calculations have a long history. The present very efficient code is the work of O'Brien. The code having been written, it seems that, despite the double exponentials appearing in the above conjectures, there would be no difficulty in extending the computations to higher values of r if it were not for the very large number of groups to be considered. Similarly, there seems to be no obstruction to using the results in [24] as a basis for proving the above conjectures for $r = 3$ other than the tedium of considering so many cases.

Thus they find that there are 82 families of 2-groups of 2-coclass at most 3, consisting of the 60 families of 2-groups coclass at most 3, and 22 other families. The groups in the other families all have a cyclic central section of bounded index. They find that, assuming that the above conjectures are true, the number of isomorphism classes of 2-groups of 2-coclass 3 and order 2^{4m+i} where $m \geq 4$ for $i = 1, 2, 3, 4$ is 2504, 2568, 8632, 1532, respectively.

It seems that a proof of the above conjectures, perhaps with worse bounds, should be possible. In particular, one would expect to be able to construct a specific isomorphism between the tree of descendants of a minimal periodic root P and the tree of descendants of a suitable descendant Q of P and to have some group-theoretic construction connecting groups that correspond under the isomorphism. Such a correspondence would arise not from forming quotients or extensions but rather from inserting or deleting a section from 'the middle' of a suitable group.

The proof of the existence of periodic roots by du Sautoy in [5] follows a completely different route (see Chapter 9). He uses model-theoretic techniques and zeta-functions to prove the following. Let G be a pro-p group of coclass r, let $k > 0$, and let $\mathcal{G}_k(G)$ be the graph whose vertices are the isomorphism classes of p-groups P of coclass r such that P/N is isomorphic to a quotient of G for some normal subgroup N of P of order at most p^k, and whose edges are defined in the usual way. Then $\mathcal{G}_k(P)$ is ultimately periodic; that is, for some P and a descendant Q of P, where both P and Q are quotients of G, the graph of descendants of P and the graph of descendants of Q are isomorphic. His theorem applies to all primes p; but in the case of $p = 2$, we know that for given r, there exists a k such that if P is a sufficiently big 2-group of coclass r, then P has a normal subgroup N of order at most 2^k such that P/N is a quotient of a pro-2 group of coclass r.

This startlingly unexpected proof does not give any explicit bounds.

Classifying the p-groups of coclass r up to isomorphism for odd primes is a much harder problem. There is no problem in principle in constructing the pro-p groups of coclass r. That is, there are only finitely many, and these can be determined algorithmically. Thus the graph $\mathcal{G}(r, p)$ can be expressed as the union of a finite number of infinite trees, or 'families', consisting of a single infinite chain and an infinite number of finite chains; together with a finite number of sporadic groups. The problem is to find a finite description of each family, bearing in mind the fact that some trees will have finite chains of unbounded length. But there is a hope that the groups in each family have a finite parametrisation that would describe the family precisely, and that the families have a sufficient amount of symmetry that such a parametric description can be obtained by computing $\mathcal{G}(r, p)$ up to groups of some order. This seems to be the case for $(p, r) = (5, 1)$ and $(p, r) = (3, 2)$. The former situation is described in [23], and the latter groups have been investigated by Leedham-Green and Newman (unpublished). These are the cases where $\mathrm{Hom}_P(\wedge^2 T, T)$ has rank at most 1 for all families. In other cases the number of groups seems too many for useful computation. Conjectures by Axel Schneider on the structure of $\mathcal{G}(r, p)$ for p odd are to be found in [25].

7 Width and obliquity

We outline the only successful approach to producing some kind of reasonable classification of the class of all finite p-groups seems to be as follows.

First, one forms a graph $\mathcal{G}$ whose vertices are the isomorphism classes of finite p-groups for the chosen prime p, and where edges join a group G (strictly its isomorphism class) to the group $G/\gamma_c(G)$ where c is the nilpotency class of G. The aim now is to understand $\mathcal{G}$. As a first step, one can try to understand the infinite chains in $\mathcal{G}$. These correspond bijectively to the isomorphism classes of countably based pro-p groups with finite lower central factors. Unfortunately there are uncountably many of these, so we have not yet seen the light.

In order to rescue the situation, we can filter the class of finite p-groups. Thus we find classes $\mathcal{X}_1 \subset \mathcal{X}_2 \subset \mathcal{X}_3 \subset \cdots$ such that:

(i) The union of these classes is the set of all isomorphism classes of finite p-groups.

(ii) Each class is quotient closed.

(iii) $\mathcal{X}_i \setminus \mathcal{X}_{i-1}$ contains, for all $i > 0$, isomorphism classes of groups of arbitrarily large class.

Now define $\mathcal{G}_i$ to be the full sub-graph of G consisting of groups in $\mathcal{X}_i$. (We will occasionally write 'group' to mean 'isomorphism class of group'.)

Locally we have gained nothing; equivalently we have lost no information. Each vertex and each edge in $\mathcal{G}$ lies in $\mathcal{G}_i$ for i big enough. But assymptotically the situation may be far simpler. Not every infinite chain in $\mathcal{G}$ lies in some $\mathcal{G}_i$; in fact we may succeed in arranging matters so that there are only finitely many infinite chains in each $\mathcal{G}_i$, and hence only countably many in their union. In fact condition (iii) above was inserted to ensure that infinite chains do arise in $\mathcal{G}_i$.

Suppose now that we can do two things. First we produce a good description of the infinite chains (or pro-p groups) arising in $\mathcal{G}_i$ for all i, and secondly we prove that for some function $f(i)$ we can move from any vertex in $\mathcal{G}_i$ to a vertex on an infinite chain by moving at most $f(i)$ steps up the graph, in other words $\mathcal{G}_i$ has *twig length* $f(i)$. We now have a classification of all finite p-groups. Of course the classification is not up to isomorphism. If $G \in \mathcal{X}_i \setminus \mathcal{X}_{i-1}$ then G has a normal subgroup N of order at most $p^{f(i)}$ such that G/N is well understood, being a lower central quotient of a pro-p group in some well understood class.

Of course this theorem may prove disappointing when applied to a specific p-group G, as G may have order less than $p^{f(i)}$. The best we can hope for is a collection of classifications of the above type that will give useful insights into many interesting p-groups.

In practice, as we have seen in the preceding sections, we have a classification of 2-groups precisely as above where $\mathcal{X}_r$ is the class of 2-groups of coclass r. Thus we can describe all 2-groups modulo a small normal subgroup, where 'small' means 'bounded' in terms of the coclass. It seems likely that we may be able to do better, and get a precise finite description of all sufficiently large 2-groups of coclass r. But all this is useless when applied to a 2-group whose order is not sufficiently large compared to its coclass.

For odd primes, a similar classification by coclass holds, except that in this case there is no bound to the twig length. However this is mitigated by the fact that using twisting we have a good description of finite chains of unbounded length descending from the infinite chains, and we can move from any vertex to a vertex on one of these branches by moving at most $g(r)$ steps up the graph for some function g of r; in other words, we can give a good description of p-groups of coclass r modulo a normal subgroup of order bounded by the coclass.

While it is very satisfying to have a non-trivial classification of the class of all p-groups (and the proof of the coclass theorems is rather hard), it must be admitted that most interesting p-groups do not have a small enough coclass to fall usefully within the coclass classification. The next step should be to produce a classification by width, where the *width* of a p-group G is w where p^w is the maximal order of a lower central factor of G. This concept extends naturally to pro-p groups. This enterprise is fraught with difficulties. We shall certainly need to use other parameters in addition to the width. Thus we will produce a filtration of the graph $\mathcal{G}$ by a family of sub-graphs indexed by a number of copies of $\mathbb{N}$, with $(m_1, m_2, \ldots, m_t) \leq (n_1, n_2, \ldots, n_t)$ if $m_i \leq n_i$ for all i, so that $G_\alpha \subset G_\beta$ if $\alpha < \beta$.

Let us now consider the pro-p groups that should appear as corresponding to our infinite chains. So what chance have we of classifying the pro-p groups of finite width?

As with finite groups the first step is the classification of simple groups, so with pro-p groups the first step is the classification of the *just infinite* groups. That is, those for which every non-trivial closed normal subgroup is open. We divide just infinite pro-p groups of finite width into four classes as follows.

Type 1. Soluble groups.
Type 2. Insoluble p-adic analytic groups.
Type 3. Non p-adic analytic groups that are linear over $\mathbb{F}_p[[t]]$.
Type 4. Groups that are not linear over $\mathbb{Z}_p$ or $\mathbb{F}_p[[t]]$.

It may be worth while making a few remarks about these four types.

Type 1. The last non-trivial term of the derived series of a soluble just infinite group is of finite index. It is now easy to see that a group of type 1 is an extension of a finite dimensional $\mathbb{Z}_p$-lattice by a finite p-group acting faithfully and irreducibly. (A finite group is said to act *irreducibly* on a lattice over an integral domain if the corresponding action on the lattice tensored with the field of fractions of the domain is irreducible.) It follows that the lattice has dimension $(p-1)p^s$ for some $s \geq 0$; or has dimension 1.

Type 2. A group G of type 2 is associated with a homogeneous semi-simple Lie algebra L over $\mathbb{Q}_p$. Given L one can construct a p-adic analytic group G_L such that G will be isomorphic to an open subgroup of G_L; but G_L is not itself a proper open subgroup of just infinite pro-p group. So groups of type 2 are indexed, up to commeasurability, by the non-abelian homogeneous semi-simple Lie algebras over $\mathbb{Q}_p$.

Type 3. There are uncountably many pro-p groups of type 3, see [11]. Examples can be constructed from semi-simple Lie algebras over local fields of characteristic p, in the same way as groups of type 2. However, in this case we do not get maximal groups corresponding to the G_L. For example any finite group (in fact any countably based pro-p group) can be embedded in the automorphism group of a local field of characteristic p, see [3], and hence any group of type 2 can be extended by any finite p-group to produce a virtually linear group over $\mathbb{F}_p[[t]]$. But 'linear' and 'virtually linear' are equivalent concepts. Generally speaking, groups of type 3 are more troublesome than groups of type 2. This is in part because groups of type 2 behave well under forming pth powers. The analogous action of t on a semi-simple $\mathbb{F}_p((t))$-Lie algebra does not correspond to a precisely defined action on a corresponding group. More seriously, there is no natural way of obtaining the Lie algebra from the group. The construction of a Lie algebra for groups of type 3 is discussed by Cartwright in [4].

Type 4. These are the 'sporadics'. The first sporadics to arise were the Nottingham groups. It was then discovered that the groups of Grigorchuk type are also just infinite. This is rather alarming, as these fascinating groups seem hard to classify. More recently Fesenko has discovered more examples arising naturally as closed subgroups of the Nottingham group. Andrei Jaikin Zapirain proves in [27] that a just infinite pro-p group that is linear over a profinite ring is linear over $\mathbb{Z}_p$ or $\mathbb{F}_p[[t]]$.

Looking at the above groups, and bearing in mind the trivial (and trivially generalised) observation that if G is just infinite then so is $G \wr C_p$, it seems sensible to adopt a stronger definition. We say that a group G is *hereditarily just infinite* if every open normal subgroup of G is just infinite. (Here 'normal' is otiose.) This is clearly equivalent to the condition that G has no non-trivial subnormal subgroups of infinite index.

Looking back over the four above types, let us consider the effect of imposing the condition that the groups should be hereditarily just infinite.

Groups of type 1 are more or less wiped out. Clearly the $\mathbb{Z}_p$-lattice is only just infinite if it has rank 1. We can refrain from weeping over the loss of these groups. Although they play a central role in the classification of groups by coclass, since every pro-p group of finite coclass is soluble, they are small beer in the present context.

For groups of type 2 we get the useful simplification that the Lie algebra is now simple rather than merely semi-simple.

A similar simplification occurs with type 3, but the real problems with these groups remains unabated.

The examples we have seen above of groups that are just infinite but not hereditarily just infinite have the property that they contain as an open subgroup a direct product of hereditarily just infinite groups. One might say that they are induced up from hereditarily just infinite groups. This is not the case with the Grigorchuk groups, so they do not return to haunt us. However the Nottingham groups and Fesenko's groups remain.

Before considering whether an even stronger condition than 'hereditarily just infinite' is needed, let us return to our original context, and ask how the property of being just infinite, or hereditarily just infinite, will be reflected in the finite quotients of a pro-p group G of finite width. If G is not (hereditarily) just infinite, it will have a non-trivial closed (sub-)normal subgroup that does not contain any term of the lower central series. Since every subgroup of a finite p-group is subnormal, we had better restrict ourselves to normal subgroups. So let P be a finite p-group, and for i less than the nilpotency class c of P, define

$$\mu_i(P) = \Big(\bigcap_{N \not\leq \gamma_{i+1}(P), N \lhd P} N \Big) \cap \gamma_{i+1}(P).$$

Now define the *obliquity* $o(P)$ of P to be $\max_{i<c} \log_p\big(\gamma_{i+1}(P) : \mu_i(P)\big)$.

This concept extends immediately to pro-p groups, the maximum (or least upper bound) being taken over all i. It turns out, see [11, p. 15], that for groups of type 2, 'of finite obliquity' and 'just infinite' are synonymous. Moreover the obliquity of a group of type 2 can be computed from a finite quotient. The obliquity of groups of type 3 is more obscure, a typical consequence of the general difficulty we have with these groups.

As for groups of type 4, the Nottingham groups have finite obliquity for odd primes, but infinite obliquity if $p = 2$. The obliquity of the other known groups of type 4 has yet to be investigated. So we may hope that the class of hereditarily just infinite pro-p groups of finite obliquity is reasonably tractable.

While the definition of obliquity is rather technical, the concept of finite obliquity in a pro-p group is very natural. A pro-p group G is of finite obliquity if and only if for any two open normal subgroups M and N of the same index in G, the index $|MN : M \cap N|$ is bounded by some constant that depends only on G. This is a natural generalisation of finite coclass, a non-cyclic pro-p group being of finite coclass if any two open normal subgroups of the same sufficiently large index in G are equal.

The class of p-groups of obliquity at most some constant is clearly quotient closed, and the reader who has followed us thus far will have correctly surmised that we propose to classify p-groups by width and obliquity. But to produce a project that can be completed in this century it seems reasonable also to bound the rank. So if $\mathcal{X}_{(w,s,o)}$ denotes the class of all finite p-groups of width at most w, of rank at most s, and of obliquity at most o, what are the pro-p groups arising from the infinite chains in the corresponding graph? If G is such a pro-p group, then the bound s requires G to be p-adic analytic. The bound o requires G to be just infinite. The bound w requires G to have bounded index in G_L in the type 2 case; if G is of type 1 then the width is already bounded by the rank.

We now come to the point.

Conjecture. Let P be a finite p-group of width w, rank s and obliquity o. Then either P has a normal subgroup of nilpotency class at most 2 (abelian if $p = 2$) and of index bounded as a function of (p, w, s, o), or P has a normal subgroup N of order bounded as a function of (p, w, s, o) such that P/N is isomorphic to a quotient of an insoluble just infinite p-adic analytic group of width w, rank s and obliquity o.

This conjecture can be re-worded to state that every p-group of given width, rank and obliquity has a quotient by a normal subgroup of bounded order that is either a quotient of a just infinite insoluble p-adic analytic group with the same width and rank, or is a constructible group that arises from a p-adic space group of the same width and rank. The point being that there is no analogue to twisting for insoluble p-adic analytic groups.

Now there are only finitely many isomorphism types of just infinite p-adic analytic groups of given rank and width; this follows from the fact that there are only finitely many semisimple p-adic Lie algebras of a given dimension, and only finitely many p-adic space groups of given rank (see [11]). Since the 'class 2 by small' groups can be constructed by twisting from the finitely many soluble just infinite pro-p groups, this conjecture, if confirmed, would give a reasonable classification of finite p-groups by width, rank and obliquity. The obliquity would play a major role in determining the order or index of the normal subgroup in the conjecture. The width also plays an obviously crucial role. Removing the rank from the list of parameters would bring all just infinite pro-p groups of finite width into play. While there may well be deep and important results to be proved classifying sets of finite p-groups of unbounded rank, it should be pointed out that there are uncountably many groups of types 3 and 4, and that linear groups over $\mathbb{F}_p[[t]]$ may be extended by arbitrary finite p-groups acting as subgroups of the Nottingham group.

References

[1] N. Blackburn, On a special class of p-groups, *Acta Math.*, **100** (1958), 49–92.

[2] R. M. Bryant and L. C. Kovács, Lie representations and groups of prime power order, *J. London Math. Soc.*, **17**(2) (1978), 415–421.

[3] R. D. Camina, Subgroups of the Nottingham group, *J. Alg.*, **196** (1997), 101–113.

[4] D . Cartwright, pro-p groups and Lie algebras over local fields of characteristic p, Ph.D. thesis, University of London, 1996.

[5] M. P. F. du Sautoy, p-groups, coclass and model theory: a proof of Conjecture P, (to appear).

[6] G.A. Fernández-Alcober, The exact lower bound for the degree of commutativity of a p-group of maximal class, *J. Alg.*, **174** (1995), 523–530.

[7] G. Higman, Enumerating p-groups, II, *Proc. London Math. Soc.*, **10** (1960), 566–582.

[8] B. Huppert, *Endliche Gruppen I*, Springer-Verlag, Berlin–Heidelberg, 1967.

[9] R. James, 2-groups of almost maximal class, *J. Austral. Math. Soc. Ser. A*, **19** (1975), 343–357.

[10] R. James, 2-groups of almost maximal class: corrigendum, *J. Austral. Math. Soc. Ser. A*, **35** (1983), p. 307.

[11] G. Klaas and C. R. Leedham-Green and W. Plesken, Linear pro-p groups of finite width, Lecture Notes in Mathematics,1674, Springer-Verlag, New York– Berlin–Heidelberg, 1997.

[12] C. R. Leedham-Green, The structure of finite p-groups, *J. London Math. Soc.*, **50** (1994), 49–67.

[13] C. R. Leedham-Green, Pro-p groups of finite coclass, *J. London Math. Soc.*, **50** (1994), 43–48.

[14] C. R. Leedham-Green and S. McKay, On p-groups of maximal class I, *Quart. J. Math. Oxford*, **27**(2) (1976), 297–311.

[15] C. R. Leedham-Green and S. McKay, On p-groups of maximal class II, *Quart. J. Math. Oxford*, **29**(2) (1978), 175–186.

[16] C. R. Leedham-Green and S. McKay, On the classification of p-groups of maximal class, *Quart. J. Math. Oxford*, **35**(2) (1984), 293–304.

[17] C. R. Leedham-Green and S. McKay, Oxford University Press, LMS Monograph, (to appear).

[18] C. R. Leedham-Green and M. F. Newman, Space groups and groups of prime power order I, *Arch. Math.*, **35** (1980), 193–202.

[19] C. R. Leedham-Green and W. Plesken, Some Remarks on Sylow groups of general linear groups, *Math. Z.*, **191** (1986), 529–553.

[20] A. Lubotzky and A. Mann, Powerful p-groups I: finite groups, *J. Alg.*, **105** (1987) 484–505.

[21] A. Lubotzky and A. Mann, Powerful p-groups II: p-adic analytic groups, *J. Alg.*, **105** (1987), 506–515.

[22] S. McKay, The precise bound to the coclass of space groups, *J. London Math. Soc.*, **50**(2) (1994), 488–500.

[23] M. F. Newman, Groups of prime-power order, *Groups-Canberra 1989*, Lecture Notes in Mathematics, 1456, Springer-Verlag, New York, Berlin, Heidelberg, 1990.

[24] M. F. Newman and E. A. O'Brien, Classifying groups by coclass, Mathematics research report No. MRR 054-96, School of Mathematical Sciences, The Australian National University, 1987.

[25] A. Schneider, Classifying p-groups by coclass, (in preparation).

[26] A. Shalev, The structure of finite p-groups: effective proof of the coclass conjectures, *Invent. Math.*, **115** (1994), 315–345.

[27] A.J. Zapirain, On linear just infinite pro-p groups, (in preparation).

School of Mathematical Sciences
Queen Mary and Westfield College
London E1 4NS
England
e-mails:C.R. Leedham-Green@qmw.ac.uk; s.mckay@qmw.ac.uk

3

Pro-p Trees and Applications

Luis Ribes and Pavel Zalesskii

0 Introduction

One of the main topics of this chapter is the study of 'combinatorial' constructions for pro-p groups. We are interested in those groups that can be defined by means of universal properties, such as free pro-p groups, free and amalgamated free products, HNN-extensions, etc. In some sense these are the basic building blocks of pro-p groups. As in the case of abstract groups, such constructions and their properties can be often best understood by studying the action of the group arising from the constructions on a natural 'tree' determined by the construction. The Bass–Serre theory of abstract groups acting on trees gives a complete and satisfying description of those groups as fundamental groups of certain graphs of groups. There is a concept of 'tree' which is appropriate for the study of pro-p groups, namely the so-called pro-p tree; we define it here and develop its main properties. By contrast with the situation for abstract groups acting on trees, the structure of a pro-p group acting on a pro-p tree is not yet fully clear. But this is one of the reasons that make this area so attractive, for it is still full of open and interesting problems. Yet, one has a considerable number of results about pro-p groups acting on a pro-p trees, and we give an ample sample of them in the first four sections of this chapter.

The other topic of this chapter is the study of certain types of results on abstract groups that can be best proved by using 'pro-p methods'. In the second part of the chapter we describe a series of these results.

In the first part of this chapter we develop the basic facts of the theory of pro-p groups acting on pro-p trees. This is a pro-p analog of the Bass–Serre theory of groups acting on trees. Knowledge of the Bass–Serre theory is not necessary, but it would be very helpful for understanding and sometimes motivating the results here. In fact, for the sake of intuition, we stress the similarities and differences between the abstract and pro-p theories in appropriate places.

In Section 1 we recall the definition of an abstract graph in a form that is convenient for us in the context of pro-p groups. We also establish some elementary facts about the fundamental group of an abstract graph. This will be used later in the chapter in our applications to produce certain algorithms related to free groups and to finite monoids. Also in this section we define profinite graphs: they can be thought of as inverse limits of finite graphs.

In Section 2 we develop the fundamental notion of pro-p tree; the guiding idea here is that the Cayley graph of a free pro-p group is the prototype of a pro-p tree.

Sections 3 and 4 are concerned with the structure of pro-p groups acting on pro-p trees. We include here what is known on this subject at the present time. There are results that provide good information related to finite subgroups (e.g., a finite subgroup of a pro-p group acting on a pro-p tree must be contained in the stabilizer of a vertex), and to abelian normal subgroups. The most general theorems give structural information in the form of several alternatives: e.g., if a pro-p group acts on a pro-p tree, then it may coincide with the stabilizer of one of the vertices (in which case we get no real information), or it contains a free pro-p subgroup missing the vertex stabilizers, or, alternatively, it contains a normal subgroup with a quotient isomorphic to either the infinite procyclic pro-p group or the infinite dihedral pro-2 group. In Section 4 these general results are applied to the concrete cases of groups that appear as free products, amalgamated free products or HNN-extensions in the category of pro-p groups.

Sections 5–8 deal with abstract groups and classes of abstract monoids. Section 5 contains basic facts about the pro-p topology on abstract groups. We include also a generalization of the theorem of M. Hall, stating that a closed finitely generated subgroup of a free group F is a free factor of some open subgroup in the pro-p topology on F.

Section 6 contains a key result for later applications to finite monoids, namely that a product of closed finitely generated subgroups of a free group F is closed in the pro-p topology on F. The proof of this theorem illustrates the use of graph theoretic methods and the inter-relation between abstract and pro-p trees.

In Section 7 we exhibit an algorithm for computing the closure of a given finitely generated subgroup of H of a free group F, yielding also an algorithm for deciding whether or not H is closed in the pro-p topology on F.

Section 8 consists of applications to finite semigroup and language theory of the results of the two preceding sections. We describe algorithms for computing the closure (in the pro-p topology) of a rational subset in a free group and in a free monoid. As a consequence a solution of a pro-p version of a well-known conjecture of J. Rhodes is given.

Finally, Section 9 surveys recent results on virtually free pro-p groups. It contains the description of the structure of a pro-p group G that contains an open subgroup of index p. Such a group is a free pro-p product whose factors are either free pro-p groups or of the form $C_p \times H$, where H is a free pro-p group. Serre proved a special case of this theorem in 1965, namely, when G is torsion free; in that case G is free pro-p. As an application of the structure of G one obtains partial answers to a series of conjectures concerning the fixed subgroup of an automorphism of a free pro-p group.

1 Abstract and profinite graphs

A *graph* is a set Γ with a distinguished subset $V = V(\Gamma)$ (the set of *vertices* of Γ), together with two maps, $d_0, d_1 : \Gamma \longrightarrow V$, that are the identity on V. The elements of $E = E(\Gamma) = \Gamma - V(\Gamma)$ are the *edges* of Γ. Let $e \in E$; then $d_0(e)$ and $d_1(e)$ are called the *initial* and *terminal* vertices of e, respectively. We sometimes refer to a graph as an *abstract graph* to emphasize the difference with the profinite graphs that are introduced later on in this section.

Let Γ be a graph; then for each edge e we introduce formal symbols $e^1 (= e)$ and e^{-1}, to be thought of as travelling along e one way or in reverse. Define $d_0(e^{-1}) = d_1(e)$ and $d_1(e^{-1}) = d_0(e)$. Set $E^{-1} = \{e^{-1} \mid e \in E\}$. A group G is said to *act on* Γ if G acts on the set Γ in such a way that $d_i(gm) = gd_i(m)$ for all $g \in G, m \in \Gamma$, $i = 0, 1$. If G acts on Γ, define $ge^{-1} = (ge)^{-1}$, for all $e \in E, g \in G$.

If v, w are vertices of the graph Γ, a *path* $p_{v,w}$ of *length* n joining v to w is a finite sequence

$$e_1^{\epsilon_1}, e_2^{\epsilon_2}, \dots, e_n^{\epsilon_n}$$

of symbols $e_i^{\epsilon_i}$, where e_i is an edge, $\epsilon_i = \pm 1$, $v = d_0(e_1^{\epsilon_1})$, $d_1(e_i^{\epsilon_i}) = d_0(e_{i+1}^{\epsilon_{i+1}})$ $(i = 1, \dots, n-1)$, and $w = d_1(e_n^{\epsilon_n})$. If for every i, $e_i^{\epsilon_i} \neq e_{i+1}^{-\epsilon_{i+1}}$, we say that this path is *reduced*. A reduced path $p_{v,w}$ is a *cycle* if $v = w$. An abstract graph Γ is *connected* if any two vertices in Γ are joined by a path. Γ is an abstract *tree* if it is connected and it contains no cycles of length ≥ 1. If in the above path $p_{v,w}$, one has $e_i^{\epsilon_i} = e_{i+1}^{-\epsilon_{i+1}}$ for some i, then we can form a new path $p'_{v,w} = e_1^{\epsilon_1}, \dots, e_{i-1}^{\epsilon_{i-1}}, e_{i+2}^{\epsilon_{i+2}}, \dots, e_n^{\epsilon_n}$ from v to w. We say that $p_{v,w}$ and $p'_{v,w}$ are *elementarily equivalent*. In general, two paths from v to w are said to be *equivalent* if one can transform one to the other by a finite sequence of elementary equivalences. It is clear that equivalence of paths is an equivalence relation on the set of paths from v to w; furthermore, each equivalence class of paths contains precisely one reduced path.

Let v be a fixed vertex of a connected abstract graph Γ. Let $\pi(\Gamma, v)$ denote the set of all cycles $p_{v,v}$ from v to v. Define a multiplication on $\pi(\Gamma, v)$ by concatenation of paths. This induces a multiplication on the set $\pi_1(\Gamma, v)$ of equivalence classes of paths in $\pi(\Gamma, v)$. It is plain that under this multiplication $\pi_1(\Gamma, v)$ becomes a group, the *fundamental group of the graph* Γ *based at* v. Observe that this definition of $\pi_1(\Gamma, v)$ is independent of the chosen vertex v, in the sense that $\pi_1(\Gamma, v) \cong \pi_1(\Gamma, w)$ if w is another vertex of Γ (recall that Γ is connected). Consequently we use the notation $\pi_1(\Gamma) = \pi_1(\Gamma, v)$ and refer to it as the *fundamental group of the graph* Γ. It is clear from this definition that one can think of $\pi_1(-)$ as a functor on the category of connected abstract graphs and their morphisms.

An easy application of Zorn's Lemma shows that every connected abstract graph Γ contains a maximal abstract subtree T. One easily verifies that $V(\Gamma) = V(T)$ (cf. Proposition I.11 of [37]).

Proposition 1.1 *Let Γ be a connected graph.*

(a) $\pi_1(\Gamma)$ *is a free abstract group.*

(b) *If Δ is a connected subgraph of Γ, then $\pi_1(\Delta)$ is a free factor of $\pi_1(\Gamma)$.*

Proof. This is well known (see, e.g., Ex. 5.3, Ch. 6 of [22]). We only give an indication of the proof. Fix $v \in V(\Gamma)$. Take a maximal tree of Γ containing v. For each $x \in E(\Gamma) - E(T)$, define a cycle c_x of Γ to consist of the unique reduced path in T from v to $d_0(x)$ followed by x, followed by the unique reduced path in T from $d_1(x)$ to v. Then c_x defines an element of $\pi_1(\Gamma)$. Let X consist of the cycles thus defined. Then one checks that $\pi_1(\Gamma)$ is free on the set X. To prove part (b), proceed as follows: pick the fixed vertex v from Δ; choose a maximal tree T' of Δ, and extend this tree to a maximal tree T of Γ; then apply the above construction. It is plain that $X \cap \pi_1(\Delta)$ is a basis for the group $\pi_1(\Delta)$, and so $\pi_1(\Delta)$ is a free factor of $\pi_1(\Gamma)$. □

If Γ and Γ' are graphs, a *morphism* $\alpha : \Gamma \longrightarrow \Gamma'$ is a map such that $d_i\alpha = \alpha d_i$ $(i = 0, 1)$. If α is surjective (respectively, injective, bijective), we say that α is an *epimorphism* (respectively, *monomorphism*, *isomorphism*). If α is an epimorphism, we shall call Γ' a *quotient graph* of Γ; if α is a monomorphism, we shall call $\alpha(\Gamma)$ a *subgraph* of Γ'.

Note that the equality $d_j\alpha(m) = \alpha d_j(m)$ implies that a morphism of graphs maps vertices to vertices. However, the next example shows that a morphism can map an edge into a vertex.

Example 1.2 Let Δ be a subgraph of a graph Γ. Consider the natural map $q : \Gamma \longrightarrow \Gamma/\Delta$ to the quotient space Γ/Δ. Define $V(\Gamma/\Delta) = q(V(\Gamma))$, $d_0(q(m)) = q(d_0(m))$, $d_1(q(m)) = q(d_1(m))$ for all $m \in \Gamma$. Then clearly Γ/Δ becomes a quotient graph of Γ and we shall say that Γ/Δ is obtained from Γ by *collapsing* Δ to a point.

If v is a vertex of Γ, define $\mathrm{Star}(\Gamma, v) = d_0^{-1}(v)$. A morphism α is a *local isomorphism* if for every v in V, α induces a bijection from $\mathrm{Star}(\Gamma, v)$ to $\mathrm{Star}(\alpha(\Gamma), \alpha(v))$.

Consider now an abstract free group F on a set X. Recall that the vertex set of the corresponding *Cayley graph* $\Gamma = \Gamma(F, X)$ is $V(\Gamma) = F$ and the edges of Γ consist of the pairs $e = (f, x)$ for all $f \in F$, $x \in X$. The incidence maps are defined by $d_0(f, x) = f$ and $d_1(f, x) = fx$ for an edge (f, x), and $d_0(f) = d_1(f) = f$ for a vertex f. It is well-known that this Cayley graph $\Gamma = \Gamma(F, X)$ is an abstract tree (cf. Proposition I.15 of [37]). Observe that F acts on the left on $\Gamma = \Gamma(F, X)$ in a natural way. The quotient graph $F\backslash\Gamma$ under this action is a "bouquet of circles": a graph with a single vertex and $|X|$ edges. If H is a subgroup of F, the natural epimorphism of graphs $\sigma : \Gamma \longrightarrow H\backslash\Gamma$ is a local isomorphism.

Lemma 1.3 *Let H be a subgroup of the free group $F = F(X)$. Then there is a natural isomorphism of groups $p_H : H \longrightarrow \pi_1(H\backslash\Gamma(F, X))$.*

Proof. We just indicate the main steps. Let h be an element of H. Consider the unique reduced path $p_{1,h}$ from 1 to h in $\Gamma(F, X)$. Define $p_H(h) = \sigma(p_{1,h})$. Clearly p_H is a homomorphism, and it is not hard to check that it is onto. Since σ is a local isomorphism, it maps reduced paths to reduced paths, and hence p_H is an injection. □

From the definition of the isomorphisms p_H one immediately deduces the following result.

Corollary 1.4 *Let $H \leq K \leq F$, where F is a free abstract group on X, and let $\Gamma = \Gamma(F, X)$ be the Cayley graph of F with respect to X. Then there exists a commutative diagram*

$$\begin{array}{ccc} \pi_1(H\backslash\Gamma) & \xrightarrow{\bar{\tau}} & \pi_1(K\backslash\Gamma) \\ {\scriptstyle p_H}\uparrow & & \uparrow{\scriptstyle p_K} \\ H & \xrightarrow{\iota} & K \end{array}$$

where $\bar{\tau}$ is induced by the natural epimorphism of graphs $\tau : H\backslash\Gamma \longrightarrow K\backslash\Gamma$. In particular, $\bar{\tau}$ is injective.

We now turn to profinite graphs. Recall first that a topological space X is a *profinite space* if it is the inverse limit of finite discrete spaces. In other words, X is profinite if it is compact, Hausdorff and totally disconnected. A graph Γ is called a *profinite graph* if Γ is a profinite space, the subset of vertices $V(\Gamma)$ is closed (and hence profinite) and the two incidence maps $d_0, d_1 : \Gamma \to V(\Gamma)$ are continuous. If $E(\Gamma)$ is closed, it is enough to define d_0 and d_1 continuously on $E(\Gamma)$. The set of edges $E(\Gamma)$ of a profinite graph Γ need not be a closed (and therefore compact) subset of Γ. Finite abstract graphs are profinite graphs.

If Γ and Γ' are profinite graphs, a *morphism of profinite graphs* $\alpha : \Gamma \longrightarrow \Gamma'$ is simply a morphism as abstract graphs which is also continuous.

Proposition 1.5 *Every profinite graph Γ can be represented as an inverse limit of finite quotient graphs of Γ.*

Proof. Let Γ be a profinite graph with set of vertices $V(\Gamma)$. Let $\Gamma = \varprojlim_{\lambda\in\Lambda} K_\lambda$ be a decomposition of Γ as an inverse limit of finite discrete spaces, with projection maps $\varphi_\lambda : \Gamma \to K_\lambda$. Define $\psi_\lambda : \Gamma \longrightarrow K_\lambda \times K_\lambda \times K_\lambda$ by setting $\psi_\lambda(m) = (\varphi_\lambda(m), \varphi_\lambda d_0(m), \varphi_\lambda d_1(m))$. Define a graph structure on $\Gamma_\lambda := \mathrm{Im}(\psi_\lambda)$ by setting $V(\Gamma_\lambda) = \psi_\lambda(V(\Gamma))$, $d_0(\psi_\lambda(m)) = \psi_\lambda d_0(m)$, $d_1(\psi_\lambda(m)) = \psi_\lambda d_1(m)$. To check that this is well defined one needs to verify that for all $m, m' \in \Gamma$ such that $\psi_\lambda(m) = \psi_\lambda(m')$, one has $\psi_\lambda d_j(m) = \psi_\lambda d_j(m')$, $j = 0, 1$. But this follows directly from the definition of ψ_λ. So Γ_λ is a quotient graph of Γ for every $\lambda \in \Lambda$, and the set $\{\Gamma_\lambda \mid \lambda \in \Lambda\}$ forms an inverse system with $\Gamma = \varprojlim \Gamma_\lambda$ as required. □

Exercise 1.6 Let $\Gamma = \varprojlim_{\lambda\in\Lambda} \Gamma_\lambda$ be an inverse limit of profinite graphs Γ_λ and let $\rho : \Gamma \longrightarrow \Delta$ be a morphism of profinite graphs, where Δ is a finite graph. Prove that there exist $\lambda \in \Lambda$ and a morphism $\rho_\lambda : \Gamma_\lambda \longrightarrow \Delta$ such that $\rho = \rho_\lambda\varphi_\lambda$, where $\varphi_\lambda : \Gamma \longrightarrow \Gamma_\lambda$ is the projection.

A profinite graph Γ is said to be *connected* if all its finite quotient graphs are connected as abstract graphs.

Proposition 1.7

(a) *Every quotient graph of a connected profinite graph is connected.*

(b) *If* $\Gamma = \varprojlim \Gamma_\lambda$ *and the* Γ_λ *are connected profinite graphs, then* Γ *is connected.*

Proof. (a) is obvious.

(b) Let Δ be a finite quotient graph of Γ. Then (cf. Exercise 1.6) there exists λ such that Δ is also a quotient graph of Γ_λ. It follows that Δ is connected as required. □

Example 1.8 A connected profinite graph need not be connected as an abstract graph. Indeed let $\Gamma = \mathbf{N} \uplus \mathbf{N} \uplus \{\infty\}$ be the one point compactification of a disjoint union of two copies of the natural numbers. Define $V(\Gamma) = \mathbf{N} \uplus \{\infty\}$, $E(\Gamma) = \mathbf{N}$, $d_0(n) = n, d_1(n) = n + 1$ for $n \in E(\Gamma)$ and d_0, d_1 to be identity maps on $V(\Gamma)$. Then Γ is a connected profinite graph (since it is the inverse limit of finite segments), but as an abstract graph it has a disconnected vertex $\{\infty\}$.

Exercise 1.9 (i) Let $\Gamma = \Gamma_1 \cup \Gamma_2$ be a union of connected profinite graphs Γ_1 and Γ_2 with $\Gamma_1 \cap \Gamma_2 \neq \emptyset$. Prove that Γ is a connected profinite graph.

(ii) Suppose a profinite graph Γ is a disjoint union of open subgraphs. Prove that Γ is not connected. (*Hint*: use the operation of collapsing from Example 1.2 and Proposition 1.7 (a).)

2 Pro-p trees

We begin this section with some facts on profinite modules which will be needed later.

Let X be a profinite space (respectively, let $(X, *)$ a pointed profinite space). Let $X = \varprojlim X_j$ be a decomposition of X as an inverse limit of finite spaces (respectively, let $(X, *) = \varprojlim (X_j, *)$ be a decomposition of $(X, *)$ as an inverse limit of finite pointed spaces). For each finite space X_j denote by $[\mathbf{F}_p X_j]$ the $\mathbf{F}_p$-vector space with basis X_j. Then $[\![\mathbf{F}_p X]\!] = \varprojlim [\mathbf{F}_p X_j]$ (respectively, $[\![\mathbf{F}_p(X, *)]\!] = \varprojlim [\mathbf{F}_p(X_j, *)]$) is called a free profinite $\mathbf{F}_p$-module on a profinite space X (respectively, on a pointed profinite space $(X, *)$). It is not hard to check that X is naturally embedded in $[\![\mathbf{F}_p X]\!]$ and that $[\![\mathbf{F}_p X]\!]$ can be characterized by the usual universal property: whenever $\varphi : X \longrightarrow M$ is a continuous map of X into an $\mathbf{F}_p$-module M, there exists a unique continuous $\mathbf{F}_p$-homomorphism $[\![\mathbf{F}_p X]\!] \longrightarrow M$ extending φ. The module $[\![\mathbf{F}_p(X, *)]\!]$ admits an analogous characterization. (Note that some authors denote a free profinite $\mathbf{F}_p$-module on a profinite space X by $\mathbf{F}_p[\![X]\!]$. We have decided not to use this notation in this chapter to avoid confusion with the usual notation for a ring of power series.)

The facts stated in the following lemma are the "inverse limits" of the corresponding facts for finite spaces, which in turn are obvious.

Lemma 2.1 *Let Y and Z be closed subspaces of the profinite pointed space $(X, *)$ such that $* \in Y$ and $* \notin Z$.*

(1) *The natural $\mathbf{F}_p$-homorphisms $[\![\mathbf{F}_pZ]\!] \longrightarrow \overline{\langle Z\rangle}$ and $[\![\mathbf{F}_p(Y,*)]\!] \longrightarrow \overline{\langle Y\rangle}$ are isomorphisms; so $\overline{\langle Z\rangle}$ can be identified with $[\![\mathbf{F}_pZ]\!]$ and $\overline{\langle Y\rangle}$ with $[\![\mathbf{F}_p(Y,*)]\!]$.*

(2) *There are natural isomorphisms $[\![\mathbf{F}_p(X,*)]\!]/[\![\mathbf{F}_pZ]\!] \cong [\![\mathbf{F}_p(X/Z,*)]\!]$ and $[\![\mathbf{F}_p(X,*)]\!]/[\![\mathbf{F}_p(Y,*)]\!] \cong [\![\mathbf{F}_p(X/Y,*)]\!]$.*

(3) *The equality $(Y,*) = \bigcap_{i\in I}(Y_i,*)$ implies $[\![\mathbf{F}_p(Y,*)]\!] = \bigcap_{i\in I}[\![\mathbf{F}_p(Y_i,*)]\!]$, where $(Y_i,*)$ are closed subsets of $(X,*)$.*

(4) *Assertions analogous to (1), (2) and (3) also hold in the nonpointed case.*

Let G be a pro-p group. Define the *complete group algebra* $[\![\mathbf{F}_pG]\!]$ to be the inverse limit

$$[\![\mathbf{F}_pG]\!] = \varprojlim_{U\in\mathcal{U}} [\mathbf{F}_p(G/U)]$$

of the ordinary group algebras $[\mathbf{F}_p(G/U)]$, where $\mathcal{U}$ is the collection of all open normal subgroups of G. Then $[\![\mathbf{F}_pG]\!]$ is a profinite ring. (Note that some authors denote a complete group algebra by $\mathbf{F}_p[\![G]\!]$.)

If a pro-p group G acts continuously on a profinite space X, then $[\![\mathbf{F}_pX]\!]$ becomes a profinite $[\![\mathbf{F}_pG]\!]$-module. The proof of the next proposition can be found in [41] or in [31].

Proposition 2.2 *Let a pro-p group G act freely on a profinite space X (respectively, on a pointed profinite space $(X,*)$). Then the module $[\![\mathbf{F}_pX]\!]$ (respectively, $[\![\mathbf{F}_p(X,*)]\!]$) is a projective $[\![\mathbf{F}_pG]\!]$-module.*

Later we shall need the following well-known fact (cf. §7, Chapter IV [20]).

Lemma 2.3 *Let $G = \langle g\rangle$ be a cyclic group of order p and M a free profinite $[\![\mathbf{F}_pG]\!]$-module. Then* $\mathrm{Im}(\nu_G) = \mathrm{Ker}(\tau_G)$ *where*

$$\nu_G : M \longrightarrow M;\ m \longrightarrow (1+g+\cdots+g^{p-1})m$$

$$\tau_G : M \longrightarrow M;\ m \longrightarrow (g-1)m.$$

Let Γ be a profinite graph. We denote by $(E^*(\Gamma),*)$ the quotient space $\Gamma/V(\Gamma)$ with the image of $V(\Gamma)$ as a distinguished point $*$.

Denote by $C(\Gamma,\mathbf{F}_p)$ the chain complex

$$0 \longrightarrow [\![\mathbf{F}_p(E^*(\Gamma),*)]\!] \stackrel{\delta}{\longrightarrow} [\![\mathbf{F}_pV(\Gamma)]\!] \stackrel{\varepsilon}{\longrightarrow} \mathbf{F}_p \longrightarrow 0$$

of profinite $\mathbf{F}_p$-modules such that $\varepsilon(v) = 1$ for every $v \in V(\Gamma)$, $\delta(\bar{e}) = d_1(e) - d_0(e)$ if $\bar{e}$ is the image in $E^*(\Gamma)$ of the edge $e \in E(\Gamma)$, and $\delta(*) = 0$. Obviously, $\mathrm{Im}(\delta) \leq \mathrm{Ker}(\varepsilon)$.

Note that if $E(\Gamma)$ is compact (which is the case for the majority of important examples), then

$$[\![\mathbf{F}_p(E^*(\Gamma),*)]\!] = [\![\mathbf{F}_p E(\Gamma)]\!].$$

In this situation the definition of the chain complex $C(\Gamma, \mathbf{F}_p)$ is completely analogous to the abstract one (see [6], Proposition 1.1).

The *homology groups* of Γ are defined as the homology groups of the chain complex $C(\Gamma, \mathbf{F}_p)$ in the usual way: $H_0(\Gamma, \mathbf{F}_p) = \mathrm{Ker}(\varepsilon)/\mathrm{Im}(\delta)$, $H_1(\Gamma, \mathbf{F}_p) = \mathrm{Ker}(\delta)$.

A profinite graph morphism $\alpha : \Gamma \longrightarrow \Delta$ defines continuous maps $\alpha_V : V(\Gamma) \longrightarrow V(\Delta)$ and $\alpha_{E^*} : (E^*(\Gamma),*) \longrightarrow (E^*(\Delta),*)$. We extend the maps α_V and α_{E^*} to homomorphisms $\tilde{\alpha}_V : [\![\mathbf{F}_p V(\Gamma)]\!] \longrightarrow [\![\mathbf{F}_p V(\Delta)]\!]$ and $\tilde{\alpha}_{E^*} : [\![\mathbf{F}_p(E^*(\Gamma),*)]\!] \longrightarrow [\![\mathbf{F}_p(E^*(\Delta),*)]\!]$. The triple $\tilde{\alpha} = (\tilde{\alpha}_{E^*}, \tilde{\alpha}_V, \mathrm{Id}_{\mathbf{F}_p})$

$$\begin{array}{ccccccccc} 0 & \longrightarrow & [\![\mathbf{F}_p(E^*(\Gamma),*)]\!] & \xrightarrow{\delta^\Gamma} & [\![\mathbf{F}_p V(\Gamma)]\!] & \xrightarrow{\varepsilon^\Gamma} & \mathbf{F}_p & \longrightarrow & 0 \\ & & {\scriptstyle\tilde{\alpha}_{E^*}}\big\downarrow & & {\scriptstyle\tilde{\alpha}_V}\big\downarrow & & {\scriptstyle\mathrm{Id}}\big\downarrow & & \\ 0 & \longrightarrow & [\![\mathbf{F}_p(E^*(\Delta),*)]\!] & \xrightarrow{\delta^\Delta} & [\![\mathbf{F}_p V(\Delta)]\!] & \xrightarrow{\varepsilon^\Delta} & \mathbf{F}_p & \longrightarrow & 0 \end{array} \qquad (1)$$

is then a homomorphism $\tilde{\alpha} : C(\Gamma, \mathbf{F}_p) \longrightarrow C(\Delta, \mathbf{F}_p)$ of complexes. Therefore, when $\Gamma = \varprojlim \Gamma_\lambda$ is an inverse limit of an inverse system of profinite graphs, the corresponding chain complexes $C(\Gamma_\lambda, \mathbf{F}_p)$ form an inverse system and $C(\Gamma, \mathbf{F}_p) = \varprojlim C(\Gamma_\lambda, \mathbf{F}_p)$ in the obvious sense.

Further, the homomorphism $\tilde{\alpha}$ of complexes induces a homomorphism of homology groups, namely $\alpha_i^* : H_i(\Gamma, \mathbf{F}_p) \longrightarrow H_i(\Delta, \mathbf{F}_p)$, $i = 0, 1$. Of course, α_1^* is just the restriction of $\tilde{\alpha}_{E^*}$ to the $\mathrm{Ker}(\delta^\Gamma)$.

Lemma 2.4

(a) *If* $\alpha : \Gamma \longrightarrow \Delta$ *is a surjective morphism of profinite graphs, then* $\alpha_0^* : H_0(\Gamma, \mathbf{F}_p) \longrightarrow H_0(\Delta, \mathbf{F}_p)$ *is surjective. If* α *is injective, then so is* $\alpha_1^* : H_1(\Gamma, \mathbf{F}_p) \longrightarrow H_1(\Delta, \mathbf{F}_p)$.

(b) *If* $\Gamma = \varprojlim \Gamma_\lambda$, *then* $H_i(\Gamma, \mathbf{F}_p) = \varprojlim H_i(\Gamma_\lambda, \mathbf{F}_p)$, $i = 0, 1$.

Proof. (a) Suppose that α is surjective. Then so is α_V, and that implies surjectivity of $\tilde{\alpha}_V$. It follows from (1) that α_0^* is surjective. Similarly, suppose that α is injective. Then so is α_{E^*}, and that implies injectivity of $\tilde{\alpha}_{E^*}$. It follows that α_1^* is injective.

(b) Let $\Gamma = \varprojlim \Gamma_\lambda$. As it was mentioned above the corresponding chain complexes $C(\Gamma_\lambda, \mathbf{F}_p)$

$$0 \longrightarrow [\![\mathbf{F}_p(E^*(\Gamma_\lambda),*)]\!] \xrightarrow{\delta_\lambda} [\![\mathbf{F}_p V(\Gamma_\lambda)]\!] \xrightarrow{\varepsilon_\lambda} \mathbf{F}_p \longrightarrow 0,$$

form an inverse system such that $C(\Gamma, \mathbf{F}_p) = \varprojlim C(\Gamma_\lambda, \mathbf{F}_p)$. This means that $\mathrm{Im}(\delta) = \varprojlim \mathrm{Im}(\delta_\lambda)$ and $\mathrm{Ker}(\varepsilon) = \varprojlim \mathrm{Ker}(\varepsilon_\lambda)$. Therefore $H_i(\Gamma, \mathbf{F}_p) = \varprojlim H_i(\Gamma_\lambda, \mathbf{F}_p)$, $i = 0, 1$, as required. □

Proposition 2.5 *A profinite graph Γ is connected if and only if $H_0(\Gamma, \mathbf{F}_p) = 0$.*

Proof. The proof is obvious for finite Γ (see, e.g., [6], Proposition 1.1.1). Now let $\Gamma = \varprojlim \Gamma_\lambda$ be infinite, where the Γ_λ are finite quotient graphs of Γ (see Proposition 1.5). If Γ is connected, then so are the Γ_λ and therefore $H_0(\Gamma_\lambda, \mathbf{F}_p) = 0$ for every λ. Consequently, by Lemma 2.4 (b), $H_0(\Gamma, \mathbf{F}_p) = 0$.

Conversely, suppose $H_0(\Gamma, \mathbf{F}_p) = 0$. Then, by Lemma 2.4 (a), $H_0(\Gamma_\lambda, \mathbf{F}_p) = 0$ for every λ. It follows that Γ_λ is connected for every λ and hence, by Proposition 1.7 (b), Γ is connected. □

A profinite graph Γ is said to be a *pro-p tree* if Γ is connected and $H_1(\Gamma, \mathbf{F}_p) = 0$. Thus Γ is a pro-p tree if and only if $C(\Gamma, \mathbf{F}_p)$ is an exact sequence.

The next lemma follows immediately from Lemma 2.4.

Lemma 2.6

(a) *Every connected subgraph of a pro-p tree is a pro-p tree.*

(b) *An inverse limit of pro-p trees is a pro-p tree.*

Proposition 2.7 *Let $\Delta_i, i \in I$, be a family of pro-p subtrees of a pro-p tree T. Then the subgraph $\Delta = \bigcap_{i\in I} \Delta_i$ is a pro-p tree (possibly empty).*

Proof. By Lemma 2.1, one has

$$[\![\mathbf{F}_p V(\Delta)]\!] = \bigcap_{i\in I} [\![\mathbf{F}_p V(\Delta_i)]\!],$$

$$[\![\mathbf{F}_p(E^*(\Delta), *)]\!] = \bigcap_{i\in I} [\![\mathbf{F}_p(E^*(\Delta_i), *)]\!].$$

Consider the exact sequence

$$0 \longrightarrow [\![\mathbf{F}_p(E^*(T), *)]\!] \xrightarrow{\delta} [\![\mathbf{F}_p V(T)]\!] \xrightarrow{\varepsilon} \mathbf{F}_p \longrightarrow 0$$

associated with T. Denote by $\varepsilon^\Delta, \varepsilon^{\Delta_i}, \delta^\Delta, \delta^{\Delta_i}$ the restrictions of ε and δ to Δ and Δ_i, respectively. Then

$$\mathrm{Ker}(\varepsilon^\Delta) = [\![\mathbf{F}_p V(\Delta)]\!] \cap \mathrm{Ker}(\varepsilon) = \bigcap_{i\in I} \mathrm{Ker}(\varepsilon^{\Delta_i}),$$

and $\mathrm{Im}(\delta^\Delta) = \bigcap_{i\in I} \mathrm{Im}(\delta^{\Delta_i})$ because δ is injective. Since Δ_i is connected, by Proposition 2.5, $\mathrm{Ker}(\varepsilon^{\Delta_i}) = \mathrm{Im}(\delta^{\Delta_i})$ for every i. It follows that $\mathrm{Im}(\delta^\Delta) = \mathrm{Ker}(\varepsilon^\Delta)$. So, by Proposition 2.5, Δ is connected, and it is a pro-p tree by Lemma 2.6. □

It follows from Proposition 2.7 that for every subset W of a pro-p tree T, there exists a smallest pro-p subtree $[W]$ containing W (in fact, $[W]$ is the intersection of all pro-p subtrees containing W). If W consists of two vertices v and w, we call $[v, w] :=$

$[\{v, w\}]$ the *geodesic* connecting v and w. Remark that an analogous definition of 'geodesic' for abstract graphs would correspond to the graph determined by the path $p_{v,w}$.

Lemma 2.8 *Let Γ be a pro-p tree, T, T' subtrees of Γ with $T \cap T' \neq \emptyset$, and let $x \in V(T)$, $x' \in V(T')$. Then $[x, x'] \cap T \cap T' \neq \emptyset$.*

Proof. By Proposition 2.7, $P = [x, x'] \cap T$ and $P' = [x, x'] \cap T'$ are pro-p trees and, by Exercise 1.9 (i), so is $T \cup T'$. Hence $[x, x'] \subseteq T \cup T'$ and $[x, x'] = P \cup P'$. Since $[x, x']$ is a connected profinite graph, its finite images are connected finite graphs. Suppose $P \cap P' = \emptyset$. Then by collapsing P and P' to vertices (see Example 1.2), we obtain a graph consisting of two vertices only, which is clearly disconnected. From this contradiction we deduce that $P \cap P' \neq \emptyset$ as desired. □

Lemma 2.9 *Let Δ be a pro-p subtree of a connected profinite graph Γ and α : $\Gamma \longrightarrow \Gamma/\Delta$ the canonical epimorphism. Then $\alpha_1^* : H_1(\Gamma, \mathbf{F}_p) \longrightarrow H_1(\Gamma/\Delta, \mathbf{F}_p)$ is an isomorphism. In particular, if Γ is a pro-p tree then so is Γ/Δ.*

Proof. Let $\beta : \Delta \longrightarrow \Gamma$ be the natural embedding. Then one has a monomorphism $\tilde{\beta} : C(\Delta, \mathbf{F}_p) \longrightarrow C(\Gamma, \mathbf{F}_p)$ and an epimorphism $\tilde{\alpha} : C(\Gamma, \mathbf{F}_p) \longrightarrow C(\Gamma/\Delta, \mathbf{F}_p)$ of the chain complexes, i.e., the following commutative diagram

$$\begin{array}{ccccccccc}
0 & \longrightarrow & [\![\mathbf{F}_p(E^*(\Delta), *)]\!] & \xrightarrow{\delta^\Delta} & [\![\mathbf{F}_p V(\Delta)]\!] & \xrightarrow{\varepsilon^\Delta} & \mathbf{F}_p & \longrightarrow & 0 \\
 & & \Big\downarrow{\scriptstyle \tilde{\beta}_{E^*}} & & \Big\downarrow{\scriptstyle \tilde{\beta}_V} & & \Big\downarrow{\scriptstyle \mathrm{Id}} & & \\
0 & \longrightarrow & [\![\mathbf{F}_p(E^*(\Gamma), *)]\!] & \xrightarrow{\delta^\Gamma} & [\![\mathbf{F}_p V(\Gamma)]\!] & \xrightarrow{\varepsilon^\Gamma} & \mathbf{F}_p & \longrightarrow & 0 \\
 & & \Big\downarrow{\scriptstyle \tilde{\alpha}_{E^*}} & & \Big\downarrow{\scriptstyle \tilde{\alpha}_V} & & \Big\downarrow{\scriptstyle \mathrm{Id}} & & \\
0 & \longrightarrow & [\![\mathbf{F}_p(E^*(\Gamma/\Delta), *)]\!] & \xrightarrow{\delta^{\Gamma/\Delta}} & [\![\mathbf{F}_p V(\Gamma/\Delta)]\!] & \xrightarrow{\varepsilon^{\Gamma/\Delta}} & \mathbf{F}_p & \longrightarrow & 0.
\end{array}$$

Clearly, we have $\mathrm{Ker}(\tilde{\alpha}_{E^*}) = \tilde{\beta}_{E^*}([\![\mathbf{F}_p(E^*(\Delta), *)]\!])$ and

$$\mathrm{Ker}(\tilde{\alpha}_V) = \mathrm{Ker}(\varepsilon^\Gamma) \cap \tilde{\beta}_V([\![\mathbf{F}_p V(\Delta)]\!]) = \tilde{\beta}_V(\mathrm{Ker}(\varepsilon^\Delta)).$$

Since δ^Δ is injective, $\mathrm{Ker}(\delta^\Gamma) \cap \tilde{\beta}_{E^*}([\![\mathbf{F}_p(E^*(\Delta), *)]\!]) = 0$. Hence $\tilde{\alpha}_{E^*}$ maps $\mathrm{Ker}(\delta^\Gamma) = H_1(\Gamma, \mathbf{F}_p)$ isomorphically onto $\mathrm{Ker}(\delta^{\Gamma/\Delta}) = H_1(\Gamma/\Delta, \mathbf{F}_p)$. Finally note that $(\tilde{\alpha}_{E^*})_{|H_1(\Gamma, \mathbf{F}_p)} = \alpha_1^*$ by definition. □

3 Pro-p groups acting on pro-p trees

Let G be a pro-p group and Γ a profinite graph. Suppose that G acts continuously on the profinite space Γ from the left. We say that *G acts on the profinite graph* Γ (from the left) if $d_i(gm) = gd_i(m)$ for all $g \in G$, $m \in \Gamma$ and $i = 0, 1$. Define $G_m = \{g \in G \mid gm = m\}$ to be the *stabilizer* of an element $m \in \Gamma$. Let $\varphi_\Gamma : G \times \Gamma \longrightarrow \Gamma$ be the map defined by $\varphi_\Gamma(g, m) = gm$ and $pr_G : G \times \Gamma \longrightarrow G$ be the projection. Then $G_m = pr_G(\varphi_\Gamma^{-1}(m) \cap (G \times \{m\}))$, so that G_m is a closed subgroup of G. Clearly, $G_m \leq G_{d_i(m)}$ for every $m \in \Gamma$, $i = 0, 1$. If the stabilizer G_m of every element $m \in \Gamma$ is trivial, we say that G acts *freely* on Γ.

If a pro-p group G acts on a profinite graph Γ, then the space $G\backslash\Gamma = \{Gm \mid m \in \Gamma\}$ of G-orbits with the quotient topology is a profinite space which admits a natural profinite graph structure as follows: $V(G\backslash\Gamma) = G\backslash V(\Gamma)$, $d_j(Gm) = Gd_j(m)$ for $j = 0, 1$. So one has a natural epimorphism $\Gamma \to G\backslash\Gamma$ of profinite graphs.

If $N \lhd G$, there is an action of G/N on $N\backslash\Gamma$ defined in the standard way: $(gN)(Nm) = Ngm$ for $g \in G, m \in \Gamma$.

Remark 3.1 Let a pro-p group G act on a profinite graph Γ and let $N \lhd G$. Consider the natural action of G/N on $N\backslash\Gamma$ defined above. Then the stabilizers in G/N are epimorphic images of the corresponding stabilizers in G.

Example 3.2 Let G be a pro-p group and X a closed subset of G containing 1. Define the *Cayley graph* $\Gamma(G, X)$ as follows: $\Gamma(G, X) = G \times X$, $V(\Gamma(G, X)) = G \times \{1\}$, $d_0(g, x) = (g, 1)$, $d_1(g, x) = (gx, 1)$ for all $g \in G, x \in X$. Then $\Gamma(G, X)$ is a profinite graph. Define a left action of G on $\Gamma(G, X)$ by setting $g' \cdot (g, x) = (g'g, x)$ for all $(g, x) \in G \times X, g' \in G$. Clearly $gd_i(m) = d_ig(m)$ for all $g \in G, m \in \Gamma(G, X)$ and $i = 0, 1$, so G acts (freely) on the Cayley graph $\Gamma(G, X)$.

If a pro-p group G acts on a pro-p tree T, then clearly

$$0 \longrightarrow [\![\mathbf{F}_p(E^*(\Gamma), *)]\!] \xrightarrow{\delta} [\![\mathbf{F}_p V(\Gamma)]\!] \xrightarrow{\varepsilon} \mathbf{F}_p \longrightarrow 0$$

is an exact sequence of G-modules.

Let X be a profinite space, F a pro-p group and $\iota : X \longrightarrow F$ a continuous mapping. We say that (F, ι) is a *free pro-p group* on the profinite space X, or simply F is a free pro-p group on X, if the following universal property is satisfied:

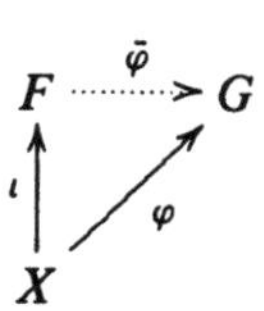

whenever $\varphi : X \longrightarrow G$ is a continuous mapping into a pro-p group G such that $\varphi(X)$ generates G, then there exists a unique continuous homomorphism $\bar{\varphi} : F \longrightarrow G$ such that the above diagram commutes: $\bar{\varphi}\iota = \varphi$. Remark that the concept of a free pro-p group on a profinite space, as we have defined it here, is equivalent to that of free pro-p group on a set (converging to 1) as used in [38] for example. This equivalence was proved in [23].

Theorem 3.3. *Let $F = F(X)$ be a free pro-p group of on a profinite space X. Then the Cayley graph $\Gamma = \Gamma(F, X \cup \{1\})$ is a pro-p tree.*

Proof. We note first that $E(\Gamma) = F \times X$ is compact. Therefore, we just have to prove that the sequence

$$0 \longrightarrow [\![\mathbf{F}_p(F \times X)]\!] \xrightarrow{\delta} [\![\mathbf{F}_p F]\!] \xrightarrow{\varepsilon} \mathbf{F}_p \longrightarrow 0,$$

where $\delta(f, x) = fx - f$ and $\varepsilon(f) = 1$ for all $f \in F, x \in X$, is exact.

We show first the exactness at the middle term, which is equivalent to the connectedness of Γ by Proposition 2.5. Clearly, $\mathrm{Ker}(\varepsilon)$ is the augmentation ideal $(\!(IF)\!)$ of the complete group algebra $[\![\mathbf{F}_p F]\!]$. So it suffices to show that $(\!(IF)\!)$ is generated by $\{x - 1 \mid x \in X\}$ as $[\![\mathbf{F}_p F]\!]$-module. Denote by X_λ the image of X in G_λ. Since $[\![\mathbf{F}_p F]\!] = \varprojlim [\mathbf{F}_p G_\lambda]$, where G_λ ranges over all finite quotient groups of F, it is enough to show that the augmentation ideal $(\!(IG_\lambda)\!)$ of G_λ is generated by $\{x - 1 \mid x \in X_\lambda\}$ as a G_λ-module. Now by induction on the size of the support of an element of $(\!(IG_\lambda)\!)$, it is easy to show that $(\!(IG_\lambda)\!)$ is generated by the $g - 1$ $(g \in G_\lambda)$ as an $\mathbf{F}_p$ module. Finally, if $g = x_1 \cdots x_n$ is a product of elements of X_λ, then $g - 1 = \sum_{i=1}^{n} x_1 \cdots x_{i-1}(x_i - 1)$ and so $g - 1$ lies in the $\mathbf{F}_p$-module generated by the $x - 1$ $(x \in X_\lambda)$.

It remains to prove that δ is a monomorphism. Since $[\![\mathbf{F}_p(F \times X)]\!]$ is a free $[\![\mathbf{F}_p F]\!]$-module on X, it suffices to prove that $(\!(IF)\!)$ is freely generated by $\{x - 1 \mid x \in X\}$. We shall show that any mapping $\varphi : X \longrightarrow M$ to any profinite $[\![\mathbf{F}_p F]\!]$-module M can be extended to a continuous $[\![\mathbf{F}_p F]\!]$-module homomorphism $\psi : (\!(IF)\!) \longrightarrow M$ such that $\psi(x - 1) = \varphi(x)$ for all $x \in X$.

Since M is an F-module, we can construct the semidirect product $M \rtimes F$. Define a homomorphism $\rho : X \longrightarrow M \rtimes F$ by $\rho(x) = (\varphi(x), x)$ and extending to F by the universal property. Define now $d : F \longrightarrow M$ by the equation $(d(f), f) = \rho(f)$, for all $f \in F$. Then d is continuous and it is a derivation, that is, $d(f_1 f_2) = d(f_1) + f_1 d(f_2)$, for all $f_1, f_2 \in F$. Since $[\![\mathbf{F}_p F]\!]$ is a free profinite $\mathbf{F}_p$-module with basis F, d can be extended to a homomorphism of profinite $\mathbf{F}_p$-modules $d : [\![\mathbf{F}_p F]\!] \longrightarrow M$. Using the fact that the set of finite $\mathbf{F}_p$-linear combinations is dense in $[\![\mathbf{F}_p F]\!]$, it is not difficult to verify that $d(mn) = \varepsilon(n)d(m) + md(n)$ for all $m, n \in [\![\mathbf{F}_p F]\!]$. Therefore, the restriction ψ of d to $(\!(IF)\!)$ is a morphism of profinite $[\![\mathbf{F}_p F]\!]$-modules. Clearly, for each $x \in X$, one has $\psi(x - 1) = d(x - 1) = d(x) - d(1) = d(x)$, because $d(1) = 0$. Since $(\varphi(x), x) = \phi(x) = (d(x), x)$, we conclude that $\psi(x - 1) = \varphi(x)$ as required. □

Theorem 3.4. *Let a pro-p group G act freely on a pro-p tree T. Then G is free pro-p.*

Proof. The short exact sequence

$$0 \longrightarrow [\![\mathbf{F}_p(E^*(T), *)]\!] \xrightarrow{\delta} [\![\mathbf{F}_p V(T)]\!] \xrightarrow{\varepsilon} \mathbf{F}_p \longrightarrow 0$$

is an $[\![\mathbf{F}_p G]\!]$-projective resolution for G by Proposition 2.2. Hence, $\mathrm{cd}(G) = 1$. Thus G is a free pro-p group (cf. [38] or [31]). □

Proposition 3.5 *Let a pro-p group G act on a pro-p tree T. Assume that $G = \overline{\langle G_m \mid m \in T \rangle}$ is generated by the stabilizers of the elements of T. Then $G \backslash T$ is a pro-p tree.*

Proof. The proof is not difficult but uses the language of homology of pro-p groups. It can be found in [41] or [32]. $\square$

Corollary 3.6 *Let a pro-p group G act on a pro-p tree T and let $H = \overline{\langle G_m \mid m \in T\rangle}$ be the subgroup generated by all stabilizers. Then H is normal in G and G/H is free pro-p.*

Proof. By Proposition 3.5, $H\backslash T$ is a pro-p tree. Clearly H is normal in G and G/H acts freely on $H\backslash T$ (cf. Remark 3.1). The result follows by Theorem 3.4. $\square$

Theorem 3.7. *Suppose that a pro-p group G acts on a pro-p tree T. Then the set*

$$T^G = \{m \in T \mid gm = m \text{ for all } g \in G\}$$

of fixed points of T is a pro-p subtree of T (possibly empty).

Proof. It is clear that T^G is a subgraph of T with vertex set $V(T^G) = (V(T))^G$, and that $E^*(T^G) = E^*(T)^G$. By Lemma 2.6(a), it is enough to check that T^G is connected. We shall distinguish several cases.

(1) We consider first the case $G \cong C_p$. By Proposition 2.5, it suffices to check that $H_0(T^G, \mathbf{F}_p) = 0$. Since every proper subgroup of G is trivial, it follows that the actions of G on the spaces $(E^*(T)/(E^*(T))^G, *)$ and $(V(T)/(V(T))^G, *)$ are free, where $*$ is the image of $E^*(T)^G$ and $V(T)^G$, respectively. By Lemma 2.1(1), we can consider the modules $[\![\mathbf{F}_p((E^*(T))^G, *)]\!]$ and $[\![\mathbf{F}_p(V(T))^G]\!]$ as submodules of $[\![\mathbf{F}_p(E^*(T), *)]\!]$ and $[\![\mathbf{F}_p V(T)]\!]$ respectively and, by Lemma 2.1(2), one has

$$[\![\mathbf{F}_p(E^*(T), *)]\!]/[\![\mathbf{F}_p((E^*(T))^G, *)]\!] = [\![\mathbf{F}_p(E^*(T)/(E^*(T))^G, *)]\!]$$

and

$$[\![\mathbf{F}_p(V(T))]\!]/[\![\mathbf{F}_p(V(T))^G]\!] = [\![\mathbf{F}_p(V(T)/(V(T))^G, *)]\!].$$

By Proposition 2.2, $[\![\mathbf{F}_p(E^*(T)/(E^*(T))^G, *)]\!]$ and $[\![\mathbf{F}_p(V(T)/(V(T))^G, *)]\!]$ are therefore projective $[\![\mathbf{F}_pG]\!]$-modules. Hence the natural epimorphisms

$$[\![\mathbf{F}_p(E^*(T), *)]\!] \longrightarrow [\![\mathbf{F}_p(E^*(T)/(E^*(T))^G, *)]\!]$$

and

$$[\![\mathbf{F}_p(V(T))]\!] \longrightarrow [\![\mathbf{F}_p(V(T)/(V(T))^G, *)]\!]$$

split and we can write $C(T, \mathbf{F}_p)$ in the following way:

$$0 \longrightarrow [\![\mathbf{F}_p((E^*(T))^G, *)]\!] \oplus M \overset{\delta}{\longrightarrow} [\![\mathbf{F}_p(V(T))^G]\!] \oplus L \overset{\varepsilon}{\longrightarrow} \mathbf{F}_p \longrightarrow 0,$$

where $M \cong [\![\mathbf{F}_p(E^*(T)/(E^*(T))^G, *)]\!]$ and $L \cong [\![\mathbf{F}_p(V(T)/(V(T))^G, *)]\!]$ are projective $\mathbf{F}_p$-modules.

We view $C(T^G, \mathbf{F}_p)$ as a subcomplex of the complex $C(T, \mathbf{F}_p)$ described above. Since T is a pro-p tree the sequence $C(T, \mathbf{F}_p)$ is exact. Then T^G is connected if and only if $\delta([\![\mathbf{F}_p(E^*(T)^G, *)]\!]) = \operatorname{Ker}(\varepsilon_{|[\![\mathbf{F}_p(V(T))^G]\!]})$. So we shall prove this equality.

Take $b \in \operatorname{Ker}(\varepsilon_{|[\![\mathbf{F}_p(V(T))^G]\!]})$. Since $C(T, \mathbf{F}_p)$ is exact, there exists $a \in [\![\mathbf{F}_p(E^*(T), *)]\!]$ with $\delta(a) = b$. Let $a = a_0 + a_1$ for some $a_0 \in [\![\mathbf{F}_p(E^*(T)^G, *)]\!]$ and $a_1 \in M$. Then $\delta(a_1) = b - \delta(a_0)$ is an element of $\delta(M) \cong M$ which is fixed by G. Let g be a generator of G and consider the endomorphisms ν_G, τ_G of the middle term of $C(T, \mathbf{F}_p)$ given by $\nu_G(m) = (1 + g + \cdots + g^{p-1})m$, $\tau_G(m) = (g - 1)m$. Then according to Lemma 2.3 the image of ν_G coincides with the kernel of τ_G on any free G-submodule of $[\![\mathbf{F}_p(V(T))]\!]$. Since $\delta(M)$ is G-free one has $\delta(a_1) = \nu_G(\delta(a_2))$, for some $a_2 \in M$. Write $\delta(a_2) = b_0 + b_1$ for some $b_0 \in [\![\mathbf{F}_p V(T)^G]\!]$ and $b_1 \in L$. Then $\delta(a_1) = \nu_G(b_0) + \nu_G(b_1)$. The element $\nu_G(b_1) = \delta(a_1) - \nu_G(b_0) = b - \delta(a_0) - \nu_G(b_0)$ lies in $L \cap [\![\mathbf{F}_p V(T)^G]\!]$, since $\nu_G(b_1) \in L$ and $b, \delta(a_0), \nu_G(b_0) \in [\![\mathbf{F}_p V(T)^G]\!]$. Therefore, $\nu_G(b_1) = 0$ and hence $\delta(a_1) = \nu_G(b_0) = pb_0 = 0$. It follows that $b = \delta(a_0)$ as required.

(2) Next we consider the case $G \cong C_{p^n}$. Consider the smallest n such that T^G is not connected. It was proved above that $n > 1$. Pick a proper nontrivial subgroup H of G. Since $|H| < n$, then T^H is a pro-p tree on which G/H acts. Hence $T^G = (T^H)^G$ is also a pro-p tree, because $| G/H |< n$, a contradiction.

(3) Suppose $G \cong \mathbf{Z}_p$. Then $G = \varprojlim_{n \in \mathbf{N}} G/G^{p^n}$, where each G/G^{p^n} is a finite cyclic p-group. Assume that $T^G \neq \emptyset$. Then the subgroups G^{p^n} lie in the stabilizer of some vertex of T. By Proposition 3.5, the quotient graph $T_n = G^{p^n} \backslash T$ is a pro-p tree, and clearly $T = \varprojlim T_n$. Furthermore, G/G^{p^n} acts on T_n in a natural way. By part (2), $T_n^{G/G^{p^n}}$ is a pro-p tree. Therefore $T^G = \varprojlim T_n^{G/G^{p^n}}$ is also a pro-p tree.

(4) Let G be arbitrary. Obviously, $T^G = \bigcap_{g \in G} T^{\langle g \rangle}$. It was proved in (3) that $T^{\langle g \rangle}$ is a pro-p tree for every $g \in G$. Hence, by Proposition 2.7, T^G is a pro-p tree as required. □

Corollary 3.8 *Suppose that a pro-p group G acts on a pro-p tree T, and let v and w be two different vertices of T. Then* $E([v, w]) \neq \emptyset$ *and* $(G_v \cap G_w) \leq G_e$ *for every* $e \in E([v, w])$.

Proof. Observe that $K = G_v \cap G_w$ acts on $[v, w]$. By Theorem 3.7, T^K is a pro-p tree. Since $v, w \in T^K$, then $E([v, w]) \subseteq E(T^K)$, and $K \leq G_e$ for every edge $e \in [v, w]$. Clearly $E[v, w] \neq \emptyset$, because otherwise $[v, w] = V([v, w])$ is not connected. □

Next we prove that a finite p-group G acting on a pro-p tree has a fixed vertex. The abstract analog of this fact is proved easily by induction on the cardinality of a minimal G-invariant subtree. In the pro-p situation one cannot use that kind of induction. So the proof below is by induction on the order of the group G.

Theorem 3.9. *Suppose that a finite p-group G acts on a pro-p tree T. Then* $G = G_v$ *for some vertex* $v \in V(T)$.

Proof. We need to prove that $T^G \neq \emptyset$. We use induction on the order of G. Suppose $| G |= p$. Then G is not free, so, by Theorem 3.4, there exists some vertex $v \in V(T)$

with nontrivial stabilizer G_v. However, a cyclic subgroup of order p does not have nontrivial proper subgroups, so $G = G_v$.

If N is a proper nontrivial normal subgroup of G, the induction hypothesis and Theorem 3.7 give that $T^N \neq \emptyset$. Since N is normal, T^N is a G-invariant subtree of T. Then G/N acts naturally on T^N and by the induction hypothesis again one has $T^G = (T^N)^{G/N} \neq \emptyset$. □

Corollary 3.10 *Let G be a pro-p group acting on a pro-p tree T and let D be a nonempty G-invariant pro-p subtree of T. Then for every vertex v in T, there exists a vertex $w \in D$ such that $G_v \leq G_w$.*

Proof. Let $G_v = \varprojlim G_v/N_\lambda$, where N_λ runs through all open normal subgroups of G_v. Set $T_\lambda = N_\lambda \backslash T$. By Proposition 3.5, the T_λ are pro-p trees on which G_v acts. Consider the natural action of G_v/N_λ on T_λ. Let D_λ be the image of D in T_λ. Since D is G-invariant, one has that D_λ is a pro-p tree on which the finite p-group G_v/N_λ acts. By Theorem 3.9, $D_\lambda^{G_v/N_\lambda} \neq \emptyset$. Since $D_\lambda^{G_v} = D_\lambda^{G_v/N_\lambda}$, one has $D_\lambda^{G_v} \neq \emptyset$. Observe that the fixed points are preserved under homomorphic images, which means that the $D_\lambda^{G_v}$ form an inverse system and $D^{G_v} = \varprojlim D_\lambda^{G_v} \neq \emptyset$. For every vertex $w \in D^{G_v}$ we obviously have $G_v \leq G_w$. □

There are some nice properties of pro-p groups acting on pro-p trees which do not have analogs in the abstract situation. The main reason for this is the compactness of pro-p trees. In the next lemma we state one such property.

Lemma 3.11 *Let G be a pro-p group acting on a pro-p tree T. Then there exists a nonempty minimal G-invariant pro-p subtree D of T. Moreover, if D contains more than one vertex, then it is unique.*

Proof. The existence of D follows from Zorn's lemma, together with the compactness of T. Let us prove the second part of the statement. Suppose that $|D| > 1$, and let D_0 be another minimal G-invariant pro-p subtree of T. We denote by T' the profinite graph obtained from T by collapsing D_0 to a point (see Example 1.2). Since D_0 is G-invariant, the action of G carries over naturally to T'. By Lemma 2.9, T' is a pro-p tree. Since D and D_0 are minimal, $D \cap D_0 = \emptyset$ (otherwise Proposition 2.7 would imply that $D \cap D_0$ is a G-invariant pro-p subtree contradicting the minimality D and D_0). This means that the pro-p tree D is isomorphic to its image D' in T', and in particular, $|D'| > 1$. We note that D' is a minimal G-invariant pro-p subtree of T'. Let v be the vertex of T' to which D_0 collapses. Then $G = G_v$. By Corollary 3.10, there is a vertex w in $V(D')$ such that $G_v \leq G_w$; that is, G fixes the vertex w. But D' is minimal, so $D' = \{w\}$, a contradiction. □

We say that the action of a pro-p group G on a pro-p tree T is *irreducible*, if T is the unique minimal G-invariant pro-p subtree of T. The action is said to be *faithful*, if the kernel $C_G(T)$ of the action is trivial. Note that if G acts faithfully on T, then $|T| > 1$.

In a sense, faithful irreducible actions carry an essential part of the information on pro-p groups acting on pro-p trees. Indeed, if a pro-p group G acts on a pro-p tree T, one can consider the action of G on a minimal G-invariant subtree D and factor out the kernel $C_G(D)$ of this action. The resulting action of $G/C_G(D)$ on D is faithful and irreducible.

In the rest of this section we shall study a faithful and irreducible action which will lead us eventually to a pro-p analog of Bass's Theorem [3] on groups acting on trees and having no nonabelian free subgroups (Theorem 3.18).

Theorem 3.12. *Let G be a pro-p group acting irreducibly on a pro-p tree T, and N a normal subgroup of G contained in the stabilizer of some vertex. Then N acts trivially on T.*

Proof. By Theorem 3.7, T^N is a pro-p subtree of T and, since N is normal in G, T^N is G-invariant. Since $N \leq G_v$, it follows that $v \in T^N$; that is, $T^N \neq \emptyset$. Irreducibility of the action implies that $T^N = T$. □

Theorem 3.13. *Let G be a pro-p group acting irreducibly on a pro-p tree T, and let N be a finite normal subgroup of G. Then N acts trivially on T.*

Proof. Follows from Theorems 3.9 and 3.12. □

Proposition 3.14 *Let G be a pro-p group acting faithfully and irreducibly on a pro-p tree T. Then every subgroup H of G containing a nontrivial subnormal subgroup N of G acts irreducibly on T.*

Proof. Let $N = N_0 \triangleleft N_1 \triangleleft \cdots \triangleleft N_k = G$ be a normal series joining N to G. Assume that k is minimal such that the proposition is false. If $k > 1$, then from the minimality of k it follows that N_1 acts irreducibly on T, and therefore $N_0 \triangleleft N_1$ acts irreducibly on T, contradicting our assumption. Therefore $k = 1$, i.e., N is normal in G. Choose a minimal N-invariant pro-p subtree D of T. We claim that D consists of more than one vertex. Indeed, if $D = \{v\}$ for some $v \in V(T)$, then $N \leq G_v$. However, by Theorem 3.13, N must then act trivially on the whole of T, contradicting the faithfulness of the action of G on T.

Thus, $|D| > 1$. Since N is normal in G, the translate gD of D by any element g of G is also a minimal N-invariant pro-p subtree of T. But, by Lemma 3.11, such a subtree is unique, so that $gD = D$ for all $g \in G$. Since G acts irreducibly on T, one concludes that $D = T$, a contradiction. □

Now we state the main theorem on pro-p groups acting faithfully and irreducibly on pro-p trees. The theorem has a 'Tits alternative' flavour: it says that a pro-p group G acting faithfully and irreducibly on a pro-p tree T either contains a free nonabelian pro-p subgroup or else is solvable (even polyprocyclic).

Theorem 3.15. *Let G be a nontrivial pro-p group acting faithfully and irreducibly on a pro-p tree T. Then one of the following holds:*

(a) *G has a free nonabelian pro-p subgroup acting freely on T;*

(b) *$G \cong \mathbf{Z}_p$;*

(c) *$G \cong \mathbf{Z}_2 \rtimes C_2$ is an infinite dihedral pro-2 group.*

We shall prove Theorem 3.15 using a sequence of auxiliary results.

Lemma 3.16 *Let G be a pro-p group acting faithfully and irreducibly on a pro-p tree T and let A be a nontrivial abelian normal subgroup of G. Then*

(a) *$A \cong \mathbf{Z}_p$;*

(b) *the centralizer $C_G(A)$ of A in G is free and acts freely on T;*

(c) *if H is a subgroup of G acting reducibly on T, then $hah^{-1} \neq a$ for all $1 \neq a \in A$ and $1 \neq h \in H$;*

(d) *if H is a subgroup of G of finite order p^n $(n > 0)$, then* (c) *is satisfied. Moreover, $p = 2$, $H \cong C_2$ and $\overline{\langle A, H\rangle} = A \rtimes H \cong \mathbf{Z}_2 \rtimes C_2$ is an infinite dihedral pro-2 group;*

(e) *if B is the subgroup of G generated by all elements of order 2, then $B = C_B(A) \rtimes \langle b\rangle$, where $b^2 = 1$.*

Proof. (c) By contradiction. Suppose that $hah^{-1} = a$ for some $1 \neq h \in H$ and $1 \neq a \in A$. Then

$$\langle h\rangle \triangleleft \langle h, a\rangle \geq \langle a\rangle \triangleleft A \triangleleft G.$$

Applying Proposition 3.14 a sufficient number of times, one gets that H acts irreducibly on T, a contradiction.

(b) By (c), $C_G(A) \cap G_v = 1$ for all $v \in V(T)$, and this means that $C_G(A)$ acts freely on T. Then, by Theorem 3.4, $C_G(A)$ is free pro-p.

(a) Follows immediately from (b).

(d) Since H is finite, by Theorem 3.9, $H = H_v$ for some $v \in V(T)$, so that the action of H on T is reducible. The natural homomorphism $\phi_p : H \longrightarrow \mathrm{Aut}(A_p)$ is injective by (c). Therefore, if $p \neq 2$ then $H = 1$ contradicting our assumption; hence $p = 2$, $\mathrm{Aut}(A_2) \cong \mathbf{Z}_2 \times C_2$, and therefore $H \cong C_2$. Now the result follows from (c).

(e) Let b_1 and b_2 be elements of order 2 in B, and a any element of A. By (d), one has $b_1ab_1^{-1} = b_2ab_2^{-1} = a^{-1}$, so that $b_1b_2 \in C_B(A)$. The index $|B : K|$ of the subgroup $K = \langle b_1b_2 \mid b_1, b_2 \in B, b_1^2 = b_2^2 = 1\rangle \leq C_B(A)$ is at most 2. By (b), $C_B(A)$ is free and hence torsion free. Thus $K = C_B(A)$ and $|B : C_B(A)| = 2$, so that $B = C_B(A) \rtimes \langle b\rangle$ for any element b of order 2 in B. □

Lemma 3.17 *Let G be a nontrivial pro-p group acting faithfully and irreducibly on a pro-p tree T. Then G can be represented as an inverse limit $G = \varprojlim G_U$, where the G_U are quotient groups of G each acting faithfully and irreducibly on a pro-p tree D_U with finite stabilizers of vertices, and having an open nontrivial normal subgroup acting freely on D_U. Moreover,*

(a) *if $G\backslash T$ is finite, so is $G_U\backslash D_U$ for every U.*

(b) *if G has no nonabelian pro-p subgroups acting freely on T, then each G_U has an open normal subgroup isomorphic to $\mathbf{Z}_p$.*

Proof. For every open normal subgroup U of G, set $\tilde{U} = \overline{\langle U_v \mid v \in V(T)\rangle}$. Since $gG_vg^{-1} = G_{gv}$, one deduces that $\tilde{U}$ is normal in G. Set $G_U = G/\tilde{U}$ and $T_U = (G/\tilde{U})\backslash T$. By Proposition 3.5, T_U is a pro-p tree for every U. Since $\tilde{U} \leq U$, one has $G = \varprojlim G_U$ and $T = \varprojlim T_U$. Consider the natural action of G_U on T_U. Since G acts irreducibly on T, $T^G = \varprojlim T_U^{G_U} = \emptyset$. This means that there exists an open normal subgroup V of G such that $T_V^{G_V} = \emptyset$. Thus, replacing the inverse systems $\{G_U\}$ and $\{T_U\}$ by the cofinal systems $\{G_U \mid U \leq V\}$ and $\{T_U \mid U \leq V\}$ respectively, we can ensure that $T_U^{G_U}$ is empty for all $U \leq V$. Then, by Lemma 3.11, there exists a unique minimal G_U-invariant subtree D_U for every $U \leq V$. Hence, for $U_1 \leq U_2$ the image of D_{U_1} in T_{U_2} contains D_{U_2}, which implies that $C_{G_{U_2}}(D_{U_2})$ contains the image of $C_{G_{U_1}}(D_{U_1})$ in G_{U_2}. Thus the $C_{G_U}(D_U)$ form an inverse system. Next we show that $\varprojlim C_{G_U}(D_U) = 1$.

Indeed, suppose that $g \in \varprojlim C_{G_U}(D_U)$. It suffices to prove that $g \in G_v$ for all $v \in V(T)$, because that would mean that g is in the kernel of the action of G on T (which is trivial, since the action of G on T is faithful). Let $\phi_U : G \longrightarrow G_U$ and $\psi_U : T \longrightarrow T_U$ be the natural epimorphisms. To prove that $g \in G_v$ one just needs to show that $g_U := \phi_U(g) \in (G_U)_{v_U}$ for any U, where $v_U = \psi_U(v)$. Let

$$\Delta_U = \overline{\bigcup_{W \triangleleft_o U} \psi_{W,U}(D_W)},$$

where $\psi_{W,U} : T_W \longrightarrow T_U$ is the canonical morphism of the corresponding inverse system. Clearly, $g_U \in C_{G_U}(\psi_{W,U}(D_W))$ for every $W \triangleleft_o U$, and therefore $g_U \in C_{G_U}(\Delta_U)$. Hence it is enough to prove that $\Delta_U = T_U$. One can see easily that for a pair of open normal subgroups W_1, W_2 of G with $W_1 \leq W_2 \leq U$ the inclusion $\psi_{W_1,W_2}(\Delta_{W_1}) \subseteq \Delta_{W_2}$ holds. This means that $\{\Delta_W \mid W \triangleleft_o U\}$ forms an inverse system. Since Δ_W is G_W-invariant, $\varprojlim(\Delta_W)$ is G-invariant. Then $\varprojlim(\Delta_W) = T$, which implies that $\Delta_U = T_U$, as desired. Thus, $\varprojlim C_{G_U}(D_U) = 1$. It follows that

$$\bigcap_{U \leq V} \phi_U^{-1}(C_{G_U}(D_U)) = 1.$$

Hence, replacing every U by $\phi_U^{-1}(C_{G_U}(D_U))$, we obtain the desired inverse system $\{G_U\}$, where each G_U acts faithfully and irreducibly on D_U.

Next, note that the image U_0 of the normal subgroup U of G in G_U acts freely on T_U. This completes the proof of the first statement of the lemma.

(a) Follows from the construction of D_U.

(b) By Corollary 3.6, U_0 is free. Then there is a subgroup U_1 of U such that the restriction $\varphi_{|U_1}$ of the natural homomorphism $\varphi : G \longrightarrow G_U$ to U_1 is an isomorphism onto U_0. This means that $U_1 \cap G_v$ is trivial for all $v \in V(T)$, and therefore that U_1 acts freely on T. By the hypothesis of (b), U_1 has no nonabelian pro-p subgroups. Since U_1 is free, it follows that $U_1 \cong \mathbf{Z}_p$ and therefore so is U_0. □

We are now in a position to prove Theorem 3.15.

Proof of Theorem 3.15. Suppose G does not satisfy (a). By Lemma 3.17 (c), G has a decomposition $G = \varprojlim G_U$, where each G_U is a quotient group of G acting faithfully and irreducibly on a pro-p tree D_U with finite stabilizers of vertices, and having an open normal subgroup A_U isomorphic to $\mathbf{Z}_p$. Therefore, since $\mathbf{Z}_p$ and $\mathbf{Z}_2 \rtimes C_2$ cannot be epimorphic images of each other (use the Hopfian property for example), it suffices to prove that every G_U is isomorphic either to $\mathbf{Z}_p$ or to $\mathbf{Z}_2 \rtimes C_2$.

Consider the action of G_U on A_U induced by conjugation, and let φ_U be the corresponding homomorphism from G_U to $\mathrm{Aut}(A_U)$. We divide the proof into two cases: $p > 2$ and $p = 2$.

Suppose that $p > 2$. By Lemma 3.16 (a), $A_U \cong \mathbf{Z}_p$, and therefore $\mathrm{Aut}(A_U) \cong \mathbf{Z}_p \times C_{p-1}$ (see Proposition 8 in [36], Chapter 2). Then, by Lemma 3.16 (d), $(G_U)_v = 1$ for all $v \in V(D_U)$. Thus G_U acts freely on D_U. By Theorem 3.4, G_U is a free pro-p group. Since $\mathbf{Z}_p \cong A_U$ is an open normal subgroup of G_U, one deduces that $G_U \cong \mathbf{Z}_p$.

Suppose now that $p = 2$; that is, $A_U \cong \mathbf{Z}_2$. If G_U acts freely on D_U, then as before, by Theorem 3.4, G_U is a free pro-2 group, and since $\mathbf{Z}_2 \cong A_U$ is open normal subgroup of G_U, one deduces that $G_U \cong \mathbf{Z}_2$.

It remains to consider the case where there exists a vertex $v \in V(D_U)$ such that $(G_U)_v \neq 1$. By Lemma 3.16 (d), all nontrivial vertex-stabilizers in D_U are isomorphic to C_2. Set $G_0 = \langle (G_U)_v \mid v \in V(D_U)\rangle$. By Corollary 3.6, G_U/G_0 is a free pro-2 group, so that $G_U \cong G_0 \rtimes P$, where $P \cong G_U/G_0$. By Lemma 3.16 (e), $G_0 = C_{G_0}(A_U) \rtimes \langle g \rangle$ where $g^2 = 1$. By Lemma 3.16 (b), $C_{G_0}(A_U)$ is a free pro-2 group and it is normal in G_U, since A_U is so. Therefore $C_{G_0}(A_U) \rtimes P$ is torsion free. Since the stabilizers $(G_U)_v$ are finite, it follows that the action of $C_{G_0}(A_U) \rtimes P$ on D_U is free. Hence, by Theorem 3.4, $C_{G_0}(A_U) \rtimes P$ is a free pro-2 group. But A_U is isomorphic to $\mathbf{Z}_2$ and open in G_U, which means that the intersection $A_U \cap (C_{G_0}(A_U) \rtimes P)$ is open in $C_{G_0}(A_U) \rtimes P$ and isomorphic to $\mathbf{Z}_2$. One infers that $C_{G_0}(A_U) \rtimes P \cong \mathbf{Z}_2$; that is, $P = \{1\}$ and $G_U = G_0$. Since g has finite order, by Theorem 3.9, it fixes some vertex, so it acts reducibly on D_U. Hence, by Lemma 3.16 (c), g acts nontrivially on A_U and therefore on $C_{G_0}(A_U) \cong \mathbf{Z}_2$, so that $G_U \cong \mathbf{Z}_2 \rtimes C_2$ as desired. □

Finally we can prove the main theorem of this section. The abstract analog of this theorem was proved by H. Bass [3]. It is curious that one of the possibilities in the abstract case (namely the case of an invariant end) does not appear in the pro-p situation. The reason for this is the compactness of a pro-p tree.

Theorem 3.18. *Let G be a pro-p group acting on a pro-p tree T. Then one of the following holds:*

(a) *$G = G_v$ is the stabilizer of a vertex v of T.*

(b) *G has a free nonabelian pro-p subgroup P such that $P \cap G_v = 1$ for every vertex v of T.*

(c) *There exists an edge e of T whose stabilizer G_e is normal in G, and the quotient group G/G_e is solvable and isomorphic to one of the following groups:*

(1) $\mathbf{Z}_p$;

(2) *an infinite dihedral pro-2 group* $\mathbf{Z}_2 \rtimes C_2 \cong C_2 \coprod C_2$.

(Here, $C_2 \coprod C_2$ denotes the free pro-2 product, defined below in Section 4.)

Proof. Choose a minimal G-invariant pro-p subtree D of T (see Lemma 3.11). Let $C_G(D)$ be the kernel of the action of G on D, and set $G_0 = G/C_G(D)$. Then G_0 acts faithfully and irreducibly on D. By Theorem 3.15, one of the following holds for G_0:

(A) $G_0 = \{1\}$;

(B) G_0 contains a nonabelian free pro-p subgroup acting freely on D;

(C) $G_0 \cong \mathbf{Z}_p$ acts freely on D;

(D) $G_0 \cong \mathbf{Z}_2 \rtimes C_2$ is an infinite dihedral pro-2 group.

We consider each of these cases in turn. If G_0 is trivial, then $|D| = 1$, since $G = C_G(D)$. This means that there is a vertex v in $V(T)$ with $G = G_v$; that is, case (a) of the theorem holds.

Consider now case (B). Let H be a nonabelian pro-p subgroup of G_0 acting freely on D. By Theorem 3.4, H is a free pro-p group. Hence G has a subgroup P mapped isomorphically onto H by the natural homomorphism $G \longrightarrow G_0$. Therefore $P \cap G_v = \{1\}$ for all $v \in V(D)$. Then, by Corollary 3.10, $P \cap G_v = \{1\}$ for all $v \in V(T)$, i.e., case (b) of the theorem holds.

If $C_G(D) = G_e$ for some $e \in E(D)$, then the cases (C),(D) correspond to items (1) and (2) of case (c) of the theorem. Thus, it remains to prove the existence of $e \in E(D)$ with $C_G(D) = G_e$.

In case (C), the fact that G_0 acts freely on D implies that $G_e = C_G(D)$ for every edge $e \in E(D)$. Consider the action of G_0 on the pro-p tree D in case (D). It is enough to show that there exists an edge $e \in E(D)$ with trivial stabilizer in G_0. We first prove that $D_0 = \{d \in D \mid (G_0)_d \neq 1\}$ is a profinite subgraph of D, i.e., that D_0 is closed in D. Let K be a subgroup of G_0 of order 2 and denote by H the maximal normal subgroup of G_0 isomorphic to $\mathbf{Z}_2$. Note that in case (D) there are exactly two conjugacy classes of a maximal finite subgroup: $K \cong C_2$ and $\langle kh \rangle$, where k and h are generators of K and H respectively. Since in this case there is an abelian open normal subgroup acting irreducibly on D (see Proposition 3.14), by Theorem 3.12 it acts freely on D, i.e., the stabilizer of every vertex in D is finite and so is at most of order 2. Denote by D^K and $D^{\langle kh \rangle}$ the sets of fixed points with respect to K and $\langle kh \rangle$ respectively. Then $D_0 = HD^K \cup HD^{\langle kh \rangle}$. Since D^K and $D^{\langle kh \rangle}$ are closed, so is D_0, and therefore it is a profinite subgraph of D.

We prove now that D_0 is disconnected. Indeed, one obviously has $HD^K \cap HD^{\langle kh \rangle} = \emptyset$, which implies that D_0 is disconnected (see Exercise 1.9). Now disconnectivity of D_0 implies the existence of an edge $e \in (D - D_0)$. Thus $(G_0)_e = \{1\}$, and the theorem is proved. □

4 Free constructions

In this section we use our earlier results to obtain some structural theorems about subgroups of free products, free products with amalgamation and HNN-extensions in the category of pro-p groups. If G is a pro-p group arising by means of one of these constructions, there is a standard pro-p tree $S(G)$, associated with the construction, on which G acts naturally. It is through this action that we shall be able to apply the results established in previous sections. We first recall the definitions of these constructions for the reader's convenience. Then we shall indicate how to build the pro-p trees $S(G)$ associated with each of the three constructions.

Let $\{G_i : i = 1, ..., n\}$ be a finite collection of pro-p groups. A free *pro-p product* consists of a pro-p group $G = \coprod_{i=1}^n G_i$ and homomorphisms $\varphi_i : G_i \longrightarrow G$, $i = 1, ..., n$, with the following universal property:

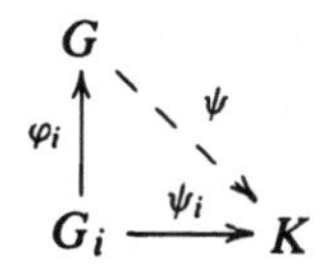

for any pro-p group K and any set of homomorphisms $\psi_i : G_i \longrightarrow K$, $i = 1, ..., n$, there exists a unique homomorphism $\psi : G \longrightarrow K$ such that $\psi_i = \psi\varphi_i$ for all $i = 1, ..., n$.

Observe that one needs to test the above universal property only for finite p-groups K; then it holds automatically for any pro-p group K, since K is an inverse limit of finite p-groups.

From the definition it is easily deduced that if a free pro-p product exists, it is unique. Existence follows from the following explicit construction of $G = \coprod_{i=1}^n G_i$. Let $G^{abs} = \star_{i=1}^n G_i$ be the free product of the G_i as abstract groups. Denote by $\varphi_i^{abs} : G_i \longrightarrow G^{abs}$ the natural embeddings. Let

$$\mathcal{N} = \{N \triangleleft_f G^{abs} \mid (\varphi_i^{abs})^{-1}(N) \triangleleft_o G_i\}.$$

Define $G := \varprojlim_{N \in \mathcal{N}} (G^{abs}/N)$ to be the completion of G^{abs} with respect to the topology defined by $\mathcal{N}$. We leave it to the reader to check the universal property for G.

Let $G = \coprod_{i=1}^n G_i$ be a free pro-p product. Then the standard graph $S = S(G)$ is defined as follows:

$$\begin{aligned} S &= \Big(\dot{\bigcup}_{i=1}^{n-1} G\Big) \bigcup \Big(\dot{\bigcup}_{i=1}^{n} G/G_i\Big), \\ V(S) &= \dot{\bigcup}_{i=1}^{n} G/G_i, \\ d_0(g) &= gG_1, \\ d_1(g) &= gG_{i+1} \text{ if } g \text{ belongs to the } i\text{-th copy of } G \text{ in } \Big(\dot{\bigcup}_{i=1}^{n-1} G\Big). \end{aligned}$$

Note that G acts naturally on $S(G)$ by left multiplication; the stabilizers of vertices are conjugate to the G_i ($i = 1, \ldots, n$) and all edge stabilizers are trivial. The quotient graph $G\backslash S(G)$ is a finite tree with $n-1$ edges.

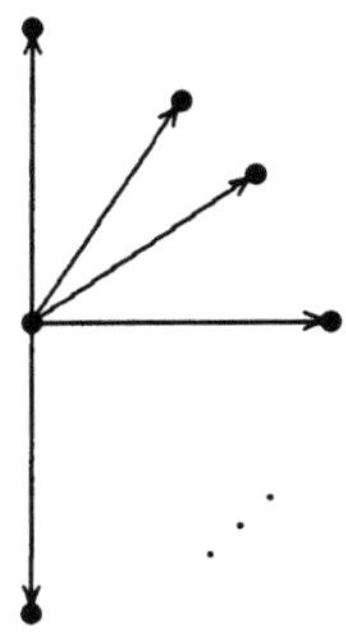

We turn now to amalgamated free products. Let G_1 and G_2 be pro-p groups and $f_i : H \longrightarrow G_i$, $i = 1, 2$, monomorphisms of pro-p groups. The *amalgamated free pro-p product* of G_1 and G_2 with amalgamated subgroup H is a pushout in the category of pro-p groups

$$\begin{array}{ccc} H & \xrightarrow{f_1} & G_1 \\ {\scriptstyle f_2}\downarrow & & \downarrow{\scriptstyle \varphi_1} \\ G_2 & \xrightarrow{\varphi_2} & G_1 \coprod_H G_2 \end{array},$$

i.e., a pro-p group $G = G_1 \coprod_H G_2$ satisfying the following universal property: for any pair of homomorphisms $\psi_1 : G_1 \longrightarrow K$, $\psi_2 : G_2 \longrightarrow K$ into a pro-p group K with $\psi_1 f_1 = \psi_2 f_2$, there exists a unique homomorphism $\psi : G \longrightarrow K$ such that the following diagram is commutative:

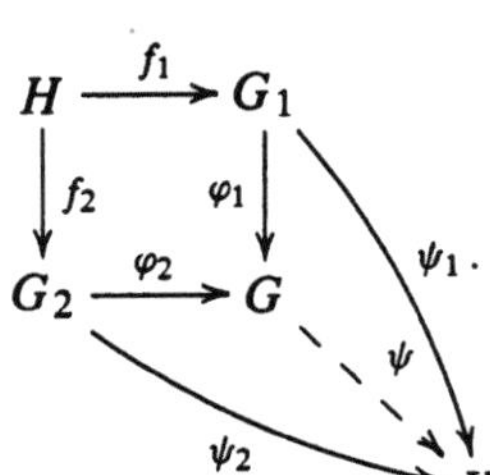

It follows from the existence and uniqueness of pushouts that an amalgamated free pro-p product exists and it is unique. As usual, it is enough to check the universal property when K is finite. As a rule, we shall consider H as a common subgroup of G_1 and G_2 and denote by the same letter H the images $f_1(H)$ and $f_2(H)$ in G_1 and G_2 respectively.

In the abstract situation the canonical homomorphisms $\varphi_i^{abs} : G_i \longrightarrow G_1 \star_H G_2$ are monomorphisms. However, in the category of pro-p groups it is not always the case (see [28]). An amalgamated free pro-p product $G = G_1 \coprod_H G_2$ will be called *proper* if the canonical homomorphisms $\varphi_i : G_i \longrightarrow G_1 \coprod_H G_2$ are monomorphisms. In this case we shall identify G_1, G_2 and H with their images in G.

If $G = G_1 \coprod_H G_2$ is not proper, one can replace G_1, G_2 and H by their canonical images in G. This operation does not change G, but the amalgamated free pro-p product $G = G_1 \coprod_H G_2$ becomes proper. So from now on we shall consider proper amalgamated pro-p products only. Of course, after this operation the amalgamated free pro-p product can become 'fictitious', in the sense that H could coincide with one of the factors, and so G would coincide with the other factor. We shall exclude such cases from our considerations.

Let $G = G_1 \coprod_H G_2$ be an amalgamated free pro-p product. Then the standard graph $S = S(G)$ is defined as follows:

$$\begin{aligned} S &= G/H \,\dot{\bigcup}\, G/G_1 \,\dot{\bigcup}\, G/G_2, \\ V(S) &= G/G_1 \,\dot{\bigcup}\, G/G_2, \\ d_0(gH) &= gG_1,\ d_1(gH) = gG_2. \end{aligned}$$

Obviously G acts naturally on $S(G)$; the stabilizers of vertices are conjugate to G_1 or G_2, and the edge stabilizers are conjugate to H. The quotient graph $G\backslash S(G)$ is a finite tree having one edge and two vertices

Next we recall the definition of an HNN-extension. Let H be a pro-p group and let $f : A \longrightarrow B$ be a continuous isomorphism between closed subgroups A, B of H. The *pro-p HNN-extension* of H with associated subgroups A, B consists of a pro-p group $G = \mathrm{HNN}(H, A, f)$, an element $t \in G$, and a homomorphism $\varphi : H \longrightarrow G$ satisfying the following universal property: for any pro-p group K, any $k \in K$ and any homomorphism $\psi : H \longrightarrow K$ satisfying $k(\psi(a))k^{-1} = \psi(f(a))$ for all $a \in A$, there is a unique homomorphism $\omega : G \longrightarrow K$ such that the diagram

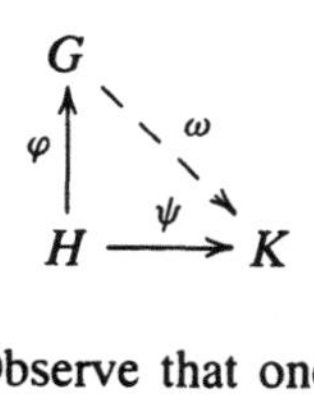

is commutative and $\omega(t) = k$. Observe that one needs to test the above universal property only for finite p-groups K, for then it holds automatically for any pro-p group K, since K is an inverse limit of finite p-groups.

From the definition it is easily deduced that if an HNN-extension exists, it is unique. The existence follows from the following explicit construction of $G =$

HNN(H, A, f). Let $G^{abs} = \text{HNN}^{abs}(H, A, f)$ be the abstract HNN-extension. Denote by $\varphi^{abs} : H \longrightarrow G^{abs}$ the natural embedding. Let

$$\mathcal{N} = \{N \triangleleft_f G^{abs} | (\varphi^{abs})^{-1}(N) \triangleleft_o H\}.$$

Define

$$G = \varprojlim_{N \in \mathcal{N}} (G^{abs}/N)$$

to be the completion of G^{abs} with respect to the topology defined by $\mathcal{N}$. We leave to the reader to check the universal property for G.

In contrast with the situation for abstract groups, the canonical homomorphism $\varphi : H \longrightarrow G = \text{HNN}(H, A, f)$ is not necessarily a monomorphism. When φ is a monomorphism, we shall call $G = \text{HNN}(H, A, f)$ a *proper* pro-p HNN-extension. If $G = \text{HNN}(H, A, f)$ is not proper, one can replace H and A by their canonical images in G (with the obvious change of f). This operation does not change G, but the pro-p HNN-extension $G = \text{HNN}(H, A, f)$ becomes proper.

Let $G = \text{HNN}(H, A, f) = \langle H, t \mid tat^{-1} = f(a), a \in A\rangle$ be a pro-p HNN-extension. Then the standard graph $S = S(G)$ is defined as follows:

$$S = G/A \uplus G/H, \ V(S) = G/H, \ d_0(gA) = gH, \ d_1(gH) = gtH.$$

The group G acts naturally on $S(G)$; vertex stabilizers are conjugate to H and edge stabilizers are conjugate to A. The quotient graph $G \backslash S(G)$ is just a loop: one edge with only one vertex:

Theorem 4.1. *In all three cases described above the graph $S = S(G)$ is a pro-p tree.*

The proof is similar to Chiswell's proof of the analogous fact in the discrete case [5]; an alternative proof will appear in the forthcoming book [32].

Theorem 4.2.

(a) *Let K be a finite subgroup of a free pro-p product $G = \coprod_{i=1}^{n} G_i$ of pro-p groups. Then $K \leq gG_ig^{-1}$ for some $g \in G$ and some $i \in \{1, \dots, n\}$;*

(b) *Let K be a finite subgroup of a proper free amalgamated pro-p product $G = G_1 \coprod_H G_2$ of pro-p groups. Then $K \leq gG_ig^{-1}$ for some $g \in G$ and $i \in \{1, 2\}$;*

(c) *Let K be a finite subgroup of a proper pro-p HNN-extension $G = \text{HNN}(H, A, f)$. Then $K \leq gHg^{-1}$ for some $g \in G$.*

Proof. Apply Theorem 3.9. □

Theorem 4.3.

(a) *Let* $G = \coprod_{i=1}^{n} G_i$ *be a free pro-p product of pro-p groups. Then* $G_i \cap gG_jg^{-1} = 1$, *whenever* $1 \leq i \neq j \leq n$ *or* $g \notin G_i$;

(b) *Let* $G = G_1 \coprod_H G_2$ *be a proper free amalgamated pro-p product of pro-p groups. Then* $G_i \cap gG_jg^{-1} \leq bHb^{-1}$ *for some* $b \in G_i$, *whenever* $1 \leq i \neq j \leq 2$ *or* $g \notin G_i$.

(c) *Let* $G = \mathrm{HNN}(H, A, f)$ *be a proper pro-p HNN-extension. Then*

$$H \cap gHg^{-1} \leq bAb^{-1}$$

for some $b \in H$, *whenever* $g \notin H$.

Proof. Apply Corollary 3.8. □

Corollary 4.4

(a) *Let* $G = \coprod_{i=1}^{n} G_i$ *be a free pro-p product of pro-p groups. Then* G_i *is self-normalizing for each* $i = 1, ..., n$.

(b) *Let* $G = G_1 \coprod_H G_2$ *be a proper (non-fictitious, i.e., such that* $G_1 \neq H \neq G_2$*) free amalgamated pro-p product of pro-p groups. Then* G_i *is self-normalizing for* $i = 1, 2$.

(c) *Let* $G = \mathrm{HNN}(H, A, f)$ *be a pro-p HNN-extension such that* $A \neq H \neq f(A)$. *Then* H *is self-normalizing.*

Theorem 4.5.

(a) *Let* $G = \coprod_{i=1}^{n} G_i$ *be a free pro-p product of non-trivial pro-p groups. Then any finite normal subgroup of* G *is trivial.*

(b) *Let* $G = G_1 \coprod_H G_2$ *be a (non-fictitious) free amalgamated pro-p product of pro-p groups. Then every finite normal subgroup of* G *is contained in* H.

(c) *Let* $G = \mathrm{HNN}(H, A, f)$ *be a pro-p HNN-extension. Then every finite normal subgroup of* G *is contained in* A.

Proof. Note that the action of G on $S(G)$ is irreducible, so that the result follows from Theorem 3.13. □

We state the interpretations of Theorem 3.18 for free products, free products with amalgamation and HNN-extensions as three separate theorems. However, we give a single proof for all the three results.

Theorem 4.6. *Let* K *be a subgroup of a free pro-p product* $G = \coprod_{i=1}^{n} G_i$ *of pro-p groups. Then one of the following holds:*

(1) $K \leq gG_ig^{-1}$ *for some* $g \in G$ *and some* $i = 1, ..., n$;

(2) K *has a nonabelian free pro-*p *subgroup* P *such that* $P \cap gG_ig^{-1} = 1$ *for every* $i = 1, ..., n$ *and every* $g \in G$.

(3) K *is solvable and isomorphic to one of the following groups:*

(a) $\mathbf{Z}_p$;

(b) *an infinite dihedral pro-2 group* $\mathbf{Z}_2 \rtimes C_2 = C_2 \coprod C_2$.

Theorem 4.7. *Let* K *be a subgroup of a proper free amalgamated pro-*p *product* $G = G_1 \coprod_H G_2$ *of pro-*p *groups. Then one of the following holds:*

(1) $K \leq gG_ig^{-1}$ *for* $i = 1$ *or* 2 *and some element* g *of* G;

(2) K *has a nonabelian free pro-*p *subgroup* P *such that* $P \cap gG_ig^{-1} = 1$ *for* $i = 1, 2$ *and every* $g \in G$;

(3) *there exists a normal subgroup* H_0 *of* K *which is contained in* H, *such that* K/H_0 *is solvable and isomorphic to one of the following groups:*

(a) $\mathbf{Z}_p$;

(b) *an infinite dihedral pro-2 group* $\mathbf{Z}_2 \rtimes C_2 = C_2 \coprod C_2$. *Moreover, in this case* $K = K_1 \coprod_{K_0} K_2$, *where* $K_1 = K \cap g_1G_1g_1^{-1}$, $K_2 = H \cap g_2G_2g_2^{-1}$ *for some elements* $g_1, g_2 \in G$ *such that* $|K_i : K_0| = 2$ *for* $i = 1, 2$.

Theorem 4.8. *Let* K *be a subgroup of a proper pro-*p *HNN-extension* $G = \mathrm{HNN}(H, A, f)$. *Then one of the following holds:*

(1) $K \leq gHg^{-1}$;

(2) K *has a nonabelian free pro-*p *subgroup* P *such that* $P \cap gHg^{-1} = 1$ *for every* $g \in G$;

(3) *there exists a normal subgroup* H_0 *of* K *which is contained in* A, *such that* K/H_0 *is solvable and isomorphic to one of the following groups:*

(a) $\mathbf{Z}_2$;

(b) *an infinite dihedral pro-2 group* $\mathbf{Z}_2 \rtimes C_2 = C_2 \coprod C_2$. *Moreover, in this case* $K = K_1 \coprod_{K_0} K_2$, *where* $K_1 = K \cap g_1Hg_1^{-1}$, $K_2 = K \cap g_2Hg_2^{-1}$ *for some elements* $g_1, g_2 \in G$ *such that* $|K_i : K_0| = 2$ *for* $i = 1, 2$.

Proof. Consider the action of G on the pro-p tree $S = S(G)$. Since K is a subgroup of G it acts naturally on S as well. Now one needs to apply Theorem 3.18 and use the fact that vertex and edge stabilizers of S in K have the form $K \cap gG_ig^{-1}$ ($K \cap H$ for an HNN-extension) and $K \cap gHg^{-1}$ ($K \cap A$ for an HNN-extension) respectively. The second part of the statements in subcase 3(b) of Theorems 4.7 and 4.8 follow directly from the universal property of a free amalgamated pro-p product. □

5 The pro-p topology of an abstract group

Let L be an abstract group and $\mathcal{N} = \mathcal{N}(L)$ be the collection of those normal subgroups of L whose index is a finite power of p. Then L can be viewed as a topological group by taking $\mathcal{N}$ to be a basis of neighborhoods of the identity element 1. We refer to the corresponding topology as the *pro-p topology* of L. Observe that this topology is Hausdorff if and only if $\bigcap_{N\in\mathcal{N}} N = 1$, that is, if and only if L is residually a finite p-group. For a subset X of L, we use $\mathrm{Cl}(X)$ to denote the closure of X in the pro-p topology of L.

Next we describe some basic properties of the pro-p topology of a group L. Let H be a subgroup of L; then the *core* H_L of H in L is the intersection of all conjugates of H in L. This is the largest subgroup of H which is normal in L. Observe that if H has finite index in L, then it has only finitely many conjugates, and so $H_L = \bigcap_{x\in L} H^x$ is a normal subgroup of L of finite index.

Example 5.1

1) Let $S_3 = \langle a, b \mid a^2, b^3, (ab)^2\rangle$ be the symmetric group of degree 3. Then $\langle b\rangle$ is open in the pro-2 topology of S_3, and $\langle a\rangle$ is not open in the the pro-3 topology of S_3.

2) If $K \triangleleft L$, then K is closed in the pro-p topology of L if and only if L/K is residually a finite p-group.

The next lemma provides concrete criteria to verify whether a subgroup of L is open, closed or dense in the pro-p topology of L.

Lemma 5.2 *Let L be an abstract group with the pro-p topology, and $H \leq L$. Then*

(i) *H is open if and only if L/H_L is a finite p-group.*

(ii) *H is closed if and only if it is an intersection of open subgroups of L.*

(iii) *$\mathrm{Cl}(H)$ is the intersection of the open subgroups of L containing H.*

(iv) *H is not dense in L if and only if there exists an open subgroup of index p in L that contains H.*

Proof. (i) If L/H_L is a finite p-group, then H_L is open, and hence so is H. Conversely, if H is open, then so is every conjugate H^x of H in L. Moreover, H has finite index in L since it contains some subgroup in $\mathcal{N}$. So H has only finitely many conjugates. Therefore H_L is open. Hence there exists some $N \in \mathcal{N}(L)$ with $N \leq H_L$. Then there is an epimorphism $L/N \longrightarrow L/H_L$, so that L/H_L is a p-group.

(ii) An open subgroup is necessarily closed; therefore the intersection of open subgroups is closed. Conversely, assume H is a closed subgroup of L, and let $x \in L - H$. Then there exists some $N \in \mathcal{N}(L)$ such that $xN \cap H = \emptyset$. Hence $x \notin HN$, so

$$H = \bigcap_{N\in\mathcal{N}(L)} HN.$$

Since each HN is open, the result follows.

(iii) This is a consequence of (ii).

(iv) If H is dense in L, $\mathrm{Cl}(H) = L$, and so H is not contained in any proper open subgroup of L. Conversely, assume that H is not dense in L. Then, by (ii), there exists a proper open subgroup T of L containing H. Since L/T_L is a finite p-group, there exists a subgroup S of index p in L such that $T \leq S$. Finally, S is open because T is open. □

The *pro-p completion* $\widehat{L}_p$ of L is defined to be

$$\widehat{L}_p = \varprojlim_{N \in \mathcal{N}} L/N.$$

Clearly $\widehat{L}_p$ is a pro-p group, and there exists a natural continuous homomorphism

$$\iota = \iota_p : L \longrightarrow \widehat{L}_p,$$

induced by the continuous epimorphisms $L \longrightarrow L/N$ ($N \in \mathcal{N}$). Observe that $\iota_p(L)$ is a dense subgroup of $\widehat{L}_p$, and that ι_p is injective if and only if L is residually a finite p-group. From now on we shall assume that L is residually a finite p-group, and we shall identify L with its image in $\widehat{L}_p$ via the injection ι, so that $L \leq \widehat{L}_p$. If X is a subset of $\widehat{L}_p$, then its closure in $\widehat{L}_p$ is denoted by $\bar{X}$.

Let $\varphi : H \longrightarrow L$ be a homomorphism of groups. One easily sees that φ induces a natural continuous homomorphism $\hat{\varphi} : \widehat{H}_p \longrightarrow \widehat{L}_p$ of the pro-p completions. This defines a functor from the category of groups to the category of pro-p groups: the *pro-p completion functor*. The following two lemmas are not hard to prove and they are left to the reader.

Lemma 5.3

(1) *The pro-p completion functor is right exact.*

(2) *Let $H \leq L$, and denote by $\iota : H \hookrightarrow L$ the inclusion map. Then $\hat{\iota} : \widehat{H}_p \longrightarrow \widehat{L}_p$ is an injection if and only if the pro-p topology of L induces on H the pro-p topology of H.*

If in the situation of part (2) above $\hat{\iota}$ is an injection, we shall identify $\widehat{H}_p$ with its image in $\widehat{L}_p$, so that $\widehat{H}_p \leq \widehat{L}_p$. Observe that in this case, $\widehat{H}_p = \bar{H}$.

Lemma 5.4 *Let L be residually a finite p-group. Then $\widehat{L}_p$ induces on L its pro-p topology. In particular if $X \subseteq L$, then* $\mathrm{Cl}(X) = L \cap \bar{X}$.

The next result provides more examples of closed subgroups.

Lemma 5.5 *Let $L = K \rtimes H$ be residually a finite p-group. Then the pro-p topology of L induces on H its pro-p topology. Moreover H is closed in the pro-p topology of L.*

Proof. Let $V \triangleleft H$ with H/V a finite p-group. Then $L/KV \cong H/V$ so that KV is open in the pro-p topology of L, and clearly $KV \cap H = V$. This proves the first assertion.

To prove that H is closed it suffices to show that if $1 \neq k \in K$, then there exists an open normal subgroup V of L such that $kV \cap H = \emptyset$. Since L is residually a finite p-group, there exists an open normal subgroup W of L such that $kW \cap W = \emptyset$. Then take $V = kW \cap K(W \cap H)$. (More conceptually: consider the identity map $\iota : L = KH \longrightarrow L$ and the projection map $\varphi : L = KH \longrightarrow L$ that sends an element of L to its projection in H; then H consists of those elements of L where the values of ι and φ coincide; since both these maps are continuous, H is closed.) □

Corollary 5.6 *Let $L = H \star R$ be the free product of residually finite p-groups H and R. Then the pro-p topology of L induces on H its pro-p topology. Moreover H is closed in the pro-p topology of L.*

Proof. Observe that $L = R^L \rtimes H$, where R^L denotes the normal closure of R. The result follows from the above lemma. □

In [12] Marshall Hall proved that every finitely generated subgroup H of a free abstract group M is a free factor of a subgroup of finite index U of M. In general there is no control on the index of U in M. However the next result shows that if one assumes H to be closed in the pro-p topology of M, then U can be chosen to have index a power of p.

Theorem 5.7. *Let M be an abstract free group, and let H be a finitely generated subgroup of M which is closed in the pro-p topology of M. Then there exists an open subgroup U of M such that $H \leq U$ and $U = H \star K$, where K is a subgroup of U. In particular, the index of U in M is a power of p.*

Proof. Let $\Gamma = \Gamma(F, X)$ be the Cayley graph of the free group F with respect to some fixed basis X (see Section 1). Let $\{U_i \mid i \in I\}$ be the collection of all open subgroups in the pro-p topology of F that contain H. For each i denote by $\tau_i : H\backslash\Gamma \longrightarrow U_i\backslash\Gamma$ the natural morphism of the corresponding quotient graphs. Since H is closed, $H = \bigcap_{i\in I} U_i$. Therefore, if Δ is a finite subgraph of $H\backslash\Gamma$, there exists some $i \in I$ such that the restriction of τ_i to Δ is an injective morphism of graphs. According to Lemma 1.3, $p_H : H \longrightarrow \pi_1(H\backslash\Gamma)$ is an isomorphism. Since H has finite rank there are finitely many cycles from $1H$ to $1H$ in $H\backslash\Gamma$ that represent generators of $\pi_1(H\backslash\Gamma)$. Let Γ_1 be the graph consisting of the vertices and edges of these cycles. Then obviously Γ_1 is a finite subgraph of $H\backslash\Gamma$, and $\pi_1(\Gamma_1) = \pi_1(\Gamma)$. Choose $j \in I$ in such a way that τ_j is injective on Γ_1. Then we have a commutative diagram

$$\begin{array}{ccc} \pi_1(\Gamma_1) = \pi_1(H\backslash\Gamma) & \xrightarrow{\bar{\tau}_j} & \pi_1(U_j\backslash\Gamma) \\ {\scriptstyle p_H}\uparrow & & \uparrow{\scriptstyle p_{U_j}} \\ H & \xrightarrow{\iota} & U_j. \end{array}$$

It follows from Proposition 1.1(b) that the image of $\pi_1(\Gamma_1) = \pi_1(H\backslash\Gamma)$ in $\pi_1(U_j\backslash\Gamma)$ is a free factor of $\pi_1(U_j\backslash\Gamma)$. Thus H is a free factor of U_j. □

Corollary 5.8 *Let F be an abstract group endowed with the pro-p topology.*

(a) *If U is an open subgroup of F, then the topology induced on U by the pro-p topology of F is the pro-p topology of U.*

(b) *Suppose that F is free, and let H be a finitely generated subgroup of F. If H is closed in the pro-p topology of F, then the topology induced on H by the pro-p topology of F is the pro-p topology of H.*

Proof. By Theorem 5.7 and Proposition 5.8, part (b) follows from part (a). By Lemma 5.2 F/U_F is a finite p-group. Let $N \triangleleft U$ with U/N a finite p-group (i.e., N is open in the pro-p topology of U). Put $K = U_F \cap N$; then U/K is a finite p-group. Clearly $K_F \leq N$ and F/K_F is a finite p-group, because both F/U_F and U_F/K_F are p-groups. Since K_F is open in F, the result follows. □

6 The product of subgroups in an abstract free group

This section contains our main application of the methods that we have introduced in previous sections. The proof of the following theorem uses heavily results in the theory of groups acting on trees, abstract as well as profinite.

Theorem 6.1. *Let G be an abstract group with a free subgroup F which is open in the pro-p topology of G. Let K and $H_1, H_2, \ldots, H_n$ be finitely generated subgroups of G. Assume that all these subgroups are closed in the pro-p topology of G. Then the set $S = H_1H_2\cdots H_nK$ is also closed in the pro-p topology of G.*

Proof. For simplicity we shall give the proof only for the case $n = 1$ (the general case offers special difficulties, but the proof we offer here contains the main ideas; the interested reader may consult [29] or [32] for a complete proof in a more general context). Put $H = H_1$. So we must prove that HK is closed.

We start with a series of reductions. Since F is open in G, the pro-p topology of F is precisely the topology induced by the pro-p topology of G (see Corollary 5.8). We shall show first that we may assume that $H, K \leq F$. Indeed, since H is finitely generated and $F \cap H$ has finite index in H, $F \cap H$ is also finitely generated; moreover, $F \cap H$ is closed in the pro-p topology of F. Say $\{h(1), \ldots, h(t)\}$ is a left transversal of $F \cap H$ in H.

Then

$$HK = \bigcup_{j=1}^{t} h(j)(F \cap H_i)K. \qquad (*)$$

Obviously $F \cap H$ is also finitely generated and closed in the pro-p topology of G. Now, if $(F \cap H)K$ is closed in G, so is the finite union $(*)$. Hence we may replace H by $F \cap H$. Similarly we may replace K by $F \cap K$, and so we may assume that the H and K are subgroups of F.

Therefore, from now on we shall assume that $G = F$ is an abstract free group.

Next we show that we may assume that one of the subgroups, say K, is a free factor of F. Since K is finitely generated and closed in the pro-p topology of F, there is an open subgroup U of F such that K is a free factor of U (see Theorem 5.7). By the argument used above, we may replace F by U. Furthermore, we may assume that the rank of F is finite. Indeed, since the subgroups H and K are finitely generated, they are contained in a free factor of F of finite rank; finally observe that the pro-p topology of F induces on this free factor its own pro-p topology (see Corollary 5.6).

Hence from now on in this proof we shall assume that $G = F$ is an abstract free group of finite rank, and that K is a free factor of F.

Choose a basis X of F in such a way that $Y = F \cap X$ is a basis of K. Consider the abstract Cayley graphs $\Gamma(F, X)$, $\Gamma(K, Y)$ and the profinite Cayley graphs $\Gamma(\widehat{F}_p, X)$, $\Gamma(\widehat{K}_p, Y)$. Then $\Gamma(F, X)$ and $\Gamma(K, Y)$ are abstract trees (cf. [37] Proposition I.15), and $\Gamma(\widehat{F}_p, X)$, $\Gamma(\widehat{K}_p, Y)$ are pro-p trees since $\widehat{F}_p$ and $\widehat{K}_p$ are the free pro-p groups on the sets X and Y, respectively (see Theorem 3.3). In addition $\Gamma(K, Y) \subseteq \Gamma(F, X)$, $\Gamma(\widehat{K}_p, Y) \subseteq \Gamma(\widehat{F}_p, X)$, $\Gamma(F, X) \subseteq \Gamma(\widehat{F}_p, X)$ and $\Gamma(K, X) \subseteq \Gamma(\widehat{K}_p, X)$; moreover these two latter embeddings are dense embeddings. Recall that if R is a subset of F, $\bar{R}$ denotes its closure in $\widehat{F}_p$. By Corollary 5.6, we may identify $\bar{K}$ with $\widehat{K}_p$.

It will suffice to prove that

$$HK = \overline{HK} \cap F.$$

Observe that since both $\bar{H}$ and $\bar{K}$ are compact, so is their product $\bar{H}\bar{K}$, by the continuity of multiplication. Hence $\bar{H}\bar{K}$ is closed in $\widehat{F}_p$ and $\overline{HK} = \bar{H}\bar{K}$. Thus we must show that

$$HK = \bar{H}\bar{K} \cap F.$$

To show this, suppose that $h \in \bar{H}$, $k \in \bar{K}$ and that $hk \in F$. One must prove that $hk \in HK$.

Since $hk \in F$, the geodesic $[1, hk]$ in $\Gamma(\widehat{F}_p, X)$ is finite, and hence so is $h^{-1}[1, hk] = [h^{-1}, k]$. Let Δ be the minimal $\bar{H}$-invariant pro-p subtree of $\Gamma(\widehat{F}_p, X)$ containing the vertex 1. Say $H = \langle r_1, \dots, r_s \rangle$. Then

$$\Delta = \bigcup_{j=1}^{s} \bar{H}[1, r_j].$$

It follows that the quotient graph $\bar{H} \backslash \Delta$ is finite.

By Lemma 2.3 and Exercise 1.9 (i), $[h^{-1}, 1] \cup \Gamma(\bar{K}, Y)$ and $\Delta \cup \Gamma(\bar{K}, Y)$ are pro-p trees. Furthermore,

$$[h^{-1}, k] \subseteq [h^{-1}, 1] \cup \Gamma(\bar{K}, Y) \subseteq \Delta \cup \Gamma(\bar{K}, Y).$$

If $h^{-1} \in \bar{K}$, then $hk \in \bar{K} \cap F = K$ and the proof is finished. Hence we may assume that $h^{-1} \notin \bar{K}$. Now, since $[h^{-1}, k]$ is finite, we may think of it as an ordered sequence of vertices and edges in $\Gamma(\widehat{F}_p, X)$. Then there must exist a first vertex of this sequence, say v, which is in $\Gamma(\bar{K}, Y)$. Let e be the last edge of $[h^{-1}, k]$ not in $\Gamma(\bar{K}, Y)$. Then v is one of the vertices of e, and e is an edge of $[h^{-1}, 1]$. So $[h^{-1}, v]$ is a finite subgeodesic of $[h^{-1}, 1]$, and $[v, k]$ is a finite geodesic in $\Gamma(\bar{K}, Y)$. Thus there exists

$t \in K$ such that $vt = k$. It follows that $hk \in F$ if and only if $hv \in F$; and $hk \in HK$ if and only if $hv \in HK$. Therefore we may replace k by v, and we shall assume from now on that

$$k \in [h^{-1}, 1] \subseteq \Delta.$$

Denote by $\pi : \Delta \longrightarrow \bar{H}\backslash\Delta$ the canonical epimorphism of graphs. Consider now the pro-p subtree $T = \Delta \cap \Gamma(\bar{K}, Y)$, and the natural action of $\bar{H} \cap \bar{K}$ on T. We shall see that the quotient graph $(\bar{H} \cap \bar{K})\backslash T$ is isomorphic to $\pi(T) \subseteq \bar{H}\backslash\Delta$; and in particular, that it is finite. Indeed, note first that π induces an epimorphism $\bar{\pi} : (\bar{H} \cap \bar{K})\backslash T \longrightarrow \pi(T)$. Now, suppose $t, t' \in T$ and $xt = t'$, for some $x \in \bar{H}$. If t is a vertex, then $t, t' \in \bar{K}$, and so $x \in \bar{H} \cap \bar{K}$. If t is an edge, then its vertices are in $\bar{K}$, and again we deduce that $x \in \bar{H} \cap \bar{K}$. Thus $\bar{\pi}$ is an isomorphism. Let $\rho : T \longrightarrow (\bar{H} \cap \bar{K})\backslash T$ be the canonical epimorphism of graphs. Since $(\bar{H} \cap \bar{K})\backslash T$ is finite, there exists a connected transversal Σ of ρ in T containing the vertex 1. Note that $\Sigma \subseteq \Gamma(F, X) \cap \Gamma(\bar{K}, Y) = \Gamma(K, Y)$. Now, k is a vertex of T. Therefore there exists $g \in \bar{H} \cap \bar{K}$ such that $gk \in \Sigma$. So $gk \in K$. Then $hg^{-1}gk = hk \in F$, and hence $hg^{-1} \in F \cap \bar{H} = H$. Thus $hk \in HK$, as desired. □

7 Algorithms in free abstract groups

In this section we establish the existence of an algorithm to compute the closure of a finitely generated subgroup H of a free abstract group F in the pro-p topology of F. This implies that there is algorithmic procedure to decide whether H is closed. Observe that since H is finitely generated, we may assume that F has finite rank.

Throughout this section we assume that F is a free abstract group of finite rank n on a fixed finite basis X. To specify or to give or to construct a finitely generated subgroup of F means to list a finite set of generators for the group. We begin with descriptions of simple algorithms to (i) construct a basis of a given subgroup M of finite index in F; (ii) decide whether a specified finitely generated subgroup of F is contained in a given subgroup of finite index of F; (iii) construct the intersection of two given subgroups of finite index of F.

Lemma 7.1 *Let M be a given subgroup of F of finite index t. Then there exists an algorithm to construct a basis of M.*

Proof. Say $m_1, \dots, m_r$ are given as words in X and that $M = \langle m_1, \dots, m_r\rangle$. Observe that the core M_F of M has at most index t^t in F. Construct every n-generated group of order at most t^t. There are only finitely many epimorphisms from F to one of these groups. Choose one such finite group G and an epimorphism $\varphi : F \longrightarrow G$ so that $|G : \tilde{M}| = t$, where $\tilde{M} = \varphi(M)$. Then $M \geq \ker(\varphi)$ and so $M = \varphi^{-1}(\tilde{M})$. Form the Cayley graph $\Gamma(F, X)$. Construct a graph $\tilde{\Gamma}(G, X)$ with vertices consisting of the elements of G, whose set of edges is $\{(g, x) \mid g \in G, x \in X\}$ and $d_0(g, x) = g$, $d_1(g, x) = g\varphi(x)$. Remark that G acts on $\tilde{\Gamma}(G, X)$ in a natural way, and that $\tilde{M}\backslash\tilde{\Gamma}(G, X) \cong M\backslash\Gamma(F, X)$. Therefore, $\pi_1(\tilde{M}\backslash\tilde{\Gamma}(G, X)) \cong \pi_1(M\backslash\Gamma(F, X)) \cong M$ (see Lemma 1.3). Note that these isomorphisms can be made explicit. Hence one can apply to the finite graph $\tilde{\Gamma}(G, X)$ the procedure indicated in Proposition 1.1 to construct a basis of M. □

Lemma 7.2 *Let M be a given subgroup of finite index t in F, and let H be a given finitely generated subgroup of F. Then there is an algorithm to decide whether $H \leq M$.*

Proof. Construct G and φ as in the proof of Lemma 7.1. Since $M \geq \ker(\varphi)$, it is clear that $M \geq H$ if and only if $\varphi(M) \geq \varphi(H)$. □

Lemma 7.3 *Let M_1 and M_2 be given subgroups of finite index t_1 and t_2 in F, respectively. Then there is an algorithm to construct $M_1 \cap M_2$.*

Proof. As in the proof of Lemma 7.1 one can construct finite groups G_1 and G_2 and epimorphisms $\varphi_1 : F \longrightarrow G_1$ and $\varphi_2 : F \longrightarrow G_2$ such that $\ker(\varphi_i) \leq M_i$ $(i = 1, 2)$. Define $\varphi : F \longrightarrow G_1 \times G_2$ by $\varphi(f) = (\varphi_1(f), \varphi_2(f))$ $(f \in F)$. Then $\ker(\varphi) \leq M_1 \cap M_2$. Therefore $M_1 \leq M_2$ if and only if $\varphi(M_1) \leq \varphi(M_2)$; and $M_1 \cap M_2 = \varphi^{-1}(\varphi(M_1) \cap \varphi(M_2))$. This gives the result. □

Lemma 7.4 *Given a free group F with a finite basis, there exists an algorithm to specify all the subgroups of a given finite index r that are open in the pro-p topology of F.*

Proof. Let $m = r^r$. Let R be a subgroup of F of index n. Then there is a homomorphism $\varphi : F \longrightarrow S_m$ and a subgroup R' of index n in S_m such that $\varphi^{-1}(R') = R$. Hence one can specify the distinct subgroups R of F of index r together with a right transversal of R in F. Therefore, using the method indicated in the proof of Lemma 7.3, one can construct the core R_F of R in F and decide whether F/F_G is a p-group; i.e., one can decide whether R is open. □

7.5 Algorithm to compute the closure $\mathrm{Cl}(H)$ of H in the pro-p topology of F

Let F be a free abstract group with a given finite basis X. Let H be a subgroup of F generated by $\{h_1, \ldots, h_r\}$. Denote by $\Gamma = \Gamma(F, X)$ the Cayley graph of F with respect to X. For a natural number m, denote by $U(m)$ the intersection of all open (in the pro-p topology) subgroups of F of index at most m containing H.

Step 1 Construct the reduced paths p_{1,h_i} in Γ from 1 to h_i $(i = 1, \ldots, r)$. Let $T = \bigcup_{i=1}^{r} p_{1,h_i}$. Set $m := p$.

Step 2 Construct $U(m)$ (see Lemmas 7.2, 7.3 and 7.4). If $U(m) = F$ (*see Lemma* 7.2), then $\mathrm{Cl}(H) = F$, by Lemma 5.2(d), and the algorithm stops.

Step 3 Construct the finite quotient graph $U(m)\backslash\Gamma$. Construct the image T_m of T in $U(m)\backslash\Gamma$, under the natural morphism of graphs $\Gamma \longrightarrow U(m)\backslash\Gamma$.

Step 4 Construct the image $p_{1,h_i}(m)$ in $U(m)\backslash\Gamma$ of the path p_{1,h_i} $(i = 1, \ldots, r)$. These images are reduced cycles in T_m since $H \leq U(m)$, and so they represent elements of

$\pi_1(T_m)$. Check whether there is an open subgroup of $\pi_1(T_m)$ of index p containing each of the elements represented by $p_{1,h_i}(m)$ $(i = 1, \ldots, r)$. *(Construct $\pi_1(T_m)$ as in Proposition 1.1; use Lemma 7.4 to construct the open subgroups of index p in $\pi_1(T_m)$ and adapt the method described in the proof of Lemma 7.1 to check whether $p_{1,h_i}(m)$ is in such open sugroups. Note that $\pi_1(T_m)$ is identified with a subgroup of F using the natural isomorphisms described in Proposition 1.1 and Lemma 1.3.)* If none exist, $\mathrm{Cl}(H) = \pi_1(T_m)$, and the algorithm stops. Otherwise, set $m := pm$ and go to Step 2.

The next task is to justify that this algorithm really computes the closure of H, and we do this in the following.

Theorem 7.6. *The algorithm 7.5 stops after a finite number of steps and its output is rm Cl(H).*

Proof. We continue with the notation described in the algorithm. The aim of this proof is to show that there exists some natural number m such that $\pi_1(T_m) = \mathrm{Cl}(H)$. We shall do this by showing that, for some m, $H \leq \pi_1(T_m) \leq \mathrm{CL}(H)$ and that $\pi_1(T_m)$ is closed.

Since $H \leq \mathrm{CL}(H) \leq U(m)$ we have natural morphisms of graphs

$$\Gamma \longrightarrow H\backslash\Gamma \xrightarrow{\tau} \mathrm{Cl}(H)\backslash\Gamma \xrightarrow{\tau_m} U(m)\backslash\Gamma.$$

Denote by Γ_1 the image of T in $H\backslash\Gamma$; then Γ_1 is a finite graph and $\pi_1(\Gamma_1) = \pi_1(H\backslash\Gamma)$ (see the proof of Lemma 1.3). Consider the commutative diagram

$$\begin{array}{ccccc}
\pi_1(\Gamma_1) = \pi_1(H\backslash\Gamma) & \xrightarrow{\bar\tau} & \pi_1(\mathrm{Cl}(H)\backslash\Gamma) & \xrightarrow{\bar\tau_m} & \pi_1(U(m)\backslash\Gamma) \\
\uparrow{\scriptstyle p_H} & & \uparrow{\scriptstyle p_{\mathrm{Cl}(H)}} & & \uparrow{\scriptstyle p_{U(m)}} \\
H & \xrightarrow{\iota} & \mathrm{Cl}(H) & \xrightarrow{\iota_m} & U(m)
\end{array}.$$

The vertical maps are isomorphisms (see Lemma 1.3) and we shall think of them as embeddings; consequently, we also think of the maps $\bar\tau$ and $\bar\tau_m$ as embeddings.

Since T_m is a subgraph of $U(m)\backslash\Gamma$, we know that $\pi_1(T_m)$ is a free factor of $\pi_1(U(m)\backslash\Gamma) = U(m)$ (see Proposition 1.1). Since $U(m)$ is closed in the pro-p topology of F, it follows from Corollary 5.6 that $\pi_1(T_m)$ is closed in F.

Denote by Γ_2 the image of Γ_1 in $\mathrm{Cl}(H)\backslash\Gamma$. Note that by Lemma 5.2, we have that $\mathrm{Cl}(H) = \bigcap_{i=1}^{\infty} U(m)$. Since Γ_2 is finite, there exists $m \geq p$ such that τ_m maps Γ_2 injectively into $U(m)\backslash\Gamma$; on the other hand, it is clear that $\tau_m(\Gamma_2) = T_m$. So $\bar\tau_m(\pi_1(\Gamma_2)) = \pi_1(T_m)$, and in particular $\pi_1(\Gamma_2)$ is closed in F. Observe that $\pi_1(\Gamma_2)$ is contained in $\mathrm{Cl}(H) = \pi_1(\mathrm{Cl}(H)\backslash\Gamma)$ (in fact, it is a free factor according to Proposition 1.1). Furthermore, the restriction to H of the injection $p_{U(m)}$ sends H into a subgroup of $\pi_1(T_{U(m)})$, so that under our identifications $H \leq \pi_1(T_m) = \pi_1(\Gamma_2) \leq \mathrm{Cl}(H)$. Thus $\pi_1(T_m) = \mathrm{Cl}(H)$, as desired. □

Corollary 7.7 *If H is a finitely generated subgroup of a free abstract group F, then its closure $\mathrm{Cl}(H)$ in the pro-p topology of F is also finitely generated (as an abstract group).*

Corollary 7.8 *Let $H_1, \ldots, H_t$ be finitely generated subgroups of a free abstract group F. Then*

$$\mathrm{Cl}(H_1 \cdots H_t) = \mathrm{Cl}(H_1) \cdots \mathrm{Cl}(H_t).$$

Proof. By Corollary 7.7 each $\mathrm{Cl}(H_i)$ is finitely generated. Hence, by Theorem 6.1, $\mathrm{Cl}(H_1) \cdots \mathrm{Cl}(H_t)$ is closed in the pro-p topology of F. On the other hand, $H_1 \cdots H_t$ is a dense subset of $\mathrm{Cl}(H_1) \cdots \mathrm{Cl}(H_t)$. □

Lemma 7.9 *Let H be a subgroup of the free group $F = F(X)$, and let $\{h_1, \ldots, h_r\}$ be a finite set of generators of H given as words on the basis X. Assume that $w \in F$. Then there is an algorithm to decide whether $w \in H$.*

Proof. See Proposition 2.21 of [19]. The idea is to establish an effective ordering of the elements of F and then use the Nielsen method. □

We can now describe an algorithm to check if a finitely generated subgroup H of the free group F is closed:

7.10 Algorithm to decide whether H is closed in the pro-p topology of F

Step 1 Apply Algorithm 7.5 to compute a set of generators of $\mathrm{Cl}(H)$.

Step 2 Apply Lemma 7.9 to decide whether $\mathrm{Cl}(H) \leq H$.

8 Algorithms for monoids

The results in this section are motivated by a famous question raised first by J. Rhodes (see [13] for a background, a history and the many interconnections of this question with other problems in theoretical computer science). The question was posed originally as to whether it is possible to describe algorithmically a certain submonoid (the so-called 'kernel') of a finite monoid. We start by describing this 'kernel'.

The identity element of a monoid is always denoted by 1. A *relational morphism* τ from a monoid M to a monoid N is a map that assigns to each element $m \in M$, a nonempty subset $\tau(m)$ of N, such that $1 \in \tau(1)$ and $\tau(mm') \supseteq \tau(m)\tau(m')$, for every $m, m' \in M$. For example, if $\rho : N \longrightarrow M$ is a epimorphism of monoids, the map $m \mapsto \rho^{-1}(m)$ is a relational morphism from M to N. Define the *p-kernel* $\mathrm{K}_p(M)$ of a finite monoid M as

$$\mathrm{K}_p(M) = \bigcap_{\tau} \tau^{-1}(1),$$

where τ ranges over all relational morphisms from M to every finite p-group (more precisely, representatives of the isomorphism classes of finite p-groups). Note that, by definition, $\tau^{-1}(1) = \{x \in M \mid 1 \in \tau(x)\}$. Then clearly $K_p(M)$ is a submonoid of M.

There is a way of describing the elements of $K_p(M)$ in terms of the pro-p topology. To explain this connection let us adopt the following notation. If X is a set, denote by X^* the free monoid on the set X. Then

$$X \subseteq X^* \subseteq \langle X \rangle = F(X),$$

where $F(X)$ is the abstract free group on X. The pro-p topology (see Section 5) of $F(X)$ induces on X^* a topology, that we call the *pro-p topology of* X^*. If $S \subseteq X^*$, we shall denote by Clm(S) the closure of S in pro-p topology of X^*. Then one has the following:

Theorem 8.1. ([26]) *Let M be a finite monoid, and let $\pi : X^* \longrightarrow M$ be a monoid epimorphism of the free monoid X^* onto M. Then $t \in K_p(M)$ if and only if* $1 \in$ Clm$(\pi^{-1}(t))$.

The following result gives a positive answer to the p-version of the question posed by Rhodes. (Incidentally, the original question posed by Rhodes refers to the computability of the kernels — not necassarily p-kernels — of finite monoids. This question also has a positive answer. See [1] for an answer using purely combinatorial methods, or [29] for an answer using the profinite topology.)

Theorem 8.2. *There exists an algorithm to compute the p-kernel* $K_p(M)$ *of a finite monoid M.*

The proof of this theorem requires the full strength of Theorem 6.1 and Algorithms 7.5 and 7.10. We come back to this proof later. We first need some considerations about rational subsets of a free group.

Let M be a monoid. If S is a subset of M, denote by S^* the submonoid generated by S. The class of *rational subsets* Rat(M) of M is the smallest class of subsets of M such that (i) the empty subset and the singleton subsets of M belong to Rat(M); (ii) if S and T are in Rat(M), then so are $S \cup T$ and ST; and (iii) if S is in Rat(M), so is the submonoid S^*. Let M be a free monoid with basis A. Then according to Kleene's theorem (see, e.g., [9] Th. 5.1), rational subsets are exactly the subsets of M recognized by finite state automata on the alphabet A; or, equivalently, the regular languages on the alphabet A.

If L is a rational subset of M, one sees easily that L can be specified in terms of finitely many singletons using finitely many times the operations union, product and $*$. We refer to any such way of expressing L as a *rational expression* (also called 'regular expressions').

Let $F = F(X)$ be a free group with finite basis X. We shall study next some properties of rational subsets of F that can be obtained as consequences of the results obtained above. First observe that if R is a rational subset of F, so is $R^{-1} = \{x \in R \mid x \in R\}$ (just consider a rational expression for R, and construct a corresponding rational expression for R^{-1}). Next note that a finitely generated subgroup H of F is rational, for if S is a finite set of generators of H, then $H = \{S \cup S^{-1}\}^*$. In fact one has

Theorem 8.3. ([2]) *A subgroup H of a free abstract group $F = F(X)$ of finite rank is a rational subset if and only if H is finitely generated.*

Lemma 8.4 *Let S be any subset of an abstract group F. Then* $\mathrm{Cl}(S^*) = \mathrm{Cl}\langle S\rangle$.

Proof. Let N be an open normal subgroup of F. Then $S^*N = \langle S\rangle N$ since F/N is a finite group. The result follows since

$$\mathrm{Cl}(S^*) = \bigcap_{N\in\mathcal{N}} S^*N, \quad \mathrm{Cl}\langle S\rangle = \bigcap_{N\in\mathcal{N}} \langle S\rangle N. \qquad \square$$

Theorem 8.5. *Let $F = F(X)$ be a free group with finite basis X endowed with the pro-p topology. Let $\mathcal{F}$ be the class of subsets of F consisting of finite unions of sets of the form $gH_1H_2\cdots H_n$, where $g \in F$, and $H_1, H_2, \ldots, H_n$ are finitely generated subgroups of F which are closed in the pro-p topology of F. Then $\mathcal{F}$ is equal to the collection of those rational subsets of F that are closed in the pro-p topology of F.*

Proof. As pointed out above, every finitely generated subgroup of F is a rational subset, and hence any set in $\mathcal{F}$ is rational and closed by Theorem 6.1.

For the converse, we first define the class $\mathcal{R}$ to consist of all rational subsets S of F whose closure $\mathrm{Cl}(S)$ is in $\mathcal{F}$. We claim that in fact $\mathcal{R}$ is the class of all rational subsets of F. Observe that the theorem follows from this claim: indeed, if the claim holds and R is a closed rational subset of F, then $R = \mathrm{Cl}(R) \in \mathcal{F}$.

To prove the claim, we must show that $\mathcal{R}$ satisfies the conditions (i)-(iii) above that characterize rational subsets (since $\mathcal{R} \subseteq \mathrm{Rat}(F)$). Obviously the empty set and singleton sets belong to $\mathcal{R}$, so (i) holds. Consider any two subsets S and T of F, then plainly $\mathrm{Cl}(S \cup T) = \mathrm{Cl}(S) \cup \mathrm{Cl}(T)$; so, if $S, T \in \mathcal{R}$, then $S \cup T \in \mathcal{R}$. To finish the verification of (ii), assume now that $S, T \in \mathcal{R}$. We must show that $ST \in \mathcal{R}$. By assumption, $\mathrm{Cl}(S), \mathrm{Cl}(T) \in \mathcal{F}$. So their product $\mathrm{Cl}(S)\mathrm{Cl}(T)$ is also a finite union of sets of the form $gH_1\cdots H_t$, with each H_i a finitely generated subgroup of F, and therefore it is in $\mathcal{F}$; in particular $\mathrm{Cl}(S)\mathrm{Cl}(T)$ is closed. Using continuity of multiplication we have

$$ST \subseteq \mathrm{Cl}(S)\mathrm{Cl}(T) \subseteq \mathrm{Cl}(ST),$$

and thus $\mathrm{Cl}(S)\mathrm{Cl}(T) = \mathrm{Cl}(ST)$. Therefore $\mathrm{Cl}(ST) \in \mathcal{F}$, i.e., $ST \in \mathcal{R}$.

To prove (iii), consider $S \in \mathcal{R}$. By Lemma 8.4, $\mathrm{Cl}(S^*) = \mathrm{Cl}\langle S\rangle$. Since S is rational, so is S^{-1}. Hence $\langle S\rangle = \{S \cup S^{-1}\}^*$ is a rational subset of F. Therefore $\langle S\rangle$ is finitely generated, according to Theorem 8.3. Then, by Corollary 7.7, $\mathrm{Cl}\langle S\rangle$ is also finitely generated; so $\mathrm{Cl}(S^*) = \mathrm{Cl}\langle S\rangle \in \mathcal{F}$, and thus $S^* \in \mathcal{R}$. $\square$

As a consequence of the above theorem we obtain

Theorem 8.6. *Let $F = F(X)$ be a free abstract group with a finite basis X. Assume that F is endowed with the pro-p topology.*

(a) *If L_1 and L_2 are closed rational subsets of F, then so is their product L_1L_2.*

(b) *The closure* $\mathrm{Cl}(R)$ *of a rational subset R of F is rational.*

(c) *There exists an algorithm to compute the closure of a rational set R of F given by a rational expression. Specifically,* $\mathrm{Cl}(R)$ *can be computed using the following rules:*

(1) $\mathrm{Cl}(T) = T$, *if T is finite;*

(2) $\mathrm{Cl}(S \cup T) = \mathrm{Cl}(S) \cup \mathrm{Cl}(T)$;

(3) $\mathrm{Cl}(ST) = \mathrm{Cl}(S)\mathrm{Cl}(T)$;

(4) $\mathrm{Cl}(S^*) = \mathrm{Cl}(\langle S\rangle) = (Y \cup Y^{-1})^*$, *where $\langle S\rangle$ denotes the subgroup of F generated by S, and where Y is a set of generators of* $\mathrm{Cl}(\langle S\rangle)$ *computed according to Algorithm 7.5.*

Proof. Part (a) follows immediately from Theorem 8.5. Part (b) is a consequence of Theorems 8.5 and 8.3. Part (c) follows again from Theorem 8.5 and Algorithm 7.5. □

Corollary 8.7 *Let S be a rational subset of the free group F given by a rational expression. Then it is decidable whether or not S is closed in the pro-p topology of F.*

Proof. According to Theorem 8.6, one can compute $\mathrm{Cl}(S)$ as a rational expression. According to Lemma 5.11 in [13], one can decide whether $S = \mathrm{Cl}(S)$, and hence whether S is closed. □

Theorem 8.8. *Let X be a finite set, X^* the free monoid on X, and $L_0, \dots, L_k$ rational subsets of X^*. Let $L = L_0^* w_1 L_1^* w_2 \cdots w_k L_k^*$, where $w_1, \dots, w_k \in X^*$. Then*

$$\mathrm{Clm}(L) = [\mathrm{Cl}(\langle L_0\rangle) w_1 \mathrm{Cl}(\langle L_1\rangle) w_2 \cdots w_k \mathrm{Cl}(\langle L_k\rangle)] \cap X^*.$$

[Recall that $\mathrm{Clm}(L)$ denotes the closure of L in the pro-p topology of X^*, and $\mathrm{Cl}(\langle L_i\rangle)$ the closure of $\langle L_i\rangle$ in the pro-p topology of the free abstract group $F = F(X)$.]

Proof. First observe that $\mathrm{Clm}(L) = \mathrm{Cl}(L) \cap X^*$. Set

$$N_i = w_k^{-1} \cdots w_{i+2}^{-1} w_{i+1}^{-1} L_i^* w_{i+1} w_{i+2} \cdots w_k.$$

Then each N_i is a submonoid of F, and

$$L = L_0^* w_1 L_1^* \cdots w_k L_k^* = w_1 \cdots w_k N_0 \cdots N_k.$$

So $\mathrm{Cl}(L) = w_1 \cdots w_k \mathrm{Cl}(N_0 \cdots N_k)$. Next we claim that $\mathrm{Cl}(N_0 \cdots N_k) = \mathrm{Cl}(N_0) \cdots \mathrm{Cl}(N_k)$. It suffices to prove that $\mathrm{Cl}(N_0) \cdots \mathrm{Cl}(N_k)$ is closed in F. Note that $\mathrm{Cl}(N_i) = \mathrm{Cl}\langle N_i\rangle$ (just observe that $N_i U = \langle N_i\rangle U$, for any open normal subgroup U of F). By Theorem 6.1, $\mathrm{Cl}\langle N_0\rangle \cdots \mathrm{Cl}\langle N_k\rangle$ is closed in F. Finally note that

$$w_1 \cdots w_k \mathrm{Cl}\langle N_0\rangle \cdots \mathrm{Cl}\langle N_k\rangle = \mathrm{Cl}\langle L_0\rangle \cdots \mathrm{Cl}\langle L_k\rangle.$$ □

It is not hard to see that a rational subset of X^* can be effectively written as a finite union of rational subsets of the form $L = L_0^* w_1 L_1^* w_2 \cdots w_k L_k^*$ with $L_0, \ldots, L_k$ rational subsets of X^* and $w_1, \ldots, w_k \in X^*$. Then the following result is a consequence of Theorems 8.6 and 8.8.

Corollary 8.9 *Let L be a rational subset of the free monoid X^*. Then there is an effective procedure to compute* $\mathrm{Clm}(L)$.

We can return now to the proof of Theorem 8.2.

Proof of Theorem 8.2. We use the notation of Theorem 8.1. It is well known that $\pi^{-1}(t)$ is a rational subset of X^* (cf. [9]). The result follows from Theorems 8.1 and Corollary 8.9. □

For other related algorithmic results concerning monoids and semigroups the reader may consult [30], [13], [21] or [32].

9 Automorphisms and p-extensions of free pro-p groups

In this section we describe some recent results on the group of fixed points of an automorphism of a free pro-p group of finite rank, as well as some structure theorems for pro-p groups that contain a free pro-p subgroup of index p. We emphasize some key points in the proofs of these results where the use of free products of pro-p groups is one of the important ingredients.

Let $\varphi : F \longrightarrow F$ be an automorphism of a free pro-p group F of finite rank n. Then

$$\mathrm{Fix}_F(\varphi) = \{x \in F \mid \varphi(x) = x\}$$

is the subgroup of fixed points of φ. It is sometimes useful to think of this subgroup as a centralizer. Consider the (relative) holomorph $G = F \rtimes \langle\varphi\rangle$. Then $\mathrm{Fix}_F(\varphi) = C_F(\varphi)$. We are interested in the rank of $\mathrm{Fix}_F(\varphi)$. (The corresponding question for abstract groups has been treated in [8], [10] and [4].)

Unlike the situation for abstract groups (cf. [10]), the order of the automorphism φ plays a decisive role in the answer to what is $\mathrm{Fix}_F(\varphi)$. It is well-known that $\mathrm{Aut}(F)$ has a natural structure of a profinite group; in fact it is virtually a pro-p group, i.e., $\mathrm{Aut}(F)$ contains a pro-p open subgroup (see for example [7], Chapter 5). It follows that the order m of φ is a supernatural number rp^t, where r is a natural number prime to p and $0 \le t \le \infty$.

Theorem 9.1. ([14]) *Assume that the rank of F is at least 2. If the order m of φ is not divisible by p (i.e., in the above notation, $t = 0$), then the rank of* $\mathrm{Fix}_F(\varphi)$ *is necessarily infinite.*

The proof of this theorem can be found in [14]. We indicate here only the main ingredients that have to do with free products of pro-p groups. We assume for simplicity that $r = q$ is a prime number. Using an argument analogous to Thompson's q-nilpotency criterion, one may assume that $\mathrm{Fix}_F(\varphi)$ is not trivial. Now, if $\mathrm{Fix}_F(\varphi)$

were finitely generated, by a theorem of Lubotzky [18], there would exist an open subgroup of U of F such that $U = \mathrm{Fix}_F(\varphi) \amalg \mathrm{T}$ (free pro-p product), for some subgroup T of U. Further, one proves that the free factor T can be chosen to be φ-invariant. The main step then is to consider the cartesian subgroup C of the free product $\mathrm{Fix}_F(\varphi) \amalg \mathrm{T}$, that is, the kernel of the natural epimorphism $\mathrm{Fix}_F(\varphi) \amalg \mathrm{T} \longrightarrow \mathrm{Fix}_F(\varphi) \times \mathrm{T}$. The group C is known to be a free pro-p group; moreover φ induces on C an automorphism, and one easily sees that $\mathrm{Fix}_C(\varphi) \neq 1$. A contradiction is reached by observing that $\mathrm{Fix}_U(\varphi) = \mathrm{C} \cap \mathrm{Fix}_F(\varphi) = 1$.

The properties of a free pro-p group are in many cases similar to those of abstract free groups (although their proofs are often very different). Theorem 9.1 can be regarded as one of the exceptions (in [10] it is shown that the rank of the subgroup of fixed points of an automorphism of an abstract free group of finite rank is always finite). However, the following conjecture restores our faith on the (quasi) perfection of the universe.

Conjecture 9.2 ([15]) *Let F be a free pro-p group of rank n, and let φ be an automorphism of F of order p^t $(0 \leq t \leq \infty)$. Then the rank of* $\mathrm{Fix}_F(\varphi)$ *is at most n.*

The answer to the above conjecture is clearly related to the study of the structure of the group $G = F \rtimes \langle\varphi\rangle$. Next we state two additional conjectures. The first one is phrased in terms of fundamental groups of a graph of groups (cf. [37] I.5, or [6] Ch. II).

Conjecture 9.3 ([16]) *Let G be a finitely generated pro-p group that is virtually free pro-p. Then G is the fundamental group of a finite graph of finite p-groups, in the category of pro-p groups. In other words, G is the pro-p completion of the abstract fundamental group of a finite graph of finite p-groups.*

Conjecture 9.4 ([16]) *Let G be a finitely generated pro-p group that contains a free pro-p subgroup of index p. Then G is a free pro-p product*

$$G \cong (\coprod_{i=1}^{t}(C_p \times H_i)) \amalg H,$$

where C_p denotes the cyclic group of order p, and H_i and H are free pro-p groups.

This conjecture has recently been proved in more generality (see Theorem 9.11 below). If ones assumes in addition that G is torsion free (not necessarily finitely generated), the proof of Conjecture 9.4 was known long time ago. In a difficult and seminal paper (for pro-p groups as well as for abstract groups) Serre proved the following:

Theorem 9.5. ([35]) *If a pro-p group G has no elements of order p and contains an open free pro-p subgroup, then G is free pro-p.*

Actually, Serre proved a more general result, namely, he showed that if G is a pro-p group with no torsion elements, and if U is an open subgroup of G, then G and U have the same p-cohomological dimension, i.e., $cd_p(G) = cd_p(U)$. As a consequence he showed:

Theorem 9.6. ([35]) *If G is a torsion free p-adic analytic group of dimension t, then $cd_p(G) = t$.*

Coming back to Conjecture 9.4, its principal motivation was the main result in [8] where an analogous result is proved for abstract groups using the Bass–Serre theory of groups acting on trees. Conjectures 9.2–9.4 are established in [15] and [16] for the special case when the rank of the free pro-p group is 2. Both the methods and the results used in these two papers are very explicit. Here are two samples.

Theorem 9.7. ([15]) *Let F be a free pro-p group of rank 2, and let φ be a nontrivial automorphism of F of finite order r. Then*

(i) *If $p \nmid r$, then $\mathrm{Fix}_F(\varphi)$ is a free pro-p group of infinite rank;*

(ii) *if $p = 2$ and $2|r$, then $\mathrm{rank}(\mathrm{Fix}_F(\varphi)) \leq 1$;*

(iii) *if $p = 3$ and $3|r$, then $\mathrm{Fix}_F(\varphi) = 1$;*

(iv) *if $p > 3$, there is no automorphism of F whose order is a multiple of p.*

Theorem 9.8. ([17]) *Let G be a pro-p group with trivial center having an open normal free subgroup F of rank 2. Then one of the following holds (the symbol $\amalg$ stands for 'free pro-p product', for the appropriate prime):*

(1) *G is a free pro-p group of finite rank.*

(2) *$p = 3$ and $G \cong C_3 \amalg C_3$.*

(3) *$p = 2$ and G has one of the following structures:*

  (a) $G \cong C_2 \amalg C_2 \amalg C_2$;
  (b) $G \cong C_2 \amalg \mathbf{Z}_2$;
  (c) $G \cong C_2 \amalg (C_2 \times \mathbf{Z}_2)$;
  (d) $G \cong C_4 \amalg C_2$;
  (e) $G \cong (C_2 \times C_2) \amalg C_2$;
  (f) $G \cong (C_2 \times C_2) \amalg_{C_2} (C_2 \times C_2) \amalg_{C_2} (C_2 \times C_2)$;
  (g) $G \cong D_4 \amalg_{C_2} (C_2 \times C_2)$, *where D_4 is the dihedral group of order* 8.

As a consequence of these explicit statements one can obtain quite accurate information about the conjugacy classes of automorphisms of a given order in $\mathrm{Aut}(F)$.

Theorem 9.9. ([16]) *Let F be a free pro-p group of rank 2. Let S denote the set of possible orders of torsion elements of $\mathrm{Aut}(F)$.*

(1) *The conjugacy classes of elements of finite order coprime to p in $\mathrm{Aut}(F)$ are in a natural one-to-one correspondence with the conjugacy classes of elements of order coprime to p in $\mathrm{GL}_2(p)$.*

(2) *Let $c(n)$ denote the number of conjugacy classes of automorphisms of order n.*

(i) *For $s \in S$ with s coprime to p, one has that s divides $p^2 - 1$. Furthermore $c(s) = \phi(s)s$, if s divides $p - 1$, and $c(s) = \phi(s)/2$, if s does not divide $p - 1$ (here ϕ denotes the Euler function).*

(ii) *If $p = 2$, then $S = \{2, 3, 4\}$. One has $c(2) = 4$, $c(3) = 1$, $c(4) = 1$.*

(iii) *If $p = 3$, then $S = \{2, 3, 4, 8\}$. One has $c(2) = 2$, $c(3) = 1$, $c(4) = 1$, $c(8) = 2$.*

(iv) *If $p > 3$, then every $s \in S$ is coprime to p.*

Part of Conjecture 9.2 has been recently established by C. Scheiderer in [34] using very elegant and beautiful methods related to group cohomology. He considers any finite p-group S of automorphisms of a free pro-p group F of finite rank. Define

$$\mathrm{Fix}_F(S) = \{x \in F \mid \varphi(x) = x, \forall \varphi \in S\}.$$

Then he shows that $\mathrm{Fix}_F(S)$ is a free factor of F, and in particular $\mathrm{rank}(\mathrm{Fix}_F(S)) \leq \mathrm{rank}(F)$. His result in turn has been extended in [17] to include free pro-p groups of any rank. We state here the latter result.

Theorem 9.10. ([34], [17]) *Suppose that F is a free pro-p group (whose rank is not necessarily finite) and let S be a finite p-group of automorphisms of F. Then the set of fixed points* $\mathrm{Fix}_F(S)$ *is a free factor of F.*

This result is a consequence of the following theorem that establishes Conjecture 9.4 (it was proved by Scheiderer when F is of finite rank, and by Herfort, Ribes and Zalesskii in full generality).

Theorem 9.11. ([34], [17]) *Let G be a pro-p group that contains a free pro-p subgroup F of index p. Then G is a free pro-p product*

$$G \cong (\coprod_{x \in X} (C_p \times H_x)) \amalg H,$$

where C_p denotes the cyclic group of order p, H is a free pro-p group and H_x are free pro-p groups indexed by a profinite space X (the space X is a quotient of the space of conjugacy classes of elements of order p in G).

The proofs of these last two theorems use two important results that we state using homology groups. Theorem 9.10, due to Scheiderer, is a generalization of the theorem of Serre mentioned above. To state it we need some notation. Given a profinite group G, denote by $\mathcal{T}$ the set of all subgroups of G of order p. Then $\mathcal{T}$ has in a natural way the structure of a profinite space. Indeed, one clearly has

$$\mathcal{T} = \varprojlim_{U \in \mathcal{U}} \mathcal{T}_U,$$

where $\mathcal{U}$ is the collection of open subgroups of G, and $\mathcal{T}_U$ denotes the (finite) set of all subgroups of G/U of order p. Observe that G acts on $\mathcal{T}$ by conjugation. Then the free $\widehat{\mathbf{Z}}$-module $[[\widehat{\mathbf{Z}}\mathcal{T}]]$ on $\mathcal{T}$ becomes a profinite G-module.

Theorem 9.12. ([34]) *Let G be a profinite group that does not contain any subgroup isomorphic to $C_p \times C_p$. Assume that H is an open subgroup of G of finite cohomological p-dimension d. If B is a profinite p-primary right G-module, the natural homomorphism $B \widehat{\otimes} [[\widehat{\mathbf{Z}}T]] \longrightarrow B$ defined by $b \widehat{\otimes} \bar{t} \mapsto b\epsilon(\bar{t})$ ($b \in B$, $\bar{t} \in [[\widehat{\mathbf{Z}}T]]$), where $\epsilon : [[\widehat{\mathbf{Z}}T]] \longrightarrow \widehat{\mathbf{Z}}$ is the augmentation map, induces isomorphisms*

$$H_n(G, B \widehat{\otimes} [[\widehat{\mathbf{Z}}T]]) \longrightarrow H_n(G, B)$$

for each $n > d$.

Another ingredient needed to prove Theorem 9.11 is a characterization of when a pro-p group can be expressed as a free pro-p product of a collection of subgroups indexed by a profinite space. The specific result used is due to O. Mel'nikov, who in turn generalizes an earlier version due to J. Neukirch [25].

Theorem 9.13. ([24]) *Let G be a pro-p group, and let $\{G_x \mid x \in X\}$ be a continuous system of subgroups of G indexed by some profinite space X. Then the following statements are equivalent:*

(1) *G is a free pro-p product*

$$G = \coprod_{x \in X} G_x \amalg H,$$

where H is a free pro-p group.

(2) *The map*

$$\mathrm{Cor} : \bigoplus_{x \in X} H_n(G_x, \mathbf{F}_p) \longrightarrow H_n(G, \mathbf{F}_p)$$

is injective for $n = 1$ and surjective for $n = 2$ (here $\mathbf{F}_p$ is the abelian group of order p with trivial G-module structure).

References

[1] C. J. Ash, Inevitable graphs: a proof of the type II conjecture and some related decision procedures, *Int. J. Alg. and Comp.*, **1** (1991), 127–146.

[2] J. Berstel, *Transductions and Context Free Languages*, Teubner, 1979.

[3] H. Bass, Some remarks on group actions on trees, *Comm. Alg.*, **4** (1976), 1091–1126.

[4] M. Bestvina and M. Handel, Train tracks and automorphisms of free groups, *Ann. Math.*, **135** (1992), 1–51.

[5] I. M. Chiswell, Exact sequences associated with a graph of groups, *J. Pure Appl. Alg.*, **8** (1976), 1–36.

[6] W. Dicks, *Trees and Projective Modules*, Lect. Notes Math., **790**, Springer, Berlin, 1980.

[7] J. D. Dixon, M. P. F. du Sautoy, A. Mann and D. Segal, *Analytic pro-p Groups*, 2nd edition, CUP, Cambridge, 1999.

[8] J. L. Dyer and G. P. Scott, Periodic automorphisms of free groups, *Comm. Alg.*, **3** (1975), 195–201.

[9] S. Eilenberg, *Automata, Languages and Machines, Vol. A*, Academic Press, New York, 1974.

[10] S. M. Gersten, Fixed points of automorphisms of free groups, *Invent. Math.*, **84** (1986), 91–119.

[11] D. Gildenhuys and L. Ribes, Profinite groups and Boolean graphs, *J. Pure Appl. Alg.*, **12** (1978), 21–47.

[12] M. Hall, A topology for free groups and related groups, *Annals of Math.*, **52** (1950), 127–139.

[13] K. Henckell, S. W. Margolis, J-E. Pin and J. Rhodes, Ash's Type II theorem, profinite topology and Malcev products. Part I, *Int. J. Alg. and Comp.*, **1** (1991), 411–436.

[14] W. N. Herfort and L. Ribes, On automorphisms of free pro-p-groups, *Proc. Amer. Math. Soc.*, **108** (1990), 287–295.

[15] W. N. Herfort, L. Ribes and P. Zalesskii, Fixed points of automorphisms of free pro-p groups of rank 2, *Can. J. Math.*, **47** (1995), 383–404.

[16] W. N. Herfort, L. Ribes and P. Zalesskii, Finite extensions of free pro-p groups of rank at most two, *Israel J. Math.*, **107** (1998), 195–227.

[17] W. N. Herfort, L. Ribes and P. Zalesskii, p-Extensions of free pro-p groups, *Forum Mathematicum*, **11** (1999), 49–61.

[18] A. Lubotzky, Combinatorial group theory for pro-p groups, *J. Pure Appl. Math.*, **25** (1982), 311–325.

[19] R. C. Lyndon and P. E. Schupp, *Combinatorial Group Theory*, Springer, Berlin, 1977.

[20] S. Mac Lane, *Homology*, Springer, Berlin, 1963.

[21] S. W. Margolis and J. E. Pin, New results on the conjecture of Rhodes and on the topological conjecture, *J. Pure Appl. Alg.*, **80** (1992), 305–313.

[22] W. S. Massey, *Algebraic Topology: An Introduction*, Harcout, Brace & World, New York, 1967.

[23] O. V. Mel'nikov, Projective limits of free profinite groups, *Dokl. Acad. Sci. BSSR*, **24** (1980), 968–970.

[24] O. V. Mel'nikov, Subgroups and homology of free products of profinite groups, *Math. USSR Izvestiya*, **34** (1990), 97–119.

[25] J. Neukirch, Freie Produkte pro-endlicher Gruppen und ihre Kohomologie, *Arch. Math.*, **22** (1971), 337–357.

[26] J. E. Pin, On a conjecture of Rhodes, *Semigroup Forum*, **39** (1989), 1–15.

[27] J. E. Pin and C. Reutenauer, A conjecture on the Hall topology for the free group, *Bull. London Math. Soc.*, **23** (1991), 356–362.

[28] L. Ribes, On amalgamated products of profinite groups, *Math. Z.*, **123** (1971), 356–364.

[29] L. Ribes and P. Zalesskii, On the profinite topology on a free group, *Bull. London Math. Soc.*, **25** (1993), 37–43.

[30] L. Ribes and P. Zalesskii, The pro-p topology of a free group and algorithmic problems in semigroups, *I. J. Alg. and Comp.*, **4** (1994), 359–374.

[31] L. Ribes and P. Zalesskii, *Profinite Groups*, Springer, Berlin, to appear, 2000.

[32] L. Ribes and P. Zalesskii, *Profinite Trees*, Springer, Berlin, to appear.

[33] C. Scheiderer, *Real and étale Cohomology*, Lect. Notes Math. **1588**, Springer, Berlin, 1994.

[34] C. Scheiderer, The structure of some virtually free pro-p groups, *Proc. Amer. Math. Soc.*, **127** (1999), 695–700.

[35] J-P. Serre, Sur la dimension cohomologique des groupes profinis, *Topology*, **3** (1965), 413–420.

[36] J-P. Serre, *A Course in Arithmetic*, Springer, Berlin, 1973.

[37] J-P. Serre, *Trees*, Springer, Berlin, 1980.

[38] J-P. Serre, *Galois Cohomology*, Springer, Berlin, 1997.

[39] J. Stallings, Topology of finite graphs, *Inv. Math.*, **7** (1983), 551–565.

[40] J. S. Wilson, *Profinite Groups*, Clarendon Press, Oxford, 1998.

[41] P. Zalesskii and O. Mel'nikov, Subroups of profinite groups acting on trees, *Math. USSR Sbornik*, **63** (1989), 405–424.

Luis Ribes
Dept. of Mathematics and Statistics
Ottawa, Ont, Canada K1V 8N2
emails: lribes@math.carleton.ca

Pavel Zalesskii
Departamento de Matemática-IE
BR-70910-900 Brasília-DF, Brazil
pz@mat.unb.br

4

Just Infinite Branch Groups

R. I. Grigorchuk

Introduction

In this chapter we define and investigate the class of branch groups. This class has only recently been defined, but the first examples of groups with a branch structure appeared at the end of the 1970s and the beginning of the 1980s ([45] [13] [14], [24]).

By a *branch group* we mean a group G which has a descending sequence of normal subgroups H_n of finite index with trivial intersection, such that $H_n = L_n \times \cdots \times L_n$ is a factorization of H_n as a product of finitely many copies of a group L_n, the factorization for H_n subdivides the factorization for H_{n-1} for each $n = 1, 2, \ldots$, and the group G acts transitively by conjugation on the set of factors. This is one version of the definition. A second version involves the action of G on an appropriate spherically homogeneous rooted tree, and the branch structure given by the pair $(\{H_n\}, \{L_n\})$ is defined in terms of this action.

Branch groups behave very similarly to the full automorphism group of such a tree. We develop in parallel both the theory of (abstract) branch groups and the theory of profinite branch groups.

The class of branch groups is important because:

1. There are many examples of groups from this class with remarkable algebraic properties.

2. It is one of three classes (two in the profinite case) that naturally partition the class of all just infinite groups.

Just infinite groups (i.e., infinite groups all of whose proper quotients are finite) are one of the main topics of this book, so we will elaborate on the last statement. It

The author thanks the Swiss National Science Foundation and the London Mathematical Society for support.

is known that every finitely generated infinite group can be mapped onto a just infinite group, and that every finitely generated infinite virtually pro-p group can be mapped onto a just infinite pro-p group (see Section 6). This means that if there is an infinite finitely generated group possessing some property which is preserved under passage to homomorphic images (such as being torsion, or of finite width etc.), then there must be a group with that property in the class of finitely generated just infinite groups.

It is of great importance that the class of just infinite groups splits into three subclasses: branch groups, groups containing a normal subgroup of finite index of the form L^d where L is a simple group, and groups containing a normal subgroup of finite index of the form L^d, where L is a hereditarily just infinite group, that is a residually finite group in which every subgroup of finite index is just infinite. In the profinite category the situation is even more straightforward because simple groups do not appear as an option.

Examples of hereditarily just infinite groups include the infinite cyclic group, the infinite dihedral group, the groups $\mathrm{SL}(n, \mathbb{Z})$ factored by their centres when $n > 2$ (a theorem of Mennicke [36]), and lattices in higher rank semisimple Lie groups factored by their centres (a theorem of Kazhdan and Margulis [35], Th. 2.6). Hereditarily just infinite profinite groups can be produced by taking profinite versions of the above examples. There are also sporadic examples such as the Nottingham group and related groups constructed by I. Fesenko (see Chapters 6 and 10). Even so, we are very far from understanding which other groups belong to this class.

The simplest examples of just infinite branch groups are produced with the help of finite automata. Another natural realization of them is as groups acting on a rooted tree.

The aim of this chapter is to give a construction of just infinite branch groups (we shall call it the *main construction*), and to investigate some concrete examples more carefully. The chapter can also be viewed as an introduction to the theory of branch groups.

As we mentioned above, besides the fact that many branch groups are just infinite, the class of branch groups contains groups with many other distinctive properties. This class is associated with the General Burnside Problem since the simplest known examples of infinite finitely generated torsion groups fall into this class ([13], [14] [24]). It is also associated with the Restricted Burnside Problem because it provides examples of residually finite torsion groups with very slow growth of element orders (a recent result of A. Rozhkov [39]).

The first examples of groups of intermediate growth were constructed in the class of branch groups ([14], [15]), thus solving Milnor's problem. This also gave new examples of amenable groups and eventually permitted the construction of a finitely presented amenable but not elementary amenable group [18].

Many branch groups have properties that classify them as fractal or self-similar groups. One of the ways to introduce the concept of fractalness in group theory is via the notion of *Hausdorff dimension*, first used in the context of profinite groups by J. Abercrombie [1] and by Y. Barnea and A. Shalev [2] (see Chapter 1). In Section 15 we compute the Hausdorff dimension of a profinite branch group.

Some of these groups have a remarkable finite width property, and thus provide the answer to a question of E. Zelmanov [3].

Branch pro-p groups, which are not virtually torsion-free, contain a copy of every countably based pro-p group [21]. This gives a new approach to the solution of some well-known problems in the theory of profinite groups.

There is some hope that branch groups can open a new page in the theory of number fields (Chapter 11).

This chapter is organized as follows. In Sections 1–3 we start with the definition of a spherically homogeneous rooted tree and its automorphism group viewed as a profinite group. In Section 2 we associate several important subgroups to every group acting on a rooted tree, for example the stabilizers and rigid stabilizers of vertices and levels. We also mention the simplest properties of these subgroups.

We provide a partial classification of automorphisms of a rooted tree, the most important types being rooted automorphisms, directed automorphisms and automatic automorphisms (automorphisms defined with the help of a finite state automaton; Section 4).

In Section 5 we define the class of branch groups and give the first examples of finitely generated branch groups. The proof of the above-mentioned trichotomy concerning the structure of finitely generated just infinite groups is given in Section 6. Section 7 contains a criterion to detect when a branch group is just infinite. With the help of this criterion we supply the first examples of just infinite finitely generated profinite branch groups.

The main construction of branch groups is described in Section 8. The principal idea is to consider a group generated by several rooted automorphisms and several directed automorphisms with a common path for the direction. At the present time all known examples of branch groups are of this type. There is some reason to hope that this is the only way to produce branch groups, and if this can be proved, it will be a great step forward in understanding the properties of branch groups.

Section 9 contains a sufficient condition and a criterion to determine when a group given by the main construction is an infinite torsion group. This provides uncountably many examples of infinite finitely generated branch p-groups for every prime p.

In Section 10 we define the congruence property for a group acting on a rooted tree and apply it to the construction of examples of just infinite profinite branch groups. These groups are profinite completions of (abstract) just infinite branch groups and the congruence property allows us to consider them as the closure of the corresponding subgroup in $\operatorname{Aut} T$.

In Section 11 we show that some just infinite branch pro-p groups have a universal embedding property: every countably based pro-p group can be embedded in a given one.

The final four sections deal with the 3-generator 2-group G constructed in [13], which has all the properties mentioned above. It has intermediate growth, is just infinite, has branch structure, enjoys the congruence property and has finite width. Its profinite completion is a just infinite branch pro-2 group with the universal embedding property.

We begin Section 12 with general information about this group. In particular some important subgroups of G are defined and investigated. Then in Section 13 we describe the derived series of G and in Section 14 we describe the lower central series of G. In particular we prove that the group G has finite width, in the sense that the ranks of the quotients $\gamma_n(G)/\gamma_{n+1}(G)$, which are elementary abelian 2-groups in this case, are uniformly bounded by 2 when $n > 1$. Further properties of this group are discussed in [28], [22], and [10].

Finally in Section 15 we show that the profinite completion $\widehat{G}$ of G has Hausdorff dimension 5/8 as a subgroup of the full automorphism group of a binary tree.

Let us conclude the introduction with some historical remarks.

The trichotomy mentioned above is based on a remarkable result of J. S. Wilson [48] (see also Chapter 5), who classified just infinite groups according to the structure of the lattice $\mathcal{L}$ of subnormal subgroups. It was proved in [48] that the so-called *structure lattice* $\mathcal{L}/\sim$ is isomorphic either to the Boolean lattice of all subsets of some finite set, or else to the lattice of open and closed subsets of a Cantor set (here $\sim$ denotes the equivalence relation of commensurability). Simple groups and hereditarily just infinite groups are the building blocks from which any just infinite group with finite structure lattice can be constructed. On the other hand, the class of just infinite groups with infinite structure lattice is precisely the class of branch groups.

Wilson unified the class of branch groups by this characterisation in terms of the structure lattice. Our aim in this chapter is to demonstrate the variety and richness of this class of groups.

During the last 20 years, the theory of (abstract) branch groups was developed in connection with the theory of groups generated by finite automata ([16], [8], [9], [20]), in relation to problems of Burnside type as in [45], [13], [14], [15], [24], [25], [26], [39] or problems on growth in groups, as in [14], [15], [17], [23].

Profinite just infinite branch groups have come under systematic investigation only recently [21].

The definition of branch groups in the context of actions on spherically homogeneous rooted trees was first given by the author at conferences on group theory in Bath, England (July 1997) and Slavyansk, Ukraine (August 1997); see also [19].

The fact that many groups with branch structure have the property of being just infinite was discovered in [25], [15].

The link between just infinite branch groups generated by finite automata and Wilson's class of just infinite groups with infinite structure lattice was indicated in [15], where the answer was given to a question of S. J. Pride ([40], Problem 5).

Results related to the subject of this chapter may be found in the following papers, as well as those cited above and later: [10], [28], [41], [42], [43], [44].

As a matter of notation, we generally write G' to denote the commutator subgroup of a group G. When G is a profinite group, G' denotes the closed subgroup generated by all commuttors in G. The symbol $\prod$ means Cartesian or unrestricted direct product; the (restricted) direct product of a family of groups $(G_i)_{i\in I}$ is denoted $\mathrm{Dr}_{i\in I}G_i$.

1 Spherically homogeneous rooted trees

Let T be a tree with a distinguished vertex $\emptyset$, called the *root vertex*. The set of vertices of the tree is written $V(T)$.

The distance of a vertex u from the root vertex is $|u|$, the *norm* of u. This is the number of edges in a geodesic path from $\emptyset$ to u. Clearly $|u| = 0$ if and only if $u = \emptyset$.

The cardinality of the set of vertices which are both adjacent to u and of norm $|u| + 1$ is called the *degree* of u. The tree is called *spherically homogeneous* if vertices of the same norm have the same degree.

Let V_n denote the set of vertices of norm n. This set is called the nth level or nth layer of T.

Clearly, a spherically homogeneous tree T is determined by the sequence $\overline{m} = \{m_n\}_{n=1}^{N}$ of natural numbers, where N may be finite or infinite depending on T, and m_n is the degree of the vertices in level $n-1$. We use the notation $T = T_{\overline{m}}$. The number m_n is the *branch degree* of the tree at level $n-1$. The sequence $\overline{m}$ is therefore called the *branch index*. We will only consider trees where m_n is finite and $m_n \geq 2$ for every n.

The case $m_n = d \geq 2$ for every n is particularly important. In this case we write $T_{\overline{m}} = T_d$. This is the rooted regular tree of degree d; when $d = 2$ it is called the *binary tree*.

For convenience of exposition, we often describe a tree by a geometric picture, for example as in Diagram 1.

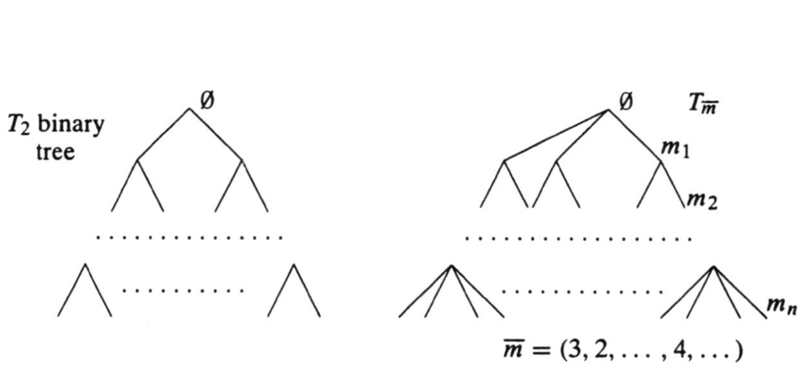

Diagram 1

The number of vertices in the nth layer of the tree $T_{\overline{m}}$ is clearly the number $N_n = m_1 m_2 \dots m_n$.

We understand a *path* in T to be a sequence of adjacent edges which starts at Ø and such that each edge occurs at most once (equivalently, there are no reversals). If T is infinite, we allow infinite paths, and the set of all of them is written ∂T and is called the *boundary* or *ends space* of T.

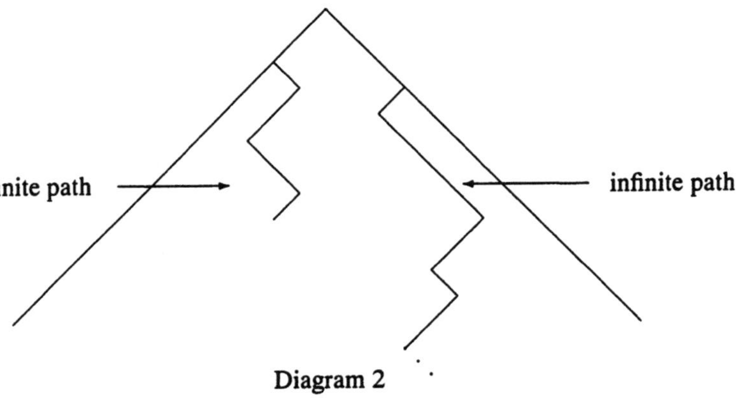

Diagram 2

To each sequence $\overline{m} = \{m_n\}$ we associate a sequence (A_n) of alphabets where $A_n = \{a_1^{(n)}, \ldots, a_{m_n}^{(n)}\}$, $n = 1, 2, \ldots$, so that $|A_n| = m_n$. A path of length n in $T_{\overline{m}}$ is identified with the sequence $(a_1, \ldots, a_i, \ldots, a_n)$ where $a_i \in A_i$, and infinite paths are identified in a natural way with infinite sequences.

In the case of a regular tree, it is more convenient to use a single alphabet $A = \{a_1, \ldots, a_d\}$ serving all levels. In the sequel we will, as a rule, use $A = \{0, 1, \ldots, d-1\}$.

Having introduced the alphabets A_n we can impose a lexicographic ordering on the set of all paths: comparison is made first by length, and if path lengths are equal, via the natural order of the first pair of non-coincident letters of the sequences defining the paths. Since the finite paths and the vertices are in a natural bijective correspondence, we have an induced linear order on the vertex set. Geometrically, this means ordering the vertices by level, and among vertices of the same level, ordering from left to right.

The subtree of T containing only the vertices from levels 0 to n is denoted $T_{[n]}$, and we write T_u for the subtree of T which has its root at $u \in V(T)$. Clearly $T_u \cong T_v$ if $|u| = |v|$. We can therefore use the notation $T_{\langle n \rangle}$ for any tree isomorphic to a subtree which begins at level n. Obviously $T_{\langle n \rangle}$ is infinite if T is infinite, and $T_{\langle n \rangle}$ is isomorphic to $T_{\sigma^n \overline{m}}$, where σ is a shift to the left in the space of sequences, in other words $(\sigma \overline{m})_n = m_{n+1}$.

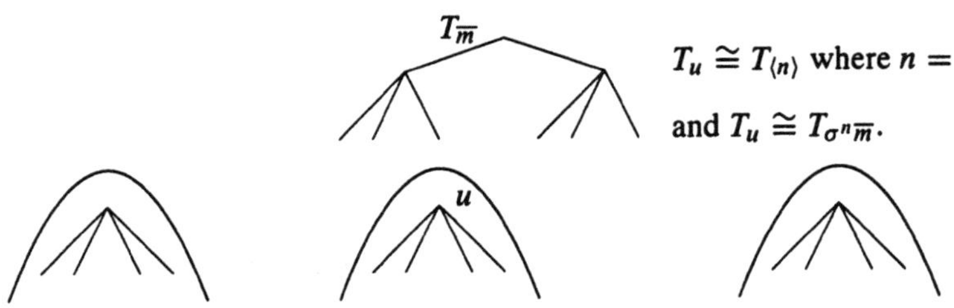

Diagram 3

2 The automorphism group of a tree and some of its subgroups

We write $\operatorname{Aut} T$ for the group of automorphisms of the rooted tree T. This is the set of bijections from $V(T)$ to $V(T)$ that preserve edge incidence and the distinguished root vertex $\emptyset$. Obviously the vertices of each level are permuted among themselves by $\operatorname{Aut} T$.

Any $g \in \operatorname{Aut} T$ can be regarded as a *labelling* of the vertices by elements of the symmetric group which acts on the edges (or vertices) below the vertex; in symbols $\{g(u)\}_{u \in V(T)}$ where for a vertex u of level n we have $g(u) \in \Sigma_{m_{n+1}}$, the symmetric group of degree m_{n+1}. The effect of $g \in G$ on a vertex u and a path p is denoted u^g and p^g, respectively. We write $\operatorname{Fix}(g)$ for the set of vertices fixed by g.

The effect of $g \in \operatorname{Aut} T$ on a vertex u, or what amounts to the same thing, on the path to that vertex $p_u = (a_1, a_2, \ldots, a_n)$, $a_i \in A_i$, is given by the formula

$$p_u^g = \left(a_1^{g(\emptyset)}, a_2^{g(a_1)}, \ldots, a_n^{g(a_1 \ldots a_{n-1})}\right) \tag{1}$$

and this formula extends in a natural way to an action on infinite paths:

$$(a_1, \ldots, a_n, \ldots)^g = \left(a_1^{g(\emptyset)}, a_2^{g(a_1)}, \ldots, a_n^{g(a_1 \ldots a_{n-1})}, \ldots\right). \tag{2}$$

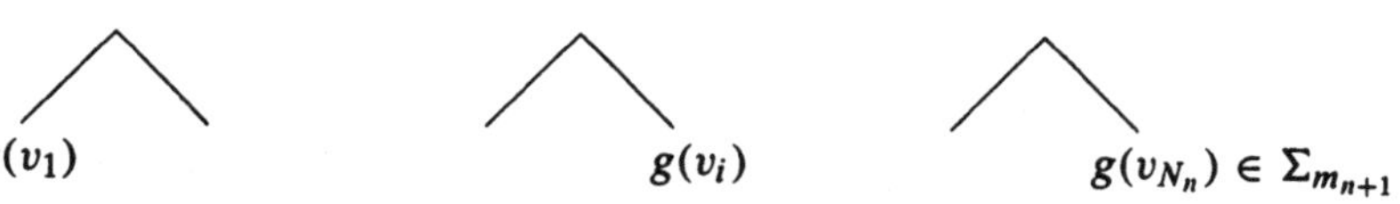

Diagram 4

The rule for composing automorphisms is given by the following statement.

Proposition 1.

(a) *Let $h = fg$, where $f, g, h \in \operatorname{Aut} T$. Then $h(u) = f(u)g(u^f)$ for all $u \in V(T)$.*

(b) *Let $f = g^{-1}$ where $g \in \operatorname{Aut} T$. Then $f(u) = (g(u^{g^{-1}}))^{-1}$ for all $u \in V(T)$.*

Proof. (a) Let $u = (a_1, a_2, \ldots, a_n)$. We have

$$\begin{aligned} u^{fg} &= \left(a_1^{f(\emptyset)}, a_2^{f(a_1)}, \ldots, a_n^{f(a_1,\ldots,a_{n-1})}\right)^g \\ &= \left(a_1^{f(\emptyset)g(\emptyset)}, \left(a_2^{f(a_1)}\right)^{g(a_1^{f(\emptyset)})}, \ldots, \left(a_n^{f(a_1,\ldots,a_{n-1})}\right)^{g(a_1^{f(\emptyset)},\ldots,a_{n-1}^{f(a_1,\ldots,a_{n-2})})}\right) \\ &= \left(a_1^{f(\emptyset)g(\emptyset)}, a_2^{f(a_1)g(a_1^{f(\emptyset)})}, \ldots, a_{n-1}^{f(a_1,\ldots,a_{n-1})g(a_1^{f(\emptyset)},\ldots,a_{n-1}^{f(a_1,\ldots,a_{n-2})})}\right), \end{aligned}$$

and this implies the relation $h(u) = f(u)g(u^f)$.

(b) This follows from (a) in an obvious manner, since if $f = g^{-1}$, then $(fg)(u) = g^{-1}(u)g(u^{g^{-1}}) = 1$. □

Intuitively, one can imagine that the automorphism g acts on T in a neighbourhood of the vertex u via the permutation $g(u)$ of the edges below u. For example, in the case of a ternary tree T_3, the automorphism b cyclically permutes the the three subtrees which begin at the first level, and is defined by the labelling in Diagram 5(a). Here $\varepsilon = (1, 2, 3) \in \Sigma_3$ is a cyclic permutation and $1 \in \Sigma_3$ is the identity element. On the other hand, the automorphism c defined in Diagram 5(b) acts via cyclic permutations at all the vertices.

In the sequel, when defining tree automorphisms by labellings, we adopt the convention that when the labelling of part of the diagram is not specified, then those vertices are to be labelled with the identity element.

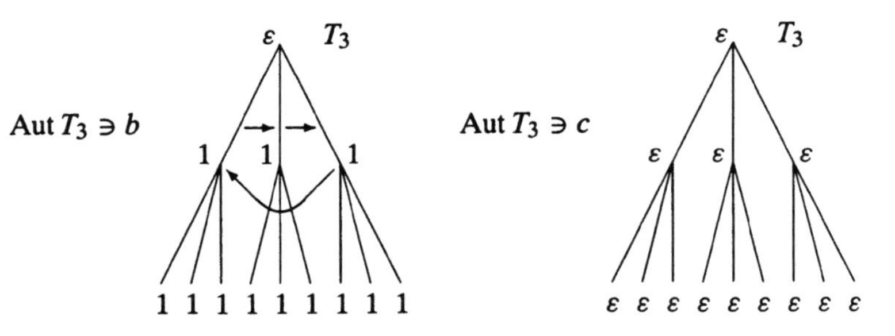

(a) Diagram 5 (b)

Let $g = \{g(u)\}_{u \in V(T)}$ be an automorphism of the tree T, and suppose that v is a vertex. Consider the automorphism $g_{\{v\}}$ of T_v whose labelling is the restriction to vertices in T_v of the labelling of g. That is $g_{\{v\}}(w) = g(w)$ if $w \in V(T_v)$. Note that v need not be a fixed point of g.

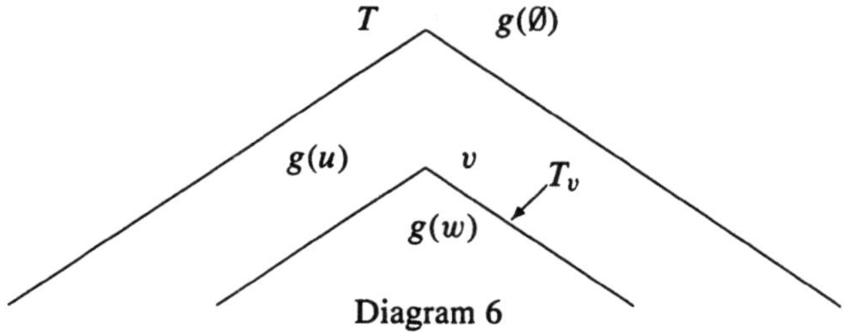

Diagram 6

The automorphism $g_{\{v\}}$ is called a *slice* or *section* of g at v. Moreover, formula (1) can be written as

$$(pq)^g = (p^g q^{g\{p\}}), \tag{3}$$

where $pq = a_1, a_2, \ldots, a_n$ is a sequence of symbols from the alphabets A_i, $i = 1, \ldots, n$ (we remind the reader that we identify the vertices with the finite paths, and the paths with the corresponding sequences of symbols).

In addition, we have the following relations:

$$(gf)_{\{v\}} = g_{\{v\}} \cdot f_{\{v^g\}}, \quad (g^{-1})_{\{v\}} = \left(g_{\{v^{g^{-1}}\}}\right)^{-1}, \tag{4}$$

viewed as holding in Aut $T_{\langle n\rangle}$ where $n = |v|$. Note that if $g \in G \leq \operatorname{Aut} T$ and T_v is isomorphic to T, then the slice $g_{\{v\}}$ is not necessarily in G.

Lemma 1. *Suppose that $g, h \in \operatorname{Aut} T$ and $u \in \operatorname{Fix} g$. Then $u^h \in \operatorname{Fix} g^h$ and $(g^h)_{\{u^h\}} = (g_{\{u\}})^{h_{\{u\}}}$.*

Proof. Using the relation (4) we get

$$\begin{aligned}(h^{-1}gh)_{\{u^h\}} &= h^{-1}_{\{u^h\}} g_{\{u\}} h_{\{u^g\}} \\ &= (h_{\{u\}})^{-1} g_{\{u\}} h_{\{u\}}.\end{aligned}$$

as claimed. □

With any group $G \leq \operatorname{Aut} T$ we associate the subgroups $\operatorname{St}_G(u) = \{g \in G \mid u^g = u\}$, the stabilizer of a vertex u, and, if T is infinite, the stabilizer $\operatorname{St}_G(e)$ of an end $e \in \partial T$.

In subsequent considerations, an important role will be played by the *level stabilizers* $\operatorname{St}_G(n)$ consisting of those $g \in G$ that fix all vertices in the nth level ($n = 1, 2, \ldots$). Clearly

$$\operatorname{St}_G(n) = \bigcap_{|u|=n} \operatorname{St}_G(u),$$

and $\operatorname{St}_G(n)$ is a normal subgroup of G of finite index.

Any $g \in \operatorname{St}_G(n)$ can be identified in a natural way with the collection $g_1, \ldots, g_{N_n}$ of elements of Aut $T_{\langle n\rangle}$, where $g_i = g\,|_{T_u}$ is the restriction of g to the subtree with root vertex u in level n having the number i in the natural ordering of the vertices in the nth level ($1 \leq i \leq N_n$). We depict this situation by writing

$$g = (g_1, \dots, g_{N_n})_n, \tag{5}$$

and often omitting the subscript to the right bracket. Relation (5) also defines an embedding

$$\psi_n : \mathrm{St}_G(n) \to \mathrm{Aut}\, T_{u_1} \times \cdots \times \mathrm{Aut}\, T_{u_{N_n}}.$$

If $n = 1$, then the subscript n attached to ψ will be omitted. Thus ψ is the embedding

$$\psi : \mathrm{St}_G(1) \to \mathrm{Aut}\, T_{\langle 1 \rangle} \times \cdots \times \mathrm{Aut}\, T_{\langle 1 \rangle}$$

with m_1 factors.

Another important class of subgroups associated with $G \leq \mathrm{Aut}\, T$ are the *vertex groups*, also called the *rigid stabilizers of vertices*, and the *rigid stabilizers of levels*.

The *vertex group* at the vertex $u \in V(T)$ is

$$\mathrm{top}_G(u) = \{g \in G \mid v^g = v \text{ for all } v \in V(T) \setminus V(T_u)\}.$$

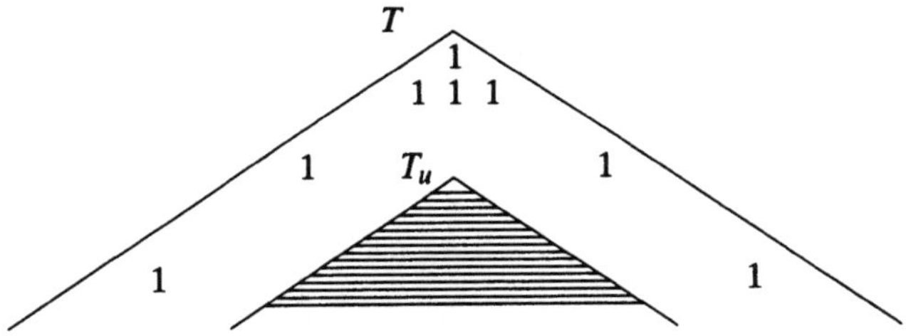

Diagram 7

(In some papers the notation $\mathrm{rist}_G(u)$ or $\mathrm{cost}_G(u)$ is used instead of $\mathrm{top}_G(u)$.)

The elements of $\mathrm{top}_G(u)$ may be characterized as those elements of G which have trivial labelling (labelling by the identity) at all vertices not in T_u. The reason for the notation should be clear from Diagram 7.

Consider the vertex groups $\mathrm{top}_G(u_1), \dots, \mathrm{top}_G(u_{N_n})$ corresponding to the vertices at the nth level. Also consider the group they generate, which is written $\mathrm{rist}_G(n)$, and is called the *rigid stabilizer* of the nth level. Clearly $\mathrm{rist}_G(n)$ is a normal subgroup of G, and we have a direct product decomposition

$$\mathrm{rist}_G(n) = \mathrm{top}_G(u_1) \times \cdots \times \mathrm{top}_G(u_{N_n}), \tag{6}$$

as can be seen from Diagram 8 below.

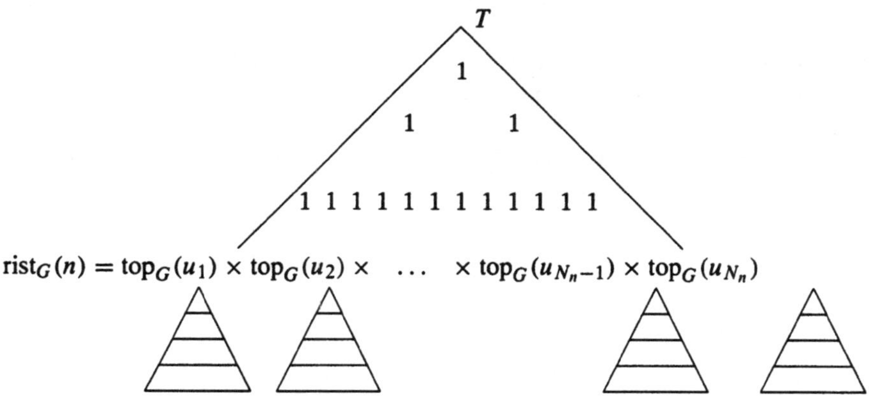

Diagram 8

Let $G \leq \mathrm{Aut}\, T$ be a group acting transitively on each layer (we will say that G acts *spherically transitively*). This condition is very important and will occur frequently throughout the chapter. Clearly, for every $n \geq 1$, the nth level vertex stabilizers $\mathrm{St}_G(u_1), \dots , \mathrm{St}_G(u_{N_n})$ are conjugate in G by Lemma 1. Let G_n denote the (common) isomorphism type of the restrictions $\mathrm{St}_G(u_i) \mid_{T_{u_i}}$ of these groups to the corresponding subtrees. The following lemma is useful for recognizing spherically transitive actions; the proof is self-evident:

Lemma A.

(a) *G acts spherically transitively if and only if it acts transitively on the first layer and for some (and hence every) vertex u of the first level, the subgroup $G_1 = St_G(u) \mid_{T_u}$ acts spherically transitively on the subtree T_u.*

(b) *G acts spherically transitively, if and only if for some k, it acts transitively on the kth layer and for some (and hence every) vertex u of the kth level, the subgroup $G_k = St_G(u) \mid_{T_u}$ acts spherically transitively on the subtree T_u.*

Similarly, we write L_n^G for the common isomorphism type of the nth level vertex groups $\mathrm{top}_G(u_1), \dots , \mathrm{top}_G(u_{N_n})$, which are also conjugate in G.

We have the following natural embeddings: $L_n^G \leq G_n$ and

$$L_n^G \times \cdots \times L_n^G \cong \mathrm{rist}_G(n) \leq \mathrm{St}_G(n) \hookrightarrow G_n \times \cdots \times G_n \tag{7}$$

(each product contains N_n factors, corresponding to the vertices in the nth layer of T).

We will call L_n^G and G_n the *lower* and *upper companion groups* of G, respectively.

Now $G \leq \mathrm{Aut}\, T$ is called *self-reproducing* if T is a regular tree and the upper companion groups G_n coincide with G for every n.

When $G = \mathrm{Aut}\, T$, the embeddings (7) become isomorphisms, and we have

$$L_n^{\mathrm{Aut} T} = (\mathrm{Aut}\, T)_n = \mathrm{Aut}\, T_{(n)};$$

that is, the lower and upper companion groups coincide with the automorphism group of a subtree $T_{\langle n\rangle}$.

We note that there are examples of groups that act transitively on each layer, but for which the lower companion groups L_n^G are all trivial. For example, this is the case for the group $G = \langle x, y\rangle$ generated by automorphisms of the binary tree defined in Diagram 9(a) and (b), which is isomorphic to the infinite dihedral group (it will be discussed in more detail in Section 6):

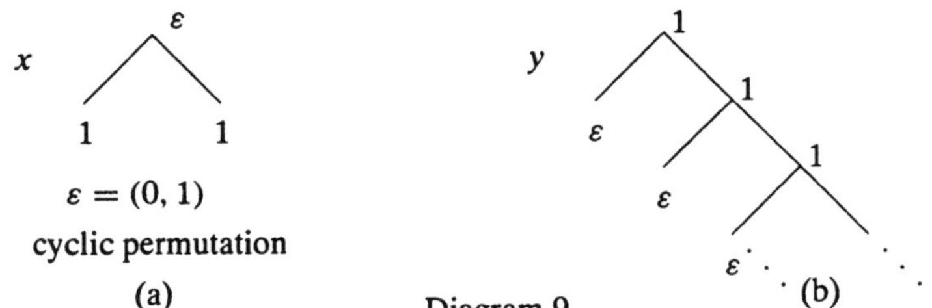

Diagram 9

We will be interested in the opposite case, namely when L_n^G has finite index in G_n for every n, and so the inclusions in (7) are inclusions of subgroups of finite index. In particular, $\mathrm{rist}_G(n)$ then has finite index in G. Such groups will be called *branch groups*; the precise definition is given in Section 5.

In the case where T is a regular tree, the groups considered later have the property that for each vertex u and each automorphism $g \in G$, the section $g_{\{u\}}$ becomes an element of G when $T_{\{u\}}$ is identified with T. In particular, this implies that the upper companion groups G_n $(n = 1, 2, \dots)$ coincide with G, which means that G is self-reproducing.

3 Aut T as a profinite group

We put a metric space structure on the boundary ∂T of the infinite tree $T = T_{\overline{m}}$. Specifically, if $p = (a_1, a_2, \dots)$ and $q = (b_1, b_2, \dots)$ are two infinite paths, let $\delta(p, q) = \frac{1}{2^k}$ where k is the largest natural number (or $+\infty$) such that $a_1 = b_1, \dots, a_k = b_k$ all hold. Then ∂T becomes a metric space, homeomorphic to the Cantor set.

The group Aut T acts on ∂T by isometries, in fact Aut T = Isom ∂T.

We now define a class of metrics on Aut T. Let $\{\lambda_n\}_{n=1}^{\infty}$ be a descending sequence of positive numbers with limit 0, and for $f, g \in \mathrm{Aut}\, T$, define $\delta(f, g) = \lambda_n$ where n is the largest number such that $f(u) = g(u)$ whenever $|u| \le n$. This metric induces a compact Hausdorff topology on Aut T, in which two elements are "close" if their actions coincide on a large subtree $T_{[n]}$. Thus Aut T is a profinite group, the topology of which does not depend on the choice of the sequence $\{\lambda_n\}$.

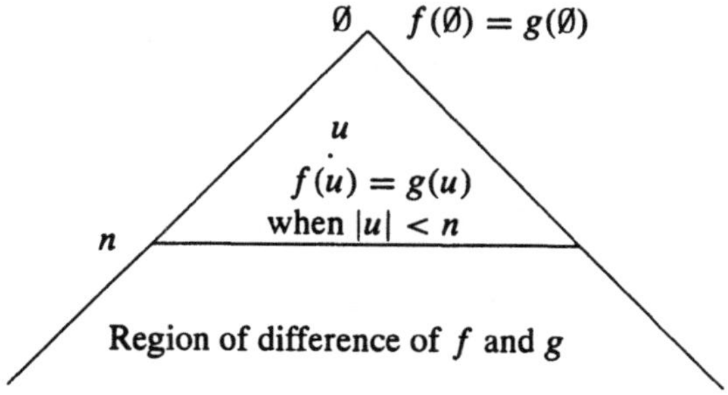

Diagram 10

The rule for the composition of automorphisms (1) shows that the metric δ is both right and left invariant.

Recall that $T_{[n]}$ is the subtree of T ending at level n. The fact that $\operatorname{Aut} T$ is a profinite group becomes more transparent if one observes that the sequence of groups $\operatorname{Aut} T_{[n]}$ $(n = 1, 2, \dots)$ forms a projective system with the homomorphisms $\operatorname{Aut} T_{[n+1]} \to \operatorname{Aut} T_{[n]}$ defined by restriction of the action of $\operatorname{Aut} T_{[n+1]}$ to $T_{[n]}$, and the group $\operatorname{Aut} T$ is the projective limit of this projective system:

$$\operatorname{Aut} T = \varprojlim_{n\to\infty} \operatorname{Aut} T_{[n]}.$$

The topology of this group defined by the system of open sets $\{\mathrm{St}_{\operatorname{Aut} T}(n)\}_{n=1}^{\infty}$ coincides with the topology defined by the metric δ.

Since $\operatorname{Aut} T$ acts transitively on the boundary, there is on ∂T a unique invariant probabilistic measure μ defined by the relations $\mu(\overline{a_1, \dots, a_n}) = 1/N_n$, where $\overline{a_1, \dots, a_n}$ is the cylindrical set consisting of all paths in ∂T which begin with the sequence $a_1, \dots, a_n$. In the sequel we will not need to use the measure μ.

The Sylow pro-p subgroups of $\operatorname{Aut} T$ are obtained as follows. For every $n \geq 1$ we fix a Sylow p-subgroup $S_n \leq \Sigma_{m_{n+1}}$. Then

$$S = \{g \in \operatorname{Aut} T : g(u) \in S_n \text{ whenever } |u| = n\}$$

is a Sylow pro-p subgroup of $\operatorname{Aut} T$, and moreover every Sylow pro-p subgroup is obtained in a similar way.

In the case of the regular tree $T = T_p$, there is a canonical way to construct a Sylow pro-p subgroup. As before, we use the alphabet $A = \{0, 1, 2, \dots, p-1\}$ for coding paths. Let $\varepsilon = (0, 1, \dots, p-1)$ be a cyclic permutation of A, and regard ε as being in the symmetric group Σ_p. Define $\operatorname{Aut}_* T_p \leq \operatorname{Aut} T_p$ to be the set of all elements whose labelling $\{g(u)\}$ takes values in $\langle \varepsilon \rangle \cong C_p$, a cyclic group of order p. Thus for every $u \in V(T)$, we have $g(u) = \varepsilon^{i(u)}$ for some $i(u)$ with $0 \leq i(u) \leq p-1$.

The group $\operatorname{Aut}_* T_p$ is isomorphic to the 'infinitely iterated wreath product'

$$\varprojlim_{n\to\infty} (C_p \wr C_p \wr \dots \wr C_p)_{n \text{ factors}},$$

and is a Sylow pro-p subgroup of Aut T.

In the sequel, most of our examples will be constructed inside this group.

It is well known that if T is finite, then Aut T is isomorphic to the iterated permutational wreath product

$$\Sigma_{m_1} \wr \Sigma_{m_2} \wr \ldots \wr \Sigma_{m_n}.$$

If also $m_1 = \cdots = m_n = p$, then every Sylow p-subgroup is isomorphic to the wreath product

$$C_p \wr C_p \wr \ldots \wr C_p.$$

The study of iterated wreath products was initiated by L. Kaloujnine in [30], [31] and continued by his students I. D. Ivanyuta [29], V. I. Sushchansky [46], Y. V. Bondarchuk and V. I. Sushchansky [7] and others, who studied restricted and unrestricted wreath products (see also P. Hall's paper [27]).

Not much is known about the structure and properties of Aut T. For some special cases the automorphism group of Aut T is described in [33]. There is a satisfactory description of the closed normal subgroups of Aut T given in [6] and [46]. We are going to mention one simple fact concerning closed subgroups of Aut T. Recall that a profinite group is said to be *countably based* if the topology of the group is defined by a countable system of neighbourhoods; these groups are characterized by the property that they are isomorphic to closed subgroups of Cartesian products of countably many finite groups. Every finitely generated profinite group belongs to this class.

Proposition 2.

(a) *Let $\overline{m}$ be an unbounded sequence of integers ≥ 2. Then every countably based profinite group is isomorphic to a closed subgroup of Aut $T_{\overline{m}}$.*

(b) *Every countably based pro-p group is isomorphic to a closed subgroup of the Sylow pro-p subgroup $\mathrm{Aut}_* T_p$ in Aut T_p.*

Proof. We first address part (a). Let G be a closed subgroup of the Cartesian product $K = \prod_{n=1}^{\infty} K_n$ of finite groups. We construct an embedding of K into Aut T where $T = T_{\overline{m}}$.

Let $(m_{i(n)})$ be a subsequence of $\overline{m}$ such that $m_{i(n)} \geq |K_n|$ for each n. For each n we fix an isomorphism of K_n with a subgroup $\tilde{K}_n$ of the symmetric group $\Sigma_{m_{i(n)}}$. For first vertex u in the $i(n)$th level, the group $\tilde{K}_n$ is realized as the subgroup of Aut T acting via permutations of the edges below u and trivially at all other vertices. Thus the Cartesian product

$$\prod_{n=1}^{\infty} \left(\tilde{K}_n \times \{1\} \times \ldots \times \{1\}\right) \cong \prod_{n=1}^{\infty} K_n \tag{8}$$

(with $m_{i(n)-1} - 2$ identity factors in the bracketed term on the left) is realized as a closed subgroup of Aut T (see Diagram 11), from which (a) follows.

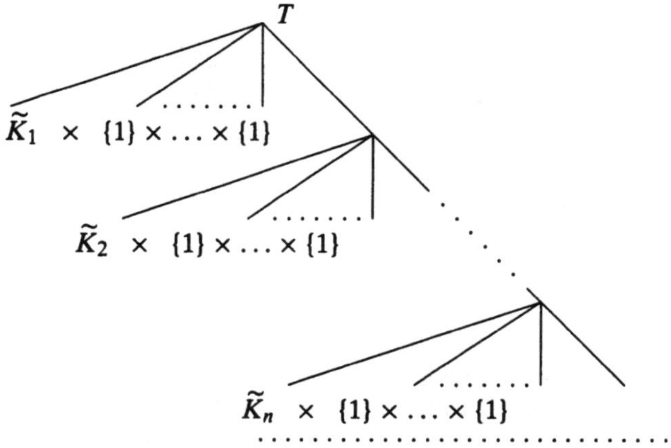

Diagram 11

Now we address part (b). The proof is analogous to that of part (a). However, we must now embed the finite p-groups K_n into iterated wreath products $W_n = C_p \wr \ldots \wr C_p$ of the appropriate number $i(n)$ of cyclic groups of order p (it is well known that such an embedding is possible).

For any $u \in V(T)$, where $T = T_p$, there is an embedding of W_n into $\operatorname{Aut} T$ which acts non-trivially only on the subtree T_u and such that for any element $g \in W_n$ and vertex $v \in V(T)$, the labelling g_v is a power ε^j of a cyclic permutation of the edges starting at v $(0 \leq j \leq p-1)$. Thus we may replace $\widetilde{K}_n$ by W_n in Diagram 11 to obtain an embedding of $\prod_{n=1}^{\infty} W_n$ into $\operatorname{Aut} T$. This completes the proof. □

In Section 11 it will be shown that some finitely generated subgroups of $\operatorname{Aut} T$ (namely some branch groups) have a similar universal embedding property.

4 Some classes of automorphisms of trees

In this section we will consider special classes of automorphisms which will be used in various constructions.

An automorphism $g \in \operatorname{Aut} T$, $T = T_{\overline{m}}$ an infinite rooted tree, is called a *rooted automorphism* if its labelling $\{g(u)\}$ is non-trivial only at the root vertex: $g(\emptyset) \neq 1$ and $g(u) = 1$ if $u \neq \emptyset$.

An automorphism $g \in \operatorname{Aut} T$ is said to be *finitary* if the support $\operatorname{supp} g(u)$ of its labelling (the set of vertices at which the labelling is non-trivial) is finite.

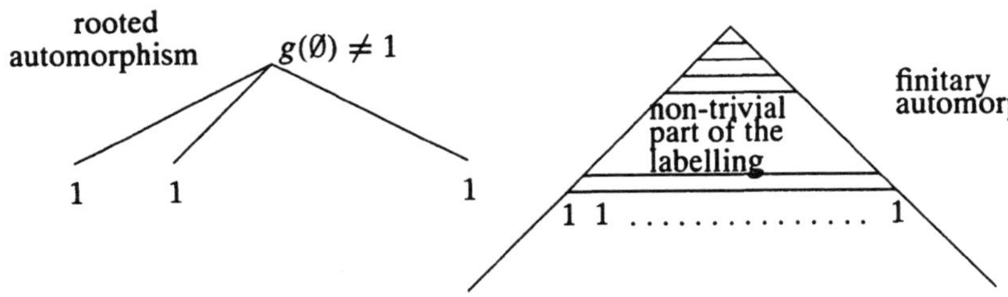

(a) (b)

Diagram 12

It is obvious that a rooted automorphism is finitary.

An automorphism $g \in \operatorname{Aut} T$ is called *directed*, with directing path $l \in \partial T$, if $g(u) = 1$ whenever u is not at distance 1 from l, and g is not finitary. Thus a directed automorphism is not finitary, leaves the path l invariant, and all its non-trivial labelling is concentrated in the neighbourhood of l.

An example of such an automorphism is given in the following diagram.

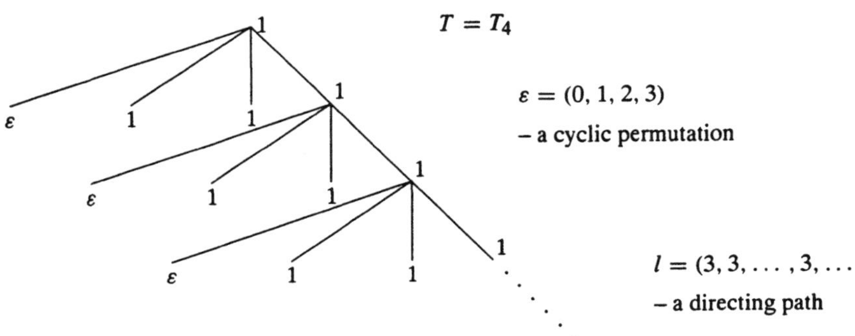

Diagram 13

Note that in future we will always use the rightmost path in the diagrams of trees as the directing path.

Especially interesting examples of groups acting on rooted trees are obtained from generators which are automorphisms of trees whose labelling is generated by finite automata. To be precise, an automorphism $g \in \operatorname{Aut} T$ of a regular tree $T = T_d$ is called *regular* or *automatic* if there is an automaton $\mathfrak{A}$ on the alphabet $A = \{0, 1, \dots, d-1\}$ with specified initial state s, and with states labelled by elements of the symmetric group Σ_d, which determines the labelling $\{g(u)\}$ in the following way: if $u = a_1 \dots a_n$, $a_i \in A$, then $g(u)$ is the label of the state of the automaton which is the end point of a path given by the input string $a_1 \dots a_n$ if it starts from the initial state.

We consider some examples. The rooted automorphism of T_d described in Diagram 14(a) (where $\varepsilon = (0, 1, \dots, d-1)$ is a cyclic permutation) is defined by the 2-state automaton shown in Diagram 14(b). The start state is distinguished by a broken circle.

(a) (b)

Diagram 14

The directed automorphism of T_d shown in Diagram 15(a) is determined by the 3-state automaton shown in Diagram 15(b).

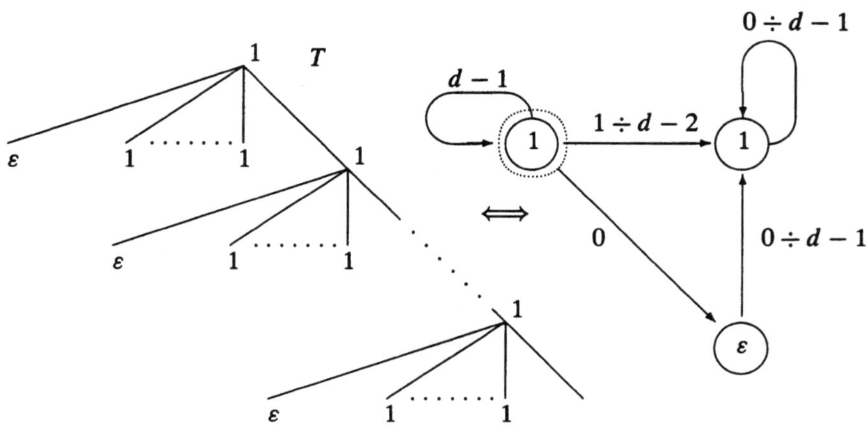

(a) Diagram 15 (b)

Similarly the directed automorphisms described in Diagrams 16(a) and 17(a) can be described by corresponding automata.

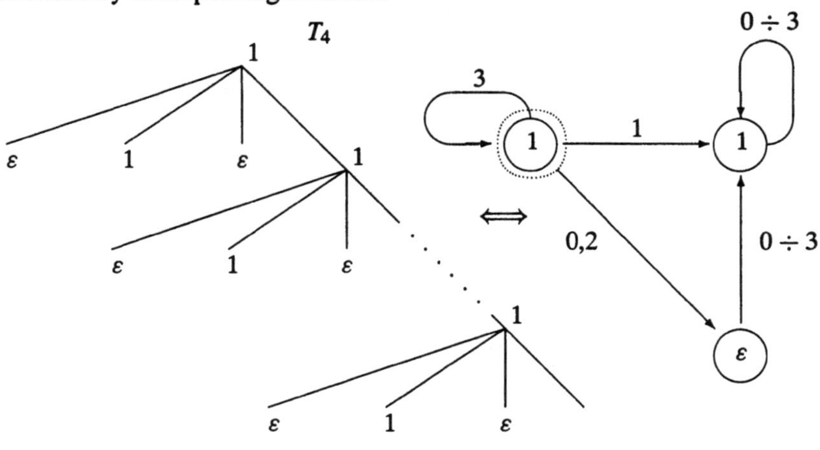

(a) (b)

Diagram 16

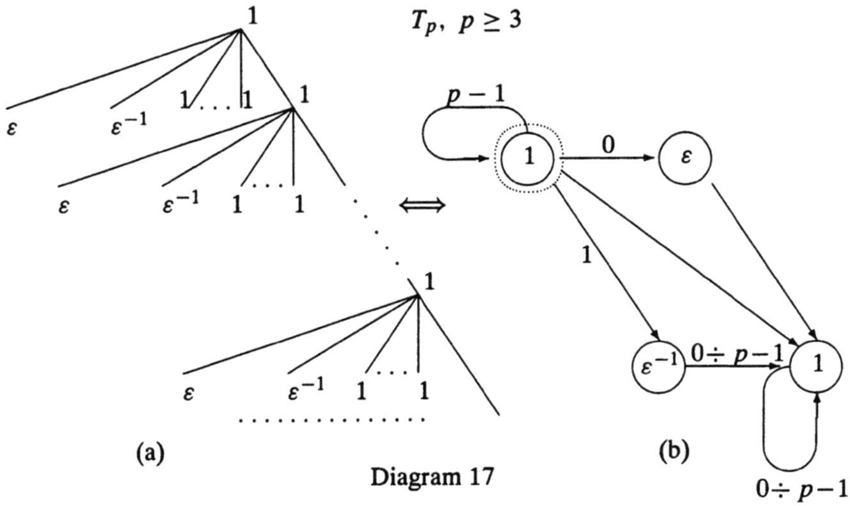

Diagram 17

Our final example is given by the directed automorphisms b, c, d specified in diagram 18; the corresponding group will be studied in Sections 12–15. These directed automorphisms of a binary tree are described simultaneously by the single automaton in Diagram 19. One chooses as the start state the state labelled by the appropriate letter.

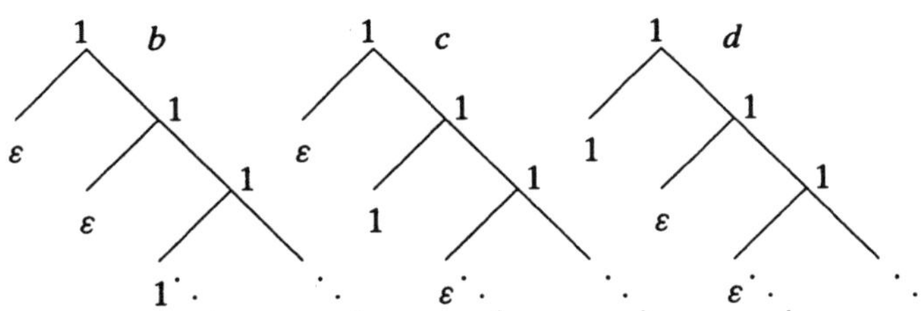

Diagram 18

(The sequences $\varepsilon\varepsilon 1$, $\varepsilon 1 \varepsilon$ and $1 \varepsilon\varepsilon$ are periodically repeated).

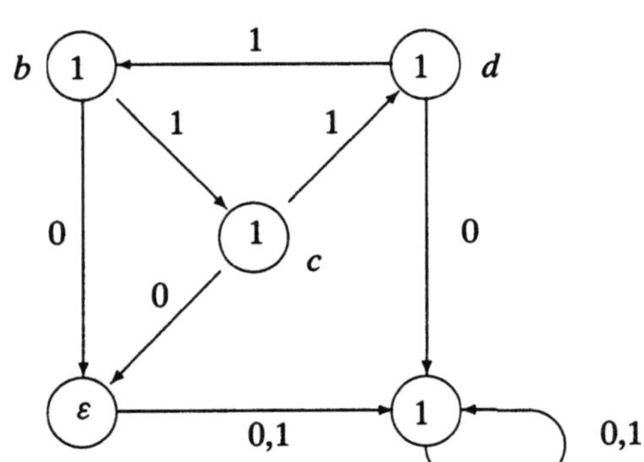

Diagram 19

The examples just given of automorphisms of trees and the corresponding automata are related to the infinite finitely generated p-groups constructed in [13] and [24]. To be precise, if we designate the rooted automorphism of the tree T_p defined by a p-cycle by a_p, and write f, g, b, c, d for the directed automorphisms defined in Diagrams 16, 17 and 18, then the following holds.

Theorem 1. *The groups $G = \langle a_2, b, c, d\rangle$ and $H = \langle a_4, f\rangle$ for $p = 2$, and $GS_p = \langle a_p, g\rangle$ where $p \geq 3$, are all infinite p-groups. In addition, the group G has intermediate growth.*

The periodicity of this group will follow from Theorem 7, proved below, which also provides many other examples of torsion groups.

The groups G and H were constructed in [13], and GS_p was constructed in [24]. They are also examples of *groups generated by finite automata*: the finite invertible automata form a group with respect to the operation of composition of automata, and the given groups are subgroups of this group.

We finish this section by constructing one more example of a tree automorphism called an *adding machine*. This plays an important role in the theory of groups acting on trees, and in the theory of dynamical systems [6].

Let $T = T_{\overline{m}}$, $\overline{m} = \{m_n\}_{n=1}^{\infty}$, and let $A_n = \{a_1^{(n)}, \dots, a_{m_n}^{(n)}\}$ be a sequence of alphabets associated with this tree. Let $\varepsilon_n = (a_1^{(n)}, a_2^{(n)}, \dots, a_{m_n}^{(n)})$ be a cyclic permutation of the letters of the alphabet A_n. The automorphism ξ defined by Diagram 20 is an adding machine.

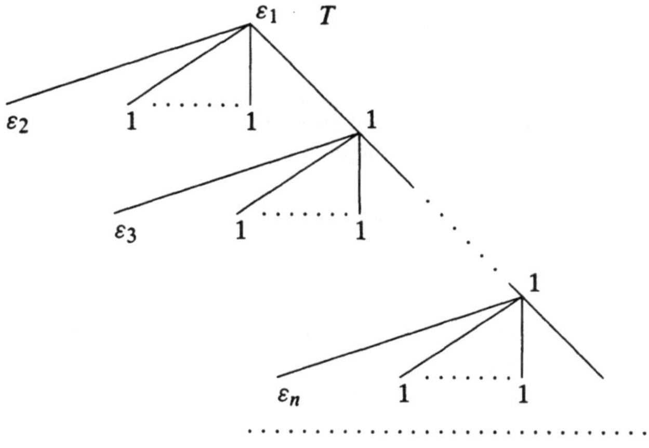

Diagram 20

It is not difficult to verify that ξ acts transitively on each layer of $T = T_{\overline{m}}$, and hence has infinite order. Moreover, any two elements $g, h \in \operatorname{Aut} T$ that act transitively on layers must be conjugate in $\operatorname{Aut} T$ (see [6]). A conjugacy criterion for elements of $\operatorname{Aut} T$ is given in [12].

The name "adding machine" can be explained by considering the ring

$$\widehat{\mathbb{Z}}_{\overline{m}} = \lim_{\overleftarrow{n \to \infty}} \mathbb{Z}/N_n\mathbb{Z}$$

of $\overline{m}$-adic integers (recall that $N_n = m_1 \dots m_n$ where $\overline{m} = \{m_n\}_{n=1}^{\infty}$), whose elements can be naturally identified with elements of the boundary ∂T, and then the action of ξ in ∂T corresponds to the map $x \mapsto x + 1$ in the ring $\widehat{\mathbb{Z}}_{\overline{m}}$.

5 Branch groups

We begin by making the key definition of this chapter.

Definition 1.

(a) The group G is said to be a *branch group* if there is a tree $T = T_{\overline{m}}$ and an embedding $G \hookrightarrow \operatorname{Aut} T$ such that

  (i) G acts transitively on each layer of the tree, and

  (ii) for each $n \geq 1$ there exists $L_n \leq \operatorname{Aut} T_{\langle n \rangle}$ such that the direct product

$$H_n = L_n^{(1)} \times \cdots \times L_n^{(N_n)} \leq \mathrm{St}_{\operatorname{Aut} T}(n) \tag{9}$$

  of $N_n = m_1 m_2 \dots m_n$ copies of L_n is normal and of finite index in G (each factor in (9) corresponds to a vertex in the nth layer and acts on the subtree hanging from that vertex).

(b) The profinite group G is a branch group if there is a tree $T = T_{\overline{m}}$ and an embedding $G \hookrightarrow \operatorname{Aut} T$ as a closed subgroup such that conditions (i) and (ii) of part (a) are satisfied, and additionally all subgroups under consideration are closed.

We remark that condition (ii) can be replaced by

(ii)′ For each $n \geq 1$, $|G : \mathrm{rist}_G(n)| < \infty$.

The pair $(\{L_n\}_{n=1}^{\infty}, \{H_n\}_{n=1}^{\infty})$ will be called a *branch structure*. It is not unique, but there is a canonical branch structure given by $(\{L_n^G\}_{n=1}^{\infty}, \{\mathrm{rist}_G(n)\}_{n=1}^{\infty})$, where L_n^G is the nth lower companion subgroup of G.

Theorem 2. *Let $G \leq \operatorname{Aut} T$ be a branch group.*

(a) *The closure $\overline{G}$ of G in $\operatorname{Aut} T$ is a profinite branch group.*

(b) *Each upper companion group G_k of G is a branch group.*

(c) *The centralizer $C_{\operatorname{Aut} T}(G)$ of G is trivial.*

Proof. Suppose that G has the branch structure $(\{L_n\}_{n=1}^{\infty}, \{H_n\}_{n=1}^{\infty})$.

(a) It is easy to see that $(\{\overline{L}_n\}, \{\overline{H}_n\})$ is a branch structure for $\overline{G}$.

(b) It is equally easy to see that $(\{L_n\}_{n=k}^{\infty}, \{H_n^*\}_{n=k}^{\infty})$ is a branch structure for G_k, where $H_n^* = L_n \times \cdots \times L_n$ with $m_k m_{k+1} \ldots m_n$ factors; note that G_k acts spherically transitively on each subtree $T_{\langle k \rangle}$, by Lemma A.

(c) Suppose $1 \neq g \in C_{\mathrm{Aut}T}(G)$, where $T = T_{\overline{m}}$. We may assume that $g \notin \mathrm{St}_{\mathrm{Aut}\,T}(1)$. Indeed, suppose $g \in \mathrm{St}_{\mathrm{Aut}\,T}(n)$ but $g \notin \mathrm{St}_{\mathrm{Aut}\,T}(n+1)$ for some n. Then there is a vertex u with $|u| = n$ such that in $g = (*, \ldots, *, g_u, *, \ldots, *)_n$, we have $g_u \notin \mathrm{St}_{\mathrm{Aut}\,T_u}(1)$.

Obviously g_u centralizes $\mathrm{St}_G(u)$, the restriction of which to the subtree T_u is a branch group by (b). Therefore we may assume that $g \notin \mathrm{St}_{\mathrm{Aut}\,T}(1)$.

Let $g = fa$ where $1 \neq a \in \Sigma_{m_1}$, $f \in \mathrm{St}_{\mathrm{Aut}T}(1)$ and f has the decomposition $f = (f_{u_1}, \ldots, f_{u_{m_1}})$, where the f_{u_i} are elements acting on the subtrees $T_{\langle 1 \rangle}$.

There are vertices v, w in the first layer such that $v^a = w$ and $v \neq w$.

Let $h \in \mathrm{St}_G(1)$ have the decomposition $h = (1, \ldots, 1, \xi, 1, \ldots, 1)$ where $\xi \in L_1$ is situated at position v. Then

$$1 = [h, g] = (1, \ldots, 1, \xi^{-1}, 1, \ldots, 1, \xi^{f_v}, 1, \ldots, 1)$$

and in this decomposition, ξ^{-1} is situated at position v and ξ^{f_v} is situated at position w. Since the element $\xi \in L_1$ can be chosen distinct from the identity, we have a contradiction. □

In the sequel, all our examples of profinite branch groups will be constructed on the basis of statement (a) in the above theorem.

Definition 1′.

(a) The group G is a *branch group* if it has trivial centre and contains descending chains of subgroups $\{H_n\}_{n=1}^{\infty}$, $\{L_n\}_{n=1}^{\infty}$, where each L_n is a subgroup of H_n and such that the following hold:

(i) H_n is normal in G, the index $|G : H_n|$ is finite and $\bigcap_{n=1}^{\infty} H_n = \{1\}$.

(ii) There is a sequence $\{N_n\}_{n=1}^{\infty}$ of natural numbers such that N_n divides N_{n+1}, and there are subgroups $L_n^{(1)}, \ldots, L_n^{(N_n)}$ of G, each isomorphic to $L_n = L_n^{(1)}$, such that H_n can be represented as the direct product

$$H_n = L_n^{(1)} \times \cdots \times L_n^{(N_n)}, \tag{10}$$

in such a way that the product decomposition (10) of H_{n+1} refines that of H_n; that is, each factor $L_n^{(i)}$ from (10) contains a product of $m_{n+1} = N_{n+1}/N_n$ factors $L_{n+1}^{(j)}$, $(i-1)m_{n+1} + 1 \leq j \leq i m_{n+1}$, from the corresponding decomposition of H_{n+1}.

(iii) When G acts on itself by conjugation, for each $n = 1, 2, \ldots$ the factors in (10) are permuted transitively among themselves.

(b) The profinite group G is a branch group if it satisfies the conditions (a), with the additional proviso that all the subgroups under discussion are closed.

Thus a group is a branch group if it has a structure similar to the structure of the whole automorphism group of a spherically homogeneous rooted tree.

If G is a branch group in the sense of the first definition, then it has trivial centre by Theorem 2, and the sequences $\{L_n\}$, $\{H_n\}$ clearly satisfy the conditions (i) – (iii) of Definition 1$'$.

Now suppose that the subgroup chains $\{L_n\}$, $\{H_n\}$ satisfy the conditions of Definition 1$'$. Let $\overline{m}$ be the sequence $\{m_n\}_{n=1}^{\infty}$ where $m_n = N_{n+1}/N_n$, and let $T = T_{\overline{m}}$ be the corresponding tree.

For each $n \geq 1$ identify the vertices of the nth level of T with the corresponding factors in the decomposition (10). Since the decomposition of H_{n+1} into factors $L_{n+1}^{(j)}$ refines the analogous decomposition for H_n, it follows that the conjugation action on the vertices in level $n+1$ agrees with the action on the vertices in level n. Thus we have created an action of G on T and the groups L_n, H_n satisfy conditions (i), (ii) of Definition 1. The only problem is that this action of G may not be faithful. This problem does not arise when G is just infinite, the case of particular interest in this chapter, and in this case the two definitions are then equivalent.

Definition 2. A branch group is called *self-similar* if, for some $d \geq 2$, it has an action on the regular tree T_d, with a branch structure in the sense of Definition 1, and such that the sequences of upper and lower companion groups $\{L_n^G\}_{n=1}^{\infty}$, $\{G_n\}_{n=1}^{\infty}$ are periodic. That is, there are constants M and k such that $L_n = L_{n+k}$ and $G_n = G_{n+k}$ for all $n \geq M$.

The nicest examples of self-similar groups are the self-reproducing groups (that is with a constant sequence $G_n = G$ of upper companion groups) which have a branch structure ($\{L_n\}$, $\{H_n\}$) for which $L_n = L$ is constant ($n = 1, 2, \dots$). In this case the embeddings (7) take the form

$$L \times \cdots \times L \lhd \mathrm{St}_G(n) \lhd G \times \cdots \times G$$

(d^n factors in each product) and L is a subgroup of finite index in G.

We also remark that a characteristic feature in the examples of self-similar groups below is that the section $g_{\{u\}}$ of any $g \in G$ at any vertex $u \in V(T)$ must be an element of G. Also there is then a normal subgroup K of finite index in G 'containing geometrically' its own dth power. This means that for arbitrary elements $k_1, \dots, k_d \in K$, the automorphism $f \in \mathrm{St}_{\mathrm{Aut}\,T}(1)$ defined in Diagram 21(a) belongs to K, and the branch structure shown in Diagram 21(b) naturally arises. We indicate this circumstance by $K \succ K \times \cdots \times K$, and we say that the group K is *self-replicating*. We also say that K *contains* $K \times \cdots \times K$ *geometrically*. This terminology will also be used in a more general setting; thus $P \succ L \times \cdots \times L$ (with N_n factors) means that P contains geometrically $L \times \cdots \times L$ in the sense that each factor L acts on a corresponding subtree $T_{\langle n \rangle}$. Similarly, for $P \leq \mathrm{St}_G(n)$ we write $P \prec L \times \cdots \times L$, with N_n factors, to indicate that the restriction of P to each subtree $T_{\langle n \rangle}$ is contained in L.

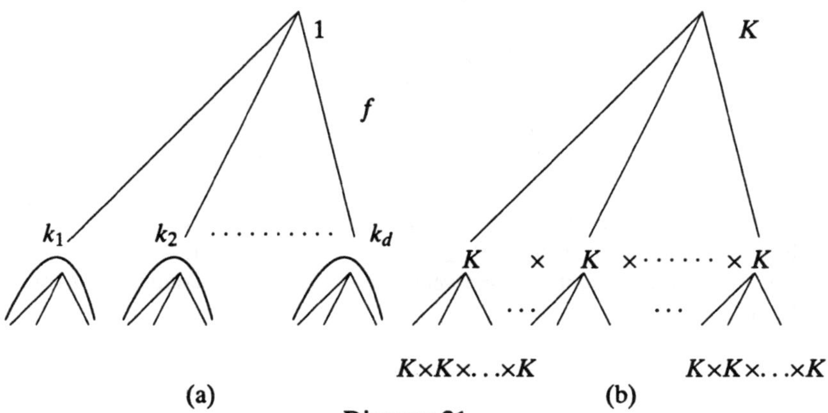

Diagram 21

We now give the simplest examples of self-similar branch groups.

Example 5.1. ($d \geq 2$) Let $G = \langle a, b \rangle \leq \operatorname{Aut} T_d$ where a is a rooted automorphism defined by the cyclic permutation $\varepsilon = (0, 1, \dots, d-1)$ and b is a directed automorphism defined by the recurrence relation $b = (a, 1, \dots, b)$ (see the convention on notation made in Section 2); alternatively, b can be defined by the labelling given in Diagram 22(b):

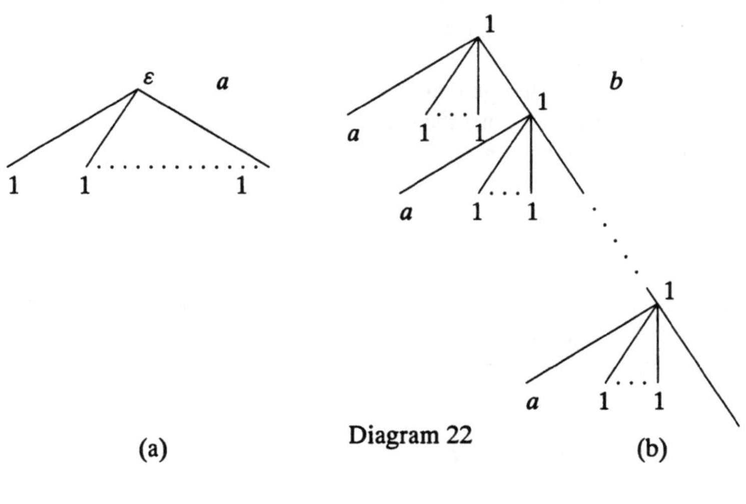

Diagram 22

The generators a and b clearly have order d. The set of elements of the form $b^{a^i}, 0 \leq i \leq d-1$, serves as a generating set for $\mathrm{St}_G(1)$ (since $G/\mathrm{St}_G(1) \cong \langle a \rangle$), and the following relations hold:

$$\begin{array}{rcl} b & = & (a, 1, \dots, 1, b) \\ b^a & = & (b, a, 1, \dots, 1) \\ \dots & \dots & \dots \\ b^{a^{d-1}} & = & (1, \dots, 1, b, a) \end{array} \tag{11}$$

(each vector is a cyclic permutation of the previous one).

Since for each co-ordinate i, $0 \leq i \leq d-1$, the elements a and b occur as entries in the ith component of some vector in (11), we may deduce the geometrical embedding

$$\mathrm{St}_G(1) \prec G \times \cdots \times G, \tag{12}$$

where the projection of $\mathrm{St}_G(1)$ on each factor of (12) is G.

From this it easily follows that G acts transitively on each layer of T. Indeed, let $u, v \in V(T)$, $|u| = |v| = n$. If $n = 1$, then obviously $u^{a^i} = v$ for some i, $0 \leq i \leq d-1$. Let $n > 1$ and suppose that u and v begin with the same letter j, $j \in \{0, 1, \dots, d-1\}$: that is, $u = ju'$, $v = jv'$. Then by inductive hypothesis there exists $f \in G$ such that $(u')^f = v'$.

On the other hand, there exists $g \in \mathrm{St}_G(1)$ having the decomposition

$$g = (*, \dots, *, f, *, \dots, *)$$

where the symbols $*$ stand for elements, and f is located in the jth position. Then evidently $u^g = v$.

Finally, if u and v have different first letters then acting on u by a suitable power of a, we can arrange that u^{a^i} and v have the same first letter, and complete the argument by reducing to the previous case.

Since the sections of either generator a or b at each $u \in V(T)$ are clearly in G, then, from Lemma 1, it follows that the section $g_{\{u\}}$ of any $g \in G$ at any $u \in V(T)$ is also in G. From this we deduce the geometrical inclusion

$$\mathrm{St}_G(n) \prec G \times \dots \times G, \tag{13}$$

and for each $n \geq 1$, the upper companion group G_n of G coincides with G.

We are going to prove that G is a branch group. To this end we consider the commutator $[b, b^a] = ([a, b], 1, \dots, 1)$. Since for each $f \in G$ there is an element $g \in \mathrm{St}_G(1)$ such that $g = (f, *, \dots, *)$, and since $[b, b^a]^g = ([a, b]^f, 1, \dots, 1)$, we have

$$[G, G] = G' \succ G' \times 1 \times \dots \times 1. \tag{14}$$

Since conjugation by a permutes the factors of (14) cyclically we conclude that

$$G' \succ G' \times \dots \times G'. \tag{15}$$

The commutator subgroup of G has finite index in G since the generators a and b each have finite order. We claim that in fact $G/G' \cong C_d \times C_d$, and the factor group is generated by the images of a and b. Indeed, since $G' \leq \mathrm{St}_G(1) \leq G$, the image of a in G/G' clearly has order d. Keeping in mind that $G' = \langle [a^i, b^j];\ 1 \leq i, j \leq d-1 \rangle$, and considering $G/\mathrm{St}_G(2)$, we can easily see that $|G : G'| = d^2$, which implies the claim.

Thus G has a branch structure in which $L_n = G'$ and $H_n = G' \times \cdots \times G'$ (d^n factors, $n = 1, 2, \ldots$). Moreover, G is a self-similar group with self-replicating subgroup $K = G'$, as follows from (15).

It can be shown that in this example, G' can be identified with the group $\mathrm{top}_G(u)$ for any $u \in V(T)$ with $|u| \geq 1$, and hence that

$$\mathrm{rist}_G(n) = G' \times \cdots \times G'$$

(d^n factors) for $n = 1, 2, \ldots$. We will see later that the group G is just infinite.

6 On the structure of just infinite groups

The purpose of this section is to establish the trichotomy mentioned in the Introduction.

Definition 3.

(a) A residually finite group G is *hereditarily just infinite* if every normal subgroup of finite index is just infinite.

(b) A profinite group G is *hereditarily just infinite* if every open subgroup is just infinite.

If a finitely generated hereditarily just infinite group G acts on a tree $T = T_{\overline{m}}$ transitively on layers, then the lower companion groups L_n^G must be trivial. Indeed, otherwise the rigid stabilizers $\mathrm{rist}_G(n) = L_n^G \times \cdots \times L_n^G$ (N_n factors) would be normal subgroups of G and hence of finite index. However they are not just infinite.

The simplest examples of abstract hereditarily just infinite groups are the infinite cyclic group and the infinite dihedral group $D = \langle x, y \mid x^2 = y^2 = 1 \rangle$. More complicated examples are given by the groups $\mathrm{SL}(n, \mathbb{Z})$ modulo their centres for $n \geq 3$, and many other lattices in Lie Groups of rank ≥ 2 factored by their centres. Examples of profinite hereditarily just infinite groups are $\mathbb{Z}_p$, $\mathrm{SL}(n, \mathbb{Z}_p)$ modulo its centre ($n \geq 3$), etc. The most extraordinary profinite examples are the so-called Nottingham groups and some similar groups constructed by I. Fesenko (see Chapters 6 and 10).

As we remarked earlier, the action of hereditarily just infinite groups on spherically homogeneous trees differs significantly from the action of branch groups. Let us demonstrate this difference using the example of the action of the dihedral group D on a binary tree where the automorphisms $x, y \in \mathrm{Aut}\, T_2$ are defined in Diagram 23.

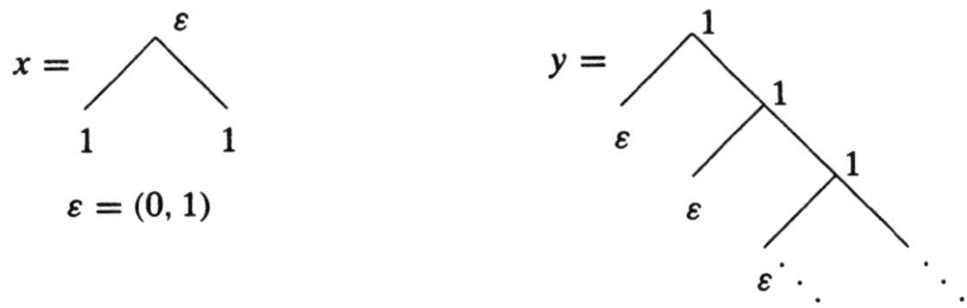

Diagram 23

Clearly $x^2 = y^2 = 1$ and $\mathrm{St}_D(1) = \langle y, y^x \rangle$ where $y = (x, y)$, $y^x = (y, x)$. Since $(xy)^2 \in \mathrm{St}_D(1)$ and we have the decomposition $(xy)^2 = (yx, xy)$, it follows that xy is an element of infinite order, and hence x, y do indeed generate the dihedral group D.

The fact that the action of D is transitive on layers and that the upper companion groups D_n coincide with D can be established just as in Example 5.1. In particular

$$\mathrm{St}_D(n) \prec D \times \cdots \times D \quad (2^n \text{ factors}).$$

However, the lower companion groups L_n^D are trivial: this follows both from the fact that D is hereditarily just infinite, and from the fact that for every element $g \in \mathrm{St}_D(n)$, there is a word $W(x, y)$ such that the decomposition of g at the nth level has the form

$$g = (W(x, y), W(y, x), \ldots, W(x, y), W(y, x)). \tag{16}$$

Since the map $x \mapsto y$, $y \mapsto x$ defines an automorphism of D, it follows from the fact that one of the coordinates of the vector (16) is equal to the identity that the whole vector is the identity.

Similarly, for an arbitrary hereditarily just infinite group G acting transitively on layers, there are automorphisms $\varphi_i \in \mathrm{Aut}\, G_n$, $1 \leq i \leq N_n - 1$, where the G_n are upper companion groups, such that for any element $g \in \mathrm{St}_G(n)$ we have the following decomposition:

$$g = \big(h, \varphi_1(h), \ldots, \varphi_{N_n-1}(h)\big)_n \in G_n \times \cdots \times G_n.$$

It would be desirable to study more carefully the actions of hereditarily just infinite groups on trees.

Now we will formulate the alternative which makes transparent the significance of the two classes of groups defined above — branch groups and hereditarily just infinite groups. It is obvious that the three classes of groups involved in the following statement are disjoint.

Theorem 3.

(a) *Let G be an abstract just infinite group. Then either G is a branch group, or G contains a normal subgroup of finite index which is isomorphic to the direct product of a finite number of copies of a group L, where L is either simple or hereditarily just infinite.*

(b) *Let G be a profinite just infinite group. Then either G is a branch group, or G contains an open normal subgroup which is isomorphic to the direct product of a finite number of copies of some hereditarily just infinite profinite group.*

Proof. Recall that the Baer radical of a group G is the subgroup generated by all cyclic subnormal subgroups of G. In [48] it is proved that a just infinite group with non-trivial Baer radical is a finite extension of a free abelian group of finite rank. Thus, when proving our statement, we may assume that the Baer radical is trivial. In this situation, the following statement due to Wielandt (see [48], Lemma 14) holds:

Lemma 2. *Let $\{L_i : i \in I\}$ be a family of subnormal subgroups of G such that $L_i \cap L_j = \{1\}$ whenever $i \neq j$. Then $\langle L_i : i \in I\rangle = Dr_{i \in I} L_i$.*

We will also need the following statement from [48], (Corollary 1 to Theorem 3B):

Let G be a just infinite group with trivial Baer radical. Let K be subnormal in G. There are only finitely many subgroups of G conjugate to K.

Let us assume now that G is an just infinite group with trivial Baer radical. If G is not hereditarily just infinite and contains no simple normal subgroup of finite index, then there are subgroups K, Q such that $K \lhd_\infty Q \lhd_f G$. For each subgroup D of G, we denote by $O(D)$ the orbit of D under conjugation by G. As K is subnormal, the orbit $O(K)$ is finite.

We may therefore choose a subset $R \subseteq O(K)$ maximal subject to

$$L_1 := \bigcap_{K^g \in R} K^g \neq \{1\}.$$

Then the intersection of any two distinct conjugates L_1^g, L_1^h is trivial, so by Lemma 2 we have

$$H_1 := \langle L_1^g \ : \ g \in G\rangle = \mathrm{Dr}_{L_1^g \in O(L_1)} L_1^g.$$

Note that there is more than one factor in this direct product, as otherwise L_1, and hence K, would be of finite index in G.

Suppose now that for some $n \geq 1$ we have

$$G \rhd_f H_n = L_n^{(1)} \times \cdots \times L_n^{(N_n)} \tag{17}$$

where $L_n^{(1)} = L_n, \ldots L_n^{(N_n)}$ are permuted transitively under conjugation by G. Then the following three cases are possible.

(1) L_n has a simple normal subgroup of finite index. In this case we have one of the conclusions of the theorem.

(2) L_n has a non-trivial normal subgroup K_n of infinite index. In this case, we operate with K_n as we did with K, above. Thus we find a suitable intersection L_{n+1} of conjugates of L_n with the property that

$$H_{n+1} := \langle L_{n+1}^f \ : \ f \in G\rangle = \mathrm{Dr}_{L_{n+1}^f \in O(L_{n+1}^f)} L_{n+1}^f$$

is normal and of finite index in G.

It is easy to see that this decomposition refines (17), and that the number of factors is strictly greater than in (17). Thus in Case (2) we can pass from H_n to H_{n+1}, making one more step in the construction of a branch structure.

(3) Where neither (1) nor (2) holds. In this case, L_n has proper subgroups of finite index, and every such subgroup is just infinite. We show that in this situation G is residually finite, hence hereditarily just infinite.

Indeed, let H_* be the intersection of all subgroups of finite index in G. If $H_* \neq 1$, then H_* is characteristic, of finite index in G, and contains H_n. Then

$$M := H_* \cap L_n$$

is a non-trivial normal subgroup of L_n, hence has finite index in L_n. Since we are not in Case (1), M is not simple, and as M is just infinite, it follows that M has a proper normal subgroup P of finite index. Then for each i the group $L_n^{(i)}$ contains a conjugate $P^{(i)}$ of P, and the subgroup

$$\mathrm{Dr}_{1 \leq i \leq N_n} P^{(i)}$$

has finite index in G, hence contains H_*. But this forces $P = M$, a contradiction.

We see that if either Case (1) or Case (3) arises at some stage of this recursive procedure, then G satisfies one of the alternative conclusions of the theorem. On the other hand, if Case (2) occurs each time, we obtain a strictly descending sequence $\{H_n\}_{n=1}^{\infty}$ of normal subgroups of G satisfying the conditions of Definition 1$'$. Thus G is a branch group provided that the intersection $\bigcap_{n=1}^{\infty}$ is trivial; and it must be, because G is just infinite.

Theorem 3(a) is proved. The proof of Theorem 3(b) is analogous, but in this case the analogues of Lemma 2 and other related statements in the profinite case are used, all considerations refer to closed subgroups, and the case of simple subgroups does not arise. Details are given in Chapter 5. □

We finish this section by proving a simple statement that establishes the existence of just infinite quotients in two important cases.

Proposition 3.

(a) *Let G be a finitely generated infinite group. Then G can be mapped onto a just infinite group.*

(b) *Let G be a finitely generated infinite profinite group which is virtually a pro-p group. Then G can be mapped onto a virtually pro-p just infinite group.*

Proof. (a) Let $\mathcal{N}$ denote the set of normal subgroups of infinite index in G. We show that every ascending chain in $\mathcal{N}$ has an upper bound in $\mathcal{N}$. Let $\{H_\alpha\}$ be such a chain, and put $H = \bigcup_\alpha H_\alpha$. Suppose that H has finite index in G. Since G is finitely generated, so is H. But then H coincides with H_β for some β, a contradiction. Thus H has infinite index in G and so $H \in \mathcal{N}$. By Zorn's lemma there exists a maximal element $M \in \mathcal{N}$. If H/M is a non-trivial normal subgroup of G/M then, by the maximality of M, H/M is of finite index in G/M. Thus G/M is just infinite.

(b) Let $\mathcal{N}$ denote the set of closed normal subgroups of infinite index in G. We use Zorn's Lemma as above. So let $\{H_\alpha\}$ be an ascending chain in $\mathcal{N}$, and now let H be the closure of $\bigcup_\alpha H_\alpha$. If H has infinite index in G then $H \in \mathcal{N}$.

Suppose that H has finite index in G. Since G is virtually pro-p and finitely generated, so is H. Hence $H/\Phi(H)$ is a finite group. Write A_α to denote the image of H_α in $H/\Phi(H)$. Since $\bigcup_\alpha H_\alpha$ is dense in H we have $\bigcup_\alpha A_\alpha = H/\Phi(H)$. It follows that $H/\Phi(H) = A_\beta$ for some β, which implies that $H_\beta = H$, a contradiction. □

It is not known whether statement (b) can be extended to all finitely generated profinite groups.

7 A criterion for a branch group to be just infinite

Let G be a branch group (discrete or profinite) acting on the tree $T = T_{\overline{m}}$, with branch structure $(\{L_n\}, \{H_n\})$.

Theorem 4. *The group G is just infinite if and only if for each $n \geq 1$, the index of the commutator subgroup L_n' in L_n is finite. Moreover if this condition holds, then every non-trivial normal subgroup P of G contains a subgroup H_n' for some $n = n(P)$.*

(Recall that in the profinite case, L_n' denotes the *closed* subgroup generated by commutators in L_n.)

Proof. One direction of the statement is obvious, since if for some k the index of L_k' in L_k is infinite, then $H_k' = L_k' \times \cdots \times L_k'$ is a normal subgroup of G of infinite index.

Now suppose that $|L_n : L_n'| < \infty$ for $n = 1, 2, \ldots$. Let $1 \neq g \in G$ and put $P = \langle g \rangle^G$. It will suffice to prove that P contains H_s' for some s, since the hypothesis implies that each of the subgroups H_s' has finite index in G.

Suppose that $g \in \mathrm{St}_G(n) \setminus \mathrm{St}_G(n+1)$. Say $g = (*, \ldots, *, g_u, *, \ldots, *)_n$, where the vertex u with $|u| = n$ is chosen so that $g_u \notin \mathrm{St}_{\mathrm{Aut}\,T_{(n)}}(1)$. Then we have $g_u = ha$ where $h \in \mathrm{St}_{\mathrm{Aut}T\langle n\rangle}(1)$ and $1 \neq a \in \Sigma_{m_{n+1}}$.

Let A_i, $i = 1, 2, \ldots$, be a sequence of alphabets corresponding to the tree T and let $x, y \in A_{n+1}$ be such that $x \neq y$ and $x^a = y$.

To an arbitrary element $\xi \in L_{n+1}$, we associate the element $f \in \mathrm{St}_G(n+1)$ defined by the following decompositions:

$$f = (1, \ldots, 1, f_u, 1, \ldots, 1)_n, \tag{18}$$

$$f_u = (1, \ldots, 1, \xi, 1, \ldots, 1)_1, \tag{19}$$

where the element f_u is in position u in (18) and the element ξ is in position x in (19).

Consider the element $[g, f]$ which belongs to P and also to $\mathrm{St}_G(n+1)$; this has the decomposition

$$\begin{aligned}[g, f] &= (1, \ldots, 1, [g_u, f_u], 1, \ldots, 1)_n \\ &= (1, \ldots, 1, a^{-1}h^{-1}f_u^{-1}haf_u, 1, \ldots, 1)_n \\ &= (1, \ldots, 1, \xi^{-h_{(x)}}, 1, \ldots, \xi, 1, \ldots, 1)_{n+1},\end{aligned}$$

where $h_{\{x\}}$ is the corresponding section of h, and the positions of the elements $\xi^{-h_{\{x\}}}$ and ξ in the last vector correspond to the vertices uy and ux respectively (if we identify vertices with string of symbols).

Let $\eta \in L_{n+1}$, and let $l \in \mathrm{St}_G(n+1)$ have the decomposition $l = (1, \dots, 1, \eta, 1, \dots, 1)_{n+1}$, where η occurs at the position of the vertex ux. Then

$$[[g, f], l] = (1, \dots, 1, [\xi, \eta], 1, \dots, 1).$$

As η was an arbitrary element of L_{n+1}, we infer that

$$P \succ 1 \times \cdots \times 1 \times L'_{n+1} \times 1 \times \dots \times 1.$$

Since the action of G is transitive on each level and P is normal in G, it follows that

$$P \succ L'_{n+1} \times \cdots \times L'_{n+1} = H'_{n+1}. \tag{20}$$

This concludes the proof. □

Corollary. *If G is a branch group acting on a tree $T_{\overline{m}}$ and G is just infinite as an abstract group, then the profinite group $\overline{G}$ which is the closure of G in $\mathrm{Aut}\, T_{\overline{m}}$ is a just infinite profinite branch group.*

Proof. It has already been mentioned that if G is a branch group, then the closure $\overline{G}$ is also a branch group. Namely, if $(\{L_n\}, \{H_n\})$ is a branch structure for G, then the pair $(\{\overline{L}_n\}, \{\overline{H}_n\})$ provides a branch structure for $\overline{G}$. As $[\overline{L}_n : \overline{L}'_n] \leq |L_n : L'_n|$ the statement follows from Theorem 4. □

Using the criterion proved above it is usually not hard to verify the fact that a given branch group is just infinite. For example, the condition $|L_n : L'_n| < \infty$ obviously holds if G is a finitely generated periodic branch group. Thus all such groups are just infinite.

Example 7.1. We are going to give an example of a non-periodic just infinite branch group. To this end we consider the group G from Example 5.1, and we will assume that $d \geq 5$.

It was established earlier that in this case, as a branch structure one can take the pair $(\{L_n\}, \{H_n\})$ where $L_n = G'$, $H_n = G' \times \cdots \times G'$ (d^n factors), $n = 1, 2, \dots$. Thus in order to apply the criterion of Theorem 4 we only have to show that $|G' : G''| < \infty$.

Recall that

$$b = (a, 1, \dots, 1, b) \tag{21}$$

and that the decomposition of the element b^{a^i} is obtained from the vector (21) by taking the ith power of a cyclic shift.

Straightforward calculations show that

$$x := [a^i, b^j] = (a^j, 1, \dots, 1, b^{-j}, a^{-j}, 1, \dots, 1, b^j)$$

and

$$y := [a^{i+1}, b^s] = (a^s, 1, \dots, 1, b^{-s}, a^{-s}, 1, \dots, 1, b^s)$$

if $2 \le i \le d-2$, where the element a^{-j} in the first vector is in the ith position and the element a^{-s} in the second vector is in the $(i+1)$th position.

Therefore

$$[x, y] = (1, \dots, 1, [a^{-j}, b^{-s}], 1, \dots, 1).$$

Thus $G'' \succ 1 \times \cdots 1 \times G' \times 1 \times \cdots \times 1$, and hence $G'' \succ G' \times \cdots \times G' = H_2$.

From this it follows that G'' has finite index in G and hence in G', so by Theorem 4 the group G is just infinite. Its closure $\overline{G}$ is a profinite just infinite branch group. When $d = p$ is prime, $\overline{G}$ is a pro-p group.

Note that there exist just infinite branch groups which are torsion-free [3].

8 The main construction of branch groups

All the examples of branch groups given in this chapter are particular cases of the following general construction.

Let $T = T_{\overline{m}}$ be an infinite spherically homogeneous rooted tree, $\{A_n\}_{n=1}^{\infty}$ an associated sequence of alphabets, $R = \{r_1, \dots, r_k\}$ a set of rooted automorphisms of the tree T, and $D = \{d_1, \dots, d_l\}$ be a set of directed automorphisms, where the directing path e is the rightmost path (in the sense of the ordering) in ∂T.

As before, we will suppose that $a_1^{(n)}, \dots, a_{m_n}^{(n)}$, where $N_n = m_1 m_2 \cdots m_n$, are the symbols of the alphabet A_n.

Geometrically, the automorphism d_j $(1 \le j \le l)$ is specified by the sequence of vectors $\omega_j = \omega_j^{(1)} \dots \omega_j^{(n)} \dots$, where $\omega_j^{(n)} = (r_{j,1}^{(n)}, \dots, r_{j,m_n-1}^{(n)})$ is a tuple of elements of the symmetric group $\Sigma_{m_{n+1}}$, $n = 1, 2, \dots$. The tuple $\omega_j^{(n)}$ here defines the action of the automorphism d_j at the $n+1$th level (see Diagram 24).

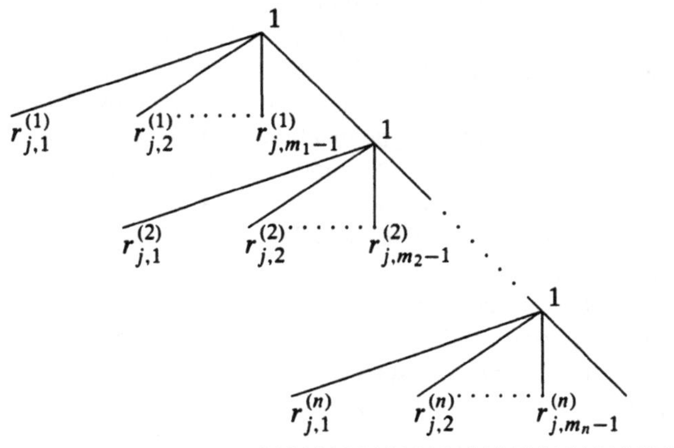

Diagram 24

Define $G = \langle R, D\rangle$ to be the group generated by the set of rooted automorphisms R and the set of directed automorphisms D. The group G is described using the sequence $\omega = \omega^{(0)}\omega^{(1)}\dots\omega^{(n)}\dots$ where $\omega^{(0)} = (r_1, \dots, r_k)$ is a tuple of rooted automorphisms with $r_i \in \Sigma_{m_1}$ for $1 \le i \le k$, and $\omega^{(n)}$ is the $l \times (m_n - 1)$ matrix

$$\omega^{(n)} = \begin{pmatrix} \omega_1^{(n)} \\ \omega_2^{(n)} \\ \vdots \\ \omega_l^{(n)} \end{pmatrix} = \begin{pmatrix} r_{1,1}^{(n)} & \dots & \dots & r_{1,m_n-1}^{(n)} \\ & & & \\ r_{l,1}^{(n)} & \dots & \dots & r_{l,m_n-1}^{(n)} \end{pmatrix}.$$

Accordingly, in this and the next section, we will denote this particular group by G_ω, call the sequence ω a *defining code* for $G = G_\omega$, and will denote the directed generators by $d_j^{(\omega)}$, $1 \le j \le l$.

Definition 4. The group $G = G_\omega = \langle R, D\rangle$ as above is called a *special group* if the following conditions hold.

1. The group $R_0 = R = \langle r_1, \dots, r_k\rangle$ and the groups $R_n \le \Sigma_{m_{n+1}}$ defined by

$$R_n := \langle r_{j,s}^{(n)},\ 1 \le j \le l,\ 1 \le s \le m_n - 1\rangle,\ n = 1, 2, \dots$$

   act transitively on the respective alphabets A_{n+1}, $n = 0, 1, \dots$.

2. If, for some M, the branch numbers m_n are equal to 2 for every $n \ge M$, then we require in addition that for each $n \ge M$, there exist j with $1 \le j \le l$ such that $r_{j,l}^{(n)} = 1$.

Among the special groups there are many examples of just infinite branch groups which often have additional interesting properties such as, for example, to be torsion, to be a group of intermediate growth, to have finite width etc.

Obviously the just infinite branch groups of Example 7.1 are special. In the next section we will construct examples of periodic branch groups. Now we are going to discuss the simplest properties of special groups. Denote by Ω the set of sequences $\omega = \omega^{(0)}\omega^{(1)}\dots\omega^{(n)}\dots$, of the type mentioned above, and satisfying the conditions of Definition 4. Let $\sigma : \Omega \to \Omega$ denote a shift to the left in this space, so that $\sigma\omega = \omega^{(1)}\omega^{(2)}\dots\omega^{n}\dots$, where the matrix $\omega^{(1)}$ is construed as a vector writing all entries of $\omega^{(1)}$ in a line. Evidently the set of sequences Ω is closed with respect to this transformation.

The action of σ on the branch index $\overline{m} = m_1m_2\dots m_n\dots$ will, as above, be understood as the standard shift, that is the transformation $\sigma\overline{m} = m_2m_3\dots m_{n+1}\dots$. The group $G_{\sigma\omega}$ defined by the sequence $\sigma\omega$ is clearly special. The group of rooted automorphisms for this group is the group R_1 generated by the elements of the vector $\omega^{(1)}$, and the directed automorphisms $d_1^{(\sigma\omega)}, \dots, d_l^{(\sigma\omega)}$ are defined by the sequence $\sigma\omega$ and act in a natural way on the tree $T_{\sigma\overline{m}}$ which coincides with a subtree $T_{(1)}$ of the tree $T = T_{\overline{m}}$.

Theorem 5.

(a) *Each special group $G = G_\omega$ is infinite and acts transitively on layers.*

(b) *For each $n \geq 1$, the groups $St_G(u)$ and $top_G(u)$ are independent of u where u ranges over the set of vertices in the nth level.*

(c) *For each $n \geq 1$, the nth upper companion group (which is equal to the restriction of $St_{G_\omega}(u)$ to T_u where $|u| = n$) coincides with the group $G_{\sigma^n\omega}$.*

Proof. Obviously $\mathrm{St}_{G_\omega}(1) = \langle (d_j^{(\omega)})^r,\ r \in R_0,\ 1 \leq j \leq l \rangle$. For the directed generators the following decompositions hold:

$$d_j^{(\omega)} = (r_{j,1}^{(1)}, \dots, r_{j,m_1-1}^{(1)}, d_j^{(\sigma\omega)}); \tag{22}$$

and the elements $(d_j^{(\omega)})^r$, $r \in R_0$ are obtained from the vector (22) by permutation of its coordinates according to the formula for the action of the element r on the set A_1 of m_1 elements. Therefore by the transitivity of the action of R_0 on A_1, the projection $\mathrm{St}_{G_\omega}(1)$ onto an arbitrary subtree beginning at an arbitrary first level vertex is the group $G_{\sigma\omega}$. This implies that the group G_ω is infinite (here we use the fact that the set Ω is closed with respect to the transformation σ).

We are going to prove the transitivity on layers of the action of special groups by induction on the level number n, simultaneously for all groups G_ω, $\omega \in \Omega$.

The fact that an arbitrary special group G_ω acts transitively on the first level follows directly from the definition, since the rooted subgroup R_0 acts on it transitively.

Suppose that an arbitrary group G_ω, $\omega \in \Omega$ acts transitively on level n. Let us prove the transitivity of the action of the groups G_ω, $\omega \in \Omega$, on the $(n+1)$th level.

Let $u, v \in V(T_{\overline{m}})$ be two vertices in the $(n+1)$th level, and assume that they begin with the same symbol $a \in A_1$, that is, $u = au'$, $v = av'$. By inductive hypothesis there is an element $g \in G_{\sigma\omega}$ acting on the subtree $T_a \cong T_{\sigma\overline{m}}$ of the tree $T = T_{\overline{m}}$ which maps the vertex u' to v'.

On the other hand, there is an element $f \in \mathrm{St}_{G_\omega}(1)$ such that the decomposition of the element f has the form

$$f = (*, \dots, *, g, *, \dots, *),$$

where the stars denote elements of the group $G_{\sigma\omega}$ and the element g is located in position a in the alphabet A_1. Since obviously $u^f = v$ we may proceed to consider the case when u and v begin with different symbols.

However, this case reduces to the previous one, since using the action of a suitable rooted automorphism $r \in R_0$ the vertex u can be mapped to a vertex beginning with the same symbol that v does. This completes the proof of (a).

As G acts transitively on layers, it acts transitively by conjugation on the set of stabilizers of vertices at level n, and also on the set of vertex groups at vertices at level n, so (b) follows.

Finally, the fact that the upper companion groups G_n of $G = G_\omega$ are equal to the groups $G_{\sigma^n\omega}$ has already been proved when $n = 1$; for arbitrary n it follows by a simultaneous induction for all the groups G_ω, $\omega \in \Omega$. □

In Section 5 it was shown that branch groups have trivial centralizers. We will now prove that special groups also have trivial centralizers in the full automorphism group of a tree.

Theorem 6. *For any special group $G = G_\omega$ acting on the tree $T = T_{\overline{m}}$, the centralizer $C_{\mathrm{Aut}T}(G)$ is trivial.*

Proof. The proof is quite similar to the proof of Theorem 2. Let $1 \neq g \in C_{\mathrm{Aut}T}(G)$; we may assume that $g \notin \mathrm{St}_{\mathrm{Aut}T}(1)$.

Suppose that $g = fr$, where $1 \neq r \in \Sigma_{m_1}$ and $f \in \mathrm{St}_{\mathrm{Aut}T}(1)$ has the decomposition

$$f = (*, \dots, *, f_u, *, \dots, *),$$

with $f_u \in \mathrm{Aut}\, T_{\sigma\overline{m}}$ and u ranging over the set of vertices in the nth level.

Let $d_1^{(\omega)}$ be the first directed generator of the group G_ω. There are vertices $u \neq w$ of the first level such that $u = w^r$. Let $c \in R_0$ be a rooted automorphism which maps the last vertex of the first level to u. Then

$$(d_1^{\omega})^c = (*, \dots, *, d_1^{(\sigma\omega)}, *, \dots, *, x, *, \dots, *),$$

where the directed element $d_1^{(\sigma\omega)}$ is located in the position of vertex u and the rooted element $x \in R_1$ is located at the position of vertex w. Then

$$1 = [(d_1^{(\omega)})^c, g] = (*, \dots, *, \left(d_1^{(\sigma\omega)}\right)^{-1} f_w^{-1} x f_w, *, \dots, *)$$

and hence we get the following relation

$$d_1^{(\sigma\omega)} = f_w^{-1} x f_w. \tag{23}$$

We claim that this is impossible. Indeed, if $x = 1$, this implies that $d_1^{(\sigma\omega)} = 1$, a contradiction (recall that by definition, a directed automorphism is not finitary). Suppose then that $x \neq 1$. Considering (23) as a relation in the group $G_{\sigma\omega}/\mathrm{St}_{G_{\sigma\omega}}(1)$, we obtain the relation $1 = x$ in $G_{\sigma\omega}/\mathrm{St}_{G_{\sigma\omega}}(1) = R_1$, again a contradiction. □

9 Torsion special groups

In this section we give necessary and with some additional assumptions sufficient conditions for a group given by the main construction to be periodic.

In the proof that certain groups are periodic, an important role is played by a simultaneous induction on the length of a word, applied to all the upper companion groups $\{G_n\}_{n=1}^\infty$ for G. This simultaneous induction on the length also plays an important role in other questions concerning the theory of branch groups.

We keep the notation of the previous section. Thus $\{A_n\}_{n=1}^\infty$ is a sequence of alphabets associated with the tree $T = T_{\overline{m}}$. Denote by s_n the last symbol of the alphabet A_n.

Recall that a directed automorphism d is defined by a sequence of vectors

$$(r_1^{(n)}, \dots, r_{m_n-1}^{(n)}), \tag{24}$$

$n = 1, 2, \dots$, with entries in the symmetric group $\Sigma_{m_{n+1}}$. For each such automorphism, when n is fixed, we define a function $\pi_d : A_n \to \Sigma_{m_{n+1}}$ by

$$\pi_d(x) = \begin{cases} r_i^{(n)}, & \text{if } x \text{ is } i\text{-th symbol in } A_n,\ 1 \leq i \leq m_n - 1 \\ 1, & \text{otherwise.} \end{cases}$$

The group generated by an arbitrary collection $\{d_1, \dots, d_l\}$ of directed automorphisms is a subgroup of the Cartesian product $\prod_{n=1}^{\infty} R_n$, where $R_n \leq \Sigma_{m_{n+1}}$ is the subgroup generated by the components of the corresponding vectors (24); it is easy to see that an arbitrary finitely generated residually finite group can be represented in this form, the tree T and directed automorphisms d_j being suitably chosen.

The following two theorems contain ideas from [13],[14], [24], [26].

Theorem 7. *Let $G = \langle R, D \rangle$ be a special group and assume that the following conditions hold:*

1. *The group $\mathcal{D} = \langle D \rangle$ generated by the set D is periodic.*
2. *For every $n \geq 1$, $r \in R_n$, $x \in A_{n+1}$ and $d \in \mathcal{D}$, the element*

$$\pi_d(x)\pi_d(x^r)\dots\pi_d(x^{r^q}) \in \Sigma_{m_{n+1}}$$

is the identity where $q+1$ is the length of the cycle containing x under the action of the element r.

Then G is an infinite torsion group.

Proof. Any element of the group G can be written in the form

$$g = r_{i_1}d_{i_1}r_{i_2}d_{i_2}\cdots r_{i_n}d_{i_n}, \tag{25}$$

where $r_{i_j} \in R_0 = \langle r_1, \dots, r_k \rangle$, $d_{i_j} \in \mathcal{D}$ and all the factors of the product are distinct from 1, except possibly for the end terms r_{i_1}, d_{i_n}.

We will call the minimal number of factors in an expression of the form (25) the length of the element $g \in G$, and denote it by $|g|$. Obviously $|g| = 1$ if and only if $1 \neq g \in R_0 \cup \mathcal{D}$ (the length of the identity element is 0).

The factors r_{i_j} we will call *r-symbols* and the factors d_{i_j} we will call *d-symbols*. A word in r- and d-symbols is called *reduced* if it does not contain any adjacent r-symbols or d-symbols.

If w is a reduced word, then obviously

$$(|w| - 1)/2 \leq |w|_d \leq (|w| + 1)/2,$$

where $|w|_x$ denotes the number of occurrences of x-symbols in w, $x \in \{r, d\}$.

If a reduced word begins with an r-symbol and ends in a d-symbol, or vice versa, then $|w| = 2|w|_r = 2|w|_d$. In this case we will call the word *cyclically reduced*.

Suppose now that $g \in \mathrm{St}_G(1)$ and

$$g = (g_1, \dots, g_{m_1}), \tag{26}$$

$g_i \in G_1$, $1 \le i \le m_1$, where G_1 is the first upper companion group for G.

As follows from previous discussions, the nth companion group G_n of G_ω ($\omega \in \Omega$) is the group $G_n = \langle R_n, d_1^{[n]}, \dots, d_l^{[n]} \rangle$, where $d_i^{[n]} = (d_i)_{\{u_n\}}$ and $u_n = s_1 s_2 \dots s_n$ is the nth section of the element d_i at the rightmost vertex u_n of level n, for $1 \le i \le l$; thus we may refer to lengths of elements of the group G_n, in the same sense as for G.

Lemma 3. *For each element $g \in St_G(1)$ the following inequalities hold:*

$$|g_i| \le (|g| + 1)/2, \ i = 1, \dots, m_1, \tag{27}$$

where the g_i are the elements specified in relation (26).

Proof. If an element g of type (25) belongs to $\mathrm{St}_G(1)$, then $r_{i_1} r_{i_2} \dots r_{i_n} = 1$ in R_0 and relation (25) can be rewritten in the following form:

$$g = d_{i_1}^{r_{i_1}^{-1}} d_{i_2}^{(r_{i_1} r_{i_2})^{-1}} \dots d_{i_n}^{(r_{i_1} \dots r_{i_n})^{-1}}. \tag{28}$$

Since for $r \in R_0$, $d \in \mathcal{D}$ the decomposition of d^r is in the form of a vector obtained from the decomposition $d = (*, \dots, *, d^{[1]})$ by the permutation r of the coordinates of this vector (here stars denote elements of R_1), the number of symbols in each g_i (considered as a word) is equal to the number of d-symbols in the product (25), and thus $|g_i| \le |g|_d \le (|g| + 1)/2$. □

Corollary. *If $g \in St_G(1)$ and $|g| > 1$, then $|g_i| < |g|$ for $1 \le i \le m_1$, where the g_i are the elements specified in relation (26).*

To complete the proof that each element of a group G satisfying the conditions of Theorem 7 has finite order, we argue by induction on the length.

If $|g| = 1$, then $g \in R_0 \cup \mathcal{D}$ and the order of g is finite, by hypothesis if $g \in \mathcal{D}$, and bounded by m_1 if $g \in R_0$. Assume now that every element g of length $\le n$ from an arbitrary group satisfying the conditions of the theorem has finite order.

If $g \in \mathrm{St}_G(1)$ and $1 < |g| \le n + 1$, then the elements g_i, from relation (26) belong to G_1 and have length $\le n$ by Lemma 3. Since every companion group G_k of G also satisfies the conditions of the theorem, it follows by inductive hypothesis that the elements g_i have finite orders for $1 \le i \le m_1$, and hence g has finite order.

Now suppose that $g \notin \mathrm{St}_G(1)$, that is $1 \ne r = r_{i_1} r_{i_2} \dots r_{i_n} \in R_0$. Let s be the order of the element r; then $g^s \in \mathrm{St}_G(1)$ and we have

$$g^s = (g_1, \dots, g_{m_1}) \tag{29}$$

for some $g_1, \dots, g_{m_1} \in G_1$. Moreover the expression (25) for g via r- and d-symbols may be taken to be cyclically reduced (otherwise g is conjugate to an element of smaller length).

The considerations used in the proof of Lemma 3 show that the number of d-symbols in each of the elements g_i, $1 \le i \le m_1$, is no more than the number of

d-symbols in the word (25) representing g. This is $|g|/2$. Therefore $|g_i| \leq |g| + 1$ for $1 \leq i \leq m_1$. Equality can hold here only if after reduction, the element g_i fails to be cyclically reduced; but in that case, g_i is conjugate to an element of length $\leq |g|$, which we will also denote by g_i. Condition 2 of the theorem then implies that $g_i \in \mathrm{St}_{G_1}(1)$ for $1 \leq i \leq m_1$ and that $g_i = (h_i)^{t_i}$ where $t_i = s/1$ and 1 is the length of the cycle containing the element $x \in A_{n+1}$ corresponding to the index i.

Let $h_i = (h_1^{(i)}, \dots, h_{m_2}^{(i)})$ with $h_j^{(i)} \in G_2$ for $1 \leq j \leq m_2$. Then by arguments similar to those used when proving Lemma 3, the following inequalities hold: $|h_j^{(i)}| \leq (|h_i|+1)/2 < |g|$ if $|g| > 1$, and by inductive hypothesis each of the elements $g_j^{(i)}$ $(1 \leq i \leq m_1,\ 1 \leq j \leq m_2)$ has finite order. Hence g has finite order and the theorem is proved. □

This theorem provides the means for constructing many examples of infinite finitely generated torsion groups. We shall now simplify the construction of torsion groups and obtain necessary and sufficient conditions for periodicity.

Let $G = \langle r, d \rangle$ be the group generated by two elements, a rooted automorphism r and a directed automorphism d. We also assume that the tree $T = T_{\overline{m}}$ is defined by the sequence $\overline{m} = \{m_n\}_{n=1}^{\infty}$ where all the numbers m_n are prime and among them, there are infinitely many which are odd.

Assume that for each $n \geq 1$, the group R_n generated by the components $r_1^{(n)}, \dots, r_{m_n-1}^{(n)}$ of the directed automorphism d is cyclic and generated by the cyclic permutation ε_n of the alphabet A_{n+1}; assume also that the rooted automorphism r is defined by a cyclic permutation ε_0 of the symbols of the alphabet A_1.

Theorem 8. *Assume that the group $G = \langle r, d \rangle$ satisfies the above conditions and d has finite order. Then G is infinite, and G is a torsion group if and only if d has finite order and for an infinite number of values of n, the element $\alpha_n = r_1^{(n)} r_2^{(n)} \dots r_{m_n-1}^{(n)}$ of the group $\Sigma_{m_{n+1}}$ is the identity.*

Proof. [sketch] Every upper companion group $G_n = \langle r^{[n]}, d^{[n]} \rangle$ ($n \geq 1$) also satisfies the conditions of the theorem, therefore G is infinite. Now let us prove the periodicity criterion.

(a) (sufficiency) Suppose that $\alpha_n = 1$ for an infinite number of values of n. The proof of the periodicity of the group G in this case is analogous to the proof of Theorem 7. However, to obtain the effect of decreasing the length of the element $g \notin \mathrm{St}_G(1)$, g should be raised not to the power $q = m_1$, but rather to the power $q = m_1 \dots m_s$ where s is the first value for which $\alpha_s = 1$. Then $g^q \in \mathrm{St}_G(s)$ and if $g^q = (g_1, \dots, g_{N_s})_s$ with $g_i \in G_s$ for $1 \leq i \leq N_s$, then $g_i \in \mathrm{St}_{G_s}(1)$ and g_i is conjugate to an element of length $\leq |g|$.

The elements g_i therefore have the following decomposition: $g_i = (g_1^{(i)} \dots g_{m_{n+1}}^{(i)})$, where $g_j^{(i)} \in G_{s+1}$ for $1 \leq j \leq m_{n+1}$, and moreover $|g_j^{(i)}| < |g_i|$ if $|g_i| > 1$; one can now use a simultaneous induction on length similar to that used in the proof of Theorem 7.

(b) (necessity) Assume that there is an N such that $\alpha_n \neq 1$ for all $n \geq N$.

We will prove that if $n \geq N$ and $1 \neq r \in R_n$, then the element $rd^{[n]} \in G_n$ has infinite order ($d^{[n]}$ denotes the nth section of the element d). From this it will follow that there are elements of infinite order in G, since G contains an element $g \in \mathrm{St}_G(n)$ which stabilizes the first vertex u of the nth level of the tree T and such that $g\,|_{T_u} = rd^{[n]}$.

Indeed, the decomposition for the element $(rd^{[n]})^{m_n} \in \mathrm{St}_{G_n}(1)$ has the form $(rd^{[n]})^{m_n} = (*, \dots, *, \varepsilon^i_{n+1} d^{[n+1]} \varepsilon^j_{n+1}, *, \dots, *)$, where $\varepsilon^{i+j}_{n+1} = \alpha_{n+1} \in R_{n+1}$ so $i + j \not\equiv 0 \pmod{m_{n+1}}$, and the stars denote elements of similar form. Since the element $\varepsilon^i_{n+1} d^{[n+1]} \varepsilon^j_{n+1}$ is conjugate to $\varepsilon^{i+j}_{n+1} d^{[n+1]}$, which has a form analogous to $rd^{[n]}$, it follows that $rd^{[n]}$ has infinite order. This completes the proof. □

We will use this theorem to construct a series of examples of infinite 2-generated p-groups, where p is any prime ≥ 3.

Example 9.1. Let $T = T_p$ be a regular tree where $p \geq 3$ is a prime number. Let r be the rooted automorphism defined by a cyclic permutation of the vertices in the first level. Let d be the directed automorphism defined by the vector $\bar{i} = (i_1, \dots, i_{p-1})$, where $0 \leq i_j \leq p - 1$, $1 \leq j \leq p - 1$ and moreover the elements $r^{(n)}_j \in R_n$, $1 \leq j \leq m_n - 1$ defining d do not depend on n and have the form $r^{(n)}_j = \varepsilon^j$, where $\varepsilon = (0, 1, \dots, p - 1)$ is a cyclic permutation. Thus d has the form shown in diagram 25.

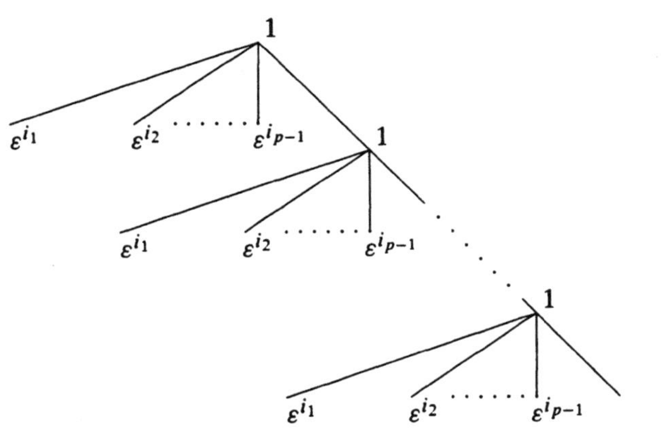

Diagram 25

The group $G = G_{\bar{i}} = \langle r, d \rangle$ is infinite if the vector $\bar{i}$ is not the zero vector, and is torsion if and only if $\sum_{j=1}^{p-1} i_j \equiv 0 \pmod{p}$, as follows from Theorem 8.

If for $\bar{i}$ we take the vector $(1, -1, 0, \dots, 0)$ we obtain the simplest example of this type, first studied in [24].

For analogous examples when $T = T_q$ and q is a prime power, more complicated necessary and sufficient conditions for periodicity were obtained by T. Vovkivskii [47]. The second example from [13] presented by Diagram 16 is a special case of this construction, with $q = 4$.

10 The congruence property and profinite completions of branch groups

Definition 5. The subgroup $G \leq \operatorname{Aut} T_{\overline{m}}$ has the *congruence property* if every subgroup of finite index in G contains the group $\operatorname{St}_G(n)$ for some n.

In this section we will shall see that for a group acting on a rooted tree and having the congruence property, its profinite completion coincides with its closure in the profinite group $\operatorname{Aut} T$ (see below, Theorem 9).

First, let us consider some examples of branch groups with the congruence property.

Example 10.1. Recall that in Example 5.1 we have $G = \langle a, b\rangle < \operatorname{Aut} T_d$, where $d \geq 3$, a is a rooted automorphism which cyclically permutes the edges of the first level, and b is a directed automorphism defined by the relation $b = (a, 1, \dots, 1, b)$. Earlier we proved that for $d \geq 3$, this is a branch group, for $d \geq 5$ it is a just infinite group, and the branch structure on G is a pair $(\{L_n\}, \{H_n\})$ where $L_n \cong G'$ for $n = 1, 2, \dots$. Suppose additionally that $d = p \geq 5$ is prime. It is clear that the upper companion groups G_n coincide with G for all n.

From Theorem 4 it follows that every non-trivial normal subgroup P of G geometrically contains the group $H_n' = L_n' \times \cdots \times L_n'$ (d^n factors) for sufficiently large n. Thus G has the congruence property if G'' contains $\operatorname{St}_G(k)$ for some k, because it is easy to see that

$$\operatorname{St}_G(n+k) \prec \operatorname{St}_G(k) \times \cdots \times \operatorname{St}_G(k) \tag{30}$$

(N_n factors respectively for each vertex of nth level).

We want to prove that the following inclusions hold: $G' \geq \operatorname{St}_G(2)$, $G'' \geq \operatorname{St}_G(3)$.

Obviously $\operatorname{St}_G(1)$ has index p in G, G' is a subgroups of $\operatorname{St}_G(1)$, and $G/G' \leq p^2$. We are going to prove that $|\operatorname{St}_G(1) : G'| = p$. Thus we have to prove that the image of G' in the factor group $\operatorname{St}_G(1)/\operatorname{St}_G(2)$ is a proper subgroup.

The group $\operatorname{St}_G(1)$ can be identified with the subgroup $G \times \cdots \times G$ (p factors) generated by the vectors

$$b^{a^i} = \tau^i(a, 1, \dots, 1, b), \tag{31}$$

$0 \leq i \leq p-1$, where τ is a cyclic permutation of the coordinates of the vector.

The group G' is generated by the elements x_{ij}, $0 < i, j \leq p-1$, where

$$x_{ij} = [a^i, b^j] = (a^j, 1, \dots, 1, b^{-j}, a^{-j}, 1, \dots, 1, b^j)$$

for $1 < i < p-1$ (the element a^{-j} is situated in the ith place), and

$$x_{1,j} = [a, b^j] = (a^j b^{-j}, a^{-j}, 1, \dots, 1, b^j),$$

$$x_{p-1,j} = [a^{p-1}, b^j] = (a^j, 1, \dots, 1, b^{-j}, a^{-j} b^j).$$

Thus the factor group $\mathrm{St}_G(1)/\mathrm{St}_G(2)$ can be identified with the subgroup of $\mathbb{Z}/p\mathbb{Z} \oplus \cdots \oplus \mathbb{Z}/p\mathbb{Z}$ generated by the images of the vectors (31), that is by the vectors $\tau^i(1, 0, \dots, 0)$ (there are p summands and we have changed to additive notation); and the image of the commutator subgroup $G' < \mathrm{St}_G(1)$ is the subgroup generated by the images of the vectors x_{i1} $(0 \le i \le p-1)$, that is, by vectors of type $(1, 0, \dots, 0, -1, 0, \dots, 0)$ where the '-1' ranges through the positions from 1 to $p-1$. In other words, the image of the commutator subgroup G' is the subgroup of index p in $\mathbb{Z}/p\mathbb{Z} \oplus \cdots \oplus \mathbb{Z}/p\mathbb{Z}$ consisting of vectors with zero-sum modulo p. It follows from this that G' has index p in $\mathrm{St}_G(1)$ and also that G' contains $\mathrm{St}_G(2)$.

The inclusion $G'' \succ G' \times \cdots \times G'$ (p factors) was established in Section 5. Therefore from (30) we may infer that

$$G'' \succ \mathrm{St}_G(2) \times \cdots \times \mathrm{St}_G(2) \succ \mathrm{St}_G(3).$$

This implies the congruence property for the groups in Example 10.1.

Example 10.2. Let $p \ge 7$ be a prime. Let $G = \langle a, b\rangle$ be the subgroup of $\mathrm{Aut}\, T_p$ generated by a rooted automorphism a given by a cyclic permutation of the symbols $\{0, 1, \dots, p-1\}$, and a directed automorphism b defined by the relation $b = (a^{i_1}, a^{i_2}, \dots, a^{i_{p-3}}, 1, 1, 1, b)$; here $i_1, \dots, i_{p-3}$ are integers in the range $0 \le i_1, \dots, i_{p-3} \le p-1$, with $i_1 \ne 0$.

Note that if $i_1 = 1$, $i_2 = p-1, i_3 = \cdots = i_{p-3} = 0$, then we obtain the Gupta-Sidki example.

We claim that a group of this type is a just infinite branch group with the congruence property. Observe that

$$x := [a^{p-2}, b^j] = (*, \dots, *, b^{-j}, a^{-i_1 j}, b^j)$$

$$y := [a^{p-3}, b^j] = (*, \cdots, *, b^{-s}, a^{-i_1 s}, *, b^s)$$

where stars denotes elements of the cyclic group $\langle a\rangle$. Therefore $G'' \ni [y, x] = (1, \dots, 1, [a^{i_1 s}, b^j], 1, 1)$ and $G'' \succ G' \times \cdots \times G'$.

From this, all the statements follow. Indeed, G has a branch structure because $|G/G'| < \infty$ and $G' \succ G' \times \cdots \times G'$. Also G is just infinite because

$$|G'/G''| < \left|G'/(G' \times \cdots \times G')\right| < |G/G'| \times \cdots \times |G/G'| < \infty.$$

Finally G has the congruence property because $G' \ge \mathrm{St}_G(2)$ and $G'' \ge \mathrm{St}_G(3)$. This can be proved just as in the previous example.

The congruence property holds for many other examples of branch groups (a further example will be considered in Section 12). For self-similar groups the verification of this property usually reduces to checking the inclusion $G'' \ge \mathrm{St}_G(k)$ for some k, which is normally feasible.

The next statement shows the significance of the congruence property.

Theorem 9. *Let G be a subgroup of $\mathrm{Aut}\, T_{\overline{m}}$ having the congruence property. Then the profinite completion $\widehat{G}$ of G is isomorphic to the closure $\overline{G}$ of G in $\mathrm{Aut}\, T_{\overline{m}}$. If in addition G is contained in a Sylow pro-p subgroup of $\mathrm{Aut}\, T_{\overline{m}}$, then $\overline{G} \cong \widehat{G} = \widehat{G}_p$, the pro-$p$ completion of G.*

Proof. Both statements follow from the fact that the normal subgroups $\mathrm{St}_{\mathrm{Aut}T_{\overline{m}}}(n)$, $(n = 1, 2, \ldots)$ form a base for the neighbourhoods of 1 in the profinite group $\mathrm{Aut}\, T_{\overline{m}}$. □

11 The universal embedding property of some branch pro-p groups

The purpose of this section is to prove that the pro-p completions of certain branch groups are universal in the class of all countably based pro-p groups with respect to embedding.

The following theorem generalises Proposition 2 in Section 3, with a similar proof:

Theorem 10. *Let $G \le \mathrm{Aut}\, T_{\overline{m}}$ be a branch group with the congruence property. Suppose that G has a branch structure $(\{L_n\}_{n=1}^{\infty}, \{H_n\})_{n=1}^{\infty}$ such that that for each $n \ge 1$, the group L_n contains a copy of every finite p-group. Then the profinite completion $\widehat{G}$ of G contains every countably based pro-p group as a closed subgroup.*

Proof. It suffices to prove that $\widehat{G}$ contains the Cartesian product of an arbitrary sequence P_n of finite p-groups.

Now consider the sequence of finite subgroups K_n of G geometrically defined as follows, where for each i, P_i denotes a fixed copy of the group P_i inside L_i:

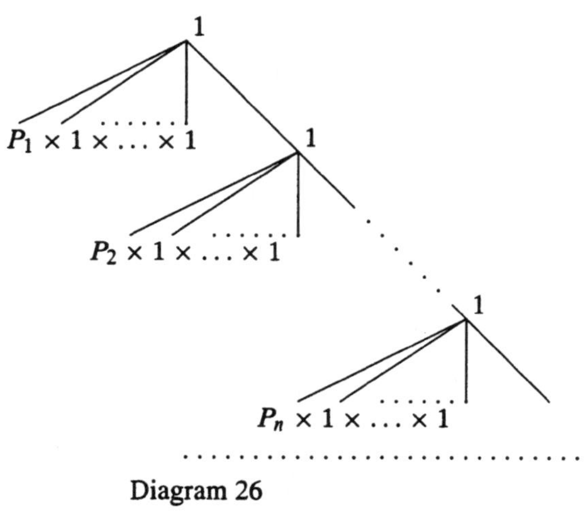

Diagram 26

Evidently $K_n \cong P_1 \times \cdots \times P_n$, and $K_n \cap \mathrm{St}_G(n) = 1$. It follows that

$$\prod_{i=1}^{\infty} P_i = \varprojlim_{n\to\infty} K_n \le \varprojlim_{n\to\infty} (G/\mathrm{St}_G(n)) \cong \widehat{G}, \qquad (32)$$

as G has the congruence property. The result follows. □

Let G be as in Theorem 10. From Theorem 2(a), we know that $\widehat{G}$ is again a branch group; and if G is just infinite, then $\widehat{G}$ is again just infinite, by the Corollary to Theorem 4. If $\widehat{G}$ is a pro-p group, then it follows that $\widehat{G}$ is universal in the class of countably based pro-p groups with respect to embedding.

As we shall see, this applies to the completions of many torsion branch groups. In particular they contain free pro-p groups, which seems surprising.

In the previous section we constructed examples of just infinite branch groups with the congruence property. We will show that they satisfy the condition of Theorem 10, namely that the groups L_n from the branch structure contain copies of every finite p-group.

Since in these examples L_n is isomorphic to the commutator subgroup G' for each $n \geq 1$, it will suffice to prove that G' contains a copy of every finite p-group. We will show that the latter property holds for the following class of torsion groups.

Example 11.1. Let $p \geq 5$ be a prime and let $G = \langle a, b \rangle \leq \operatorname{Aut} T_p$, where a is a rooted automorphism defined by cyclic permutation of the edges of the first level, and b is a directed automorphism defined by $b = (a^{i_1}, \dots, a^{i_{p-2}}, 1, b)$ where $i_1 \neq 0$, $0 \leq i_1, \dots, i_{p-2} \leq p-1$ and $\sum i_j \equiv 0 \pmod{p}$. Recall that the latter condition implies that the group G is periodic (see Theorem 8).

We have

$$b^{a^{p-1}} = (a^{i_2}, \dots, a^{i_{p-2}}, 1, b, a^{i_1})$$

and

$$G' \ni [b^{a^{p-1}}, b] = (1, \dots, 1, [b, a^{i_1}]).$$

Therefore $G' \succ G' \times \cdots \times G'$ (p factors). This shows that G is a branch group.

If $p > 6$ and $i_{p-1} = i_{p-2} = 0$, then we are in the situation of Example 10.2 and therefore G has the congruence property. Thus $\widehat{G} \cong \overline{G}$; as G is contained in a Sylow pro-p subgroup of $\operatorname{Aut} T$, we find that in this case $\widehat{G}$ is a pro-p group.

We will now prove that every finite p-group can be embedded in G'. For this we need the following proposition, which is a slight modification of the corresponding statement from [25].

Proposition 4. *Let $G = \langle a, b, \dots, d \rangle \leq \operatorname{Aut} T_n$, where $n \geq 2$, be a self-similar group generated by a rooted automorphism a which cyclically permutes the vertices of the first level, together with some directed automorphisms $b, \dots, d$. Assume that G contains a self-replicating subgroup K, that is $|G : K| < \infty$ and $K \succ K \times \cdots \times K$ (n factors). Finally assume that K contains an element k of order n which stabilizes a vertex $u \in V(T_n)$ and such that $k\,|_{T_u} = a$.*

Then for each $N \geq 1$, the group K contains the iterated wreath product of N copies of the cyclic group of order n.

Proof. Let us use the alphabet $\{0, 1, \dots, n-1\}$ for coding the paths and vertices of the tree T. We define a sequence of elements $\{k_i\}_{i=1}^{\infty}$ inductively: $k_1 = k$ and the element

$k_{i+1} \in \mathrm{top}_K((u0)^i)$ is defined by the relation $k_{i+1}|_{T_{(u0)^i}} = k$. We claim that the group $Q_N = \langle k_1, \dots, k_N \rangle$, is isomorphic to the N-fold wreath product that we want, with the ith factor in the product being generated by k_i.

We will conduct the proof by induction on N. The base case $N = 1$ is clear. Assume that Q_N is an iterated wreath product of N copies of the cyclic group of order n. We will prove the analogous fact for Q_{N+1}.

By inductive hypothesis

$$P := \langle k_2, \dots, k_{N+1} \rangle \cong C_n \wr \cdots \wr C_n$$

(N copies) where P is a subgroup of $\mathrm{St}_G(|u|+1)$.

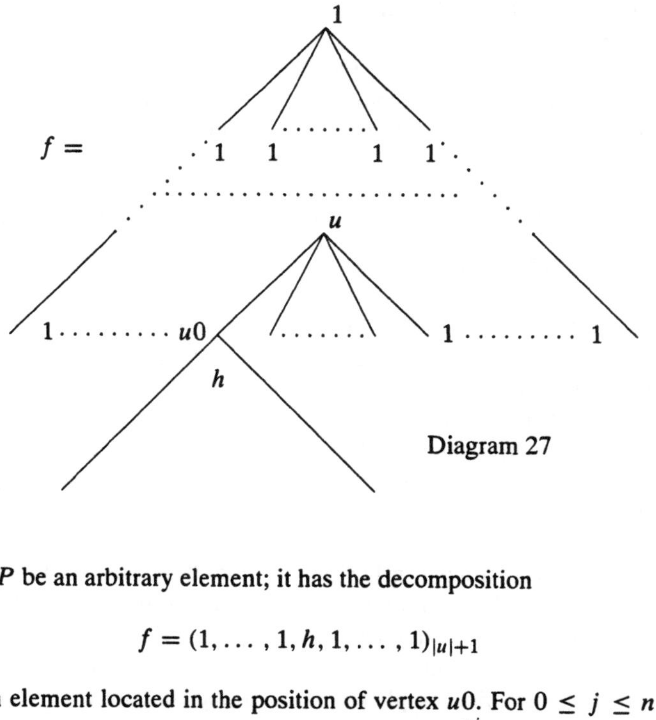

Diagram 27

Let $f \in P$ be an arbitrary element; it has the decomposition

$$f = (1, \dots, 1, h, 1, \dots, 1)_{|u|+1} \tag{33}$$

where h is an element located in the position of vertex $u0$. For $0 \le j \le n-1$, The decomposition at the $(|u|+1)$th level of the element f^{k^j} is obtained from (33) by the jth power of the cyclic permutation of the coordinates which are numbered using the vertices $u0, u1, \dots, u(n-1)$. Thus $S = \langle f^{k^j}, f \in P, 0 \le j \le n-1 \rangle \cong P \times \cdots \times P$ (n factors) and the group generated by $\langle k \rangle$ acts by conjugation on S cyclically permuting the factors of P. Thus $\langle k, S \rangle \cong P \wr C_p$ when acting on a subtree T_u. As f acts trivially on $T \setminus T_u$ and k has order n, it follows that $\langle k, S \rangle$ is isomorphic to $P \wr C_p$ as a group acting on the whole tree T.

The proposition is proved. □

Since an arbitrary finite p-group can be embedded into an iterated wreath product of N copies of the cyclic group C_p, using a sufficiently large N, the embedding problem reduces to verifying the conditions of Proposition 4 for the groups under consideration. Moreover, since in our case $L_i = G'$ for each i, it suffices to construct an element $k \in G'$ satisfying the conditions of the proposition.

We now return to Example 11.1. Consider the element $x = [a, b] = (a^{i_1}b, *, \dots, *, b)$, where the stars denote elements of the group $\langle a \rangle$, and $i_n \neq 0$. Then $x^p = ((a^{i_1}b)^p, 1, \dots, 1)_1 = (a^j ba^t, 1, \dots, 1)_2$ where j, t are numbers such that $0 \leq j, t \leq p - 1$ and $j + t \equiv 0 \pmod{p}$.

Thus $x^p \in \mathrm{St}_G(3)$, x^p is an element of order p, and

$$x^p = (\tau^{-j}(a^{i_1}, \dots, a^{i_{p-2}}, 1, b), 1, \dots, 1)_3$$

where τ, as before, denotes the cyclic permutation of the vector coordinates. Hence there exists a vertex in the third level $u = 00j$ such that $x^p \mid_{T_u} = a^{i_1}$ with $i_1 \neq 0$, and we can use Proposition 4.

The universal embedding property which we have been discussing can be extended to all branch pro-p groups which are not virtually torsion-free. For the details, see [21]. Related result have been obtained by J. S. Wilson (unpublished).

12 A just infinite branch 2-group of finite width (preliminary data)

Until now, all examples of branch groups which we have considered were related to the case of a regular tree T_d where $d \geq 3$ and so one might think that the case $d = 2$ is exceptional. Although in some situations the minimal possible branch number of the tree, namely 2, does indeed create certain difficulties, nevertheless there are many examples of branch groups acting on the tree T_2.

The remaining sections of this chapter will be dedicated to one such group, in fact the one from which the theory of branch groups began. An exposition of some other properties of the group can be found in [28] and [38].

This group was constructed in [13] as an example of an infinite 3-generator 2-group. Then in [14], [15] it was established that it is a group of intermediate growth (which answered Milnor's question on the existence of such groups), and also G has a branch structure, is just infinite, and has the congruence property. It follows from Theorem 10 that the profinite completion of this group contains every countably based pro-2 group as a closed subgroup.

In [17] it was shown that G has bounded average width and in [38] it was established that the width of G is finite. So far, this has been the only known example of a branch group with bounded width.

Our main goal is to prove the finiteness of the width of the group G (see Theorem 13 below). Along the way we shall establish most of the properties of G listed above (except for periodicity, the proof of which is analogous to the proof of Theorem 7 – although in one place there is a minor difference. See [13]). Nor do we prove the intermediate growth property, which can be found in [15] or [18]). Also we will describe the upper and lower central series of G. Our proof of the finiteness of the width differs from the one given in [38] (although both follow the same strategy).

Let us turn to the definition of the group G. It acts on a regular tree $T = T_2$ and is generated by automorphisms a, b, c, d where a is a rooted automorphism permuting the vertices in the first level, and b, c, d are the directed automorphisms defined in Diagram 18.

It is easy to see that the generators satisfy the relations $a^2 = b^2 = c^2 = d^2 = 1$, $[b, c] = [b, d] = [c, d] = 1$, $bcd = 1$ that is $bc = d$, $bd = c$, $cd = b$, and also many other relations which do not follow from those above. For example $(ad)^4 = (ac)^8 = (ab)^{16} = 1$.

In fact G is not finitely related, but can be given by the following presentation first obtained by I. Lysionok in [34]:

$$G = \left\langle a, b, c, d \;\middle|\; \begin{array}{rl} 1 & = a^2 = b^2 = c^2 = bcd \\ & = \sigma^k((ad)^4) = \sigma^k((adacac)^4),\ k = 0, 1, \dots \end{array} \right\rangle, \tag{34}$$

where the substitution σ is defined by

$$\sigma : \begin{cases} a \to aca \\ b \to d \\ c \to b \\ d \to c. \end{cases}$$

The question of the finite presentability of G is considered in [13], [15] and [20]. The latter article also describes a method for obtaining presentations of type (34) for special groups.

Since $d = bc$, G is in fact a three generator group, but it is more convenient to use the four generators a, b, c, d. Note that the elements $1, b, c, d$ form a subgroup of G isomorphic to Klein's group $C_2 \times C_2$. Because the generators have order 2 we may just use words in the positive alphabet $\{a, b, c, d\}$ for representing the elements of G; indeed any element $g \in G$ can be expressed in the form $g = *a * a * \cdots * a*$, where the stars denote elements of the set $\{b, c, d\}$ and the end stars may additionally take the value 1.

The stabilizer $\mathrm{St}_G(1)$, which we denote by H, is clearly generated by the elements b, c, d, b^a, c^a, d^a (in fact four generators suffice, for example c, d, c^a, d^a). Since these generators have the following decompositions

$$\begin{array}{ll} b = (a, c) & b^a = (c, a) \\ c = (a, d) & c^a = (d, a) \\ d = (1, b) & d^a = (b, 1), \end{array} \tag{35}$$

it follows that the map $\psi : H \to G \times G$ given by

$$\psi : \begin{cases} b & \to \quad (a, c) \\ c & \to \quad (a, d) \\ d & \to \quad (1, b) \\ b^a & \to \quad (c, a) \\ c^a & \to \quad (d, a) \\ d^a & \to \quad (b, 1) \end{cases} \tag{36}$$

defines an embedding of H into $G \times G$. Denote by φ_i the composition of ψ with the projection $\pi_i : G \times G \to G$ on the appropriate factor ($i = 0, 1$). Then $\psi = \varphi_0 \times \varphi_1$ and $\varphi_i : H \to G$ is an epimorphism (since in (36) each generator of G is present both in the position of the first coordinate of the image, and of the second).

The homomorphisms φ_i may be conveniently used when showing that certain elements of G given by words are equal to 1. For example, for the element $(ab)^{16}$ this verification is accomplished as follows (Diagram 28):

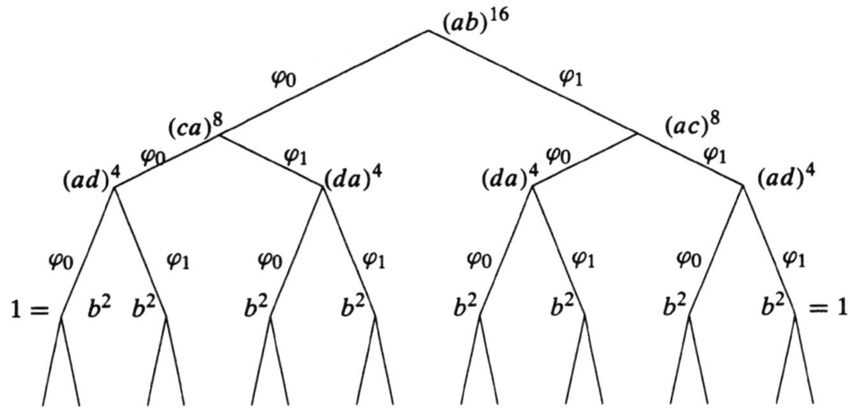

Diagram 28

From this diagram it is clear that $(ab)^{16} = (ac)^8 = (ad)^4 = 1$; in fact the method here illustrated is an algorithm which solves the word problem in G. (Concerning algorithmic properties of the group G and other branch groups, see [15], [18], [20], [22] and literature quoted there.)

We specify certain subgroups of G, namely $B = \langle b \rangle^G$, $K = \langle (ab)^2 \rangle^G$, $D = \langle a, d \rangle$, the final one being a copy of the dihedral group of order 8. Additionally, we define two subgroups of $G \times G$, namely $D_1 = \langle (a, 1), (d, 1) \rangle$ and $D_2 = \langle (a, d), (d, a) \rangle$. Obviously $D \cong D_1 \cong D_2$.

We also fix the elements $x = (ab)^2$, $y = (bd^a)^2$, $z = (b^a d)^2$. note that $x = (ca, ac)$, $y = (x, 1)$, $z = (1, x)$.

Proposition 5. *The group G is equal to the semidirect product $G = B \rtimes D$ and hence $|G : B| = 8$, $B \cap D = \{1\}$.*

Proof. Clearly the factor group G/B is generated by the images $\overline{a}, \overline{d}$ of elements a, d and hence is a dihedral group of order 2, 4 or 8. In fact the order of this group is 8.

To establish this fact, it suffices to find the index of the image $\overline{B}$ of B in the factor group $\overline{G} = G/\mathrm{St}_G(3)$.

An elementary verification yields that $\overline{G} \cong C_2 \wr C_2 \wr C_2$ is the automorphism group of the binary tree up to the third level and that $|\overline{B}| = 16$. Thus $|\overline{G} : \overline{B}| = 2^7/2^4 = 2^3$ and hence $G/B \cong D$, $B \cap D = \{1\}$ and $G = B \rtimes D$. □

Proposition 6. *The group B is generated by the elements b, b^a, y, z.*

Proof. That $y, z \in B$ follows from the fact that $y = [b, d^a]$, $z = y^a$ and B is normal in G.

Denote by B_* the subgroup of B generated by elements b, b^a, y, z. To prove the statement, it suffices to check that B_* is normal in G. To do this, it suffices to verify that B_* is invariant under conjugation by d.

Now

$$\begin{aligned} db^a d &= b^a(b^a d)^2 \in B_* \\ dyd &= ((ab)^2, 1) = y \\ dzd &= (1, (ba)^2) = (1, (ab)^{14}) = z^7 = z^{-1} \end{aligned}$$

since $(ab)^{16} = 1$. The result follows. □

Proposition 7.

(a) *$G \times G$ is equal to the semidirect product $(B \times B) \rtimes (D \times D)$.*

(b) *The group $\psi(H)$ is equal to the semidirect product $(B \times B) \rtimes D_2$. It is a subgroup of index 8 in $G \times G$ and has complement D_1.*

Proof. a) The subgroup $B \times B$ is normal in $G \times G$ and has index 64. From Proposition 5 it follows that $(B \times B) \cap (D \times D) = \{(1, 1)\}$, which completes the proof.

b) First note that D_1 is a normal subgroup of $D \times D$ and that there is a semidirect product decomposition $D \times D = D_1 \rtimes D_2$

As $(1, b) \in \psi(H)$ the inclusion $1 \times B \subset \psi(H)$ holds, and similarly we also have $B \times 1 \subset \psi(H)$. Indeed for any element $g \in G$ there is an element $h \in \mathrm{St}_G(1)$ such that $h = (*, g)$, where star denotes an element of the group G. Since $(1, b)^h = (1, b^g)$, this shows that $1 \times B \subset \psi(H)$.

Since $\psi(c) = (a, d)$ and $\psi(c^a) = (d, a)$, it follows that $\psi(H)$ is the group generated by $B \times B$ and D_2. But $B \times B$ is normal in $\psi(H)$ and $(B \times B) \cap D_2 = \{(1, 1)\}$. Hence $\psi(H) = (B \times B) \rtimes D_2$ and D_1 is a complementary subgroup in $G \times G$. □

Proposition 8. *The group K is generated by the elements x, y, z, has index 16 in G, and is self-replicating, that is $K \succ K \times K = K_1$ where $K/K_1 \cong C_4$ is the cyclic group generated by the image of $(ab)^2$.*

Also $(ac)^4 \in K$ and the following relations hold:

$$(ac)^2 \equiv (ad)^2 \bmod K \tag{37}$$

$$(ab)^4 \equiv (ac)^4 \bmod K_1. \tag{38}$$

Proof. The element $y = (bd^a)^2$ is in K since

$$y = ab[(ab)^2, d]ba.$$

Hence also $z = y^a \in K$.

Let K_* be the group generated by x, y, z. Then K_* is normal and hence coincides with K. Indeed,

$$\begin{aligned} a(ab)^2a &= (ba)^2 \in K_*, \\ b(ab)^2b &= (ba)^2 \in K_*, \\ d(ab)^2d &= (b^ad)^{-2}(ab)^2 \in K_*, \\ b(bd^a)^2b &= (bd^a)^{-2} \in K_*, \end{aligned}$$

and the other relations may be verified similarly.

Using the information about the generators of B, it is easy to conclude that $B/K \cong \langle b\rangle$ and hence that K is of index 16 in G. In fact $G/K \cong C_2 \times D$ where C_2 is generated by the image of b and D is generated by the images of a and d, which may be verified by looking at the images of G and K in $G/\mathrm{St}_3(G)$.

Since $(bd^a)^2 = ((ab)^2, 1)$, we have $K \succ K \times 1$. Similarly $K \succ 1 \times K$ and hence $K \succ K \times K$. Since $y, z \in K_1$, the factor group K/K_1 is generated by the image of the element $(ab)^2$. We will prove that $(ab)^8 \in K_1$ but $(ab)^4 \notin K_1$.

Note first that $(ad)^2 \notin K$ since otherwise in the factor group G/K the images $\overline{a}, \overline{d}$ of the elements a, d would commute, contradicting the isomorphism $G/K \cong C_2 \times D$.

The relation (37) holds because the element

$$(ac)^2(ad)^{-2} = acacdada$$

is conjugate to $(ab)^2 \in K$. Thus $(ac)^2 \notin K$ and since $(ab)^4 = ((ca)^2, (ac)^2)$ we have $(ab)^4 \notin K_1$. Also since $(ac)^4 = ((da)^2, (ad)^2)$ it follows that $(ab)^4 = (ac)^4 \bmod K_1$. Hence $(ac)^4 \in K$. Finally $(ab)^8 = ((ca)^4, (ac)^4) \in K_1$. □

Corollary. *G is a branch group.*

For each $m \geq 1$ we shall denote by K_m the subgroup of $\mathrm{St}_G(m)$ which is the product of 2^m copies of K acting on the subtrees of the binary tree which begin at the the mth level.

Proposition 9. *The equality $K' = K_2$, holds as do the inclusions $K \geq St_G(3)$, $K' \geq St_G(5)$.*

Proof. A direct calculation shows that

$$\begin{aligned} \mathrm{St}_G(3) \ni [x, y] &= ((ba)^2, 1, 1, 1)_2 \\ &= (ac, ca, 1, 1, 1, 1, 1, 1,)_3. \end{aligned}$$

Therefore $K' \succ K \times 1 \times 1 \times 1$ and so $K' \succ K \times K \times K \times K$. However since $[x, y] \in K_2$, $[x, z] \in K_2$, and y commutes with z, in fact the equality $K' = K_2$ is established.

Since the index of the image $\overline{K}$ of K in the factor group $G/\mathrm{St}_G(3)$ coincides with $|G : K| = 16$, it follows that $K \geq \mathrm{St}_G(3)$. Finally, by (30) we obtain that $K' \geq \mathrm{St}_G(5)$. □

Corollary. *The group G is just infinite.*

This is an application of Theorem 4; that $|K/K'| < \infty$ follows from the facts that G and hence K are torsion, and that K is finitely generated.

Proposition 10. *G has the congruence property. The profinite completion of G is a pro-2 group and contains as a closed subgroup every countably based pro-2 group.*

Proof. The fact that G has the congruence property follows from Theorem 4 and the previous proposition. Thus by the corollary to Theorem 4, $\widehat{G}$ coincides with the closure of G in Aut T_2, and we can use Theorem 10. Since now $L_n = K$ for each n, it remains to construct the element $k \in K$ mentioned in Proposition 4. Such an element is, for example, $k = (ab)^8 \in \mathrm{St}_G(4)$ for which we have the decomposition

$$(ab)^8 = (a, c, a, c, a, c, a, c, \dots, a, c)_4.$$

Evidently k has order 2, and if the vertex u is defined by the sequence (0, 0, 0, 0), then $k\,|_{T_u} = a$. □

Theorem 11. *The following relations hold:*

1) $x^4 \equiv (x^{-2}, x^2) \bmod K_2$

2) $x^a = x^b = x^{-1},\ x^d = x^{-c} = (1, x^{-1})x$

3) $[x, a] = [x, b] = x^{-2}$

4) $[x, c] = x^{-2}(1, x)$

5) $[x, d] = (1, x^{-1})^x$

6) $[x^2, a] = [x^2, b] = x^4$

7) $[x^2, c] \equiv (x^2, 1) \bmod K_2$

8) $[x^2, d] \equiv (1, x^2) \bmod K_2$.

Proof. 1). Using the relation (37) we obtain the following relations mod K_2:

$$\begin{aligned} x^4 &= ((ca)^4, (ac)^4)_1 = ((ad)^2, (da)^2, (da)^2, (ad)^2)_2 \\ &= ((ac)^2, (ca)^2, (ca)^2, (ac)^2)_2 = ((ba)^4, (ab)^4) = (x^{-2}, x^2). \end{aligned}$$

2). The first group of relations is obvious. The second follows from these together with the relations $[d, x^{-1}] = x^d x^{-1} = (1, x^{-1})$.

3). The verification of these is obvious.

4). $x^2[x, c] = (ab)^2 c (ab)^2 c = (1, x)$.

5). $x[x, d]x^{-1} = d(ab)^2 d(ba)^2 = (1ca1ac, bacbca) = (1, (ba)^2) = (1, x^{-1})$.

6). $[x^2, a] = (ba)^8 = x^{-4} = x^4$ because $x^8 = 1$. Similarly $[x^2, b] = x^{-4} = x^4$.

7). $[x^2, c] = (ba)^4 c (ab)^4 c = ((ac)^4, (ca)^2 d (ac)^2 d)$. But $(ac)^4 \equiv (ab)^4 \bmod K_1$, and $(ca)^2 d (ac)^2 d$ is conjugate to $(acad)^2 = (1, (ab)^2)$ which belongs to K_1. Thus $[x^2, c] \equiv (x^2, 1) \bmod K_2$.

8). $[x^2, d] = (ba)^4 d(ab)^4 d = ((ac)^2 1(ca)^2 1, (ca)^2 b(ac)^2 b) = (1, cacabacabc)$. However the following relations hold modulo K_1: $c(acab)^2 c = ((da)^2, d(ac)^2 d) = ((da)^2, d(ad)^2 d) = ((da)^2, (ad)^2) = ((ca)^2, (ac)^2) = (ab)^4 = x^2$.

The theorem is proved. □

13 The derived series of G

The results of this section are based on [37]. We recall that the derived series $G^{(n)}$, $n \geq 1$ of G is defined as follows: $G^{(1)} = G' = [G, G]$ and $G^{(n+1)} = [G^{(n)}, G^{(n)}]$, $n \geq 1$. The goal of this section is to describe the terms of the derived series and also to compute their indices in G.

We keep the notation $x = (ab)^2$, $y = (bd^a)^2$, $z = (b^a d)^2$ from the preceding section, and recall that the following relations hold: $y = (x, 1)$, $z = (1, x)$; these will be used on many occasions in what follows. As usual we denote by C_n the cyclic group of order n.

Theorem 12. *The following relations hold:*

(a) $G^{(1)} = \langle [a, d], K \rangle$ and $G^{(1)}/K \cong C_2$.

(b) $G^{(2)} = \langle y^2, yz, K^{(1)} \rangle$ and $G^{(2)}/K^{(1)} \cong C_2 \times C_4$.

(c) $K_1 = \langle y, G^{(2)} \rangle$ and $K_1/G^{(2)} \cong C_2$.

(d) $G^{(n)} = K_{2n-3}$ if $n \geq 3$.

The indices satisfy

$$[G : G^{(n)}] = \begin{cases} 2^3 & \text{if } n = 1 \\ 2^7 & \text{if } n = 2 \\ 2^{2^{2n-2}+2} & \text{if } n \geq 3. \end{cases} \tag{39}$$

Proof. We will first prove the statements (a) – (d).

(a) Since $(ab)^2 = [a, b]$, the inclusion $G^{(1)} \geq P := \langle [a, d], K \rangle$ is obvious. On the other hand $G^{(1)} = \langle [a, b], [a, d] \rangle^G$ and therefore to establish the reverse inclusion, it suffices to show that P is a normal subgroup of G. However $[a, d]^a = [a, d]^d = [d, a]$ and $[a, d]^b = [b, a][b, a]^2[a, d] \in P$. Therefore $G^{(1)} = P$ because K is normal. The relation $G^{(1)}/K \cong C_2$ is obvious because $[a, d]^2 = 1$.

(b) From (a) and the fact that $K = \langle x, y, z \rangle$ it follows that

$$G^{(2)} = K^{(1)} \cdot \langle [[a, d], x], [[a, d], y] \rangle^G.$$

Since $[[a, d], x]^b = y^{-1} z$ and $[[a, d], y] = y^2$, we have $G^{(2)} \geq Q := \langle y^2, yz, K^{(1)} \rangle$. However, the group Q is normal in G: this follows from the inclusions $[y^2, \xi]$, $[yz, \xi] \in Q$, where $\xi \in \{a, b, c\}$, which can be verified directly. Therefore $G^{(2)} = Q$.

The elements y and z commute, and also the following relations hold:

$$\begin{aligned} y^4 &= (x^4, 1) \equiv (1, 1) = 1 \bmod K^{(1)} \\ (yz)^4 &= (x^4, x^4) \equiv (1, 1) = 1 \bmod K^{(1)}, \end{aligned}$$

because $x^4 \in K_1$ and $K^{(1)} = K_2$ (see Proposition 9). From this we conclude that $G^{(2)}/K^{(1)} \cong C_2 \times C_4$.

(c) The group K_1 contains the elements y and z and hence contains the group $G^{(2)}$. Since $K = \langle x, K^{(1)}\rangle$ we have $K_1 = \langle (x,1), (1,x), K_1\rangle$; from the relation proved in the previous item, we infer the equality $K_1 = \langle y, G^{(2)}\rangle$, and as $y^2 \in G^{(2)}$ we see that $K_1/G^{(2)} \cong C_2$.

(d) From the expression for $G^{(2)}$ given in (b), we conclude that $G^{(3)} = K^{(2)}\langle [y^2,\eta], [yz,\eta], [y^2,\xi], [yz,\xi]\rangle^G$ where

$$\eta = (x,1,1,1) \in \mathrm{St}_K(2), \; \xi = (x,1,1,1,1,1,1,1) \in \mathrm{St}_K(3),$$

because $K_1 = K_2$ and $K = \langle x, y, z\rangle$, $y = (x, 1)$, $z = (1, x)$. However,

$$\begin{aligned}[y^2,\eta] &= [(x^2,1),(x,1,1,1)] = ([(ca)^2,(ab)^2],1,1,1) \\ &= (x^{-1},x,1,1,1,1,1,1) \in K_3,\end{aligned}$$

$$\begin{aligned}[yz,\eta] &= [(x,x),(x,1,1,1)] = ([ca,(ab)^2],1,1,1) \\ &= (y^{-1},1,1,1) = (x^{-1},1,1,1,1,1,1,1),\end{aligned}$$

and therefore $G^{(3)} \geq K_3$. Similarly one can verify that the commutators $[y^2,\xi]$, $[yz,\xi]$ belong to K_3. Since $K^{(1)} = K_2$ we have $K^{(2)} = K_4$, which leads to $G^{(3)} = K_3$.

Using this relation and the fact that $K^{(1)} = K_2$, we obtain the relations $G^{(n)} = K_{2n-3}$, $n = 3, 4, \ldots$ by induction on n.

After the items (a) – (d) have been proved, the question of indices $|G : G^{(n)}|$ is easily solved. Indeed the value of $|G : G^{(1)}|$ follows from the isomorphism $G/G^{(1)} \cong C_2^3$, and the equality $|G : G^{(2)}| = 2^7$ follows from the previous one, together with the sequence

$$G^{(1)} \underset{2}{\triangleright} K \underset{4}{\triangleright} K_1 \underset{2}{\triangleright} G^{(2)}$$

of nested subgroups, where the subscripts indicate the relative indices. Finally since $|K : K_1| = 4$ we have $|K_n : K_{n+1}| = 4^{2^n}$, from which the equality (39) for $n \geq 3$ follows immediately. □

14 The lower central series of G

As usual, $\gamma_n(G)$ denotes the nth term of the lower central series of G. Thus $\gamma_1(G) = G$ and $\gamma_{n+1}(G) = [G, \gamma_n(G)]$.

Since G is generated by elements of order 2, each of the factor groups $\gamma_n(G)/\gamma_{n+1}(G)$ is an elementary abelian 2-group.

The goal of this section is to give a full description of the terms of the lower central series, and consequently to compute the ranks of the groups $\gamma_n(G)/\gamma_{n+1}(G)$, which (as will be seen) are positive and bounded by 3. Thus G has *finite width*, a property that plays an important role in group theory and its applications ([17],[32]) .

Let $S = \langle (ab)^4 \rangle^G$ and for $m \geq 0$ let S_m be direct product of 2^m copies of the group S acting on the subtrees beginning at the vertices of the mth level. Recall that K_m is the subgroup similarly defined with $(ab)^2$ in place of $(ab)^4$. Now for $m \geq 1$ we put

$$N_m = S_{m-1} \cdot K_m.$$

Since S_m and K_m are normal in G, each of the groups N_m is normal in G. They play a primary role in the study of the lower central series of G.

We can now state the main result. It is due to A. Rozhkov [38] (having roots in [17]). Our exposition is quite different from that in [38], though it is based on the same idea. For more details and related results see [3].

Theorem 13. *The following hold:*

$$\mathrm{rk}(\gamma_n(G)/\gamma_{n+1}(G)) = \begin{cases} 3 \text{ if } n = 1 \\ 2 \text{ if } n = 2 \text{ or } 2^m < n \leq 3 \cdot 2^{m-1} \\ 1 \text{ if } 3 \cdot 2^{m-1} < n \leq 2^{m+1}, m = 1, 2, \ldots \end{cases}$$

Also $\gamma_{2^m+1}(G) = N_m$ *for each* $m \geq 1$.

Proof. [Sketch] Let $u = x^2$ where, as before, $x = (ab)^2$. Using the relations of Theorem 11, it is easy to prove that $N_1/N_2 \cong C_2^3$, a basis for this factor group being given by the images of the elements $(x, 1), (1, x) \in K_1$ and the element $u = x^2 \in S_0$. Denote these images by $(\overline{x}, 1), (1, \overline{x})$ and $\overline{u}$ respectively. Thus the relations $\overline{x}^2 = 1, \overline{u}^2 = 1$ both hold, and we will use these in the sequel.

For each $m \geq 1$ we have

$$N_m/N_{m+1} \cong N_1/N_2 \times \cdots \times N_1/N_2 \cong C_2^{3 \cdot 2^{m-1}} \tag{40}$$

(with 2^m factors in the middle term). Denote by W_m and $\overline{W}_m$ the vector spaces of dimension 2^m over the field $\mathbb{F}_2$ whose bases are the following sets of vectors:

$$\{(1, \ldots, 1, \overline{x}, 1, \ldots, 1)_m \mid \overline{x} \text{ in the } i\text{th position}, 1 \leq i \leq 2^m\}$$

$$\{(1, \ldots, 1, \overline{u}, 1, \ldots, 1)_m \mid \overline{u} \text{ in the } i\text{th position}, 1 \leq i \leq 2^m\}$$

respectively (we use multiplicative notation, identifying the elements of C_2 with the additive group of $\mathbb{F}_2$). Then $N_1/N_2 = W_1 \oplus \overline{W}_0$, and we can rewrite (40) as

$$N_m/N_{m+1} \cong W_m \oplus \overline{W}_{m-1}. \tag{41}$$

To understand the following discussion, it is convenient to identify the given basis elements of W_m with the vertices of the mth level of the binary tree T labelled by the symbol $\overline{x}$, and to identify the given basis elements of $\overline{W}_m$ with the vertices of the mth level labelled by the symbol $\overline{u}$.

We already know that $G/G' \cong C_2^3$ and $G' = \langle [a, d], K \rangle$. Since $x \in K$, the subgroup $\gamma_3(G)$ contains the elements $x^2 = [a, x], y = [x, d]^{ba} = (x, 1)$, and thus $\gamma_3(G) \geq N_1$.

Considering the commutators of the generators $[a, d], x, y, z$ of G' with the elements a, b, c, d, one can easily verify that they all belong to the group N_1. Thus we arrive at the equality $\gamma_3(G) = N_1$.

The factor group $\gamma_2(G)/\gamma_3(G) = G'/N_1$ is generated by the images of the elements $[a, d]$ and x, since the generators y and z of K belong to K_1 and hence to N_1. Since $x^2 \in N_1$ but $[a, d]x \notin N_1$, it follows from this that $\gamma_2(G)/\gamma_3(G) \cong C_2^2$.

Assuming that for some $m \geq 1$ we have $\gamma_{2^m+1}(G) = N_m$, we shall show that then $\gamma_{2^{m+1}+1}(G) = N_{m+1}$, and establish the other statements of Theorem 13. To this end, we introduce vectors of type $p = (p_1, \dots, p_{2^m})$ where $p_i \in \{1, x\}$, which we will call *x-vectors of dimension m*. Such a vector is called non-trivial if it has at least one non-identity component.

We will need the following technical statement, the proof of which can be found in [38].

Lemma 4. *Let $p \in St_G(m)$ where $p = (p_1, \dots, p_{2^m})$ is a non-trivial x-vector of dimension m. Then $[\langle p\rangle, G] \geq N_{m+1}$.*

Corollary. *If $p \in \gamma_n(G)$ then $\gamma_{n+1}(G) \geq N_{m+1}$.*

Recall that we agreed to identify basis elements of the spaces $W_m, \overline{W}_m$ with the vertices of the mth level of T labelled by the corresponding symbols.

Any group Γ acting on the tree T by automorphisms has a natural action on each of the spaces $W_m, \overline{W}_m, m = 1, 2, \dots$ (moreover the Γ-modules $W_m, \overline{W}_m$ are clearly isomorphic).

For the Γ-module W we define the derived module $[\Gamma, W]$ as the submodule of W generated by elements of the form $w - hw$ where $w \in W, h \in \Gamma$. For each number $m = 1, 2, \dots$, define $W_m^0 = W_m, W_m^1 = [\Gamma, W_m]$ and inductively $W_m^{i+1} = [\Gamma, W_m^i], i = 1, 2, \dots$. Similarly for each $m = 1, 2, \dots$ the sequence $\overline{W}_m^i$ $(i \geq 1)$ of Γ-modules is defined and is isomorphic to the previous one.

One can show that for an arbitrary group $\Gamma \leq \operatorname{Aut} T$, the sequence of subspaces

$$W_m = W_m^0 \supset W_m^1 \supset \dots \supset W_m^i \supset \dots$$

is strictly decreasing and terminates in zero.

Definition. The group Γ is called *tame* if for each $m \geq 1$ the following relations hold: $\dim(W_m^i / W_m^{i+1}) = 1, i = 0, 1, \dots, 2^m - 1$.

A Γ-module W with the property that $\dim(W^i / W^{i+1}) = 1$ for $i = 0, \dots, \dim(W) - 1$ is called uniserial. Such modules play an important role in the theory of pro-p groups of finite coclass (see [11], Chapter 10).

Proposition 11. ([3]) *Let $\Gamma \leq \operatorname{Aut} T$ act transitively on the vertices of each level. Suppose that for each $m = 1, 2, \dots$ there is an element $h_m \in \Gamma$ the labelling of which at the vertices in the mth level is as follows:*

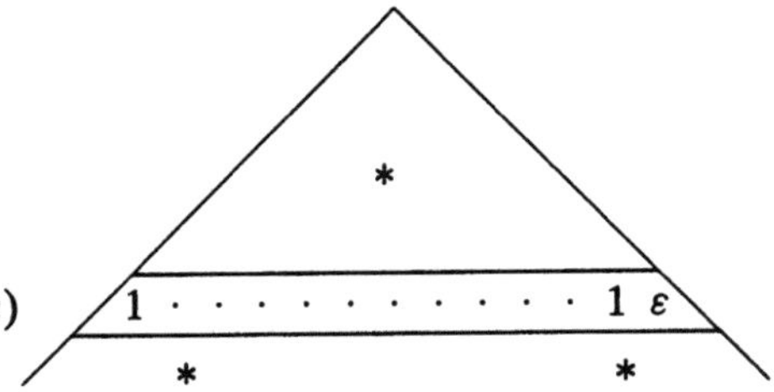

Diagram 30

and is otherwise arbitrary (here ε denotes a permutation). Then Γ is tame.

We omit the proof, and return now to our group G. It is easy to see that G satisfies the conditions of Proposition 11, taking for the elements h_m, $m = 1, 2, \ldots$ one of the elements $\{b, c\}$, depending on m.

Now we fix $m \geq 1$, and make the inductive hypothesis that $\gamma_{2^m+1}(G) = N_m$. Let $2^m + 1 \leq n \leq 2^{m+1}$ and suppose that $\gamma_n(G) \geq N_{m+1}$. Putting $i = n - (2^m + 1)$, we see that the isomorphism (41) identifies $\gamma_n(G)/N_{m+1}$ with the submodule $W_m^i \oplus \overline{W}_{m-1}^i$ of $W_m \oplus \overline{W}_{m-1}$. Since G is tame, W_m^i has dimension $2^m - i$, and it follows that $\gamma_n(G)$ contains $2^m - i \geq 1$ independent x-vectors.

Applying the corollary to Lemma 4, we infer that $\gamma_{n+1}(G) \geq N_{m+1}$. It follows that

$$\gamma_n(G)/\gamma_{n+1}(G) \cong (W_m^i \oplus \overline{W}_{m-1}^i)/(W_m^{i+1} \oplus \overline{W}_{m-1}^{i+1}).$$

When $i < 2^{m-1}$ this vector space has dimension 2 over $\mathbb{F}_2$, while if $2^{m-1} \leq i < 2^m$ it has dimension 1, since $\overline{W}_{m-1}^{2^{m-1}} = 0$.

Thus we obtain the assertions of Theorem 13 regarding $\mathrm{rk}(\gamma_n(G)/\gamma_{n+1}(G))$ for n in the range $2^m < n \leq 2^{m+1}$. We also see, by induction on n, that $\gamma_{2^{m+1}+1}(G) \geq N_{m+1}$; that this is an equality now follows on comparing the indices $|N_m : \gamma_{2^{m+1}+1}(G)|$ and $|N_m : N_{m+1}|$.

This completes the proof. □

15 The Hausdorff dimension of $\widehat{G}$

In this final section, we compute the Hausdorff dimension of the closure $\overline{G}$ of the group G considered in Sections 12–14, with respect to a natural metric on Aut T where T is the binary tree (this is one of the metrics considered in Section 3). Recall that $\overline{G}$ is isomorphic as a profinite group to the profinite completion $\widehat{G} = \widehat{G}_2$.

As a result of this computation, we find that Aut T contains a (topologically) finitely generated closed subgroup of positive Hausdorff dimension. This result is perhaps rather unexpected since Aut T itself requires infinitely many topological generators. Analogous computations can be carried out for the other branch groups considered in this chapter, again yielding positive results. This shows that branch groups are 'large' subgroups of Aut T. In contrast, the hereditarily just infinite subgroups of Aut T have Hausdorff dimension zero.

Let P be a profinite group filtered by a descending chain $\{P_n\}_{n=0}^{\infty}$ of open normal subgroups.

The formula

$$d(g,h) = \inf_n \left\{ \frac{1}{|P/P_n|} : g^{-1}h \in P_n \right\}$$

defines a metric on P, and any closed subgroup C of P has a Hausdorff dimension $\operatorname{Dim} C$ w.r.t. this metric. According to Abercrombie [1] and Barnea and Shalev [2], $\operatorname{Dim} C$ can be computed using the formula

$$\operatorname{Dim} C = \liminf_{n\to\infty} \frac{\log |CP_n/P_n|}{\log |P/P_n|}, \tag{42}$$

and takes a value in the interval [0, 1]; see also Chapter 1.

In the case $P = \operatorname{Aut} T$ where T is a spherically homogeneous rooted tree, the most natural filtration is given by

$$P_n = \operatorname{St}_P(n), \ n \geq 0.$$

For this filtration and the binary tree T, formula (42) takes the form

$$\operatorname{Dim} C = \liminf_{n\to\infty} \frac{1}{2^n} \log_2 |C/\operatorname{St}_C(n)|, \tag{43}$$

which is what we shall use for our computation.

Theorem 14. *Let $G = \langle a, b, c, d\rangle$ be the subgroup of* Aut T *defined in Section 12, where T is the binary rooted tree. Then*

$$\operatorname{Dim} \overline{G} = 5/8.$$

Proof. Since $|\overline{G}/\operatorname{St}_{\overline{G}}(n)| = |G/\operatorname{St}_G(n)|$, to apply formula (43) we only have to determine the orders of the finite quotients $G_n = G/\operatorname{St}_G(n)$, which can be naturally identified with groups acting on the subtrees $T_{[n]}$.

It was indicated in the proof of Proposition 4 that $G_3 \cong C_2 \wr C_2 \wr C_2$, and it is easy to see that $G_1 \cong C_2$, $G_2 \cong C_2 \wr C_2$. Let

$$\psi : H = \operatorname{St}_G(1) \to G \times G$$

be the homomorphism defined by (36) in Section 12, let $H_n \leq G_n$ be the image of H in G_n, and write

$$\psi_n : H_n \to G_{n-1} \times G_{n-1}$$

for the mapping induced by ψ.

Lemma 4. *If $n \geq 4$, then ψ_n is an injective homomorphism, and* Im ψ_n *has index* 8 *in $G_{n-1} \times G_{n-1}$.*

The proof of this lemma is analogous to those of Propositions 5 and 7, and we omit it. Since H_n has index 2 in G_n it follows from Lemma 5 that that

$$|G_n| = \frac{1}{4}|G_{n-1}|^2$$

when $n \geq 4$, and as $|G_3| = 2^7$ an easy calculation leads to the conclusion

$$|G_n| = 2^{5 \cdot 2^{n-3}+2}.$$

With (43) this gives the result. □

The following result has recently been obtained by B. Klopsch and C. Roever (unpublished):

Theorem. *Let G be a profinite branch group. Then the Hausdorff spectrum of G (relative to the natural congruence filtration) is the full interval* $[0, 1]$*; that is, for each real number* $\alpha \in [0, 1]$ *there exists a closed subgroup of Hausdorff dimension* α *in G.*

They also observe that many of the subgroups of positive Hausdorff dimension are in fact finitely generated (topologically); it is not known whether *every* value in $[0, 1]$ can be realised by a finitely generated subgoup.

Acknowledgements

I express deep gratitude to L. Bartholdi, N. Gupta, P. de la Harpe, A. Lubotzky, A.Machì, A. Mann, S. Mozes, V. N. Remeslennikov, A.V. Rozhkov, A. Shalev, S. Sidki, J. S. Wilson, and V. I. Sushchansky with whom I discussed many topics of this chapter at different periods.

Also I am happy to acknowledge the help of T. Ceccherini-Silberstein, G. C. Smith, P. A. Zalesskii, B. Klopsch, C. Roever, F. Scarabotti, F. Tolli, and other mathematicians who read a preliminary version of this chapter and made many useful comments.

I thank G. C. Smith and O. Tabachnikova for the translation into English and for technical help in the preparation of the text.

Finally, I would like to thank the editors for their excellent idea of producing this book, and in particular D. Segal for his painstaking editorial work, and for moral and other help that enabled me to finish the writing of this chapter.

References

[1] J. L. Abercrombie, Subgroups and subrings of profinite rings, *Math. Proc. Cambridge Phil. Soc.* **116** (1994).

[2] Y. Barnea and A. Shalev, Hausdorff dimension, pro-p groups and Kac–Moody algebras, *Trans. Amer. Math. Soc.* **349** (1997), 5073–5091.

[3] L. Bartholdi and R. I. Grigorchuk, Lie methods in growth of groups and groups of finite width, Computational and Geometric Aspects of Modern Algebra (Michael Atkinson et al., ed.), *London Math. Soc. Lect. Note Ser.*, vol. 275, Cambridge Univ. Press, Cambridge, 2000, pp. 1–27

[4] L. Bartholdi and R. I. Grigorchuk, On parabolic subgroups and Hecke algebras of some fractal group, preprint of FIM, ETH Zürich, 1999.

[5] L. Bartholdi and R. I. Grigorchuk, On the spectrum of Hecke-type operators related to some fractal groups, preprint of Max-Planck Institute in Bonn, 1999.

[6] H. Bass, M. V. Otero-Espinar, D. N. Rockmore and C. P. L. Tresser, *Cyclic Renormalization and Automorphism Groups of Rooted Trees*, Springer Lecture Notes in Mathematics **1621** (1996).

[7] Y. V. Bodnarchuk, V. I. Sushchansky, Computation of the Sylow 2-subgroup of a symmetric group of degree which is a power of 2, in *Computations in Algebra and Combinatorics*, Kiev, 1978.

[8] A. M. Brunner and S. Sidki, The generation of $GL_n(\mathbb{Z})$ by finite state automata, *Internat. J. Algebra and Comput.* **8**:1 (1998), 127–139.

[9] A. M. Brunner, S. Sidki and A. C. Viera, A just-nonsolvable torsion-free group defined on the binary tree, *J. Alg.* **211** (1999), 99–114.

[10] T. Ceccherini-Silberstein, A. Machì and F. Scarabotti, Il gruppo di Grigorchuk di crescita intermedia, (in Italian), *Rend. Circ. Mat. Palermo*, (to appear).

[11] J. D. Dixon, M. P. F. du Sautoy, A. Mann and D. Segal, *Analytic pro-p Groups*, 2nd ed., Cambridge Studies in Advanced Maths. **61**, C.U.P., Cambridge, 1999.

[12] P. Gawron, V. I. Sushchansky and V. Nekraskevych, Conjugacy classes of a group of automorphisms of a tree, *Math Notes*, to appear.

[13] R. I. Grigorchuk, On the Burnside problem for periodic groups, *Func. Anal. Appl.* **14** (1980), 41–43.

[14] R. I. Grigorchuk, On Milnor's problem on group growth, *Soviet Math. Dokl.* **28** (1983), 23–26.

[15] R. I. Grigorchuk, The growth degrees of finitely generated groups and the theory of invariant means, *Izv. Akad. Nauk SSR, Ser. Matem.* **48**:5 (1984) 939–985.

[16] R. I. Grigorchuk, On torsion groups generated by finite automata, *Abstracts of the All-Union Algebra Conference*, Kishinev, 1985.

[17] R. I. Grigorchuk, On the Hilbert-Poincaré series of graded algebras associated with groups, *Matem. Sbornik* **180**:2 (1989), 207–225.

[18] R. I. Grigorchuk, An example of a finitely presented amenable group which does not belong to the class EG, *Matem. Sbornik* **189**:1 (1998), 75–98.

[19] R. I. Grigorchuk, On branch and just infinite groups, *Abstracts of International Algebra Conference in Memory of Kurosh*, Moscow, 1998.

[20] R. I. Grigorchuk, On the system of defining relations and the Schur multiplier of periodic groups generated by finite automata. In *Proceedings of Groups St Andrews 1997 in Bath*, ed. C. M. Campbell, E.F. Robertson, N. Ruskuc, G. C. Smith. L.M.S. Lecture Notes (1999), 290–317.

[21] R. I. Grigorchuk, W. N. Herfort and P. A. Zaleskii, The profinite completion of certain torsion p-groups, Algebra. Proc. Int. Conf. of the 90th Birthday of A.G. Kurosh, Walter de Gruyter, Berlin, New York, 2000, pp. 113—123.

[22] R. I. Grigorchuk and J. S. Wilson, The conjugacy problem for certain branch groups, to appear.

[23] N. Gupta and J. Fabrykowski, On groups with sub-exponential growth function II, *J. Indian Math. Soc. (N. S.)* **56**:1-4 (1991), 217–228.

[24] N. Gupta and S. Sidki, On the Burnside problem for periodic groups, *Math. Z.* **182** (1983), 385–388.

[25] N. Gupta and S. Sidki, Some infinite p-groups, *Algebra i Logika* (1983), 584–589.

[26] N. Gupta and S. Sidki, Extension of groups by tree automorphisms, Contributions to group theory, *Cont. Math.* **33**, 232–246. A.M.S., Providence, Rhode Island, 1984.

[27] P. Hall, Wreath products and characteristically simple groups, *Proc. Cambridge Phil. Soc.* **58** (1962), 170–182.

[28] P. de la Harpe, *Topics in Geometric Group Theory*, University of Chicago Press, Chicago, 2000.

[29] I. D. Ivanyuta, Sylow p-subgroups of the countable symmetric group, *Ukrainian Math. J.* **15**:3 (1963), 240–249.

[30] L. Kaloujnine, Sur les p-groupes de Sylow du groupe symétrique de degré p^m, *C.R. Acad. Sci. Paris*, **221** (1945), 222–224.

[31] L. Kaloujnine, Sur les p-groupes de Sylow des groupes symétriques finis, *Ann. Sci. Ecole Norm. Sup.* **65** (1948), 239–279.

[32] G. Klaas, C. R. Leedham-Green and W. Plesken, *Linear pro-p Groups of Finite Width*, Lecture Notes in Math. **1684**, Springer, Berlin, 1997.

[33] P. Lentulis et J. Tits, Sur le groupe des automorphismes de certains produits en couronne, *C.R. Acad. Sci. Paris* **305** (1987), 847–852.

[34] I. G. Lysionok, A system of defining relations for the Grigorchuk group, *Matem. Zametki* **38** (1985), 503–511.

[35] G. Margulis, *Discrete Subgroups of Semisimple Lie Groups*, Springer-Verlag, Berlin, 1989.

[36] J. L. Mennicke, Finite factor groups of the unimodular group, *Annals of Math.* **31** (1965), 31–37.

[37] A. V. Rozhkov, Centralizers of elements in some groups of automorphisms of a tree, *Izv. RAN, Ser. Matem.* **57**:6 (1993), 82–105.

[38] A. V. Rozhkov, The lower central series of some groups of automorphisms of trees, *Matem. Zametki* **60**:2 (1996), 225–237.

[39] A. V. Rozhkov, Finiteness conditions in groups of automorphisms of trees, *Algebra i logika* **37**:2 (1998), 568–605.

[40] S. J. Pride, The concept of "largeness" in group theory, in *Word Problems II* (Conf. on decision problems in algebra, Oxford, 1976). Studies in Logic and Foundations of Math. **95**, pp. 299–335, North-Holland, Amsterdam–New York, 1980.

[41] S. Sidki, On a 2-generated infinite 3-group: the presentation problem, *J. Alg.* **110** (1987), 13–23.

[42] S. Sidki, On a 2-generated infinite 3-group: subgroups and automorphisms, *J. Alg.* **110** (1987), 24–55.

[43] S. Sidki, Regular trees and their automorphisms, *Monografias de matematica*, v. 56, IMPA, Rio de Janeiro, 1998.

[44] S. Sidki, Automorphisms of 1-rooted trees: growth, circuit structure and acyclicity, *J. of Math. Sci.* **100**:1 (2000), 1925–1943.

[45] V. I. Suschansky, Periodic p-groups of permutations and the unrestricted Burnside problem, *Dokl. AN SSSR* **247** (1979), 557–561.

[46] V. I. Suchshansky, Normal Structure of isometry groups of semifinite Baire metrics in *Infinite Groups and Mixed Algebraic Structures*, Kyev. Institute of Mathematics NAN of Ukraine, (1993), 270–289.

[47] T. Vovkivsky, Infinite torsion groups arising as generalisations of the second Grigorchuk group, Algebra. Proc. Int. Conf. of the 90th Birthday of A.G. Kurosh, Walter de Gruyter, Berlin, New York, 2000.

[48] J. S. Wilson, Groups with every proper quotient finite, *Proc. Camb. Phil. Soc.* **69** (1971), 373–391.

Steklov Institute of Mathematics
Gubkina Street 8
Moscow 117966
Russia
e-mail: grigorch@mi.ras.ru

5

On Just Infinite Abstract and Profinite Groups

John S. Wilson

1 Introduction

An infinite group G is called *just infinite* if all non-trivial normal subgroups have finite index; if G is profinite it is merely required that all non-trivial closed normal subgroups have finite index. Just infinite groups have arisen in a variety of contexts. The abstract just infinite groups having non-trivial abelian normal subgroups are precisely the space groups whose point groups act rationally irreducibly on the abelian normal subgroups (see McCarthy [7]). Many arithmetic groups are known to be just infinite modulo their centres; examples are $\mathrm{SL}_n(R)$ for $n \geqslant 3$ and $\mathrm{Sp}_{2n}(R)$ for $n \geqslant 2$, where R is the ring of integers of an algebraic number field (see [1]). The Nottingham group over $\mathbb{F}_p$ (described in Chapter 6) is a just infinite pro-p group. Grigorchuk [3], [4] and Gupta and Sidki [5] have introduced and studied some infinite finitely generated p-groups which act on trees, and many, together with their pro-p completions, are just infinite. Using Zorn's Lemma it is easy to see that if Γ is either an infinite finitely generated abstract group or an infinite finitely generated pro-p group, then Γ has a just infinite quotient group. Therefore to decide whether a group-theoretic property implies finiteness, it is sometimes sufficient to consider just infinite groups.

A group is called *hereditarily just infinite* if each of its subgroups of finite index is just infinite. Among the examples of just infinite groups listed above, all are hereditarily just infinite, except for those arising in the work of Grigorchuk and Gupta and Sidki. It is easy to prove that if H is a non-abelian hereditarily just infinite group and F is a finite transitive permutation group, then the permutational wreath product $H \operatorname{wr} F$ is just infinite; clearly it is not hereditarily just infinite unless $F = 1$.

In the paper [10], which was essentially the second half of my PhD dissertation, a structure theory for abstract just infinite groups was developed. This theory is valid and useful also for profinite just infinite groups. Here we give a simultaneous self-contained treatment for both the abstract and profinite cases. The theory describes a division of these groups into two classes. The groups G in the first class resemble the wreath products mentioned above; like them, they have a hereditarily just infinite subgroup H whose normal closure is the direct product of the finitely many distinct conjugates of H. Evidently this normal closure has finite index in G, and so for many purposes the study of the groups in this class reduces to the study of hereditarily just infinite groups.

The groups in the second class have very rich structure. When [10] was written, the only known examples were closely related to wreath powers; the more recent examples of Grigorchuk [3], [4] and Gupta and Sidki [5] point up the interest of this class. The groups in this class are certainly too large and diverse to admit the possibility of a classification; instead one may be able to prove general results that testify to their largeness. We illustrate this by proving that all just infinite profinite groups in this class have non-abelian abstract free subgroups. Discovery of examples in the context of actions on trees is no accident: if G is any abstract or profinite just infinite group in the second class, then trees can be defined within the structure of the group G itself on which G and its automorphism group act faithfully, and indeed the groups in this second class are just the groups called branch groups in Chapter 4.

2 Subnormal subgroups

We begin with some remarks concerning the profinite case. When studying profinite groups, we are interested only in subgroups which are profinite, and so we normally use the term 'subgroup' to mean 'closed subgroup'. Thus if X is a subset of a profinite group X, then $\langle X\rangle$ denotes the *closed* subgroup generated by X; this is the closure of the abstract subgroup generated by X.

In much of the work which follows (and in particular in Lemmas 2.1, 2.2), the abstract and profinite subgroups generated by families of subgroups under consideration do in fact coincide. This is because these subgroups are often built up in finitely many steps from products of subgroups. The product of two closed subsets of a profinite group G is always closed (see [11], (0.3.1)), and so if H, K are (closed) subgroups of G and if $HK = KH$, then HK is both the abstract and the profinite group generated by H and K. It is essentially this which makes possible a simultaneous treatment of abstract and profinite just infinite groups.

To simplify notation, for a subgroup H and subset T of a group G, we write H^T for $\langle H^t \mid t \in T\rangle$.

Let G be a group. We write $H \lhd^n G$ if there is a chain of subgroups

$$H = G_n \lhd \cdots \lhd G_1 \lhd G. \qquad (1)$$

If $H \lhd^n G$ for some n, we say that H is a subnormal subgroup of G and we write H sn G; the *defect* of a subnormal subgroup is the least n such that $H \lhd^n G$.

The terms H_r of the *normal closure series* of H are defined recursively as follows: $H_0 = G$ and $H_{r+1} = H^{H_r}$ for $r \geqslant 0$. Thus $H_{r+1} \lhd H_r$ for each r. If there is a series (1), then clearly $H_r \leqslant G_r$ for each r. Therefore $H \lhd^n G$ if and only if $H_n = H$.

Let H be a (closed) subgroup of a profinite group G. Using the fact that the normalizer of a (closed) subgroup is closed, it is not hard to see that $H \lhd^n G$ with G regarded as an abstract group if and only if $H \lhd^n G$ with G regarded as a profinite group. Indeed, if K is an abstract subgroup containing H, with topological closure $\bar{K}$, then the closed subgroups generated by the sets H^K, $H^{\bar{K}}$ coincide, so that the normal closure series for H in G regarded as a profinite group is obtained from the normal closure series in G as an abstract group by taking topological closures.

In the following two lemmas, and the rest of the chapter, we consider the abstract and profinite cases together.

Lemma 2.1 *Let H, K sn G and suppose that $H \lhd \langle H, K\rangle$. Then $\langle H, K\rangle$ sn G.*

Proof. This is a well-known result of Wielandt in the abstract case. Let

$$H = H_n \lhd \cdots \lhd H_1 \lhd G \tag{2}$$

be the normal closure series of H in G. Since K normalizes both G and H, it is clear that K normalizes each term of (2). Therefore for each r, we have $H_{r+1} \lhd H_r K$ and K sn $H_r K$, so that $H_{r+1}K/K$ sn $H_r K/K$ and hence $H_{r+1}K$ sn $H_r K$. Thus

$$\langle H, K\rangle = H_n K \text{ sn } H_{n-1}K \text{ sn } \cdots \text{ sn } H_1 K \text{ sn } G.$$

Lemma 2.2 *Suppose that each subnormal subgroup of G has only finitely many conjugates in G. If H, K sn G then $\langle H, K\rangle$ sn G.*

Proof. (Of course, in the profinite case, only closed subnormal subgroups are considered in the hypothesis.) Let $J = \langle H, K\rangle$ and let

$$H = H_n \lhd \cdots \lhd H_1 \lhd H_0 = J$$

be the normal closure series of H in J. We argue by induction on n, the case when $n \leqslant 1$ being given by Lemma 2.1. Assume that $n > 1$. Then H has only finitely many conjugates in H_{n-2}; they are all normal in H_{n-1} and H_{n-1} is their product. Therefore from finitely many applications of Lemma 2.1, we have H_{n-1} sn G. The result follows now by induction.

3 B-closed and N_0-closed classes

We write $H \leqslant_f G$ (resp. $H \lhd_f G$) to indicate that H is a subgroup (resp. a normal subgroup) of finite index in the group G.

Let $\mathcal{X}$ be a class of groups. We say that $\mathcal{X}$ is B-closed if the following holds: whenever H, K are normal subgroups of their join and both $H \cap K$ and HK are in $\mathcal{X}$, then $H \in \mathcal{X}$.

For any class $\mathcal{X}$ of groups, we define $\mathcal{X}'$ to be the class of groups all of whose normal subgroups are in $\mathcal{X}$, and we define $\mathcal{X}^{(n)}$ recursively by setting $\mathcal{X}^{(0)} = \mathcal{X}$ and $\mathcal{X}^{(n+1)} = (\mathcal{X}^{(n)})'$ for $n \geqslant 0$. Clearly $\mathcal{X}^{(n)}$ is the class of groups G such that $H \in \mathcal{X}$ for all $H \lhd^n G$.

Lemma 3.1 *If $\mathcal{X}$ is B-closed, then so is $\mathcal{X}^{(n)}$ for each $n \geqslant 0$.*

Proof. It suffices to prove that $\mathcal{X}'$ is B-closed. Suppose that $H, K \lhd \langle H, K\rangle$ and that both $H \cap K$ and HK are in $\mathcal{X}'$. Let $L \lhd H$ and write $L_1 = L^K$. Then $L_1^H = L_1$ since $HK = KH$, so that $L_1 \lhd HK$. Hence $L_1 \in \mathcal{X}$. But $L \leqslant L_1 \leqslant LK$ and so $L(L_1 \cap K) = L_1 \in \mathcal{X}$. Moreover $L \cap K \lhd H \cap K$ and so $L \cap (L_1 \cap K) = L \cap K \in \mathcal{X}$. Since L and $L_1 \cap K$ are both normal subgroups of L_1, it follows that $L \in \mathcal{X}$. Thus $H \in \mathcal{X}'$, as required.

Lemma 3.2 *The class of just infinite groups is B-closed.*

Proof. Suppose that $H, K \lhd \langle H, K\rangle$ and that HK and $H \cap K$ are just infinite. Then $H \cap K \neq 1$ and so $H \cap K \lhd_f HK$.

Let $1 \neq L \lhd H$. Suppose that $L \cap K = 1$. Then $L \cong LK/K$, which is finite since HK/K is finite, and so L is finite. Since $L \lhd H$ and $|HK : H|$ is finite, L has only finitely many conjugates in HK, all of which are normal in H, and so their join L_1 is a finite normal subgroup of HK. Since HK is just infinite, this is a contradiction. Thus if $1 \neq L \lhd H$, then $1 \neq L \cap K \lhd H \cap K$, and so $|H \cap K : L \cap K|$ is finite. It follows that $|H : L|$ is finite. Therefore H is just infinite, as required.

Let $H \lhd^2 G$ and let T be a finite subset of G. Then $H^t \lhd H^G$ for each $t \in T$, and the group H^T is the product of these subgroups. The following result is clear.

Lemma 3.3 *Let $H \lhd^2 G$ and let T' be any finite set of elements of G. Set $T = T' \cup \{1\}$ and $H^* = H \cap H^{T'}$. Then*

$$(H^{*T'}H)(H^{T'}) = H^T \quad \text{and} \quad (H^{*T'}H) \cap (H^{T'}) = H^{*T}.$$

In [8] the following result was proved for abstract groups.

Proposition 3.4 *Suppose that $\mathcal{X}$ is B-closed. Let $G \in \mathcal{X}'$ and let $G_0 \lhd_f G$. Then $G_0 \in \mathcal{X}'$.*

We recall the argument; it applies in the profinite setting as well as the abstract setting.

If $H \lhd G_0$, then $H^S \lhd G$ and so $H^S \in \mathcal{X}$, for any transversal S to G_0 in G. Pick a non-empty subset T of smallest cardinality such that $H^T \in \mathcal{X}$ for all $H \lhd G_0$. Replacing T by a translate Tt_0^{-1} with $t_0 \in T$, we may suppose that $1 \in T$ and we must prove that $T' = T \setminus \{1\}$ is empty. Suppose that this is not the case. Choose $H \lhd G_0$ and define $H^* = H \cap H^{T'}$. We have $H^T \in \mathcal{X}$ and $H^{*T} \in \mathcal{X}$. Since $\mathcal{X}$ is B-closed, it follows from Lemma 3.3 that $H^{T'} \in \mathcal{X}$. However this is a contradiction to the definition of T; and the result follows.

Corollary 3.5 *Let $\mathcal{X}$ be a B-closed class. If G is just infinite and every normal subgroup of G is in $\mathcal{X}$, then every subnormal subgroup of G is in $\mathcal{X}$.*

Proof. The hypothesis is that $G \in \mathcal{X}'$. Suppose that $G \in \mathcal{X}^{(n)}$ for some $n \geqslant 1$. By Lemma 3.1, $\mathcal{X}^{(n-1)}$ is B-closed, and so by Proposition 3.4 each normal subgroup of G is in $\mathcal{X}^{(n)}$; in other words, $G \in \mathcal{X}^{(n+1)}$. The result follows by induction.

In [8] it was proved that if G is an abstract group satisfying max-n, the maximal condition for normal subgroups, then so is each normal subgroup G_0 of finite index. The argument is rather like that in Proposition 3.4: one considers a minimal set T such that for each ascending chain (H_i) of normal subgroups of G_0 the chain (H_i^T) is finite, and one uses Lemma 3.3. Again, the argument remains valid in the profinite setting; the main steps are indicated in [11], (12.6.7). The class of groups satisfying max-n is easily seen to be B-closed. Therefore we have the following fundamental result; the second assertion is an easy consequence of the first.

Theorem 3.6 *Let G be a just infinite group. Then every subnormal subgroup of G satisfies* max-n, *and for each n the group G satisfies the maximal condition on subgroups K satisfying $K \triangleleft^n G$.*

A class $\mathcal{X}$ of groups is called N_0-closed if $HK \in \mathcal{X}$ whenever H, K are normal subgroups of a group and $H, K \in \mathcal{X}$. Examples are the following classes:
- the class of finite groups;
- for each prime p, the class of p-adic analytic groups;
- the class of nilpotent groups.

Lemma 3.7 *Let G be a just infinite group, let H* sn *G and let $\mathcal{X}$ be an N_0-closed class containing H. Then the normal subgroup of G generated by H is in $\mathcal{X}$.*

Proof. Let $(H_r)_{r=0}^n$ be the normal closure series of H in G. We argue by induction on n, the result being clear if $n \leqslant 1$. Let $n \geqslant 2$. Since H_{n-1} satisfies max-n, it has a subgroup U which is maximal subject to being a product of finitely many conjugates of H in H_{n-2}. Thus $U \in \mathcal{X}$ because $\mathcal{X}$ is N_0-closed, and clearly $U \triangleleft H_{n-2}$ so that $U = H_{n-1}$. Since $H_{n-1}^G = H_1$, the result follows by induction.

Corollary 3.8 *Let G be a just infinite group.*

(a) *G has no non-trivial finite subnormal subgroups.*

(b) *If G has a non-trivial subnormal subgroup which is p-adic analytic for some prime p, then G has an open normal p-adic analytic subgroup.*

(c) *If G has a non-trivial abelian subnormal subgroup then G has a finitely generated torsion-free abelian normal subgroup of finite index.*

Proof. Using Lemma 3.7 for the N_0-closed classes listed above, we obtain (a), (b). We also see that if the hypothesis of (c) holds, then G has a non-trivial nilpotent normal subgroup, whose centre Z is a non-trivial abelian normal subgroup of G; moreover Z satisfies the maximal condition for (normal) subgroups and so it is finitely generated.

The torsion subgroup of Z is therefore finite and, being normal in G, trivial. (Thus Z is free abelian in the abstract case; in the profinite case, since its Sylow subgroups are normal in G, it is a free abelian pro-p group of finite rank for some prime p.)

The virtually abelian just infinite groups have been studied extensively in the abstract and pro-p cases. In what follows, we shall restrict attention to just infinite groups which have no non-trivial abelian subnormal subgroups.

4 Near complements for subnormal subgroups

Throughout this section, G denotes a just infinite group having no non-trivial abelian subnormal subgroups.

Definition. Let H be a subnormal subgroup of G. A *near complement* to H in G is a subnormal subgroup D of G such that

$$\langle H, D\rangle = H \times D \quad \text{and} \quad |G : H \times D| < \infty.$$

Lemma 4.1 *Let* $1 \neq H \lhd^2 G$. *Then* H *has a near complement* D *in* G *satisfying* $D \lhd H^G$.

Proof. If H has finite index in G, we may take $D = 1$. Suppose then that H has infinite index. Since G satisfies the maximal condition for subgroups K such that $K \lhd^2 G$, we may further assume that each such subgroup K satisfying $H < K$ has a near complement E with $E \lhd K^G$. Since $|G : H^G|$ is finite, H has finitely many conjugates in G, and their intersection is 1. Thus there is an intersection $V \neq 1$ of conjugates of H such that $H \cap V = 1$. Therefore $HV = H \times V$. We have $H < HV \lhd H^G$ so that HV has a near complement E satisfying $E \lhd \langle H^G\rangle$. Clearly VE is a near complement for H and $VE \lhd H^G$.

Proposition 4.2 *Let* H sn G. *Then there is a subgroup* $K \lhd_f H$ *with* $K \lhd^2 G$.

Proof. Suppose that $H \lhd^n G$. We assume $n \geqslant 3$ and we make the obvious induction hypothesis. Since G satisfies the maximal condition on subgroups L with $L \lhd^n G$, we can also assume that all such subgroups larger than H satisfy the conclusion.

Set $H_1 = H^G$, $H_2 = H^{H_1}$. First suppose that $H_2 \leqslant_f G$ and write $N = \bigcap_{g\in G} H_2^g$. Then $H \cap N \leqslant_f H$ and $H \cap N \lhd^{n-2} N \lhd G$, so that by induction there is a subgroup $L \lhd^2 G$ with $L \leqslant_f H \cap N \leqslant_f H$. Clearly the subgroup $K = \bigcap_{h\in H} L^h$ has the required properties.

Assume then that H_2 has infinite index. By Lemma 4.1, H_2 has a near complement $D \neq 1$ which is normal in H_1. Then

$$H \times D \lhd^{n-2} H_2 \times D \lhd H_1 \lhd G,$$

and by the maximal choice of H there is a subgroup $L \lhd_f H \times D$ with $L \lhd L^G$. Since G has no non-trivial abelian subnormal subgroups, we have $C_L(D) \leqslant C_{HD}(D) = H$. Moreover L, D are normalized by L^G and H, and so $C_L(D)$ is normal in $\langle L^G, H\rangle$.

Finally, $C_L(D)$ contains $H \cap L$ which has finite index in H. Therefore $K = C_L(D)$ has the required properties.

For subgroups A, B of a group G we write

$$[A, B] = \langle [a, b] \mid a \in A, b \in B \rangle.$$

Using the identity $[aa', b] = [a, b]^{a'}[a', b]$, one sees that $[A, B]$ is normalized by A, and similarly by B.

Lemma 4.3 *Let K sn H sn G and let $|H : K|$ be finite. Then any subnormal subgroup D of G satisfying $[D, K] = 1$ also satisfies $[D, H] = 1$.*

Proof. By induction on $|H : K|$, it suffices to consider the case when $K \lhd H$. Write $C = C_G(K)$. Now $C \cap K = C_K(K) \lhd K$ sn G, so that $C \cap K = 1$. Therefore $C \cap H$ is finite and $C \cap H \lhd H$ sn G, so that $C \cap H = 1$ by Corollary 3.8.

Let H be subnormal of defect n in $\langle H, D \rangle$, let $(H_r)_{r=0}^n$ be the normal closure series of H in $\langle H, D \rangle$ and define Γ_r recursively by setting $\Gamma_0 = D$ and $\Gamma_r = [\Gamma_{r-1}, H]$ for each $r > 0$. Since H normalizes C and $D \leqslant C$, we have $\Gamma_1 \leqslant C$. By the remark preceding this lemma, $\langle H, \Gamma_{r-1} \rangle$ normalizes Γ_r for $r \geqslant 1$, and an easy induction shows that $\Gamma_{r+1} \leqslant \Gamma_r$ and $H_r = H\Gamma_r$ for each $r \geqslant 1$.

Suppose that $n \geqslant 2$ and set $F = \Gamma_{n-1}$. Thus $[F, H] = \Gamma_n \leqslant H \cap C = 1$. Since $[[F, H], \Gamma_{n-2}] = [[F, \Gamma_{n-2}], H] = 1$, by the three subgroups lemma we have $[[\Gamma_{n-2}, H], F] = 1$. Therefore F is abelian. Choose $u \in H_{n-2}$ such that $H^u \nleqslant H$. Because H and H^u are subnormal in G and normalize each other, we have HH^u sn G by Lemma 2.1. Thus since $F \lhd \langle H, \Gamma_{n-2} \rangle = H_{n-2}$ we have

$$HH^u \cap F \lhd HH^u \text{ sn } G.$$

But $HH^u \leqslant H_{n-1} = HF$ and so

$$H < HH^u = H(HH^u \cap F).$$

Therefore $HH^u \cap F$ is a non-trivial abelian subnormal subgroup of G and we have a contradiction.

We conclude that $n = 1$, so that H is normal in $\langle H, D \rangle$. Hence $[H, D] \leqslant H \cap C = 1$ and the result is proved.

Corollary 4.4 *If H sn G then H has a near complement in G.*

This follows immediately from 4.1–4.3.

Corollary 4.5

(a) *Each subnormal subgroup H of G has only finitely many conjugates in G.*

(b) *The join of two subnormal subgroups of G is again subnormal.*

(c) *Each subnormal subgroup H is a direct factor of a subnormal subgroup of finite index.*

Proof. Assertion (a) follows since H is normalized by $H \times D$, where D is a near complement to H in G, and (b) follows from (a) and Lemma 2.2. Finally, if D is a near complement to H, then $H \times D$ sn G.

Corollary 4.6 *Let H, K sn G. Then $H \cap K = 1$ if and only if $[H, K] = 1$.*

Proof. If $[H, K] = 1$, then $H \cap K$ is an abelian subnormal subgroup of G, and so $H \cap K = 1$.

Now suppose that $H \cap K = 1$. We can find normal subgroups H_0, K_0 of finite index in H, K such that $H_0 \triangleleft^2 G$ and $K_0 \triangleleft^2 G$; say H_0, K_0 are both normalized by N where $N \triangleleft_f G$. We replace H_0, K_0 by their intersections with N. Thus we now have $H_0, K_0 \triangleleft N$ and $H_0 \cap K_0 = 1$, so that $[H_0, K_0] = 1$, and two applications of Lemma 4.3 show that $[H, K] = 1$.

As a consequence of Corollary 4.6, we obtain

Corollary 4.7 (a) *If H_1, H_2, K are subnormal subgroups and $H_1 \cap K = H_2 \cap K = 1$ then $\langle H_1, H_2 \rangle \cap K = 1$.* (b) *If $H_1, \ldots, H_n$ are subnormal subgroups such that $H_i \cap H_j = 1$ for $i \neq j$ then*

$$\langle H_1, \ldots, H_n \rangle = H_1 \times \cdots \times H_n.$$

5 The structure lattice

We can now follow the development in [10]. Throughout this section, G denotes a just infinite group having no non-trivial abelian subnormal subgroups.

Lemma 5.1 *If K sn G, there is a greatest L sn G such that $K \leqslant L$ and $|L : K| < \infty$. (By greatest, we mean containing all others.)*

Proof. Let D be a near complement to K in G. Since $|G : K \times D| < \infty$, we can choose L sn G, with $K \leqslant L$ and $L \cap D = 1$, and with $|G : L \times D|$ as small as possible. Using Corollary 4.7 and Lemma 4.3, it is easy to see that L has all the required properties.

For each subnormal subgroup K of G, we write $\hat{K}$ for the greatest subnormal subgroup of G containing K as a subgroup of finite index.

Lemma 5.2 *The following conditions are equivalent, for two given subnormal subgroups H and K of G:*

(a) $|H : H \cap K|\,|K : H \cap K| < \infty$;

(b) $|\langle H, K \rangle : H|\,|\langle H, K \rangle : K| < \infty$;

(c) *H and K have a common near complement.*

Proof. If $|\langle H, K \rangle : H| < \infty$, then

$$|K : H \cap K| \leqslant |\langle H, K \rangle : H| < \infty.$$

So (b) implies (a).

If $|H : H \cap K|\,|K : H \cap K| < \infty$, and if D is a near complement to H, then it is also a near complement to $H \cap K$. So by Lemma 4.3, D is a near complement to K. Therefore (a) implies (c).

Finally, if D is a near complement to both H and K, then D is also a near complement to $\langle H, K\rangle$, by Corollary 4.7, so that

$$|\langle H, K\rangle : H| = |\langle H, K\rangle \times D : H \times D| < \infty.$$

Thus (c) implies (b).

For subnormal subgroups H and K of G, we write $H \sim K$ if any of the conditions of Lemma 5.2 is satisfied.

Lemma 5.3 (a) $\sim$ *is an equivalence relation.* (b) *If K_1, K_2, L_1 and L_2 are subnormal subgroups of G such that $K_1 \sim K_2$ and $L_1 \sim L_2$, then*

$$\text{(i)}\ \ K_1 \cap L_1 \sim K_2 \cap L_2 \quad \textit{and} \quad \text{(ii)}\ \ \langle K_1, L_1\rangle \sim \langle K_2, L_2\rangle.$$

Proof. All statements except (ii) are obvious consequences of the definitions. To prove (ii), let D be a near complement to $\langle K_1 \cap K_2, L_1 \cap L_2\rangle$. By Lemma 4.3,

$$D \cap K_1 = D \cap K_2 = D \cap L_1 = D \cap L_2 = 1,$$

so that, by Corollary 4.7, D is a near complement to both $\langle K_1, L_1\rangle$ and $\langle K_2, L_2\rangle$.

Lemma 5.3 shows that $\sim$ is a congruence on the lattice of subnormal subgroups of G. Thus, in the quotient $\mathcal{L} = \mathcal{L}(G)$ of the lattice of subnormal subgroups of G by the relation $\sim$, the join $a \vee b$ and meet $a \wedge b$ of any two elements a and b of $\mathcal{L}$ are well defined, and $\mathcal{L}$ is a lattice. We call $\mathcal{L}$ the *structure lattice* of G. Its elements are congruence classes $[H]$ of subnormal subgroups of G. By Proposition 4.2, every $a \in \mathcal{L}$ contains a subgroup $H \lhd^2 G$, and Lemma 5.1 shows that each $a = [H] \in \mathcal{L}$ contains a greatest element, namely $\hat{H}$. For any class $a = [H]$, we write $\hat{a} = \hat{H}$, so that the map $a \mapsto \hat{a}$ is an injective order-preserving map from $\mathcal{L}$ into the lattice of subnormal subgroups of G. The class $[1]$ contains only one element by Corollary 3.7. The lattice $\mathcal{L}$ has greatest element $[G]$ and least element $[1]$, and if G is hereditarily just infinite, then $\mathcal{L}$ has only these two elements.

Lemma 5.4 *Every element of $\mathcal{L}$ has a unique complement in $\mathcal{L}$.*

Proof. If D is a near complement to H in G, then $[H]$ and $[D]$ are complements in L; if $[D_1]$ and $[D_2]$ are both complements to $[H]$, then D_1 and D_2 both have H as a near complement in G, so that $[D_1] = [D_2]$.

Lemma 5.5 *$\mathcal{L}$ is a modular lattice.*

Proof. Let $h, k, l \in \mathcal{L}$, with $h \leqslant k$. Write $\hat{h} = H, \hat{k} = K$ and $\hat{l} = L$. Let D be a near complement to H in G and set $M = (H \times D) \cap L$. Then $H^M = H$, so that

$HM = MH$. Also $H \leqslant K$. Therefore $(HM) \cap K = H(M \cap K)$. Since $M \cap L \in l$, we have $(h \vee l) \wedge k = h \vee (l \wedge k)$, as required.

It is well known that every modular lattice with unique complementation is distributive, and so is a Boolean lattice; see for example Sikorski [2], p. 35. Therefore we have

Theorem 5.6 *$\mathcal{L}$ is a Boolean lattice.*

It follows that if $\mathcal{L}$ is finite, then $\mathcal{L}$ is isomorphic to the lattice of all subsets of a finite set. This case is elucidated by the next result.

Proposition 5.7 *Suppose that $\mathcal{L}$ has an atom (i.e., an element $a \neq [1]$ such that $[1] < b \leqslant a$ implies that $b = a$). Then $\mathcal{L}$ is finite, and G has a hereditarily just infinite subgroup $H = \hat{H} \lhd^2 G$ such that H^G is the direct product $H_1 \times \cdots \times H_n$ of all the distinct conjugates of H under G. For any K sn G, we have $K \sim H_{i1} \times \cdots \times H_{ir}$ for some subset $\{i_1, \ldots, i_r\}$ of the set $\{1, 2, \ldots, n\}$. So $\mathcal{L}$ is isomorphic to the lattice of all subsets of $\{1, 2, \ldots, n\}$.*

Proof. Let a be an atom of $\mathcal{L}$, and let $H = \hat{a}$. For every non-trivial subnormal subgroup K contained in H, we have $[K] = [H]$, and so $|H : K| < \infty$. Therefore H is hereditarily just infinite. Clearly $\widehat{H^g} = H^g$ is hereditarily just infinite for each $g \in G$.

If for some g and g' in G we have $H^g \cap H^{g'} \neq 1$, then $H^g \sim H^{g'}$, and since $\widehat{H^g} = H^g$ and $\widehat{H^{g'}} = H^{g'}$, we have $H^g = H^{g'}$. Thus distinct conjugates of H have trivial intersection, and so, by Corollary 4.7, $\langle H^G \rangle$ is the direct product $H_1 \times \cdots \times H_n$ of all the distinct conjugates of H under G.

Now let K sn G. Relabelling the subgroups H_i if necessary, we may assume that $H_i \cap K \neq 1$ for $i \leqslant r$ and $H_i \cap K = 1$ for $i > r$. We shall show that $K \sim H_1 \times \cdots \times H_r$. Write $D = H_{r+1} \times \cdots \times H_n$. Since for each $i \leqslant r$ we have $H_i \sim H_i \cap K$, it follows that

$$H_1 \times \cdots \times H_r \sim (H_1 \cap K) \times \cdots \times (H_r \cap K).$$

Therefore the subgroup $(H_1 \cap K) \times \cdots \times (H_r \cap K)$ of K has D as a near complement, so that, since $D \cap K = 1$ by Corollary 4.7, K also has D as a near complement. Hence

$$K \sim (H_1 \cap K) \times \cdots \times (H_r \cap K) \sim H_1 \times \cdots \times H_r.$$

The lattice isomorphism is now obvious. Map each subset $\{i_1, \ldots, i_r\}$ of $\{1, 2, \ldots, n\}$ to the element $[H_{i_1} \times \cdots \times H_{i_r}]$ of $\mathcal{L}$. This map preserves joins and complements, and so is a Boolean lattice homomorphism. The map is clearly injective, and the argument above shows that it is surjective.

We can now list some equivalent conditions for the finiteness of $\mathcal{L}$:

Theorem 5.8 *The following conditions for G are equivalent:*

(a) *$\mathcal{L}$ is finite;*

(b) *$\mathcal{L}$ satisfies the maximal condition;*

(c) *$\mathcal{L}$ has an atom;*

(d) *G satisfies the maximal condition for subnormal subgroups;*

(e) *G has a non-trivial subnormal hereditarily just infinite subgroup $H = \hat{H} \lhd^2 G$ such that H^G is the direct product of all the distinct conjugates of H under G.*

Proof. For any Boolean lattice $\mathcal{L}$, the conditions (a) and (b) are equivalent and imply (c). Proposition 5.7 shows that (c) implies (b) and (e). Obviously (e) implies (d). Finally, since the map $a \mapsto \hat{a}$ from $\mathcal{L}$ into the lattice of subnormal subgroups of G is injective and order-preserving, (d) implies (b).

We give two further properties which imply that the structure lattice of G is finite.

Proposition 5.9 (a) *If there is a bound on the defects of the subgroups $\hat{a}$ with $a \in \mathcal{L}$ then $\mathcal{L}$ is finite.* (b) *If there is a bound on the numbers $|G : N_G(\hat{a})|$ with $a \in \mathcal{L}$, then $\mathcal{L}$ is finite.*

Proof. (a) Suppose that $\hat{a} \lhd^n G$ for each $a \in \mathcal{L}$. By Theorem 3.5, the group G satisfies the maximal condition on subgroups K satisfying $K \lhd^n G$, and since the map $a \mapsto \hat{a}$ from $\mathcal{L}$ is injective and order-preserving, it follows that $\mathcal{L}$ satisfies the maximal condition. Therefore the result follows from Theorem 5.8.

(b) If H sn G and if H has defect n, then the terms with indices $1, 2, \ldots, n$ in the normal closure series of H in G are distinct and so H has at least n conjugates. The result follows.

The case considered in 5.8 and 5.9, in which $\mathcal{L}$ is finite, is illustrated by taking G to be a wreath product $H \text{ wr } \Sigma_n$, where H is hereditarily just infinite and not virtually abelian and Σ_n is the symmetric group of degree n, and taking G to be any subgroup of finite index which permutes transitively the conjugates of H in the wreath product. For this example, $\mathcal{L}$ is isomorphic to the set of subsets of $\{1, \ldots, n\}$.

Proposition 5.10 *Suppose that G is a group for which $\mathcal{L}$ is finite and of order 2^n. Then G can be embedded as a subgroup of finite index in a permutational wreath product H wr Σ_n, where H is hereditarily just infinite and not virtually abelian.*

Proof. Let L be a hereditarily just infinite subgroup such that L^G is the direct product of the conjugates $L_1, \ldots, L_n$ of L in G. The action of G on L^G by conjugation gives an embedding of G in $\text{Aut}(L_1 \times \cdots \times L_n)$, which may be identified with $\text{Aut}(L) \text{ wr } \Sigma_n$; the image of L^G is the base group of $\text{Inn}\,(L) \text{ wr } \Sigma_n$ and so, since $|G : L^G|$ is finite, the image of G has finite index in $H \text{ wr } \Sigma_n$ for some subgroup H of $\text{Aut}(L)$ containing $\text{Inn}\,(L)$ as a subgroup of finite index. Each such subgroup H is hereditarily just infinite and not virtually abelian.

6 Characterization of infinite structure lattices

We continue to follow the treatment in [10].

Proposition 6.1 *Let G be a group satisfying* max-n. *Then G has only finitely many subgroups of index n, for each integer n. In particular, G has only countably many subgroups of finite index.*

This result was proved in [9], and an inspection of the proof shows that it remains valid (and becomes simpler) for profinite groups satisfying max-n. We do not include the proof, noting merely that if G is a pro-p group with max-n, then $G/\Phi(G)$ must be finite, so that G is finitely generated; of course it is well known that finitely generated profinite groups have only finitely many subgroups of index n, for each integer n (see [11], (4.1.2)).

For the remainder of this section, G denotes a just infinite group having no non-trivial abelian subnormal subgroups and $\mathcal{L}$ denotes its structure lattice. Our first aim is to prove that G has only countably many subnormal subgroups.

Lemma 6.2 *Let $N \lhd G$ and let*

$$\mathcal{L}_N = \{a \mid a \in \mathcal{L} \text{ and } a \text{ contains a normal subgroup of } N\}.$$

Then $\mathcal{L}_N$ is a finite Boolean sub-lattice of $\mathcal{L}$.

Proof. If $H, K \lhd N$, then $HK \lhd N$ and $H \cap K \lhd N$. Therefore $\mathcal{L}_N$ is a sub-lattice of $\mathcal{L}$. Let $H \lhd N$ and let D be a near complement to H in G. Then D^u is a near complement to H^u for each $u \in G$, and so $[D^u] = [D]$ for each $u \in N$, since complementation in $\mathcal{L}$ is unique. Therefore $\bigcap_{u \in N} D^u$ is a near complement to H and so the complement of $[H]$ in $\mathcal{L}$ is in $\mathcal{L}_N$. Therefore $\mathcal{L}_N$ is a Boolean lattice.

Let

$$[H_1] \leqslant [H_2] \leqslant \cdots \leqslant [H_n] \leqslant \cdots \qquad (3)$$

be an ascending chain in $\mathcal{L}_N$ with $H_n \lhd N$ for each n. Then

$$H_1 \leqslant H_1 H_2 \leqslant \cdots \leqslant H_1 H_2 \cdots H_n \leqslant \cdots$$

is an ascending chain of normal subgroups of N. This chain must break off by Theorem 3.6, and since

$$[H_1 H_2 \ldots H_n] = [H_1] \vee [H_2] \vee \cdots \vee [H_n] = [H_n],$$

so does the chain (3). Thus the Boolean lattice $\mathcal{L}_N$ satisfies the maximal condition and so it is finite.

Proposition 6.3 *G has only countably many subnormal subgroups. Hence $\mathcal{L}$ is countable.*

Proof. First we note that each element a of $\mathcal{L}$ contains only countably many subgroups: this follows from Proposition 6.1 since if b is the complement to a in $\mathcal{L}$, the map $H \mapsto H \times \hat{b}$ from a to $[G]$ is clearly injective. Therefore it suffices to prove that $\mathcal{L}$ is countable. Every element of $\mathcal{L}$ contains a subgroup $H \lhd^2 G$ by Corollary 4.2, and so, in the notation of Lemma 6.2, we have $\mathcal{L} = \bigcup_{N \lhd G} \mathcal{L}_N$. Each lattice $\mathcal{L}_N$ is finite, and there are only countably many normal subgroups N by Proposition 6.1. The result follows.

Theorems 5.8 and 6.3 together show that *if $\mathcal{L}$ is infinite, then it is a countable atomless Boolean lattice.* These properties characterize $\mathcal{L}$ completely as a Boolean lattice; this is left as an exercise for the reader in [2]. A concrete description of $\mathcal{L}$ can be given by considering its Stone space X. By the Stone representation theorem, $\mathcal{L}$ is isomorphic to the lattice of closed and open subsets of X. All Stone spaces are compact, Hausdorff and totally disconnected, and since $\mathcal{L}$ is countable and atomless, the space X is second countable and has no isolated points. It follows that X is homeomorphic to Cantor's ternary set. Therefore we have

Proposition 6.4 *If $\mathcal{L}$ is infinite, then it is isomorphic to the lattice of closed and open subsets of Cantor's ternary set.*

The first examples to be found of just infinite groups with infinite structure lattice were wreath products with factors indexed by the set $\mathbb{N}^-$ of negative integers. They may be described as follows.

For each $n \in \mathbb{N}$ let V_n be a non-empty set. We consider the set T of finite strings $v_1 \dots v_n$ with $n \in \mathbb{N} \cup \{0\}$ and $v_i \in V_i$ for $i \leqslant n$ and the set T^∞ of infinite strings $v_1 v_2 \dots$ with $v_i \in V_i$ for each $i \in \mathbb{N}$. We may regard T as a rooted tree, with edges all pairs of the form $(v_1 \dots v_n, v_1 \dots v_n v_{n+1})$, and with root vertex the empty string $\emptyset$; then T^∞ is the set of ends of T. For each $v \in T$, the set of strings in T beginning with v is a rooted subtree of T, with root vertex v. We fix a 'trivial' element $1_n \in V_n$ for each n.

Suppose that H_n is a transitive permutation group on V_n for each $n \in \mathbb{N}$. We define an action of H_n on T as follows. Let $\xi \in H_n$ and $v = v_1 \dots v_r \in T$. If $r \geqslant n$ and $v_i = 1_i$ for $i < n$ we let $v\xi$ be the string with ith entry v_i for $i \neq n$ and nth entry $v_n\xi$; otherwise we set $v\xi = v$. In other words, ξ fixes all elements not in the rooted tree having root $1_1 \dots 1_{n-1}$ and permutes in the obvious fashion the rooted trees having roots $1_1 \dots 1_{n-1} v_n$ with $v_n \in V_n$. Thus we have a faithful action of H_n on T, consisting of tree automorphisms. We identify each H_n with its image in $\mathrm{Aut}(T)$ and we write

$$W = \langle H_n \mid n \in \mathbb{N} \rangle.$$

Wreath products with infinitely many factors were first considered in Hall [6]. In Hall's notation, W is the wreath product $\mathrm{Wr}_{m \in \mathbb{N}^-} H_{-m}$. To describe its natural structure as a permutation group, we consider its induced action on T^∞: the set of infinite strings having only finitely many non-trivial entries is an orbit and the (faithful) action on this orbit defines W as a permutation group.

If each set V_n is finite, then $\mathrm{Aut}(T)$ has finite orbits on T (namely, for each n, the set of strings of length n). Therefore $\mathrm{Aut}(T)$ is a profinite group. It is not hard to prove that $\mathrm{Aut}(T)$ is the profinite completion of the subgroup $W = \langle H_n \mid n \in \mathbb{N} \rangle$ where $H_n = \mathrm{Sym}\,(V_n)$ for each n.

Let $n \in \mathbb{N}$ and write S_n for the set of strings in T of length exactly n and T_n for the set of ends of the rooted tree with root $1_1 \dots 1_n$. Then $W_n^r = \langle H_i \mid i \leqslant n \rangle$ and $W_n^l = \langle H_i \mid i > n \rangle$ act as transitive permutation groups on S_n, T_n, respectively. The nature of the actions of W and W_n^l on T^∞, T_n makes it transparent that if T^∞ is identified with $S_n \times T_n$ in the obvious fashion (by deletion of $1_1 \dots 1_n$ and concatenation), then W becomes the permutational wreath product $W_n^l \operatorname{wr} W_n^r$. Clearly $W = \bigcup_{n \in \mathbb{N}} W_n^r$ so that W is locally finite.

We note that if H, K are arbitrary transitive permutation groups and if N is a normal subgroup of $H \operatorname{wr} K$ satisfying $N \cap K \neq 1$, then N contains the derived group of the base group B of $H \operatorname{wr} K$ (see, for example, [6], Lemma 4), and that if in addition H is a non-abelian simple group, then B is the unique minimal normal subgroup of $H \operatorname{wr} K$.

Now suppose that each H_i is a non-trivial perfect transitive group. Then $(W_n^l)_{n\in\mathbb{N}}$ is a descending sequence of subnormal subgroups of defect 2 in W with each term of infinite index in its predecessor. Let $1 \neq N \lhd W$. Then $N \cap W_n^r \neq 1$ for some n; choose n minimal with this property. From above, N contains the derived group of the base group B of $W_n^l \operatorname{wr} W_n^r$; however each group H_i is perfect, and so B is perfect and $B \leqslant N$. Therefore W/N is finite, and N cannot be abelian. We conclude that W is a just infinite group which is not virtually abelian, and consideration of the chain $(W_n^l)_{n\in\mathbb{N}}$ shows that W has infinite structure lattice. If each H_i is simple and N, n are as above, then N must equal the base group of $W_{(n-1)}^l \operatorname{wr} W_{(n-1)}^r$; if moreover all groups H_i are isomorphic as permutation groups, it follows that each subnormal subgroup of W is isomorphic to a finite direct power of W.

7 Structure trees

Throughout this section, G is a just infinite group with no non-trivial abelian subnormal subgroups and with infinite structure lattice $\mathcal{L}$.

Proposition 7.1 *The action of* Aut (G) *on the subgroup lattice of G induces a faithful action with finite orbits on $\mathcal{L}$.*

Proof. For $\theta \in \operatorname{Aut}(G)$ and for subnormal subgroups H, K with $H \sim K$ we clearly have $H\theta \sim K\theta$. Therefore we obtain a well-defined action of Aut (G) on $\mathcal{L}$ by setting $[H]\theta = [H\theta]$ for all $[H] \in \mathcal{L}$, $\theta \in \operatorname{Aut}(G)$. For each non-trivial normal subgroup N of G, the set

$$\mathcal{L}_N = \{[H] \mid H \lhd N\}$$

is finite by Proposition 6.2. But if N acts trivially on $\mathcal{L}$, then for all $a \in \mathcal{L}$ and all $u \in N$, we must have $\hat{a}^u = \hat{a}$, so that $\hat{a} \cap N \lhd N$ and $a \in \mathcal{L}_N$. Thus G acts faithfully on $\mathcal{L}$, and hence so does Aut (G) since each non-trivial normal subgroup of Aut (G) contains non-trivial inner automorphisms. By Proposition 6.1, for each integer $r \geqslant 1$, there are only finitely many normal subgroups of index r, and their intersection V_r is a non-trivial characteristic subgroup of G. Since each non-trivial normal subgroup N contains V_r for some r and since $\mathcal{L} = \bigcup_N \mathcal{L}_N$, we have that $\mathcal{L} = \bigcup_{r\geqslant 1} \mathcal{L}_{V_r}$. Each set $\mathcal{L}_{V_r}$ is finite and clearly it is a union of orbits for Aut (G).

The above action of Aut (G) also yields a faithful action on the Stone space of $\mathcal{L}$. However all groups with infinite structure lattices have isomorphic structure lattices, and actions on objects reflecting more of the properties of G would clearly be useful. We shall isolate, as subsets of $\mathcal{L}$, trees on which G and Aut (G) act faithfully.

Definition. A basal subgroup of G is a non-trivial subnormal subgroup A such that $A \cap A^g = 1$ whenever $A \neq A^g$.

By Corollary (4.7) (b), A is a basal subgroup if and only if A^G is the direct product of the distinct conjugates of A, and if A is a basal subgroup, then $A \lhd^2 G$.

Write

$$\mathcal{B} = \{[A] \mid A \text{ a basal subgroup of } G\}.$$

Thus $\mathcal{B}$ is a countable partially ordered set; it satisfies the maximal condition since G satisfies the maximal condition for subgroups $K \lhd^2 G$. Clearly $\mathcal{B}$ is invariant under the action of Aut (G) on $\mathcal{L}$.

We may regard $\mathcal{B}$ as a graph by taking as its edges all pairs (a, b) of elements of $\mathcal{B}$ such that b is a maximal element of $\{c \mid c \in \mathcal{B}, c < a\}$. Using some of the results below, it is not hard to prove that $\mathcal{B}$ is a connected graph, and clearly Aut (G) acts on it with finite orbits. With mild restrictions on G (for example, the requirement that each subgroup of finite index has finitely many maximal subgroups of finite index) one can prove that this graph has finite valency.

Instead of proving these facts, we abstract from this graph certain subtrees on which the action of G is more easily understood. First we collect some properties of basal subgroups.

Lemma 7.2 *Let A, B be basal subgroups.*

(a) *If $A \cap B \neq 1$ then $A \cap B$ is a basal subgroup.*

(b) *If $N_G(A) \leqslant L \leqslant G$ then A^L is basal and $N_G(A^L) = L$.*

(c) *The stabilizer of $[A]$ in G is $N_G(A)$.*

Proof. (a) Suppose that $A \cap B \neq (A \cap B)^g$. Then $A \neq A^g$ or $B \neq B^g$, so that either $A \cap A^g = 1$ or $B \cap B^g = 1$ and the result follows.

(b) Suppose that $A^L \cap (A^L)^g \neq 1$. Since A^G is the direct product of the groups A^t with t in a transversal to $N_G(A)$ in G and since $N_G(A) \leqslant L$, we must have $L \cap Lg \neq \emptyset$, so that $g \in L$ and $A^L = (A^L)^g$. Now let $g' \in N_G(A^L)$. Then $A^L = (A^L)^{g'} = A^{Lg'}$, so that $Lg' = L$ and $g' \in L$.

(c) Clearly $N_G(A)$ stabilizes $[A]$; and if g is an element of G stabilizing $[A]$ then $[A] = [A^g]$ so that $A \cap A^g \neq 1$ and hence $A = A^g$.

Lemma 7.3 (a) *Let H be a non-trivial subnormal subgroup of G. Then H contains a basal subgroup.* (b) *Let $h \in \mathcal{L}$ and $h \neq [1]$. Then there is an element $a \in \mathcal{B}$ with $a \leqslant h$.* (c) *If $\mathcal{L}$ is infinite then $\mathcal{B}$ has no minimal elements.*

Proof. (a) Let V be a maximal intersection of conjugates of H subject to $1 < V \leqslant H$. Clearly V is a basal subgroup.

(b) This follows directly from (a).

(c) Suppose that b is an element of $\mathcal{B}$. By Proposition 5.7, $\mathcal{L}$ has an element h with $[1] < h < b$, and from (b) there is an element $a \in \mathcal{B}$ with $a \leqslant h$. Therefore $a < b$ in $\mathcal{B}$.

Lemma 7.4 *Let M, N be normal subgroups of G such that $1 < N < M$.*

(a) *Let A_1 be a non-trivial subnormal subgroup of G normalized by N. Then there are basal subgroups $A \leqslant A_1$ with $[A]$ minimal in $\mathcal{L} \setminus \{[1]\}$ with respect to being stabilized by N.*

(b) *Suppose that A is as in* (a). *Then $[A^M]$ is a minimal element of $\mathcal{L} \setminus \{[1]\}$ with respect to being stabilized by M.*

Proof. (a) Consider the lattice $\mathcal{L}_N$ defined in Proposition 6.2; it contains $[A_1 \cap N] = [A_1]$ and since it is finite, we can choose a minimal element $[H] \neq [1]$ with $H \lhd N$ and $H \leqslant A_1$. Choose A to be an interesction of as many conjugates of H as possible with respect to $A \neq 1$. Clearly A is a basal subgroup with the required properties.

(b) From 7.2 (b), the subgroup A^M is basal with normalizer $MN_G(A)$. Suppose that b is an element of $\mathcal{L}$ stabilized by M and satisfying $[1] < b \leqslant [A^M]$. Set $K = \hat{b}$. Then $K \cap A^M \leqslant_f K$, and so by Corollary 4.7 we have $K \cap A^u \neq 1$ for some $u \in M$. Since M stabilizes b, it normalizes K and hence $K \cap A \neq 1$. Moreover $[K \cap A]$ is stabilized by N and hence $K \cap A \leqslant_f A$ by our choice of A. Thus $(K \cap A)^t \leqslant_f A^t$ for each t in a transversal to $N_M(A)$ in M. We conclude that $(K \cap A)^M \leqslant_f A^M$ and that $[A^M] = [(K \cap A)^M] \leqslant b$. The result follows.

Now we fix a filtration $\mathcal{F} = (G_i)_{i \geqslant 0}$ of G by normal subgroups; thus

$$G = G_0 > G_1 > \cdots > G_i > \cdots \qquad \text{and} \qquad \bigcap G_i = 1.$$

We call an element b of $\mathcal{B}$ a *vertex* if there is an i such that b is a minimal element of $\mathcal{L} \setminus \{[1]\}$ stabilized by G_i; the smallest such i is called the *$\mathcal{F}$-depth* of b. Let $\mathcal{T}$ be the set of all vertices; by Lemma 7.4 (a), $\mathcal{T}$ is non-empty. We make $\mathcal{T}$ into a directed graph by taking as edges all pairs (a, b) with $a, b \in \mathcal{T}$ such that b is a maximal element of $\{v \in \mathcal{T} \mid v < a\}$. Clearly the action of G on $\mathcal{L}$ induces an action on $\mathcal{T}$, and if $\mathcal{F}$ consists of characteristic subgroups then Aut (G) acts on $\mathcal{T}$.

Lemma 7.5 *Let $a, b \in \mathcal{T}$ and suppose that $b < a$. The following are equivalent:*

(i) *(a, b) is an edge;*

(ii) *$a = [B^{G_i}]$ and $b = [B]$ for some basal subgroup B normalized by G_{i+1}, where $i + 1$ is the $\mathcal{F}$-depth of b.*

Proof. Suppose that (a, b) is an edge. Write $a = [A], b = [B_1]$, where A, B_1 are basal subgroups, and write $B = A \cap B_1$. Since $b < a$ we have $B \leqslant_f B_1$, so that $b = [B]$, and B is basal by Lemma 7.2. By Lemma 7.4, $[B^{G_i}]$ is in $\mathcal{T}$ and has $\mathcal{F}$-depth at most i. Since $b < a$ and a is a vertex, a is stabilized by G_i. Hence

$$b = [B] < [B^{G_i}] \leqslant [A^{G_i}] = [A] = a.$$

Since (a, b) is an edge it follows that $a = [B^{G_i}]$.

Conversely, suppose that (ii) holds, and that $[B] < [C] \leqslant [B^{G_i}]$ with C basal and $[C] \in \mathcal{T}$; let $[C]$ have $\mathcal{F}$-depth j. The minimal property of vertices implies that $j < i + 1$ and that $[C] = [B^{G_i}]$. Therefore (a, b) is an edge.

Theorem 7.6

(a) *$\mathcal{T}$ is a tree.*

(b) *If a, b are vertices of $\mathcal{T}$, of $\mathcal{F}$-depths i, j, and if $b < a$ in $\mathcal{L}$, then $j > i$, and there is a unique (directed) path in $\mathcal{T}$, of length at most $j - i$, from a to b. The vertices in this path form a decreasing sequence with respect to the order in $\mathcal{L}$.*

Proof. (a) Suppose that there is a cycle in $\mathcal{T}$, and choose a minimal vertex b in this cycle. Then $\mathcal{T}$ must have two distinct edges (a_1, b), (a_2, b), and we obtain a contradiction from Lemma 7.5. Connectivity of $\mathcal{T}$ will follow from (b), which implies that there are paths from $[G]$ to all vertices of $\mathcal{T}$.

(b) Write $a = [A]$, $b = [B]$, with A, B basal; replacing B by the basal subgroup $A \cap B$ we may suppose that $B \leqslant A$. By Lemma 7.4 we have $[B^{G_i}] \in \mathcal{T}$. Moreover $B^{G_i} \leqslant A^{G_i} = A$, and hence $[A] = [B^{G_i}]$ since $[A] \in \mathcal{T}$. Therefore $[A]$, like $[B]$, is stabilized by G_j, and hence $i < j$, again since $[A] \in \mathcal{T}$. Now it follows from Lemma 7.5 that each pair of distinct vertices in the sequence

$$[A] = [B^{G_i}] \geqslant [B^{G_{i+1}}] \geqslant \cdots \geqslant [B^{G_{j-1}}] \geqslant [B]$$

is connected by an edge. Therefore there is a path with the required properties, and there are no other paths by (a).

Proposition 7.7 *Let a, b be vertices of $\mathcal{T}$, of $\mathcal{F}$-depths i, j with $i \leqslant j$, and write $a = [A]$, $b = [B]$, with A, B basal subgroups. The following are equivalent:*

(i) *$b \leqslant a$ in $\mathcal{L}$;*

(ii) *the path in $\mathcal{T}$ from $[G]$ to b passes through a;*

(iii) *A and B do not centralize each other.*

Proof. If $b \leqslant a$ then we obtain a (directed) path from $[G]$ to b by concatenating the paths from $[G]$ to a and from a to b, and so (ii) holds. Clearly (ii) implies (i).

If (i) holds then $A \cap B$ is certainly non-trivial and so non-abelian, and (iii) holds. Suppose that (iii) holds. Since $[A, B] \leqslant A \cap B$ we have $A \cap B \neq 1$, and so $A \cap B$ is a basal subgroup by Lemma 7.2; moreover G_j normalizes both B and $A \cap B$, and from the definition of vertices we have $[B] = [A \cap B] \leqslant A$, so that (i) holds.

Theorem 7.8

(a) *Let a be a vertex and suppose that there is an edge (a, b) in $\mathcal{T}$; let b have $\mathcal{F}$-depth $i + 1$. Then the set of vertices c such that (a, c) is an edge consists of vertices of $\mathcal{F}$-depth $i + 1$ and it is permuted transitively and non-trivially by G_i. Thus the cardinality of this set is a divisor greater than 1 of $|G_i/G_{i+1}|$.*

(b) *G has at most one orbit on the set of vertices of $\mathcal{F}$-depth i, for each $i \geqslant 0$.*

(c) *$\mathcal{T}$ has only finitely many vertices of $\mathcal{F}$-depth i, for each $i \geqslant 0$.*

Proof. (a) Write $b = [B]$ with B basal. We have $a = [B^{G_i}]$ by Lemma 7.5, and so $(a, [B^g])$ is an edge for each $g \in G_i$; these edges are fixed by G_{i+1} and there is more than one of them since $[B] \neq [B^{G_i}]$. We show now that each edge (a, c) in $\mathcal{T}$ is one of these edges. Write $c = [C]$ with C basal, and suppose that c has $\mathcal{F}$-depth $j+1$. Since $C \cap B^{G_i} \neq 1$ we have $C \cap B^g \neq 1$ for some $g \in G_i$ by Corollary 4.7. Since both $[C]$, $[B^g]$ are stabilized by G_{k+1}, where $k = \max(i, j)$, so is $[C \cap B^g]$. It follows from Lemma 7.2 and the minimal property of vertices that either $[C] = [C \cap B^g]$ or $[B^g] = [B]^g = [C \cap B^g]$. Therefore $[C] \leqslant [B]^g$ or $[B]^g \leqslant [C]$, and so since both $(a, [B]^g)$, $(a, [C])$ are edges we have $[C] = [B]^g$.

(b) Let a_1, a_2 be vertices of $\mathcal{F}$-depth i and write $a_1 = [A_1]$, $a_2 = [A_2]$ with A_1, A_2 basal. By Corollary 4.7 we may choose $g \in G$ such that $A_1^g \cap A_2 \neq 1$, and $A_1^g \cap A_2$, like A_1^g, A_2, is basal and is normalized by G_i. We conclude that

$$[A_2] = [A_1^g \cap A_2] = [A_1^g] = [A_1]^g.$$

(c) Since G_i stabilizes vertices of $\mathcal{F}$-depth i this follows immediately from (b).

In order to ensure that the tree $\mathcal{T}$ is reasonably large we shall strengthen our hypothesis on the filtration $\mathcal{F}$. We assume that the following condition holds:

(F) *each normal subgroup of finite index in G contains some term of $\mathcal{F}$.*

Condition F holds for all filtrations of G if G is profinite (see [11], (0.3.1)). In general, filtrations satisfying condition F can be constructed easily; for example, one can enumerate the non-trivial normal subgroups of G and take G_i to be the intersection of the first i subgroups, for each $i \geqslant 1$.

Proposition 7.9 *Suppose that $\mathcal{F}$ satisfies condition* F. *Then $\mathcal{T}$ has no minimal vertices.*

Proof. Let a be a vertex and write $a = [A]$, with A basal. By Lemma 7.3 (c) there is a basal subgroup B_1 with $[B_1] < [A]$, and $N_G(B_1)$ has finite index, so that $G_j \leqslant N_G(B_1)$ for some j by condition F. By Lemma 7.4 (a) we can find a vertex $[B]$ of $\mathcal{F}$-depth j with B basal and with $B \leqslant B_1$. Clearly $[B] < [A]$.

The above two results show that if condition F holds then for each i the tree $\mathcal{T}$ has vertices which fail to be stabilized by G_i. Thus G acts faithfully on $\mathcal{T}$; and if $\mathcal{F}$ consists of characteristic subgroups then Aut (G) acts faithfully on $\mathcal{T}$. Moreover, the orbits of G are the sets of elements of equal $\mathcal{F}$-depth. There is an additional important property of the action of G on $\mathcal{T}$ which distinguishes it from actions on trees obtained in other ways.

We require a notion introduced by Grigorchuk and discussed more fully in Chapter 4. Let T be a tree with a distinguished vertex v_0 and let H be a group which acts on T and fixes v_0. The *rigid stabilizer* $\mathrm{Rst}_H(u)$ in H of a vertex u is the subgroup of H which fixes every vertex w such that the path from v_0 to w does not pass through u. Since the orbit of u under H consists of vertices with equal distance from v_0, the subgroup $\mathrm{Rst}_H(u)$ also fixes u. In the case of interest to us, the vertex $[G]$ of $\mathcal{T}$ plays the role of the vertex v_0.

Lemma 7.10 *Suppose that $\mathcal{F}$ satisfies condition* F. *Let $[A]$ be a vertex of $\mathcal{T}$, with A basal.*

(a) $A \leqslant \mathrm{Rst}_G([A])$.

(b) *If $h \in A$ and h fixes every vertex b with $b \leqslant [A]$, then $h = 1$.*

Proof. (a) This follows from Proposition 7.7.

(b) Clearly h also fixes all vertices $b = [B]$ with $B \leqslant A^g \neq A$, and so the normal subgroup N generated by h fixes every vertex b with $b \leqslant [A]$. If $N \neq 1$, then $G_i \leqslant N$ for some i and it follows that each b with $b \leqslant [A]$ has $\mathcal{F}$-depth at most i, in contradiction to Proposition 7.9.

Proposition 7.11 *Suppose that $\mathcal{F}$ satisfies condition* F. *Let $a \in \mathcal{T}$ and let $R = \mathrm{Rst}_G(a)$. Then R is a basal subgroup and $[R] = a$.*

Proof. Let R^{g_1}, R^{g_2} be distinct conjugates of R. Then R^{g_1}, R^{g_2} are the rigid stabilizers of ag_1, ag_2, and no vertex of $\mathcal{T}$ is moved by both R^{g_1}, R^{g_2}. Therefore $[R^{g_1}, R^{g_2}]$ stabilizes all vertices of $\mathcal{T}$ and so $[R^{g_1}, R^{g_2}] = 1$. It follows that R is a basal subgroup.

Choose a basal subgroup A such that $a = [A]$. By Lemma 7.10 we have $A \leqslant R \cap A^G$. Let $g \in R \cap A^G$. Let A_1 be a conjugate of A with $A_1 \neq A$ and let h be the projection of g in A_1. Since g fixes all vertices b with $b \leqslant [A_1]$ so does h, and hence $h = 1$ by Lemma 7.10. Thus $g \in A$. It follows that $R \cap A^G = A$ and hence that $|R : A| < \infty$. We conclude that $a = [A] = [R]$, as required.

We single out one special case. Suppose that all proper quotient groups of G are finite p-groups, for some prime p. Adding further subgroups to a filtration $\mathcal{F}$ satisfying condition F if necessary, we obtain a maximal filtration, in which $|G_i/G_{i+1}| = p$ for each $i \geqslant 0$. The tree $\mathcal{T}$ is therefore isomorphic to the rooted p-ary tree T_p, in which p edges descend from each vertex.

We summarize our conclusions for this special case:

Theorem 7.12 *Suppose that all proper quotient groups of the (abstract or profinite) group G are finite p-groups, where p is a prime, and that no normal subgroup of G is a direct product of finitely many hereditarily just infinite groups. Then associated with each maximal filtration satisfying condition* F *there is a faithful action on a tree isomorphic to T_p whose orbits on vertices are the sets of vertices of equal depth and in which the rigid stabilizers of vertices are basal subgroups.*

Thus for each such group G we recover actions on trees isomorphic to T_p. How just infinite groups G may be obtained, starting from the tree T_p, is explained in Chapter 4.

8 Free subgroups

As an application of the theory of the previous sections, we shall prove that each profinite just infinite group with infinite structure lattice has many non-abelian abstract free subgroups. The existence of torsion groups like those discussed at the end of Section 6 and in Chapter 4 shows that the restriction to profinite groups cannot be omitted, and the existence of pro-p just infinite groups with infinite structure lattice shows that we cannot hope to establish the existence of free profinite subgroups. We do not know whether just infinite pro-p groups with infinite structure lattice necessarily have non-abelian free pro-p subgroups.

In this section, F_r denotes a free abstract group of rank r, with basis $x_1, \ldots, x_r$, $F_r^\sharp$ denotes $F_r \setminus \{1\}$, and $H^{(r)}$ denotes the direct product of r copies of a group H. For $w \in F_r^\sharp$ and $\mathbf{h} = (h_1, \ldots, h_r) \in H^{(r)}$ we write $w(\mathbf{h})$ for $w(h_1, \ldots, h_r)$, and if $S \subseteq H^{(r)}$ we write $w(S) = \{w(s) \mid s \in S\}$. If H is profinite, then we say that *many* elements of $H^{(r)}$ have a property if every non-empty open subset of $H^{(r)}$ has an element with the property. We shall show that if G is a profinite just infinite group with infinite structure lattice, then for each $r \in \mathbb{N}$ many r-tuples of elements of G freely generate abstract free subgroups.

Lemma 8.1 *Let G be a profinite group, let S be a (closed) subgroup of G and let $r \in \mathbb{N}$. Then either*

(i) *for many $\mathbf{g}$ in $G^{(r)}$ there is no $w \in F_r^\sharp$ such that $w(\mathbf{g}) \in S$, or*

(ii) *there exist $N \lhd_O G$, $\mathbf{u} \in G^{(r)}$ and $w \in F_r^\sharp$ such that $w(\mathbf{u}N^{(r)}) \subseteq S$.*

Proof. Suppose that (i) fails to hold and let U be an open subset of $G^{(r)}$ illustrating the fact. Then U contains a coset C of an open normal subgroup of $G^{(r)}$ (see [11], p. 8). For each $w \in F_r^\sharp$, let V_w be the closed set $\{\mathbf{g} \in C \mid w(\mathbf{g}) \in S\}$. Thus $\bigcup_{w \in F_r^\sharp} V_w = C$. Since $F_r^\sharp$ is countable, some set V_w has non-empty interior by Baire's category theorem (see [11], (2.7.10)) and so contains a coset of the form $\mathbf{u}N^{(r)}$ with $N \lhd_O G$. The result follows.

Lemma 8.2 *Let $w \in F_r^\sharp$ be cyclically reduced and suppose that w ends with a positive power of x_1. Then*

$$w(x_1 t, x_2, \ldots, x_r) = w(x_1, x_2, \ldots, x_r)(\textstyle\prod_{\lambda \in \Lambda} t^{\epsilon_\lambda w_\lambda})t$$

where $\epsilon_\lambda = \pm 1$ for each λ and the elements $w_\lambda = w_\lambda(x_1, \ldots, x_r)$ are non-trivial proper right segments of w.

This follows from an easy calculation.

Now let G be an abstract or profinite just infinite group with infinite structure lattice. With G we associate the tree $\mathcal{T}$ corresponding to a filtration satisfying condition F, and we consider also the set Ω of ends of this tree; the elements α of Ω can be thought of as infinite paths from $[G]$ without repeated edges. The action of G on $\mathcal{T}$ induces an action on Ω. If $\alpha \in \Omega$ and $[A] \in \mathcal{T}$ we write $\alpha \leqslant [A]$ if $[A]$ is a vertex of the path α. The following result is clear from 7.2 (c) and the nature of the action of G on $\mathcal{T}$.

Lemma 8.3 (a) *If $[A_1]$, $[A_2]$ are vertices of $\mathcal{T}$ and $[A_1] \leqslant [A_2]$, then $N_G(A_1) \leqslant N_G(A_2)$.* (b) *The stabilizer in G of an end α is $\bigcap_{\alpha \leqslant [A] \in \mathcal{T}} N_G(A)$.*

Lemma 8.4 *Let $(w_\lambda)_{\lambda \in \Lambda}$ be a finite family of elements of $F_r^\sharp$. Fix $\mathbf{u} \in G^{(r)}$ and $N \lhd_f G$. For $\mathbf{k} \in N^{(r)}$, each basal subgroup A and each end α let*

$$\Lambda(A, \mathbf{k}) = \{\lambda \mid A^{w_\lambda(\mathbf{uk})} = A\} \quad \textit{and} \quad \Lambda(\alpha, \mathbf{k}) = \{\lambda \mid \alpha\, w_\lambda(\mathbf{uk}) = \alpha\}.$$

Let $[A_1]$, $[A_2]$ *be vertices of* $\mathcal{T}$ *and let* α *be an end.*

(a) *If* $[A_1] \leqslant [A_2]$, *then* $\Lambda(A_1, \mathbf{k}) \subseteq \Lambda(A_2, \mathbf{k})$. *If* $\alpha \leqslant [A_1]$, *then* $\Lambda(\alpha, \mathbf{k}) \subseteq \Lambda(A_1, \mathbf{k})$. *Moreover* $\Lambda(\alpha, \mathbf{k}) = \Lambda(A, \mathbf{k})$ *for some basal subgroup* A *with* $\alpha \leqslant [A] \in \mathcal{T}$.

(b) $\Lambda(A, \mathbf{k})$ *depends only on the coset of* $\mathbf{k}$ *modulo* $M^{(r)}$ *where* $M = N \cap \bigcap_{g \in G} N_G(A^g)$.

(c) *Suppose that* $\Lambda(\alpha, \mathbf{k}) \neq \emptyset$ *for each* $\mathbf{k} \in N^{(r)}$. *Then there exist* $\lambda \in \Lambda$, $\mathbf{k} \in N^{(r)}$ *and* $M \lhd_f G$ *such that* α *is fixed by* $w_\lambda(\mathbf{uk}M^{(r)})$.

Proof. Assertion (a) follows from 8.3 and (b) is clear. Suppose that the hypothesis of (c) holds. Choose $\mathbf{k}$ with $|\Lambda(\alpha, \mathbf{k})|$ as small as possible. By (a) we have $\Lambda(\alpha, \mathbf{k}) = \Lambda(A, \mathbf{k})$ for some basal subgroup A with $\alpha \leqslant [A] \in \mathcal{T}$. Let $M = N \cap \bigcap_{g \in G} N_G(A^g)$. For all $\mathbf{k}' \equiv \mathbf{k}$ modulo $M^{(r)}$, we have

$$\Lambda(\alpha, \mathbf{k}') \subseteq \Lambda(A, \mathbf{k}') = \Lambda(A, \mathbf{k}) = \Lambda(\alpha, \mathbf{k}),$$

and our choice of $\mathbf{k}$ shows that $\Lambda(\alpha, \mathbf{k}') = \Lambda(\alpha, \mathbf{k})$, so that $\alpha\, w_\lambda(\mathbf{uk}') = \alpha\, w_\lambda(\mathbf{uk}) = \alpha$ for each $\lambda \in \Lambda(\alpha, \mathbf{k})$. Thus α is fixed by $w_\lambda(\mathbf{uk}M^{(r)})$ for each such λ.

Proposition 8.5 *Let* G *be an abstract or profinite just infinite group with infinite structure lattice and let* $\mathcal{T}$ *be its structure tree with respect to a filtration satisfying condition* F. *Let* $w \in F_r^\sharp$ *and let* α *be an end of* $\mathcal{T}$. *Then there is no coset* $\mathbf{u}N^{(r)}$ *with* $\mathbf{u} \in G^{(r)}$ *and* $N \lhd_f G$ *such that* $w(\mathbf{u}N^{(r)})$ *fixes* α.

Proof. Suppose that there exist $w, \alpha, \mathbf{u}, N$ such that $w(\mathbf{u}N^{(r)})$ fixes α; choose w of minimal length. Exchanging x_1, x_i for some i and replacing x_1 by x_1^{-1} if necessary, and making corresponding changes to $\mathbf{u}$, we may assume that w ends with a positive power of x_1. Clearly w is cyclically reduced since if $w = x_1^{-1}w'x_1$ then $w'(\mathbf{u}N^{(r)})$ fixes αu_1^{-1}. We write $w(x_1 t, x_2, \dots, x_r)$ in the form given by 8.2. Thus for all $\mathbf{v} = (v_1, \dots, v_r) \in \mathbf{u}N^{(r)}$ and $b \in N$ the element

$$w(v_1 b, v_2, \dots, v_r) = w(\mathbf{v})(\textstyle\prod_{\lambda \in \Lambda} b^{\epsilon_\lambda w_\lambda(\mathbf{v})})b$$

fixes α, and so therefore does $(\prod_{\lambda \in \Lambda} b^{\epsilon_\lambda w_\lambda(\mathbf{v})})b$.

Let A be a basal subgroup satisfying $\alpha \leqslant [A] \in \mathcal{T}$ and $A \leqslant N$. For all $\mathbf{k} \in N^{(r)}$, we define $\Lambda(A, \mathbf{k})$ as in 8.4. Let $b \in A$. Each element $b^{\epsilon_\lambda w_\lambda(\mathbf{uk})}$ lies in a conjugate of A; the elements in conjugates not equal to A normalize all subgroups of A, and hence fix α. Therefore $(\prod_{\lambda \in \Lambda(A,\mathbf{k})} b^{\epsilon_\lambda w_\lambda(\mathbf{uk})})b$ fixes α. However $b^{\pm 1}$ cannot fix α for all $b \in A$, since then both A and A^G would normalize all basal subgroups B with $\alpha \leqslant [B] \leqslant [A]$, and hence $\Lambda(A, \mathbf{k})$ cannot be empty. Therefore by 8.4 we have $\Lambda(\alpha, \mathbf{k}) \neq \emptyset$ for all $\mathbf{k} \in N^{(r)}$, and we can find $\lambda \in \Lambda$, $\mathbf{v} = \mathbf{uk} \in \mathbf{u}N^{(r)}$ and some $M \lhd_f G$ such that $w_\lambda(\mathbf{v}M^{(r)})$ fixes α. But this is a contradiction since w_λ has smaller length than w.

Theorem 8.6 *Let* G *be a just infinite group with infinite structure lattice.*

(a) *G does not satisfy any (abstract) group law.*

(b) *If G is profinite, then for each $r \in \mathbb{N}$, there are many r-tuples of elements of G which freely generate abstract free subgroups.*

(c) *If G is a pro-p group, then G has dense abstract free subgroups of finite rank.*

Proof. Assertion (a) follows from 8.5, and assertion (b) from 8.1 and 8.5. Moreover, if G is profinite and α is an end of a structure tree for G, then taking S in 8.1 to be the stabilizer of α we see that, for each r, many r-tuples of elements of G generate free subgroups which act freely on their orbits containing α.

Now suppose that G is a pro-p group. Then G can be generated by d elements, where p^d is the order of the Frattini quotient $\overline{G}$ of G. Moreover d elements generate G as a profinite group if and only if their images in $\overline{G}$ generate $\overline{G}$. Thus the set of d-tuples of elements of G which generate G as a profinite group is a non-empty closed and open set in $G^{(d)}$, and from (b), many d-tuples in this set generate abstract free subgroups of rank d.

References

[1] H. Bass, J. Milnor and J.-P. Serre, Solution of the congruence subgroup problem for SL_n ($n \geqslant 3$) and Sp_{2n} ($n \geqslant 2$). *Inst. Hautes Études Sci. Publ. Math.* **33** (1967), 59–137.

[2] P. Crawley and R.P. Dilworth. *Algebraic Theory of Lattices*, Prentice Hall, 1973.

[3] R.I. Grigorchuk, The growth degrees of finitely generated groups and the theory of invariant means. *Izv. Akad. Nauk. SSSR. Ser. Mat.* **48** (1984), 939–985 (English translation: *Math. SSSR. Izv.* **25** (1985)).

[4] R.I. Grigorchuk, On the growth degrees of p-groups and torsion-free groups. *Math. Sb.* **126** (1985), 194–214 (English translation: *Math. USSR Sbornik* **54** (1986), 185–205).

[5] N. Gupta and S. Sidki, On the Burnside problem for periodic groups. *Math. Z.* **182** (1983), 385–388.

[6] P. Hall, Wreath powers and characteristically simple groups. *Proc. Cambridge Philos. Soc.* **58** (1962), 170–184.

[7] D. McCarthy, Infinite groups whose proper quotient groups are finite. *Comm. Pure Appl. Math.* **21** (1968), 545–562.

[8] J.S. Wilson, Some properties of groups inherited by normal subgroups of finite index. *Math. Z.* **114** (1970), 19–21.

[9] J.S. Wilson, Groups satisfying the maximal condition for normal subgroups. *Math. Z.* **118** (1970), 107–114.

[10] J.S. Wilson, Groups with every proper quotient finite. *Proc. Cambridge Philos. Soc.* **69** (1971), 373–391.

[11] J.S. Wilson, *Profinite Groups*, Clarendon Press, Oxford, 1998.

School of Mathematics and Statistics
University of Birmingham
Edgbaston, Birmingham B15 2TT, England
e-mail:jsw@for.mat.bham.ac.uk

6

The Nottingham Group

Rachel Camina

1 Introduction

As interest in pro-p groups has grown, so has interest in the Nottingham group. Known to number theorists as the group of wild automorphisms of the local field $\mathbb{F}_q((t))$ (where $q = p^e$), this finitely generated pro-p group $\mathcal{N} = \mathcal{N}(\mathbb{F}_q)$ was introduced to the group theory community in the work of D. Johnson [13] (himself inspired by an article of S. Jennings [12]) and his Ph.D. student I. York [26] [27]. Viewing $\mathcal{N}$ as a group of formal power series under substitution, D. Johnson and I. York proved many of the initial structural results. Some of the proofs are reproduced below and the reader will see the ease with which one can calculate within the group. This, alongside the group's interesting properties, has led to the Nottingham group becoming a favourite test case for conjectures concerning pro-p groups.

For $p \neq 2$ straightforward calculations yield that the Nottingham group is not soluble, is not p-adic analytic (equivalently not of finite rank), is of finite width, that is $|\gamma_i(\mathcal{N})/\gamma_{i+1}(\mathcal{N})|$ is bounded, and is hereditarily just infinite, that is any non-trivial closed normal subgroup of an open subgroup is of finite index and therefore open. Roughly speaking, the commutator structure of $\mathcal{N}$ is tight and well behaved, whereas the pth powers drop quickly, not yielding any structure. When $p = 2$, the commutator structure is not so transparent and many facts are still unknown for this case. See the last section of this chapter for the present state of knowledge.

It was viewing the Nottingham group as a group of field automorphisms, as opposed to formal power series, that led C. Leedham-Green and A. Weiss to prove that every finite p-group can be embedded in $\mathcal{N}$, thereby establishing that the Nottingham group is not linear. This result was then extended to show that every countably-based pro-p group can be embedded in $\mathcal{N}$ [3], a property the Nottingham group shares with the Grigorchuk group, another interesting pro-p group of finite width; see Chapter 4 and [9]. More recently it has been shown that the Nottingham group is not analytic over a pro-p ring (see Chapter 1, §5 for details).

The purpose of this chapter is two-fold: to provide an easy reference point for many of the basic techniques used to study the Nottingham group and to survey the current state of knowledge of this group (apologies to any authors whose work has unwittingly been omitted).

Notation. For a ring R, $R[[t]]$ denotes the ring of formal power series in one variable t over R. If F is a field, $F((t))$ denotes the field of finite-tailed Laurent series over F; this is the field of fractions of $F[[t]]$.

2 Defining the Nottingham group

Definition 1. Let R be a commutative ring with identity. The Nottingham group over R is denoted by $\mathcal{N}(R)$ and defined to be the group of formal power series of the form

$$f = t(1 + \sum_{k=1}^{\infty} \alpha_k t^k) \in R[[t]]$$

under formal substitution: given $g \in \mathcal{N}(R)$ set $fg = g(1 + \sum_{k=1}^{\infty} \alpha_k g^k)$.

That this defines a group was proved by S. Jennings [12].

Equivalently one can consider $\mathcal{N}(R)$ as a group of automorphisms of the ring $R[[t]]$. Define $f \in \mathrm{Aut}(R[[t]])$ by setting $(t)f := f$ and extending the action of f to the whole of $R[[t]]$. A simple calculation shows that $(t)fg = fg$ and we thus have $\mathcal{N}(R)$ embedded as a subgroup of $\mathrm{Aut}(R[[t]])$.

Although these are no doubt interesting groups, and many of the references given consider the general case, in this chapter we are going to restrict ourselves to considering the 'classic' Nottingham group, that is, when $R = \mathbb{F}_q$, a finite field. Throughout this chapter, $\mathcal{N}$ will denote $\mathcal{N}(\mathbb{F}_q)$ and (according to the author's preference) the elements of $\mathcal{N}$ will be thought of as automorphisms of $\mathbb{F}_q[[t]]$.

Let $R = \mathbb{F}_q$ be a finite field of characteristic p, where $q = p^e$. Define $A(\mathbb{F}_q)$ to be the group of formal power series of the form

$$F = \sum_{k=1}^{\infty} \alpha_k t^k \in \mathbb{F}_q[[t]] \text{ with } \alpha_1 \neq 0$$

under substitution. Call $A(\mathbb{F}_q)$ the 'full' group of formal power series over $\mathbb{F}_q$ under substitution. Again, $A(\mathbb{F}_q)$ can be considered as a group of automorphisms of $\mathbb{F}_q[[t]]$. When $q = p$, we have the following stronger result.

Theorem 1.

$$\begin{aligned} A(\mathbb{F}_p) &\cong \mathrm{Aut}(\mathbb{F}_p[[t]]) \\ &\cong \mathrm{Aut}^c(\mathbb{F}_p((t))) \\ &\cong \mathrm{Aut}(\mathbb{F}_p((t))) \end{aligned}$$

where Aut^c *denotes continuous field automorphisms.*

Clearly a ring automorphism σ of $\mathbb{F}_p[[t]]$ can be extended to a field automorphism of $\mathbb{F}_p((t))$ by setting

$$\sigma((\sum_{i=0}^{\infty}\alpha_i t^i)^{-1}) = (\sigma(\sum_{i=0}^{\infty}\alpha_i t^i))^{-1}.$$

The converse depends on the fact that the valuation of $\mathbb{F}_p((t))$ defined by

$$v(\sum_{i=-k}^{\infty}\alpha_i t^i) = -k, \quad \alpha_{-k} \neq 0,$$

is the only normalised valuation of $\mathbb{F}_p((t))$ with respect to which $\mathbb{F}_p((t))$ is complete. Let σ be an automorphism of $\mathbb{F}_p((t))$; then the valuation v_σ defined by

$$v_\sigma(\sum_{i=-k}^{\infty}\alpha_i t^i) = v(\sigma(\sum_{i=-k}^{\infty}\alpha_i t^i))$$

is also a normalised valuation of $\mathbb{F}_p((t))$ with respect to which $\mathbb{F}_p((t))$ is complete, thus $v_\sigma = v$. Consequently σ is continuous, and hence restricts to an automorphism of $\mathbb{F}_p[[t]]$. Note that the continuity of the automorphism and the lack of field automorphisms of $\mathbb{F}_p$ imply that each element of $\mathrm{Aut}(\mathbb{F}_p((t)))$ is determined by its effect on t.

Similar reasoning shows that $A(\mathbb{F}_q)$ is the group of continuous automorphisms of $\mathbb{F}_q((t))$ that fix the finite subfield $\mathbb{F}_q$.

Next we define a chain of subgroups of the Nottingham group. First note that

$$A_n(\mathbb{F}_q) := \{f : (t)f = \sum_{k=1}^{n}\alpha_k t^k \text{ where } \alpha_k \in \mathbb{F}_q \text{ and } \alpha_1 \neq 0\}$$

is a subgroup of $\mathrm{Aut}(\mathbb{F}_q[[t]]/t^{n+1}\mathbb{F}_q[[t]])$ of order $(q-1)q^{n-1}$. Consider the natural maps

$$\Phi_n : A(\mathbb{F}_q) \longrightarrow A_n(\mathbb{F}_q)$$

$$(f : t \mapsto \sum_{i=1}^{\infty}\alpha_i t^i) \mapsto (f_n : t \mapsto \sum_{i=1}^{n}\alpha_i t^i).$$

Let $\mathcal{N}_n = \mathrm{Ker}(\Phi_n)$. Then $\mathcal{N}_1 = \mathcal{N}$ and we have defined a chain of closed normal subgroups of the Nottingham group satisfying $|\mathcal{N}_n : \mathcal{N}_{n+1}| = q$. It can be seen that $\mathcal{N} \cong \varprojlim(\mathcal{N}/\mathcal{N}_n)$ is a pro-p group and that the $\mathcal{N}_n$ form a base for the neighbourhoods of the identity in $\mathcal{N}$. Also, note that $|A(\mathbb{F}_q) : \mathcal{N}| = q - 1$.

To study the Nottingham group, it is useful to introduce the following specific elements.

Definition 2. For $n \geqslant 1$ and $\alpha \in \mathbb{F}_q$ define $e_n[\alpha] \in \mathcal{N}_n$ by

$$(t)e_n[\alpha] = t(1 + \alpha t^n).$$

It is easy to see that $(t)e_n[\alpha]e_n[\beta] \equiv t(1+(\alpha+\beta)t^n) \bmod \langle t^{2n+1}\rangle$ and consequently that $(t)e_n[\alpha]^k \equiv t(1+k\alpha t^n) \bmod \langle t^{2n+1}\rangle$ for all integers k. Choose $\gamma_i \in \mathbb{F}_q$ for $1 \leqslant i \leqslant e$ so that $\mathbb{F}_q$ is additively generated by $\{\gamma_1, \gamma_2, \dots, \gamma_e\}$. Then it follows that $\langle e_n[\gamma_1]\mathcal{N}_{n+1}, e_n[\gamma_2]\mathcal{N}_{n+1}, \dots, e_n[\gamma_e]\mathcal{N}_{n+1}\rangle = \mathcal{N}_n/\mathcal{N}_{n+1}$ for all $n \geqslant 1$. Thus

$$\mathcal{N}/\mathcal{N}_n = \langle e_i[\gamma_j]\mathcal{N}_n : 1 \leqslant i \leqslant n-1,\ 1 \leqslant j \leqslant e\rangle.$$

As $\mathcal{N} \cong \lim_{\leftarrow}(\mathcal{N}/\mathcal{N}_n)$, it follows that $\mathcal{N}$ is topologically generated by $\{e_i[\gamma_j] : 1 \leqslant i, 1 \leqslant j \leqslant e\}$. Although we will see later that $\mathcal{N}$ is topologically generated by just $2e$ elements ($e_1[\gamma_j]$ and $e_2[\gamma_j]$ where $1 \leqslant j \leqslant e$, will do) it is often useful to think of this infinite set of topological generators.

The following definition is also important.

Definition 3. If $1 \neq g \in \mathcal{N}$, then there exists an integer $n \geqslant 1$ such that $g \in \mathcal{N}_n \setminus \mathcal{N}_{n+1}$. Define this n to be the *depth* of g, $D(g)$. Further, define the depth of the identity to be ∞.

3 Commutator structure

The following proposition is fundamental.

Proposition 1. *Let $g, h \in \mathcal{N}$ with $D(g) = m$ and $D(h) = n$. Then*

$$\begin{array}{rcll} D([g,h]) & = & m+n & \text{if } \ m \not\equiv n \bmod p \\ D([g,h]) & > & m+n & \text{if } \ m \equiv n \bmod p. \end{array}$$

More precisely, if $(t)g = t(1+\sum_{i=m}^{\infty}\alpha_i t^i)$ and $(t)h = t(1+\sum_{j=n}^{\infty}\beta_j t^j)$, for some $\alpha_i, \beta_j \in \mathbb{F}_q$ with $\alpha_m \neq 0 \neq \beta_n$, then

$$(t)[g,h] \equiv t(1+\alpha_m\beta_n(m-n)t^{m+n}) \quad \bmod \langle t^{m+n+2}\rangle.$$

Proof. First note that

$$\begin{aligned} (t)gh &= t(1+\sum_{j=n}^{\infty}\beta_j t^j)(1+\sum_{i=m}^{\infty}\alpha_i t^i(1+\sum_{j=n}^{\infty}\beta_j t^j)^i) \\ &\equiv t(1+\sum_{j=n}^{m+n}\beta_j t^j + \sum_{i=m}^{m+n}\alpha_i t^i + (m+1)\alpha_m\beta_n t^{m+n}) \quad \bmod \langle t^{m+n+2}\rangle. \end{aligned}$$

Similarly

$$(t)hg \equiv t(1+\sum_{i=m}^{m+n}\alpha_i t^i + \sum_{j=n}^{m+n}\beta_j t^j + (n+1)\beta_n\alpha_m t^{m+n}) \quad \bmod \langle t^{m+n+2}\rangle.$$

Define $s \in \mathcal{N}$ as follows:

$$(t)s = t(1+\alpha_m\beta_n(m-n)t^{m+n}).$$

Then a quick calculation yields

$$(t)hgs \equiv (t)gh \bmod \langle t^{m+n+2} \rangle$$

and the result follows. □

I. York calculated the next term of a commutator in $\mathcal{N}$ [26, Lemma 1.2.9], [16].

A slightly more sensitive calculation than that in Proposition 1 yields the following for $\alpha, \beta \in \mathbb{F}_q$:

$$[e_m[\alpha], e_n[\beta]] \equiv e_{m+n}[\alpha\beta]^{m-n} \bmod \mathcal{N}_{m+2n}\mathcal{N}_{2m+n}. \tag{1}$$

This yields a connection with the first Witt algebra W. This Lie algebra W is defined over $\mathbb{F}_p$, with basis $\{f_0, f_1, \dots, f_{p-1}\}$ and multiplication given by $[f_i, f_j] = (i - j)f_{i+j \bmod p}$. (This is clearly reminiscent of Equation 1.) For $p > 2$, the first Witt Lie algebra is simple. W has a natural graded structure $W = \bigoplus_{i=0}^{p-1} W_i$, where $W_i = \langle f_i \rangle$. This C_p-grading allows us to define the *loop algebra* $L(W, p, \alpha)$ of W, where α is the p-tuple $(W_0, \dots, W_{p-1})$, as

$$L(W, p, \alpha) = \bigoplus_{n \in \mathbb{Z}} W_{n \bmod p} \otimes t^n.$$

The 'positive part' of this loop algebra is

$$\begin{aligned} L^+(W, p, \alpha) &= L(W, p, \alpha) \cap (W \otimes_{\mathbb{F}_p} t\mathbb{F}_p[t]) \\ &= \bigoplus_{n \geqslant 1} W_{n \bmod p} \otimes t^n, \end{aligned}$$

which is isomorphic to the graded Lie algebra $L(\mathcal{N}(\mathbb{F}_p))$ of $\mathcal{N}(\mathbb{F}_p)$ given by $\bigoplus_{n \geqslant 1} \mathcal{N}_n(\mathbb{F}_p)/\mathcal{N}_{n+1}(\mathbb{F}_p)$. For further discussion of this Lie algebra see [17].

The first Witt Lie algebra is just one of a family of Witt Lie algebras. In a similar manner, just infinite pro-p groups can be associated with all the Witt Lie algebras. In this way the Nottingham group becomes part of a family of just infinite pro-p groups; see [21] for more details.

A. Caranti has considered the graded Lie algebra associated to $\mathcal{N}(\mathbb{F}_p)$, for odd p, with respect to the lower central series. Thus his grading is a C_{p-1} grading; this follows from Remark 1(ii), below. He has proved that this Lie algebra is not finitely presented [5]. However, in doing this he proved that a certain central extension of the Lie algebra is finitely presented. It is thought that this result could be used to prove that the Nottingham group is finitely presented as a pro-p group. Strong computational evidence for this conjecture has been given by E. O'Brien.

We now return to the commutator structure of $\mathcal{N}$.

Theorem 2. *For* $p \neq 2$

$$[\mathcal{N}_r, \mathcal{N}_s] = \overline{[\mathcal{N}_r, \mathcal{N}_s]} = \begin{cases} \mathcal{N}_{r+s} & \text{if} \quad r \not\equiv s \bmod p \\ \mathcal{N}_{r+s+1} & \text{if} \quad r \equiv s \bmod p. \end{cases}$$

Proof. Let $n \geqslant r+s$ if $r \not\equiv s \bmod p$ and $n > r+s$ otherwise. Since $p \neq 2$ there exist $e_u[1] \in \mathcal{N}_r$ and $e_v[1] \in \mathcal{N}_s$ with $u \not\equiv v \bmod p$ and $u+v=n$. Thus, by Equation 1, it follows that $\langle [e_u[\gamma_j], e_v[1]]\mathcal{N}_{n+1} : 1 \leqslant j \leqslant e\rangle = \mathcal{N}_n/\mathcal{N}_{n+1}$. This yields the result for $\overline{[\mathcal{N}_r, \mathcal{N}_s]}$. However, since it now follows that $\mathcal{N}_n$ is a finitely generated pro-p group for $n \geqslant 1$, we may deduce that the abstract subgroup $[\mathcal{N}_r, \mathcal{N}_s]$ is closed [6, Chap. 1]. □

Remark 1. Assume that $p \neq 2$. The following are now clear from Theorem 2.
(i) Let $\{\mathcal{N}^{(n)}\}_{n\geq 0}$ be the derived series of $\mathcal{N}$. Then $\mathcal{N}^{(n)} = \mathcal{N}_{2^{n+1}-1}$ and $|\mathcal{N} : \mathcal{N}^{(n)}| = q^{2^{n+1}-2}$.
(ii) The lower central series of $\mathcal{N}$ is given by $\gamma_m(\mathcal{N}) = \mathcal{N}_{r_m}$ where $r_m = m+1+[(m-2)/(p-1)]$. Thus

$$\left|\frac{\gamma_m(\mathcal{N})}{\gamma_{m+1}(\mathcal{N})}\right| \leqslant p^{2e}$$

and $\mathcal{N}$ is of finite width.
(iii) A similar analysis of commutators shows that, for $p \neq 2, 3$, $\mathcal{N}(\mathbb{F}_p)$ is the closure of the derived group of $A(\mathbb{F}_p)$ [4]. This might help to explain why some results are only valid for $p \geqslant 5$ [15].

The following proposition illustrates the tight structure of normal subgroups within the Nottingham group. In particular it shows that the Nottingham group is *just infinite*, that is any non-trivial closed normal subgroup is of finite index. It is proved by repeated applications of Equation 1, as in the proof of Theorem 2:

Proposition 2. *Suppose $p \neq 2$ and let $g \in \mathcal{N}$ with $D(g) = n$. Then*

$$\overline{\langle g\rangle^{\mathcal{N}}} = \begin{cases} \langle g, \mathcal{N}_{n+1}\rangle & \text{if} \quad n \not\equiv 1 \bmod p \\ \langle g, \mathcal{N}_{n+2}\rangle & \text{if} \quad n \equiv 1 \bmod p. \end{cases}$$

Thus $\mathcal{N}$ is just infinite.

Note that if $q = p$ and $D(g) = n$, then $\langle g, \mathcal{N}_{n+1}\rangle = \mathcal{N}_n$.

In order to understand the connection between the normal subgroup structure and the lower central series of a pro-p group of finite width, the notion of obliquity has been introduced [14, III.15].

Definition 4. Let G be a pro-p group of finite width. Put

$$\mu_i(G) = \gamma_{i+1}(G) \cap \bigcap \{N \lhd G : N \not\leqslant \gamma_{i+1}(G)\}.$$

The *obliquity* of G is defined to be

$$o(G) = \sup_{i\in\mathbb{N}}(\log_p|\gamma_{i+1}(G) : \mu_i(G)|) \in \mathbb{N} \cup \infty.$$

From Proposition 2 and Remark 1(ii) it is clear that, for odd p, each normal subgroup of $\mathcal{N}$ lies between some $\gamma_j(\mathcal{N})$ and $\gamma_{j+1}(\mathcal{N})$. Thus when p is odd, $\mathcal{N}$ has obliquity zero. M. du Sautoy has used this fact to calculate the normal zeta function of $\mathcal{N}$. The *normal zeta function* of a pro-p group G is defined as

$$\zeta_G^{\triangleleft}(s) = \sum a_{p^n}^{\triangleleft}(G)p^{-ns}$$

where $a_{p^n}^{\triangleleft}(G) = |\{N : N \triangleleft G \text{ and } |G : N| = p^n\}|$.

Theorem 3. ([7]) *For $q = p \geqslant 3$*

$$\zeta_{\mathcal{N}}^{\triangleleft}(s) = \zeta_p(s) + p^{-s}\zeta_p(ps-1),$$

where $\zeta_p(s) = 1/(1-p^{-s})$.

In Section 5 the number of subgroups (as opposed to normal subgroups) of a given index is considered.

B. Klopsch has proved the following.

Theorem 4. ([16]) *Suppose $p \neq 2$ and $g \in \mathcal{N}$. Then*

$$\langle g \rangle^{\mathcal{N}} = \overline{\langle g \rangle^{\mathcal{N}}}.$$

Thus $\mathcal{N}$ is just infinite as an abstract group.

A topological group G is *hereditarily just infinite* if every open subgroup of G is just infinite.

Proposition 3. *$\mathcal{N}$ is hereditarily just infinite for $p \neq 2$.*

Proof. Let H be an open subgroup of $\mathcal{N}$ and let $h \in H$ with $D(h) = m$. Since H is open $H \geqslant \mathcal{N}_n$ for some n. Hence $[h, e_i[\alpha]] \in \langle h \rangle^H$ for all $i \geqslant n$ and all $\alpha \in \mathbb{F}_q$. Thus, by Proposition 1, for $i \not\equiv m \mod p$ there are elements of depth $m+i$ in $\langle h \rangle^H$ which generate $\mathcal{N}_{m+i}/\mathcal{N}_{m+i+1}$. For $i \equiv m \mod p$, and i sufficiently large, take commutators of length 3 of the form $[h, e_s[1], e_{i-s}[\alpha]]$ with $e_s[1]$ and $e_{i-s}[\alpha] \in \mathcal{N}_n$. In this way we conclude that $\overline{\langle h \rangle^H} \geqslant \mathcal{N}_{m+2n+2}$. □

4 Power structure

To study the power structure of the Nottingham group it is useful to introduce an infinite matrix, $M(g)$, for each $g \in \mathcal{N}$ [26, Chapter 5]. Define $M(g) = \{m_{ij}\}$ by

$$m_{ij} = \text{coefficient of } t^j \text{ in } (t^i)g.$$

(We are thinking of g as an automorphism of $t\mathbb{F}_q[[t]]$, which is regarded as a topological vector space with basis $t, t^2, t^3, \ldots$) Clearly $M(g) = I + U(g)$ where $U(g)$ has zero entries on and below the diagonal. Since the characteristic of $\mathbb{F}_q$ is p, it follows that $M(g^p) = (I + U(g))^p = I + U(g)^p$.

The following analysis of the power structure of $\mathcal{N}$ is taken from [17].

Lemma 1. *If $D(g) = n$ then*

$$\begin{aligned} D(g^p) &= np \quad \text{if } n \equiv 0 \bmod p \\ D(g^p) &> np \quad \text{if } n \not\equiv 0 \bmod p. \end{aligned}$$

Proof. Let $h \in \mathbb{F}_q[[t]]$ and suppose that $h \equiv \alpha t^j \bmod \langle t^{j+1} \rangle$ for some $\alpha \in \mathbb{F}_q$. Also suppose that $(t)g = t(1 + \sum_{k=n}^{\infty} \beta_k t^k)$. Then it is easy to see that $hU(g) \equiv j\alpha\beta_n t^{n+j} \bmod \langle t^{n+j+1} \rangle$. Thus

$$hU(g)^p \equiv j(j+n)(j+2n)\cdots(j+(p-1)n)\alpha\beta_n^p t^{pn+j} \bmod \langle t^{pn+j+1} \rangle.$$

The result follows. □

Remark 2. (i) It is now clear that $\mathcal{N}_n^p \leqslant \mathcal{N}_{np}$.
(ii) It follows that if $g \in \mathcal{N}$ has depth a multiple of p, then g has infinite order.

Theorem 5. *If p is odd and $n \equiv 0 \bmod p$, then $\mathcal{N}_n^p = \overline{\mathcal{N}_n^p} = \mathcal{N}_{np}$.*

Proof. By the proof of Lemma 1 it can be seen that for $n \equiv 0 \bmod p$

$$e_n^p[\alpha] \equiv e_{np}[\alpha^p] \mod \mathcal{N}_{np+1}.$$

The result follows from Remark 2(i), Proposition 2 and Theorem 4 since

$$\mathcal{N}_{np} = \langle e_n^p[\gamma_i] : 1 \leqslant i \leqslant e \rangle^{\mathcal{N}} \leqslant \mathcal{N}_n^p \leqslant \mathcal{N}_{np}.$$ □

Note that, since g^p commutes with g, it is no surprise that $D(g^p) > np$ if $n \not\equiv 0 \bmod p$. In fact, from Lemma 1 and Proposition 1, it follows that $D(g^p) \geqslant np + \bar{n}$, where $n \equiv \bar{n} \bmod p$ and $0 \leqslant \bar{n} \leqslant p - 1$.

Theorem 6. *If p is odd, then $\mathcal{N}_n^p = \overline{\mathcal{N}_n^p} = \mathcal{N}_{np+\bar{n}}$, where $n \equiv \bar{n} \bmod p$ and $0 \leqslant \bar{n} \leqslant p - 1$.*

Proof. By Theorem 5 we may assume that $n \not\equiv 0 \bmod p$. The previous comment makes clear that $\mathcal{N}_n^p \leqslant \mathcal{N}_{np+\bar{n}}$, so we need just consider the other inclusion. First we prove the following claim. *Claim.* $e_n[\alpha]^p \in \mathcal{N}_{(p+1)n}$ if $n \not\equiv 0 \bmod p$.

We write the matrix $M(e_n[\alpha])$ as $I + V(e_n[\alpha]) + W(e_n[\alpha])$ where $V(e_n[\alpha])$ maps t^j to $j\alpha t^{n+j}$, and $W(e_n[\alpha])$ maps t^j into $\langle t^{2n+j} \rangle$. Then, as in the proof of Lemma 1, since $n \not\equiv 0 \bmod p$, it follows that $V(e_n[\alpha])^p = 0$. Thus it is clear that $U(e_n[\alpha])^p = (V(e_n[\alpha]) + W(e_n[\alpha]))^p$ maps t into $\langle t^{(p+1)n+1} \rangle$ and the claim follows.

Let $g = e_n[1]e_{n+\bar{n}}[\gamma]$ for some $\gamma \in \mathbb{F}_q^*$. We show that $D(g^p) = np + \bar{n}$. Now $g^p = e_n[1]^p e_{n+\bar{n}}[\gamma]^p rs$, where r is a product of pth powers of commutators in $e_n[1]$ and $e_{n+\bar{n}}[\gamma]$ and s is a product of commutators in $e_n[1]$ and $e_{n+\bar{n}}[\gamma]$ of weight $\geqslant p$. Consider the commutator $c = [e_{n+\bar{n}}[\gamma],_{p-1} e_n[1]]$ which occurs as a factor of s with exponent 1. All other factors of s will have depth greater than $np + \bar{n}$. Repeated applications of Proposition 1 show that $D(c) = np + \bar{n}$. By the claim, Lemma 1 and Proposition 1, it follows that $D(g) = D(c)$.

A more careful analysis using Equation 1 shows that

$$[e_{n+\bar{n}}[\gamma],_{p-1} e_n[1]] \equiv e_{pn+\bar{n}}[\gamma]^k \mod \mathcal{N}_{np+\bar{n}+1}$$

where $k \equiv (p-1)!\bar{n}^{p-1} \equiv -1 \bmod p$. Thus for $\bar{n} \not\equiv 1$, by Proposition 2 and Theorem 4 we have

$$\mathcal{N}_{np+\bar{n}} = \langle (e_n[1]e_{n+\bar{n}}[\gamma_i])^p : 1 \leqslant i \leqslant e \rangle^{\mathcal{N}} \leqslant \mathcal{N}_n^p \leqslant \mathcal{N}_{np+\bar{n}}.$$

Similarly, $(e_n[1]e_{n+\bar{n}}[\gamma]^x)^p \equiv e_{np+\bar{n}}[\gamma]^{-x} \bmod \mathcal{N}_{np+\bar{n}+1}$ for any integer x. Suppose $\bar{n} = 1$ and $(e_n[1]e_{n+1}[\gamma]^x)^p \equiv e_{np+1}[\gamma]^{-x} e_{np+2}[\beta] \bmod \mathcal{N}_{np+3}$. Since an element commutes with its pth power, it follows from Equation 1 that $\beta \equiv -x^2\gamma^2 \bmod p$. As p is odd we can choose integers x_1 and x_2 which are congruent neither to each other nor to 0 modulo p. Then by Proposition 2 and Theorem 4 it follows that

$$\mathcal{N}_{np+1} \leqslant \langle (e_n[1]e_{n+1}[\gamma_i]^{x_j})^p : 1 \leqslant i \leqslant e, 1 \leqslant j \leqslant 2 \rangle^{\mathcal{N}} \leqslant \mathcal{N}_n^p \leqslant \mathcal{N}_{np+1}. \qquad \square$$

Remark 3. (i) It now follows that for $p \neq 2$,

$$\Phi(\mathcal{N}_n) = [\mathcal{N}_n, \mathcal{N}_n]\mathcal{N}_n^p = [\mathcal{N}_n, \mathcal{N}_n] = \mathcal{N}_{2n+1}.$$

Hence

$$d(\mathcal{N}_n) = (n+1)e.$$

So, in particular, $\mathcal{N} = \mathcal{N}_1$ is a $2e$-generator pro-p group and

$$\mathcal{N} = \overline{\langle e_1[\gamma_i], e_2[\gamma_i] : 1 \leqslant i \leqslant e \rangle}.$$

(ii) I. York noted that $f : t \mapsto t/(1-t)$ is a p-element of $\mathcal{N}$ [26, Example 5.5.1]. This can be generalised to $f_i[\alpha] : t \mapsto t/(1-\alpha t^i)^{1/i}$ where $i \not\equiv 0 \bmod p$ and $\alpha \in \mathbb{F}_q^*$ (this was pointed out to the author by M. Barakat). Thus $\mathcal{N}(\mathbb{F}_p)$ for $p \neq 2$ can be generated by two elements of order p, namely $f_1[1]$ and $f_2[1]$.B. Klopsch proved that these elements give a complete set of representatives for the conjugacy classes of elements of order p in $\mathcal{N}$ [15]; see Section 6.
(iii) By Remark 2(i) and Proposition 1

$$\Phi(\mathcal{N}(\mathbb{F}_{2^n})_m) \leqslant \mathcal{N}(\mathbb{F}_{2^n})_{2m}.$$

(iv) $\mathcal{N}$ is not p-adic analytic since, by (i) and (iii), $\mathcal{N}$ does not have finite rank.
(v) It has recently been shown, using Remark 1(i) and Proposition 3, that $\mathcal{N}$ is not analytic over any pro-p ring (see Chapter 1, Section 5). For the $p = 2$ versions of Remark 1(i) and Proposition 3, see the last section.

5 Open subgroups

In this section we consider the open subgroups of $\mathcal{N}(\mathbb{F}_p)$ for $p \geqslant 3$. This work is previously unpublished and is due to C. Leedham-Green and A. Shalev.

Theorem 7. *For $q = p \geqslant 5$ the lower rank of $\mathcal{N}$ is 2; that is $\mathcal{N}$ has a basis of neighbourhoods of the identity consisting of 2-generator (but not 1-generator) subgroups.*

Proof. As the $\mathcal{N}_n$ are not abelian the lower rank of $\mathcal{N}$ is clearly greater than 1. Let H be an open subgroup of $\mathcal{N}$; we prove that it contains a 2-generator open subgroup. As H is open $H \geqslant \mathcal{N}_n$ for some $n \equiv 0 \bmod p$. Hence $H \geqslant \overline{\langle e_{n+1}, e_{n+2}\rangle}$, where $e_{n+1} = e_{n+1}[1]$ and $e_{n+2} = e_{n+2}[1]$. *Claim.* $\overline{\langle e_{n+1}, e_{n+2}\rangle}$ is an open subgroup of $\mathcal{N}$. To prove the claim we introduce the notion of an allowable partition. Let $m \in \mathbb{N}$. An *allowable partition* of m is a sequence $\{b_i\}_{i=1}^r$ satisfying the following conditions:
(i) every b_i is either $n+1$ or $n+2$
(ii) if $s_k = \sum_{i=1}^k b_i$ then $s_k \not\equiv b_{k+1}$
(iii) $s_r = m$.
First we prove that there exist integers M and l such that for every $m > M$, m has a partition satisfying (i) and (iii), with $|u - v| \leqslant l$, where exactly u of the b_i are equal to $n+1$ and v are equal to $n+2$.

Now, $m = x(n+1) + d$ where $0 \leqslant d < n+1$ and $x \in \mathbb{N}$. Also $d = d(n+2) - d(n+1)$. Putting these together yields $m = (x-d)(n+1) + d(n+2)$. So for m sufficiently large we have a decomposition $m = u(n+1) + v(n+2)$ with $u \geqslant v \geqslant 0$. Replacing u with $u - (n+2)$ and v with $v + (n+1)$, we can ensure that $0 \leqslant u - v < 2n+3$.

Given this partition for m we can now construct an allowable partition for m. Construct a sequence according to (i) and (ii) above, choosing $n+1$ whenever possible. The choice will go as follows:

$$n+1, n+2, n+1, n+1, \ldots, n+2, n+1, n+1, \ldots$$

i.e., every choice of $n+2$ will be followed by at least two choices of $n+1$. Thus after a bounded number of steps the number of $(n+1)$s will exceed the number of $(n+2)$s by $u - v$. As $u - v$ is bounded independently of m, by choosing m sufficiently large we may assume that the sum of the b_i so far is less than m. Now complete the sequence by taking alternate $(n+1)$s and $(n+2)$s and ignoring (ii). Finally, adjust the partition by swapping pairs $n+1$ and $n+2$ so that (ii) is satisfied. This is possible, since if $x \equiv n+1 \bmod p$ or $x + n + 1 \equiv n+2 \bmod p$, then $x \not\equiv n+2$ and $x + n + 2 \not\equiv n+1 \bmod p$.

Thus there exists $N \in \mathbb{N}$ such that m has an allowable partition for all $m \geqslant N$. Using Proposition 1 it now follows that for all $m \geqslant N$ there exists $g \in \overline{\langle e_{n+1}, e_{n+2}\rangle}$ with $D(g) = m$. Thus $\overline{\langle e_{n+1}, e_{n+2}\rangle} \geqslant \mathcal{N}_N$. □

More generally, the above proof can be used to show that $\overline{\langle e_q[1], e_r[1]\rangle}$ is an open subgroup of $\mathcal{N}(\mathbb{F}_p)$ for $p \geqslant 5$ whenever
(1) $q \not\equiv r \bmod p$,
(2) $q, r \not\equiv 0 \bmod p$, and
(3) q and r are coprime.

To prove that $\mathcal{N}(\mathbb{F}_3)$ has lower rank 2, instead of considering the subgroup $\overline{\langle e_{n+1}[1], e_{n+2}[1]\rangle}$ of H take $\overline{\langle e_{n+1}[1], g\rangle}$ where $D(g) = n+2$ and $D(g^3) = 3n+8$; this is possible by Theorem 6.

The rest of this section is devoted to giving an upper bound for the number of subgroups of a given index in $\mathcal{N} = \mathcal{N}(\mathbb{F}_q)$. Denote the number of subgroups of index

p^k in a group H by $a_{p^k}(H)$. Recall that a finitely generated pro-p group H is p-adic analytic if and only if there exists an integer c such that $a_{p^k}(H) \leqslant p^{kc}$ for all k, that is if and only if H has polynomial subgroup growth [6].

Theorem 8. *Assume that p is odd. Let H be an open subgroup of index p^k in $\mathcal{N}$. Then H contains $\mathcal{N}_n$ where $n = \lceil 2kp/(p-1) \rceil$.*

Proof. A positive integer m has $[m/2]$ distinct decompositions of the form $m = u + v$ with $u, v > 0$. If we also require that $u \not\equiv v \bmod p$, then m has at least $[m/2 - m/2p]$ such decompositions. Thus if $m \geqslant 2(k + \frac{k}{p-1})$, then m has at least

$$k + \frac{k}{p-1} - \frac{k + k/(p-1)}{p} = k$$

decompositions.

We say H *covers* $\mathcal{N}_w/\mathcal{N}_{w+1}$ if $H\mathcal{N}_{w+1} \geqslant \mathcal{N}_w$. *Claim.* H fails to cover at most k quotients $\mathcal{N}_w/\mathcal{N}_{w+1}$.

First note that since H is open, $H \geqslant \mathcal{N}_r$ for some $r \in \mathbb{N}$, and hence H covers $\mathcal{N}_w/\mathcal{N}_{w+1}$ for all $w \geqslant r$. Let $1 \leqslant i < r$. Then

$$\left|\frac{H \cap \mathcal{N}_i}{H \cap \mathcal{N}_{i+1}}\right| = \left|\frac{(H \cap \mathcal{N}_i)\mathcal{N}_{i+1}}{\mathcal{N}_{i+1}}\right| \quad \text{divides } p^e,$$

and this index is equal to p^e just when H covers $\mathcal{N}_i/\mathcal{N}_{i+1}$. Now $|\mathcal{N} : \mathcal{N}_r| = p^{(r-1)e}$ and $|\mathcal{N} : H| = p^k$, so $|H : \mathcal{N}_r| = p^{(r-1)e-k}$. Also

$$|H : \mathcal{N}_r| = |H : H \cap \mathcal{N}_2||H \cap \mathcal{N}_2 : H \cap \mathcal{N}_3| \ldots |H \cap \mathcal{N}_{r-1} : \mathcal{N}_r|.$$

So at most k of these quotients are less than p^e, and hence H fails to cover at most k of the quotients $\mathcal{N}_w/\mathcal{N}_{w+1}$ for $1 \leqslant w \leqslant r - 1$. Hence the claim is true.

Suppose, for a contradiction, that $H \not\geqslant \mathcal{N}_n$. Then H does not cover a quotient $\mathcal{N}_m/\mathcal{N}_{m+1}$ for some $m \geqslant n$ and hence H fails to cover at most $k-1$ of the quotients $\mathcal{N}_u/\mathcal{N}_{u+1}$ with $1 \leqslant u < m$. So, in particular, there exist two such quotients $\mathcal{N}_u/\mathcal{N}_{u+1}$ and $\mathcal{N}_v/\mathcal{N}_{v+1}$ each covered by H and such that $u + v = m$, $1 \leqslant u, v < m$ and $u \not\equiv v \bmod p$. Hence there exists $h \in H$ with $h \equiv e_u[1] \bmod \mathcal{N}_{u+1}$, and for $\alpha \in \mathbb{F}_q$ there exists $g_\alpha \in H$ with $g_\alpha \equiv e_v[\alpha] \bmod \mathcal{N}_{v+1}$. By Proposition 1 it follows that there exists $l_\alpha \in H$ with $l_\alpha \equiv e_{u+v}[\alpha]^{u-v} \bmod \mathcal{N}_{u+v+1}$. Thus H covers $\mathcal{N}_m/\mathcal{N}_{m+1}$, a contradiction. □

Lemma 2. *Let P be a group of order p^m, and let $0 \leqslant k \leqslant m$. Then $a_{p^k}(P) \leqslant C(p)p^{k(m-k)}$, where $C(p) = \prod_{i \geqslant 1}(1 - p^{-i})^{-1}$.*

Proof. It is enough to consider the case where P is elementary abelian [22]. This case is standard and can be found in [11, III.8.5]. □

Theorem 9. *For $p \neq 2$*

$$a_{p^k}(\mathcal{N}) \leqslant C(p)p^{k((2k+\lceil\frac{2k}{p-1}\rceil-1)e-k)}.$$

Proof. By Theorem 8 an open subgroup of index p^k contains $\mathcal{N}_n$, where $n = \lceil 2k(\frac{p}{p-1}) \rceil$. Therefore, it can be considered as a subgroup of $\mathcal{N}/\mathcal{N}_n$ of index p^k. Since $|\mathcal{N}/\mathcal{N}_n| = p^{(n-1)e}$ the result follows from the preceding lemma. □

6 Closed subgroups and automorphisms

Viewing the Nottingham group as a group of automorphisms led to the following result.

Theorem 10. (C. Leedham-Green, A. Weiss) [3] *Every finite p-group can be embedded in $\mathcal{N}$.*

Proof. We prove that every finite p-group can be embedded in $\mathcal{N}(\mathbb{F}_p)$ and then the result follows. Let H be a finite p-group. According to E. Witt [24] [25], there exists a totally ramified extension field K of $\mathbb{F}_p((t))$ such that $H \cong \mathrm{Gal}(K/\mathbb{F}_p((t)))$. As the extension is totally ramified $K \cong \mathbb{F}_p((t))$. Thus H is isomorphic to a subgroup of $\mathrm{Aut}(\mathbb{F}_p((t)))$, which is isomorphic to $A(\mathbb{F}_p)$ by Theorem 1. Since $|A(\mathbb{F}_p) : \mathcal{N}(\mathbb{F}_p)| = p-1$ it follows that H is isomorphic to a subgroup of $\mathcal{N}(\mathbb{F}_p)$. □

Corollary 1. ([3]) *$\mathcal{N}$ is not linear over any field.*

Proof. This follows from the previous theorem, since the derived length of a soluble linear group in a given dimension is bounded [23, 3.7]. □

An analysis of Witt's work led to the following theorem (see Chapter 10 for details).

Theorem 11. ([3]) *Every countably-based pro-p group can be embedded as a closed subgroup in $\mathcal{N}$.*

However, explicitly writing down not-open subgroups of the Nottingham group has proved difficult. An element of order p^2 is still not known. The rest of this section is devoted to those closed subgroups of the Nottingham group that are easily described.

An interesting set of closed subgroups of the Nottingham group can be defined as follows:

$$\mathcal{N}_{(r)} = \{f : t \mapsto t(1 + \sum_{i=1}^{\infty} \alpha_i t^{ri})\} \leqslant \mathcal{N}.$$

If $r \not\equiv 0 \bmod p$ then $\mathcal{N}_{(r)} \cong \mathcal{N}$. However, if $r \equiv 0 \bmod p$, then $\mathcal{N}_{(r)}$ is a torsion free pro-p group, see Remark 2(ii). I. Fesenko has proved that $\mathcal{N}(\mathbb{F}_p)_{(p^n)}$ is hereditarily just infinite and not p-adic analytic for $p \neq 2$. He then uses these subgroups to answer a question posed by J. Coates and R. Greenberg [8] (see Chapter 10 for details). The fact that these subgroups are equal to their normalisers [4] suggested the likelihood that $\mathcal{N}$ has no outer automorphisms of order p. This has since been proved for $p \geqslant 5$ [15].

In particular B. Klopsch has proved the following theorem. Recall, $A(\mathbb{F}_q)$ denotes the 'full' group of formal power series over $\mathbb{F}_q$ under substitution. $A(\mathbb{F}_q)$ is a finitely generated profinite group and $\mathcal{N}(\mathbb{F}_q)$ is its unique Sylow pro-p subgroup. In fact $A(\mathbb{F}_q) \cong \mathbb{F}_q^* \ltimes \mathcal{N}(\mathbb{F}_q)$. Clearly an automorphism σ of $\mathbb{F}_q$ acts naturally on $A(\mathbb{F}_q)$

by sending $\sum_{i=1}^{\infty} \alpha_i t^i$ to $\sum_{i=1}^{\infty} \alpha_i^{\sigma} t^i$. An automorphism of $A(\mathbb{F}_q)$ is said to be *standard* if it is the composition of an inner automorphism and an automorphism induced by some field automorphism of $\mathbb{F}_q$. An automorphism of $\mathcal{N}(\mathbb{F}_q)$ is standard if it is the restriction of a standard automorphism of $A(\mathbb{F}_q)$.

Theorem 12. ([15]) *Let $q = p^e$ and $p \geqslant 5$. Then all automorphisms of $A(\mathbb{F}_q)$ and all automorphisms of $\mathcal{N}(\mathbb{F}_q)$ are standard. Furthermore*

$$\mathrm{Aut}(A(\mathbb{F}_q)) \cong \mathrm{Aut}(\mathcal{N}(\mathbb{F}_q)) \cong \mathrm{Aut}(\mathbb{F}_q) \ltimes A(\mathbb{F}_q).$$

A corresponding result for substitution groups of formal power series over infinite fields was proved by B. Muckenhoupt [18].

To prove the previous theorem B. Klopsch considers the conjugacy classes of p-elements in $\mathcal{N}$. (Once again denote $\mathcal{N}(\mathbb{F}_q)$ simply by $\mathcal{N}$.) He proves that for $p \geqslant 5$,

$$\{F(n,\lambda) \mid n \in \mathbb{N} \text{ with } n \not\equiv 0 \bmod p, \lambda \in \mathbb{F}_q^*\}$$

where

$$F(n,\lambda) := t(1 - n\lambda t^n)^{-1/n} \in \mathcal{N}$$

is a complete set of representatives for the conjugacy classes of elements of order p in $\mathcal{N}$. He also gives a description of the centralisers of such elements.

Previously, Pálfy P. [19] had determined the abundance k, and hence the number of conjugacy classes of the quotients $\mathcal{N}(\mathbb{F}_p)/\mathcal{N}(\mathbb{F}_p)_n$ for $1 \leqslant n \leqslant 3p+2$. There had been some speculation that the natural quotients of $\mathcal{N}(\mathbb{F}_p)$ would provide an infinite series of p-groups (for fixed p) satisfying $k(G) < c\log|G|$ for some constant c. However, the results of Pálfy do not substantiate this expectation. More recently, J. Sangroniz has studied the conjugacy classes and characters of $\mathcal{N}(\mathbb{F}_q)/\mathcal{N}(\mathbb{F}_q)_n$ for $p+2 < n \leqslant 2p+2$ [20]. The cases $1 \leqslant n \leqslant p+2$ can be found in [26].

The subgroups $\mathcal{N}_{(r)}$ are examples of index subgroups. Let $\Lambda \subseteq \mathbb{N}$. We define the subset

$$\mathcal{N}[\Lambda] := \{f : t \mapsto t(1 + \sum_{\lambda \in \Lambda} \alpha_\lambda t^\lambda)\} \subseteq \mathcal{N}.$$

A subgroup H of $\mathcal{N}$ is called an *index subgroup* if there exists $\Lambda \subseteq \mathbb{N}$ such that $H = \mathcal{N}[\Lambda]$. Clearly, every index subgroup of $\mathcal{N}$ is closed. I. York [26, Proposition 3.4.5] gave sufficient conditions on Λ so that $\mathcal{N}[\Lambda]$ is a subgroup of $\mathcal{N}$. Y. Barnea and B. Klopsch have recently proved the following result.

Theorem 13. ([1]) *Let $\Lambda \subseteq \mathbb{N}$. Then $\mathcal{N}(\mathbb{F}_p)[\Lambda]$ is a closed subgroup of $\mathcal{N}(\mathbb{F}_p)$ if and only if Λ satisfies*

$$\forall \lambda, \mu \in \Lambda,\ \forall k \leqslant \lambda + 1 \ :\ \binom{\lambda+1}{k} \equiv 0 \bmod p \quad \text{or} \quad \lambda + k\mu \in \Lambda.$$

That this condition is necessary can be easily seen by considering the element $e_\lambda[1]e_\mu[1]$. The following examples of such Λ are given in [1].

(i) $\Lambda = r\mathbb{N}$ for some $r \in \mathbb{N}$,

(ii) $\Lambda = p\mathbb{N} \cup (p\mathbb{N} - 1)$,

(iii) $\Lambda = p^i\mathbb{N} - 1$ for some $i \in \mathbb{N}$,

(iv) $\Lambda = \{p^i - 1 | i \in \mathbb{N}\}$.

Example (i) leads to the subgroups $\mathcal{N}_{(r)}$ which have already been discussed. Example (iv) defines a subgroup of $\mathcal{N}(\mathbb{F}_p)$ isomorphic to $U_1 = 1 + t\mathbb{F}_p[[t]]$, the multiplicative group of one-units of the valuation ring $\mathbb{F}_p[[t]]$, a free abelian pro-p group of infinite rank.

Their motivation in proving this result was to calculate further the *Hausdorff spectrum* of $\mathcal{N}(\mathbb{F}_p)$. For the definition and further details on this, see Chapter 1. The Hausdorff dimension is a way of measuring closed subgroups of a pro-p group G. Briefly, each closed subgroup is given a value in the closed interval $[0, 1]$, open subgroups have the value 1 and finite subgroups the value 0; the Hausdorff spectrum $\mathrm{Spec}(G)$ is the set of all values taken by closed subgroups of G. Although the complete Hausdorff spectrum of the Nottingham group is not known, for $p \geqslant 5$ it is known that

$$[0, p^{-1}] \cup \{p^{-1} + p^{-j}\}_{j=1}^{\infty} \cup \{d^{-1}\}_{d=1}^{\infty} \subseteq \mathrm{Spec}(\mathcal{N}(\mathbb{F}_p)) \subseteq \{d^{-1}\}_{d=1}^{\infty} \cup [0, 3p^{-1}];$$

see [2] and [1]. An analysis of the index subgroups yielded the interval $[0, p^{-1}]$.

Clearly if $q > p$ then $\mathbb{F}_q$ has non-trivial subfields and $\mathcal{N}$ has corresponding subgroups. Furthermore, if R is a commutative ring with 1 and I is an ideal of R then $\mathcal{N}(I)$ is a closed normal subgroup of $\mathcal{N}(R)$. B. Klopsch [16] has studied the abstract normal subgroup structure of $\mathcal{N}(R)$ in order to answer the question as to whether the profinite completion $\widehat{\mathcal{N}(R)}$ of the group $\mathcal{N}(R)$ is equal to the group $\mathcal{N}(\widehat{R})$. He concludes that this is true if R is *saturated* with ideals of finite index, i.e. if every additive subgroup of finite index in R contains an ideal of finite index in R; but it can be false if R is not saturated, for example if $R = \mathbb{F}_p[[t]]$.

7 $p = 2$

The structure of $\mathcal{N}(\mathbb{F}_{2^n})$ differs from the case of odd characteristic. In particular $\mathcal{N}(\mathbb{F}_{2^n})$ can be written as a product of two non-intersecting closed subgroups, the subgroup of 'even' elements and the subgroup of 'odd' elements (this was first noted in [17]). That the 'even' elements of the Nottingham group form a subgroup has already been seen. Recall that

$$\mathcal{N}_{(2)} = \{f \in \mathcal{N} : (t)f = t(1 + \sum_{i=1}^{\infty} \alpha_i t^{2i})\} \leqslant \mathcal{N}.$$

However, for $p = 2$ the 'odd' elements also form a subgroup (see Theorem 13)

$$\{f \in \mathcal{N}(\mathbb{F}_{2^n}) : (t)f = t(1 + \sum_{i=0}^{\infty} \alpha_i t^{2i+1})\} \leqslant \mathcal{N}(\mathbb{F}_{2^n})$$

and we have the required decomposition of $\mathcal{N}(\mathbb{F}_{2^n})$.

Many of the proofs outlined in the previous sections (in particular in Sections 3 and 4) are no longer valid when $p = 2$. This is generally because if g and h are elements of $\mathcal{N}(\mathbb{F}_{2^n})$ with $D(g) \equiv D(h) \bmod 2$, then g and h commute modulo $\mathcal{N}(\mathbb{F}_{2^n})_{D(g)+D(h)+1}$. Therefore when $p = 2$, one has to work harder to generate elements of even depth. With this aim the notions of Even and Odd Depth are introduced.

Let $f \in \mathcal{N}(\mathbb{F}_{2^n})$ be given by $(t)f = t(1+\sum_{i=1}^{\infty} \alpha_i t^i)$. The *Odd Depth* of f is the least odd i such that α_i is non-zero; the *Even Depth* is defined analogously. Clearly the minimum of the Even Depth and the Odd Depth equals the depth of f. Equivalently, note that f can be written uniquely as hk where h is an 'even' element of $\mathcal{N}(\mathbb{F}_{2^n})$ and k an 'odd' element. The Even Depth of f is then $D(h)$ and the Odd Depth is $D(k)$.

The following simplified formulae for multiplication and commutation in $\mathcal{N}(\mathbb{F}_{2^n})$ were given by Hegedüs P.

Let $f, g \in \mathcal{N}(\mathbb{F}_{2^n})$ with $(t)f = t(1 + \sum_{i \geqslant m} \alpha_i t^i)$ and $(t)g = t(1 + \sum_{j \geqslant n} \beta_j t^j)$. Then

$$\begin{aligned}(t)fg &\equiv t(1+\sum_{j\geqslant n}\beta_j t^j + \sum_{i \geqslant m}\alpha_i t^i + \sum_{i\geqslant m,\, j\geqslant n}\alpha_i\beta_j(i+1)t^{i+j}\\ &+ \sum_{i\geqslant m,\, j\geqslant n}\alpha_i\beta_j^2\frac{i(i+1)}{2}t^{i+2j}) \quad \bmod \langle t^{m+3n+1}\rangle.\end{aligned}$$

This yields the following commutator formula:

$$\begin{aligned}(t)[f,g] &\equiv t(1+\sum_{k+l=s}\alpha_k\beta_l s t^s + \sum_{k+2l=s}\alpha_k\beta_l^2\frac{k(k+1)}{2}t^s\\ &+ \sum_{2k+l=s}\alpha_k^2\beta_l\frac{l(l+1)}{2}t^s + \sum_{k+l_1+l_2=s}\alpha_k\beta_{l_1}\beta_{l_2}(l_1+1)(k+l_2)t^s\\ &+ \sum_{k_1+k_2+l=s}\alpha_{k_1}\alpha_{k_2}\beta_l(k_1+1)(l+k_2)t^s) \bmod \langle t^{m+n+2\min(n,m)+1}\rangle.\end{aligned}$$

Letting $f = g$ in the original formula gives

$$(t)f^2 \equiv t(1+\sum_{i,\, j\geqslant m}\alpha_i\alpha_j(i+1)t^{i+j} + \sum_{i,\, j\geqslant m}\alpha_i\alpha_j^2\frac{i(i+1)}{2}t^{i+2j}) \bmod \langle t^{4m+1}\rangle.$$

C. Leedham-Green and S. McKay have calculated the lower central series for $\mathcal{N}(\mathbb{F}_2)$. Denote $e_i[1]$ simply by e_i.

Theorem 14. ([17]) *(i) The lower central series for* $\mathcal{N} = \mathcal{N}(\mathbb{F}_2)$ *is as follows:*

$\gamma_2(\mathcal{N}) = \langle e_3e_4e_6, e_5, \mathcal{N}_7\rangle$,

$\gamma_{2i}(\mathcal{N}) = \langle e_{4i-1}e_{8i-4}, \{e_{2j+1} : j \geqslant 2i\}, \mathcal{N}_{8i-1}\rangle$ *for* $i > 1$,

$\gamma_{2i+1}(\mathcal{N}) = \langle e_{4i+1}e_{8i+2}, e_{8i}, \{e_{2j+1} : j \geqslant 2i+1\}, \mathcal{N}_{8i+3}\rangle$ *for* $i \geqslant 1$.

(ii) The derived quotient of $\mathcal{N}$ is the direct product of cyclic groups of orders 2, 2 and 4. For $i > 1$ the lower central quotients $\gamma_i(\mathcal{N})/\gamma_{i+1}(\mathcal{N})$ are elementary abelian, of rank 2 if i is even, and rank 4 if i is odd.

Thus $\mathcal{N}(\mathbb{F}_2)$ is also of finite width. C. Leedham-Green and S. McKay have also proved the following.

Theorem 15. ([17]) *$\mathcal{N}(\mathbb{F}_2)$ is just infinite.*

That $\mathcal{N}(\mathbb{F}_2)$ is not 2-adic analytic was noted in Remark 3(iv). The following results, which are due to Hegedűs P., are used to prove that $\mathcal{N}(\mathbb{F}_2)$ is not analytic over any pro-p ring; see Chapter 1, Section 5 for details.

Theorem 16. ([10]) *$\mathcal{N}(\mathbb{F}_2)$ is hereditarily just infinite.*

Theorem 17. ([10]) *Let $\{\mathcal{N}^{(n)}\}_{n\geqslant 0}$ be the derived series of $\mathcal{N} = \mathcal{N}(\mathbb{F}_2)$. Then $\mathcal{N}^{(n)} \supseteq \mathcal{N}_{[4c^n]}$, where $c = \frac{1+\sqrt{17}}{2} < 3$, and $|\mathcal{N} : \mathcal{N}^{(n)}| \leqslant 2^{4c^n-1}$.*

References

[1] Y. Barnea and B. Klopsch, Index subgroups of the Nottingham group, preprint.

[2] Y. Barnea and A. Shalev, Hausdorff dimension, pro-p groups, and Kac–Moody algebras, *Trans. Amer. Math. Soc.*, **349** (1997), 5073–5091.

[3] R. D. Camina, Subgroups of the Nottingham group, *J. Algebra*, **196** (1997), 101–113.

[4] R. D. Camina, Some natural subgroups of the Nottingham group, *Proc. Edin. Math. Soc.*, **42** (1999), 333–339.

[5] A. Caranti, Presenting the graded Lie algebra associated to the Nottingham group, *J. Algebra,* **198** (1997), 266–289.

[6] J. D. Dixon, M. P. F. du Sautoy, A. Mann and D. Segal, *Analytic pro-p Groups, 2nd ed.* Cambridge studies in advanced math., **61** CUP, 1999.

[7] M. P. F. du Sautoy, The zeta function of $sl_2(\mathbb{Z})$, *Forum Mathematicum,* to appear.

[8] I. Fesenko, On just-infinite pro-p groups and arithmetically profinite extensions of local fields, *J. reine angew. Math*, **517** (1999), 61–80.

[9] R. I. Grigorchuk, On the Burnside problem for periodic groups, *Funktsional. Anal. i Prilozhen,* **14** (1980), 53–54, [Russian], *Functional Anal. Appl.*, **14** (1980), 41–43, [English transl.].

[10] P. Hegedűs, The Nottingham group for $p = 2$, preprint.

[11] B. Huppert, *Endliche Gruppen I,* Springer-Verlag, 1967.

[12] S. A. Jennings, Substitution groups of formal power series under substitution, *Canad. J. Math.*, **6** (1954), 325–340.

[13] D. L. Johnson, The group of formal power series under substitution, *J. Austral. Math. Soc.*, **45** (1988), 296–302.

[14] G. Klaas, C. R. Leedham-Green and W. Plesken, *Linear Pro-p-Groups of Finite Width,* Springer Lecture Notes **1674**, Springer-Verlag, 1997.

[15] B. Klopsch, Automorphisms of the Nottingham group, *J. Algebra*, **223** (2000), 37–56.

[16] B. Klopsch, Normal subgroups in substitution groups of the formal power series, *J. Algebra,* to appear.

[17] C. Leedham-Green and S. McKay, preprint.

[18] B. Muckenhoupt, Automorphisms of formal power series under substitution, *Trans. Amer. Math. Soc.*, **99** (1961), 373–383.

[19] Pálfy P. P., The number of conjugacy classes in some quotients of the Nottingham group, *Proc. Edinburgh Math. Soc.*, **41** (1998), 369–384.

[20] J. Sangroniz, Conjugacy classes and characters in some quotients of the Nottingham group, *J. Algebra*, **211** (1999), 26–41.

[21] A. Shalev, Finite p-groups, in *Finite Groups,* eds.: B. Hartley, G. M. Seitz, A. V. Borovik and R. M. Bryant, Kluwer Academic Publishers, 1995.

[22] G. E. Wall, Some applications of the Eulerian functions of a finite group, *J. Austral. Math. Soc.*, **II** (1961-62), 35–59.

[23] B. A. F. Wehrfritz, *Infinite Linear Groups,* Springer-Verlag, 1973.

[24] E. Witt, Der Existenzsatz für abelsche Funktionenkörper, *J. für Mathematik,* **173** (1935), 43–51.

[25] E. Witt, Konstruktion von galoisschen Körpern der Charakteristik p zu vorgegebener Gruppe der Ordnung p^k, *J. für Mathematik,* **174** (1936), 237–245.

[26] I. O. York, The group of formal power series under substitution, Ph.D. thesis, Nottingham University, 1990.

[27] I. O. York, The exponent of certain p-groups, *Proc. Edinburgh Math. Soc.*, **33** (1990), 483–490.

DPMMS, Center for Mathematical Sciences
Wilberforce Road
Cambridge CB3 0WB
England
e-mail:r.d.camina@dpmms.cam.ac.uk

7

On Groups Satisfying the Golod–Shafarevich Condition

E. Zelmanov

In [2], E. S. Golod and I.R. Shafarevich found a sufficient condition for a pro-p group presented by generators and relators to be infinite. In the same volume of *Izvestia*, this criterion was applied to the solution of two outstanding problems: the construction of an infinite tower of class fields (by I. R. Shafarevich) and the construction of a counterexample to the General Burnside Problem (by E. S. Golod).

Let p be a prime number, let K be the field with p elements and let $\widehat{F}_p$ be the free pro-p group on $m \geqslant 2$ generators. Consider the group algebra $K\widehat{F}_p$ and its fundamental ideal (augmentation ideal) $\omega(K\widehat{F}_p)$. The i-th term of the Zassenhaus filtration $\widehat{F}_p = (\widehat{F}_p)_1 > (\widehat{F}_p)_2 > \dots$ is defined as

$$(\widehat{F}_p)_i = \{g \in \widehat{F}_p \mid 1 - g \in \omega(K\widehat{F}_p)^i\}.$$

It is known that $\bigcap_{i\geqslant 1}(\widehat{F}_p)_i = (1)$ (see [4]). Hence, an arbitrary nonidentity element $g \in \widehat{F}_p$ falls into some gap $g \in (\widehat{F}_p)_i \setminus (\widehat{F}_p)_{i+1}$. Let us call i the *degree* of g.

For a subset R of $(\widehat{F}_p)_2$ containing r_i elements of degree i $(i \geqslant 2)$, we write $H_R(t) = \sum_{i=1}^{\infty} r_i t^i$.

Consider a pro-p group $G = \langle x_1, \dots, x_m \mid R = 1\rangle$ presented by generators x_i and relators R in the category of pro-p groups. This means that $G = \widehat{F}_p/N$, where N is the closed normal subgroup of $\widehat{F}_p$ generated by R.

Again consider the group algebra $A = KG$ and the fundamental ideal $\omega =$

Partially supported by NSF grant DMS-9622870.

$\omega(KG)$. Denote $H_A(t) = 1 + \sum_{i\geqslant 1} \dim(\omega^i/\omega^{i+1})t^i$. Golod and Shafarevich [2] (see also Koch [3], Roquette [6], Vinberg [9]) proved that

$$\frac{H_A(t)}{1-t}(1 - mt + H_R(t)) \geqslant \frac{1}{1-t},$$

where inequality of formal series is understood in the formal way: $\sum a_i t^i \geqslant \sum b_i t^i$ if $a_i \geqslant b_i$ for each i.

Now, if there exists a number $0 < t_0 < 1$ such that $H_R(t)$ converges at t_0 and $1 - mt_0 + H_R(t_0) < 0$, then the series $H_A(t)$ cannot converge at t_0. Hence, the group G is infinite. If such a number t_0 exists, we will say that the group G (or, more precisely, the presentation $G = \langle x_1, \dots, x_m \mid R = 1\rangle)$ satisfies the *Golod–Shafarevich condition*. In particular, if $|R| < \frac{m^2}{4}$, then $1 - mt + H_R(t) \leqslant 1 - mt + |R|t^2$ is negative at some $0 < t_0 < 1$. Hence, an arbitrary m generated group, presented by $< \frac{m^2}{4}$ relators from $(\widehat{F}_p)_2$, satisfies the Golod–Shafarevich condition.

To avoid trivial cases, we will exclude the group of p-adic integers ($m = 1, R = \emptyset$) from groups satisfying the Golod–Shafarevich condition.

Important examples of pro-p groups satisfying the Golod–Shafarevich condition are pro-p completions of 3-manifold groups (see A. Lubotzky, [5]).

In [10], J. Wilson proved that certain infinite groups cannot have a presentation satisfying the Golod–Shafarevich condition and conjectured that a pro-p group satisfying the Golod–Shafarevich condition contains a free pro-p group of rank $\geqslant 2$. In this paper we prove this conjecture.

Theorem. *A pro-p group satisfying the Golod–Shafarevich condition contains a free nonabelian pro-p group.*

Remark that in [11] pro-p groups satisfying the Golod–Shafarevich condition were shown to contain an abstract nonabelian free group.

A finite p-group P is said to be *involved* in a pro-p group G if there exist an open subgroup K in G and a closed normal subgroup L of K such that $P \cong K/L$.

Corollary. *If G is a pro-p group satisfying the Golod–Shafarevich condition, then an arbitrary finite p-group is involved in G.*

Let us discuss similar notions for profinite algebras. Let $K\langle Y\rangle$ be the free associative K-algebra in m free generators $Y = \{y_1, \dots, y_m\}$ and let $K\langle\langle Y\rangle\rangle$ be the algebra of infinite series. Let I_i denote the ideal of $K\langle Y\rangle$ consisting of all sums $\sum \alpha_v v$, where $\alpha_v \in K$ and the length of each word v is $\geqslant i$. Clearly, $K\langle\langle Y\rangle\rangle$ is the completion of $K\langle Y\rangle$ with respect to the topology defined by the system of ideals $I_i, i \geqslant 1$. Let $\widehat{I_i}$ denote the closure of the ideal I_i in $K\langle\langle Y\rangle\rangle$.

From $\bigcap_{i\geqslant 1} \widehat{I_i} = (0)$, it follows that an arbitrary nonzero element $a \in K\langle\langle Y\rangle\rangle$ falls into some gap $a \in \widehat{I_i}\backslash\widehat{I_{i+1}}$. We say that $\deg(a) = i$.

Let R' be a subset of $\widehat{I_2}$ such that R' contains r_i elements of degree i for every $i \geqslant 2$. As above, denote $H_{R'}(t) = \sum_{i=2}^{\infty} r_i t^i$.

Let $I(R')$ be the closed ideal of the algebra $K\langle\langle Y\rangle\rangle$ generated by R'. We say that the algebra $A = K\langle\langle Y\rangle\rangle/I(R') = \langle y_1, \dots, y_m \mid R' = (0)\rangle$ is presented by generators Y and relators R'. The algebra A has a natural filtration $A = A_1 \supset A_2 \supset \dots$, where $A_i = \widehat{I_i} + I(R')/I(R')$. Let $H_{gr(A)}(t) = 1 + \sum_{i \geqslant 1} \dim(A_i/A_{i+1})t^i$ be the Hilbert series of the associated graded algebra. In [2], [3], [6], [9] it was proved that

$$\frac{H_{gr(A)}(t)}{1-t}(1 - mt + H_{R'}(t)) \geqslant \frac{1}{1-t}.$$

This result, in fact, implies the above cited result for pro-p groups. Indeed, $1 + \widehat{I_1}$ is a subgroup of the group of invertible elements of $K\langle\langle Y\rangle\rangle$. The quotient groups $1+\widehat{I_1}/1+\widehat{I_i}$ are finite p-groups. Hence, $1+\widehat{I_1}$ is a pro-p group. The closed subgroup of $1+\widehat{I_1}$ generated by $1 - y_1, \dots, 1 - y_m$ is isomorphic to $\widehat{F}_p$.

Let $R \subseteq (\widehat{F}_p)_2$ and let $G = \langle x_1, \dots, x_m \mid R = 1\rangle$. For an arbitrary element $f \in R$ we have $f(1 - y_1, \dots, 1 - y_m) = 1 + f'(y_1, \dots, y_m)$, $f' \in \widehat{I_1}$. It is easy to see that the degree of the element f' in $K\langle\langle Y\rangle\rangle$ is equal to the degree of the element f in $\widehat{F}_p$.

Denote $R' = \{f' \mid f \in R\}$. The algebra $\langle y_1, \dots, y_m \mid R' = (0)\rangle$ is isomorphic to the completion of the group algebra KG (see [4]).

In [1], E.S. Golod proved the following lemma.

Lemma (E. S. Golod). *For an arbitrary $\varepsilon > 0$ and an arbitrary $0 < t_0 < 1$, there exists a subset $M \subseteq I_2$ such that*

(1) $H_M(t_0) < \varepsilon$,

(2) the algebra $I_1 + I(M)/I(M)$ is nil.

We will start with a slight generalization of Golod's lemma.

Lemma 1. *For an arbitrary $\varepsilon > 0$ and an arbitrary $0 < t_0 < 1$, there exist a subset $M \subseteq I_2$ and a function $h : \mathbb{N} \to \mathbb{N}$ such that*

(1) $H_M(t_0) < \varepsilon$,

(2) for any $k \geqslant 1$, each set of k elements from $I_{h(k)} + I(M)/I(M)$ generates a nilpotent subalgebra.

Proof. Choose a sequence (ε_k) of positive numbers such that $\sum_{k=1}^{\infty} \varepsilon_k < \varepsilon$, and let $\delta_k = \frac{\varepsilon_k}{1+\varepsilon_k}$.

Let $h : \mathbb{N} \to \mathbb{N}$ be an increasing function such that for an arbitrary $k \geqslant 1$, we have $k \cdot t_0^{h(k)} < \delta_k$. The Cartesian product $I_{h(k)}^{(k)} = \underbrace{I_{h(k)} \times \dots \times I_{h(k)}}_{k}$ is a countable set. Let $\gamma_k : I_{h(k)}^{(k)} \to \mathbb{N}$ be a bijection. For an arbitrary k-tuple $\alpha = (a_1, \dots, a_k) \in I_{h(k)}^{(k)}$, let M_α consist of all products $a_{i_1} \dots a_{i_{\gamma_k(\alpha)}}$, $1 \leqslant i_1, \dots, i_{\gamma_k(\alpha)} \leqslant k$. The total number of elements in M_α is not greater than $k^{\gamma_k(\alpha)}$ and the degree of each element is greater than or equal to $h(k) \cdot \gamma_k(\alpha)$. Let

$$M(k) = \bigcup_{\alpha} M_\alpha, \quad M = \bigcup_{k \geqslant 1} M(k).$$

Then

$$\begin{aligned} H_{M(k)}(t_0) &\leqslant \sum_{\alpha \in I_{h(k)}^{(k)}} k^{\gamma_k(\alpha)} t_0^{h(k)\gamma_k(\alpha)} \\ &= \sum_{\alpha} (k t_0^{h(k)})^{\gamma_k(\alpha)} < \sum_{\alpha} \delta_k^{\gamma_k(\alpha)} \\ &\leqslant \frac{\delta_k}{1 - \delta_k} = \varepsilon_k \end{aligned}$$

and therefore $H_M(t_0) < \sum_{k=1}^{\infty} \varepsilon_k \leqslant \varepsilon$. The lemma is proved. □

Let $G = \langle x_1, \dots, x_m \mid R = 1 \rangle$ be a pro-p group satisfying the Golod–Shafarevich condition, where $R \subseteq (\widehat{F}_p)_2$, $0 < t_0 < 1$, $H_R(t)$ converges at t_0 and $1 - mt_0 + H_R(t_0) < 0$.

As above, consider the subset $R' = \{f' \mid f \in R\}$ of $\widehat{I_2} \lhd K\langle\langle Y \rangle\rangle$, and the compact algebra $K\langle\langle Y \rangle\rangle / I(R')$. The elements $1 - y_i + I(R')$, $1 \leqslant i \leqslant m$, generate a closed pro-$p$ subgroup of $1 + \widehat{I_1}/I(R')$ which is isomorphic to G. Choose $\varepsilon > 0$ such that $1 - mt_0 + H_R(t_0) + \varepsilon < 0$. By Lemma 1, there exist a subset M of I_2 and a function $h : \mathbb{N} \to \mathbb{N}$ such that $H_M(t_0) < \varepsilon/2$ and, for each $k \geqslant 1$, k arbitrary elements from $I_{h(k)} + I(M)/I(M)$ generate a nilpotent subalgebra.

Suppose that contrary to the assertion of the theorem, G does not contain a nonabelian free pro-p group. This means that for an arbitrary pair of elements $a, b \in G$ there exists a nonidentity element w of the free pro-p group of rank 2 such that $w(a, b) = 1$. Consider the Zassenhaus filtration of G:

$$G = G_1 > G_2 >, \dots, G_i = \{g \in G \mid 1 - g \in \omega(KG)^i\}.$$

Choose a sufficiently big number k so that $2t_0^k < \frac{\varepsilon}{2}$.

Let $F(2)$ be the abstract free group on two free generators x_1, x_2. For a nonidentity element $w \in F(2)$, consider the subset $(G_k \times G_k)_w$ of the Cartesian product $G_k \times G_k$ consisting of those pairs (a, b) for which there exists an element $w' \in \widehat{F(2)}_p$ such that $\deg w' > \deg w$ and $w(a, b) = w'(a, b)$. Since, according to our assumption, G_k does not contain a nonabelian free pro-p group, it follows that

$$G_k \times G_k = \bigcup_{1 \neq w \in F(2)} (G_k \times G_k)_w.$$

It is easy to see that each $(G_k \times G_k)_w$ is a closed subset of $G_k \times G_k$. By Baire's Category theorem (see [8]), there exist an element $1 \neq w \in F(2)$, a number $\ell \geqslant k$ and a pair of elements $a, b \in G_k$ such that for arbitrary elements $h', h'' \in G_\ell$, we have $w(ah', bh'') = w'(ah', bh'')$, $\deg w' > \deg w$.

We shall identify G with the subgroup of $1+\widehat{I_1}/I(R')$ generated by $\{1-y_i+I(R') : 1 \leqslant i \leqslant m\}$. Since $a, b \in G_k$, it follows that there exist $a', b' \in \widehat{I_k}$ such that $a = 1 + a' + I(R'), b = 1 + b' + I(R')$.

Now let $R'' = R' \cup M \cup \{a', b'\}$. Then

$$\begin{aligned} 1 - mt_0 + H_{R''}(t_0) &\leqslant 1 - mt_0 + H_{R'}(t_0) + H_M(t_0) + 2t_0^k \\ &\leqslant 1 - mt_0 + H_R(t_0) + \varepsilon < 0, \end{aligned}$$

so the algebra $K\langle\langle Y\rangle\rangle/I(R'')$ still satisfies the Golod–Shafarevich condition.

For $1 \leqslant i \leqslant m$ put $a_i = y_i + I(R'')$, $g_i = 1 - a_i$. The closed subgroup G' of $1+\widehat{I_1}/I(R'')$ generated by $g_1, \dots, g_m$ is a homomorphic image of G. Hence, for arbitrary elements $g, h \in G'_\ell$, there exists an element $w' \in \widehat{F(2)_p}$ such that $\deg w' > \deg w$ and $w(g, h) = w'(g, h)$.

Let A be the abstract subalgebra of $K\langle\langle Y\rangle\rangle/I(R'')$ generated by $a_1, \dots, a_m$. For any $i \geqslant 1$, i arbitrary elements from $A^{h(i)} = I_{h(i)} + I(R'')/I(R'')$ generate a nilpotent subalgebra. In particular, A is a nil algebra because every element from A is nilpotent modulo $A^{h(1)}$.

The abstract group $G(A)$ generated by $g_1, \dots, g_m$ lies in $1 + A$.

Let the degree of the element $w \in F(2)$ be n. This means that $w = \rho_1^{p^{e_1}\cdot k_1} \cdots \rho_s^{p^{e_s}\cdot k_s} \cdot \tilde{w}$, where $\rho_1, \dots, \rho_s$ are distinct left-normed group commutators in x_1, x_2 and for each $1 \leqslant i \leqslant s$, we have $(k_i, p) = 1$, $\text{length}(\rho_i) \cdot p^{e_i} = n$, $\deg \tilde{w} > n$.

Considering the commutator $(x_1, w) = x_1^{-1} w^{-1} x_1 w$ instead of w, if necessary, we can assume that $e_i = 0$ for $1 \leqslant i \leqslant s$, and therefore that $\text{length}(\rho_i) = n$.

Let $\tilde{\rho}_i$ be the Lie commutator corresponding to ρ_i. Consider the element $\tilde{f} = \sum_{i=1}^{s} k_i \tilde{\rho}_i$ of the free Lie algebra. For arbitrary elements $g, h \in G(A)_\ell$ there exists an element $w' \in \widehat{F(2)_p}$, of degree $> n$, such that $\rho_1(g, h)^{k_1} \cdots \rho_s(g, h)^{k_s} = w'(g, h)$. This implies that $\tilde{f}(1-g, 1-h)$ is equal to an infinite (but converging in $K\langle\langle Y\rangle\rangle/I(R'')$) sum of products of $1-g, 1-h$, each product having length $\geqslant n+1$.

Let $f(y_1, \dots, y_n)$ be the full linearization of $\tilde{f}$ (see [7]). Choose n arbitrary elements $h_1, \dots, h_n$ from $G(A)_s$, where $s \geqslant \max(\ell, h(n))$. Then $h_1, \dots, h_n \in 1 + A^{h(n)}$. Therefore the subalgebra generated by $1-h_1, \dots, 1-h_n$ is nilpotent. Hence

$$f(1-h_1, \dots, 1-h_n) = \sum \pm P,$$

where each P is a product of elements $1-h_1, \dots, 1-h_n$ of length $\geqslant n+1$, and the sum on the right hand side is finite.

From now on our aim will be to prove the following proposition.

Proposition. *Let $A = \langle a_1, \dots, a_m\rangle$ be an m generated associative K-algebra. Suppose that*

(1) there exists a function $h : \mathbb{N} \to \mathbb{N}$ such that for each i, i arbitrary elements from $A^{h(i)}$ generate a nilpotent subalgebra. In particular, A is a nil algebra.

(2) Let $G(A)$ be the subgroup of $1+A$ generated by $g_1 = 1-a_1, \dots, g_m = 1-a_m$. Suppose further that there exist a multilinear polynomial $f(x_1, \dots, x_n)$ and an

integer $s \geqslant h(n)$ such that for arbitrary elements $h_1, \dots, h_n \in G(A)_s$ we have $f(1-h_1, \dots, 1-h_n) = \sum \pm P$, where each P on the right hand side is a product of elements $1-h_i$, $1 \leqslant i \leqslant n$, of length $\geqslant n+1$.

Then the algebra A is nilpotent.

Let us show that the proposition implies the theorem. Indeed, we proved above that the algebra $K\langle\langle Y\rangle\rangle/I(R'')$ has a dense subalgebra that satisfies the assumptions of the proposition. Hence, it follows from the proposition that the algebra $K\langle\langle Y\rangle\rangle/I(R'')$ is nilpotent. This contradicts the fact that the presentation $\langle Y \mid R' = (0)\rangle$ satisfies the Golod–Shafarevich condition, and the theorem is proved.

A non-nilpotent algebra B is said to be *just non-nilpotent* if every proper homomorphic image of B is nilpotent. The following lemma is well known.

Lemma 2. *Every finitely generated non-nilpotent algebra has a just non-nilpotent homomorphic image.*

Proof. Let B be a finitely generated non-nilpotent algebra. Consider the set $\mathcal{P}$ of ideals of B whose quotients are not nilpotent. This set is inductive. Indeed, let $I_1 \subset I_2 \subset \dots$ be an ascending chain of ideals lying in $\mathcal{P}$, $I = \bigcup_\alpha I_\alpha$. If the quotient algebra B/I is nilpotent, then I is a finitely generated algebra. Hence I coincides with some member of the chain, a contradiction. By Zorn's Lemma, $\mathcal{P}$ contains a maximal (in $\mathcal{P}$) ideal I. The algebra B/I is just non-nilpotent. The lemma is proved. □

Suppose that an algebra A satisfies the assumptions of the proposition, but is not nilpotent. Passing to a homomorphic image, if necessary, we will assume that the algebra A is just non-nilpotent.

Lemma 3. $\bigcap_{i\geqslant 1} A^i = (0)$.

Proof. If $\bigcap_{i\geqslant 1} A^i \neq (0)$, then the quotient algebra $A/\bigcap_{i\geqslant 1} A^i$ is nilpotent. Hence there exists $k \geqslant 1$ such that $A^k = A^{k+1} = \dots$. The algebra A^k is finitely generated; it can be generated by d elements. Since $A^k \subseteq A^{h(d)}$, it follows that the algebra A^k is nilpotent. Hence, the algebra A is nilpotent as well, a contradiction. The lemma is proved. □

The multilinear polynomial $f(x_1, \dots, x_n)$ can be represented as

$$f(x_1, \dots, x_n) = x_n f_{n-1}(x_1, \dots, x_{n-1}) + \sum_{i=1}^{n-1} x_i(\dots).$$

We will assume that n, the degree of the polynomial, is minimal among integers satisfying the assumptions of the proposition.

In other words, for any integer $t \geqslant h(n-1)$, there exist elements $h_1, \dots, h_{n-1} \in G(A)_t$ such that $f_{n-1}(1-h_1, \dots, h_{n-1})$ does not lie in $\langle 1-h_1, \dots, 1-h_{n-1}\rangle^n$, where $\langle 1-h_1, \dots, 1-h_{n-1}\rangle$ is the subalgebra of A generated by $1-h_1, \dots, 1-h_{n-1}$.

Let us fix elements $h_1, \dots, h_{n-1} \in G(A)_s$ such that $f_{n-1}(1-h_1, \dots, 1-h_{n-1}) \notin \langle 1-h_1, \dots, 1-h_{n-1}\rangle^n$. Since the algebra $\langle 1-h_1, \dots 1-h_{n-1}\rangle$ is finite dimensional,

Lemma 3 implies that there exists an integer $q \geqslant s$ such that $\langle 1-h_1, \dots, 1-h_{n-1}\rangle \cap A^q = (0)$.

Lemma 4. *For arbitrary elements $g, h \in G(A)$ we have $1-(g,h) \equiv -(1-g)(1-h) \mod (A(1-g)+A(1-g)A)$.*

Proof. $1-(g,h) = 1-g^{-1}h^{-1}gh = g^{-1}h^{-1}(hg-gh) = g^{-1}h^{-1}((1-h)(1-g)-(1-g)(1-h)) \equiv -(1-g)(1-h) \mod (A(1-g)+A(1-g)A)$. The lemma is proved. □

Choose a commutator $\sigma \in G(A)_q$. Let V denote the K-linear span of the elements $1-\sigma, 1-(\sigma, g_{i_1}, \dots, g_{i_r})$, $r \geqslant 1, i_j \geqslant 1$, where $(\sigma, g_{i_1}, \dots, g_{i_r}) = ((\dots(\sigma, g_{i_1}), g_{i_2}), \dots, g_{i_r})$ is a left normed commutator. Clearly, $V \subseteq A^q$. Let I denote the ideal $A(1-\sigma)+A(1-\sigma)A$ of the algebra A. For arbitrary coefficients $1 \leqslant k_1, \dots, k_t \leqslant p-1$ we have

$$k_1(1-(\sigma, g_{i_1}, g_{i_2}, \dots)) + \dots + k_t(1-(\sigma, g_{j_1}, g_{j_2}, \dots)) \\ \equiv 1-(\sigma, g_{i_1}, g_{i_2}, \dots)^{k_1} \dots (\sigma, g_{j_1}, g_{j_2}, \dots)^{k_t} \mod I.$$

Indeed, by Lemma 4

$$(\sigma, g_{i_1}, g_{i_2}, \dots)^{k_1} = (((\sigma, g_{i_1}, g_{i_2}, \dots) - 1) + 1)^{k_1} \\ \equiv k_1((\sigma, g_{i_1}, g_{i_2}, \dots) - 1) + 1 \mod I.$$

Lemma 5. *For an arbitrary element $v \in V$ there exists an element $u \in \langle 1-h_1, \dots, 1-h_{n-1}\rangle^n$ such that $vf_{n-1}(1-h_1, \dots, 1-h_{n-1}) \equiv vu \mod I$.*

Proof. Let $v = k_1(1-\sigma, g_{i_1}, g_{i_2}, \dots)) + \dots + k_t(1-(\sigma, g_{j_1}, g_{j_2}, \dots))$ and let $h = (\sigma, g_{i_1}, g_{i_2}, \dots)^{k_1} \dots (\sigma, g_{j_1}, g_{j_2}, \dots)^{k_t}$. Then $v \equiv 1-h \mod I$. We have

$$f(1-h_1, 1-h_2, \dots, 1-h_{n-1}, 1-h) \\ \equiv (1-h)f_{n-1}(1-h_1, \dots, 1-h_{n-1}) \mod I.$$

On the other hand

$$f(1-h_1, \dots, 1-h_{n-1}, 1-h) \in \langle 1-h_1, \dots, 1-h_{n-1}, 1-h\rangle^{n+1}.$$

Hence

$$f(1-h_1, \dots, 1-h_{n-1}, 1-h) \equiv P + (1-h)u \mod I,$$

where $P \in \langle 1-h_1, \dots, 1-h_{n-1}\rangle$, $u \in \langle 1-h_1, \dots, 1-h_{n-1}\rangle^n$.

The elements $f(1-h_1, \dots, 1-h_{n-1}, 1-h)$, $(1-h)u$ and the ideal I lie in A^q. Therefore

$$P \in \langle 1-h_1, \dots, 1-h_{n-1}\rangle \cap A^q = (0).$$

The lemma is proved. □

The algebra $\langle 1-h_1, \dots, 1-h_{n-1}\rangle^n$ is finite. Let $d = |\langle 1-h_1, \dots, 1-h_{n-1}\rangle^n|$. For an arbitrary element $u \in \langle 1-h_1, \dots, 1-h_{n-1}\rangle^n$ the set

$$V_u = \{v \in V \mid vf_{n-1}(1-h_1, \dots, 1-h_{n-1}) \equiv vu \mod I\}$$

is a subspace in V. From Lemma 5 it follows that

$$V = \bigcup_u V_u.$$

This implies that at least one subspace V_{u_0}, $u_0 \in \langle 1-h_1, \dots, 1-h_{n-1}\rangle^n$, has codimension $\geqslant d$.

For an arbitrary element $u \in \langle 1-h_1, \dots, 1-h_{n-1}\rangle^n$, let $J(u)$ denote the ideal of A generated by the nonzero element $f_{n-1}(1-h_1, \dots, 1-h_{n-1}) - u$. Since the algebra A is just non-nilpotent it follows that there exists $r \geqslant 1$ such that

$$A^r \subseteq \bigcap J(u),$$

where u runs over $\langle 1-h_1, \dots, 1-h_{n-1}\rangle^n$.

Let V_i denote the subspace of V spanned by all elements of the type $1 - (\sigma, g_{j_1}, \dots, g_{j_t})$, where $1 \leqslant j_1, \dots, j_t \leqslant m$, $(i-1)r \leqslant t < ir$.

Since $\dim_K V/V_{u_0} \leqslant d$, it follows that there exists $1 \leqslant t \leqslant d$ such that

$$V_t \subseteq V_{t+1} + \cdots + V_{d+1} + V_{u_0}.$$

Lemma 6. $(1-\sigma)A^{(d+1)r} \subseteq (1-\sigma)A^{(d+1)r+1} + I$.

Proof. Consider the product $(1-\sigma)a_{i_1}\dots a_{i_{(d+1)r}}$ and its subproduct $a_{i_{(t-1)r+1}}\dots a_{i_{tr}} \in A^r$. Because of the choice of the integer r, we have

$$a_{i_{(t-1)r+1}}\dots a_{i_{tr}} = \sum v'(f_{n-1}(1-h_1, \dots, 1-h_{n-1}) - u_0)v'',$$

where v', v'' are words in $a_1, \dots, a_m$, possibly empty.

Let $v' = a_{j_1}\dots a_{j_\mu}$. Without loss of generality, we can assume that $\mu < r$, otherwise

$$a_{i_1}\dots a_{i_{(t-1)r}} v'(f_{n-1}(1-h_1, \dots, 1-h_{n-1}) - u_0)v'' a_{i_{tr+1}}\dots a_{i_{(d+1)r}} \in A^{(d+1)r+1}.$$

By Lemma 4

$$\begin{aligned}\pm(1-\sigma)a_{i_1}\dots a_{i_{(t-1)r}} v' &\equiv 1 - (\sigma, g_{i_1}, \dots, g_{i_{(t-1)r}}, g_{j_1}, \dots, g_{j_\mu}) \mod I\\ &= 1 - \sigma'\end{aligned}$$

where $1 - \sigma' \in V_t$.

Hence, $1 - \sigma' = v_0 + v_{t+1} + \dots + v_{d+1}$, where $v_0 \in V_{u_0}$ and $v_k \in V_k$ for $t+1 \leqslant k \leqslant d+1$. Again by Lemma 4, $v_k \in (1-\sigma)A^{(k-1)r} + I$. Therefore

$$\begin{aligned}&v_k(f_{n-1}(1-h_1, \dots, 1-h_{n-1}) - u_0)v'' a_{i_{tr+1}}\dots a_{i_{(d+1)r}} \in\\ &\qquad A^{(k-1)r} \cdot A \cdot A^{(d+1-t)r} \subseteq (1-\sigma)A^{(d+k-t)r+1}\end{aligned}$$

and it remains to notice that $k - t \geqslant 1$. Finally, $v_0 \in V_{u_0}$ implies

$$v_0(f_{n-1}(1-h_1, \dots, 1-h_{n-1}) - u_0) \in I.$$

The lemma is proved. □

Let $w_1, \dots, w_\mu$ be all words in $a_1, \dots, a_m$ of length $< (d+1)r$, including the empty word.

Denote $A^0 = K \cdot 1 + A$.

Lemma 7.

$$(1-\sigma)A \subseteq A^0(1-\sigma)(\sum_{j=1}^{\mu} Kw_j) + A^k.$$

Proof. Consider the set of pairs $(i, j), 0 \leqslant i, j$ with the reverse lexicographical order: if $i < i'$, then $(i', j') < (i, j)$. If $j < j'$, then $(i, j') < (i, j)$. We will use induction on (i, j) to prove that

$$A^i(1-\sigma)A^j \subseteq U_k = A^0(1-\sigma)(\sum_{\nu=1}^{\mu} Kw_\nu) + A^k.$$

If $i \geqslant k$ or $j \geqslant k$, then clearly $A^i(1-\sigma)A^j \subseteq U_k$.

Suppose that for all pairs less than (i, j) the inclusion holds. Choose an arbitrary word w of length j (if $j = 0$ then w is the empty word). We need to show that $A^i(1-\sigma)w \subseteq U_k$.

If $j < (d+1)r$, then $w \in \{w_1, \dots, w_\mu\}$ and the assertion is clear. If $j \geqslant (d+1)r$, then by Lemma 6 we have $(1-\sigma)w \in (1-\sigma)A^{j+1} + A(1-\sigma)A^0$. Hence, $A^i(1-\sigma)w \subseteq A^i(1-\sigma)A^{j+1} + A^{i+1}(1-\sigma)A^0$. By the induction assumption, the right hand side lies in U_k. The lemma is proved. □

The choice of the integers d and r did not depend on the choice of the commutator σ. Hence, Lemma 7 is valid for an arbitrary $\sigma \in G(A)_q$.

Lemma 8. *For an arbitrary element $\sigma \in G(A)_{\max(q,h(\mu))}$, the element $1-\sigma$ generates a nilpotent ideal in A.*

Proof. The elements $(1-\sigma)w_1, \dots, (1-\sigma)w_\mu$ lie in $A^{h(\mu)}$. Hence, the subalgebra $\langle(1-\sigma)w_1, \dots, (1-\sigma)w_\mu\rangle$ is nilpotent. Say $\langle(1-\sigma)w_1, \dots, (1-\sigma)w_\mu\rangle^\ell = (0)$. We will show that $((1-\sigma)A^0)^\ell = (0)$. Indeed, by Lemma 7, for an arbitrary $k \geqslant 1$ we have

$$((1-\sigma)A^0)^\ell \subseteq A^0((1-\sigma)(\sum_{j=1}^{\mu} Kw_j))^\ell + A^k = A^k.$$

Hence, $((1-\sigma)A^0)^\ell \subseteq \bigcap_{k=1}^{\infty} A^k = (0)$ by Lemma 3. The lemma is proved. □

Now we are ready to finish the proof of the proposition. Since, according to our assumptions, the algebra A is just non-nilpotent it follows that A does not contain nonzero nilpotent ideals.

Hence by Lemma 8 the group $G(A)$ is nilpotent. Hence the center Z of $G(A)$ is nontrivial. Let $1 \neq z \in Z$. The element $1-z \in A$ commutes with the generators $a_i = 1-g_i, 1 \leqslant i \leqslant m$, of the algebra A. Hence $1-z$ lies in the center of A. We have already remarked that the algebra A is nil. Hence, the ideal $(1 - z)A$ of A nilpotent, a contradiction. The proposition and the theorem are proved.

The author is grateful to Jamshid Derakhshan who carefully read the manuscript.

References

[1] E.S. Golod, On nil-algebras and residually finite p-groups, *Izv. Akad. Nauk SSSR*, **28** (1964), 273–276.

[2] E.S. Golod, I.R. Shafarevich, On the class field tower, *Izv. Akad. Nauk SSSR*, **28** (1964), 262–272.

[3] Zum Satz von Golod-Šafarevič, *Math. Nachr.*, **42** (1969), 321–333.

[4] H. Koch, *Galoissche Theorie der p-Erweiterungen*, Springer-Verlag, Berlin, 1970.

[5] A. Lubotzky, Group presentations, p-adic analytic groups and lattices in $SL_2(\mathbf{C})$, *Ann. Math.*, **118** (1983), 115–130.

[6] P. Roquette, On class field towers, *Algebraic Number Theory*, Academic Press, London, 1967, 231–249.

[7] L. Rowen, *Polynomial Identities in Ring Theory*, Academic Press, New York, 1985.

[8] W. Rudin, *Functional Analysis*, McGraw-Hill, New York, 1973.

[9] E. B. Vinberg, On the theorem concerning the infinite dimension of an associative algebra, *Izv. Akad. Nauk SSSR*, **29** (1965), 209–214.

[10] J. S. Wilson, Finite presentations of pro-p groups and discrete groups, *Invent. Math.*, **205** (1991), 177–183.

[11] J.S. Wilson, E. Zelmanov, Identities for Lie algebras of pro-p groups, *J. Pure Appl. Algebra*, **81** (1992), 103–109.

Department of Mathematics
Yale University
10 Hillhouse Avenue/P.O. Box 208823
New Haven, CT 06520-8283, USA
email:zelmanov@pascal.math.yale.edu

8

Subgroup Growth in pro-p Groups

Avinoam Mann

Let G be any group. We write $a_n(G)$ for the number of subgroups of G of index n. If G is a profinite group, we consider only open subgroups.[1] If G is finitely generated, either as an abstract or as a profinite group, then $a_n(G)$ is finite. The study of the series $a_n(G)$, a subject that is known as *subgroup growth*, was begun by Hurwitz in the 19th century, with geometric motivation, and has become an active area of research in recent years. Let us mention, e.g., that a pro-p group G is p-adic analytic if and only if its subgroup growth is *polynomial*, i.e., there exist constants C and s such that $a_n(G) \leq Cn^s$ for all n (see Section 4 below and [2]). This characterisation of p-adic analytic groups was an instrumental stage in the characterisation of all groups of polynomial subgroup growth (known for short as PSG groups). Other applications of subgroup growth are found in [12] and [4].

In this chapter we will discuss the subgroup growth of pro-p groups in general. We first determine the subgroup growth of free groups and of free pro-p groups. Next we pass to the number of normal subgroups of a free pro-p group. This is intimately related to questions about the number of groups of a given order. Then we discuss the number of subgroups with a bounded number of generators, and the number of normal subgroups with a bounded number of normal generators. Again, the latter problem is related to the number of groups of given order and number of defining relations. Here simple probabilistic ideas turn out to be very useful. We will define some analytic functions related to this growth, and prove a functional equation which holds between these functions when our group is p-adic analytic. We also give several characterisations of p-adic analytic groups by their subgroup growth, and mention some related results. As

[1] In a finitely generated pro-p group, all finite index subgroups are open. It is still an open problem whether this holds in all finitely generated profinite groups.

an application of the functional equation mentioned above, we will derive a marvellous formula of P. Hall.

Some proofs that are easily available in the literature are omitted.

Since we will consider almost exclusively finitely generated groups, we will usually not mention that explicitly. We denote by $d(G)$ the minimal number of generators of G.

1 Free groups

By way of comparison, we start by calculating the growth of $a_n(F)$ when F is a free group. Since a free profinite group is the completion of an abstract free group, it does not matter whether we consider an abstract free group or a free profinite group. This subgroup growth provides an upper bound for the subgroup growth of all groups. For more on subgroup growth in abstract groups, see Chapter 9 and the survey papers [10], [11], [19].

First, let G be any group with d generators, and let H be any subgroup of G of index n. We can represent G as a permutation group on the cosets of H, and this defines a homomorphism of G onto a transitive subgroup of the symmetric group S_n. Identifying the coset H itself with the numeral 1, this homomorphism in turn determines H as the stabiliser of 1. Since the numbering of the other cosets of H by the numerals $2, \dots, n$ is arbitrary, there are $(n-1)!$ homomorphisms which correspond to H in this way. Since a homomorphism is determined by its effects on the generators, there are at most $(n!)^d$ homomorphisms of G into S_n. To get a bound for the number of subgroups, we have to divide by the number of homomorphisms which correspond to a single subgroup, obtaining

$$a_n(G) \leq n.(n!)^{d-1}.$$

Now let $G = F = F_d$ be free of rank d (*i.e* free on d generators. This is *not* the rank of G in the sense used in §4 below). Then $(n!)^d$ is the true number of homomorphisms of F into S_n, but not all these homomorphisms are onto transitive subgroups, so not all of them determine subgroups of index n. Let ϕ be a homomorphism onto a non-transitive subgroup K of S_n; then we can write $\{1, \dots, n\}$ as the union of two K-invariant subsets, of sizes k and $n-k$, where we may assume that $k \leq n-k$, and ϕ is determined by specifying first the two invariant subsets, and then the induced homomorphism into $S_k \times S_{n-k}$. The number of such homomorphisms is then at most $\binom{n}{k}(k!)^d((n-k)!)^d = n!(k!(n-k)!)^{d-1}$, and an upper bound for the number of all homomorphisms with a non-transitive image is obtained by summing the above expression over all relevant values of k.

Now for $1 \leq k \leq n-k$ we have

$$k!(n-k)!/n! = k!/((n-k+1)...n) \leq (1/n)(1/2)^{k-1},$$

and so $\sum_{k=1}^{n-1} n!(k!(n-k)!)^{d-1} \leq (2/n)(n!)^d$. Subtracting this last expression from $(n!)^d$, we see that as $n \to \infty$, the number of homomorphisms with a transitive image is asymptotic to the number of all homomorphisms, and dividing by $(n-1)!$ we obtain

Theorem 1. $a_n(F_d) \sim n(n!)^{d-1}$.

We now return to our proper subject, pro-p groups. By contrast with abstract groups, the subgroup growth of finitely generated free pro-p groups turns out to be exponential, and this provides an upper bound for the growth of all finitely generated pro-p groups. Recall that every open subgroup of a pro-p group G has index equal to a power of p, so $a_n(G) = 0$ if n is not a power of p. Let $R = R_d$ be a free pro-p group of rank d.

Theorem 2. $\lim_{k\to\infty}(a_{p^k}(R))^{1/p^k} = p^{(d-1)/(p-1)}$.

We first prove a well-known fact:

Lemma 3. *Let G be any group.*

(a) Let H be a subgroup of finite index in G. Then $d(H) - 1 \le |G : H|(d(G) - 1)$. In particular, if G is finitely generated, then so is H.

(b) If $G = R$ is a finitely generated free pro-p group, then equality is obtained in (a).

Of course, the equality also holds if G is an abstract free group; this is Schreier's formula.

Proof. (a). Let $G = \langle x_1, \dots, x_d\rangle$. Choose representatives $a_1 = 1, \dots, a_n$ for the right cosets of H in G. Let $x \in G$, and write $x = y_1 \dots y_r$, where each y_i is one of the generators x_j or its inverse. We rewrite this expression as follows: $x = y_1 a_{k_1}^{-1}.a_{k_1} y_2 a_{k_2}^{-1}.a_{k_2} y_3 \dots y_r a_{k_r}^{-1}.a_{k_r}$, where the indices k_i are chosen so that $y_1 a_{k_1}^{-1}$ and $a_{k_i} y_{i+1} a_{k_{i+1}}^{-1}$ lie in H. If $x \in H$, then $a_{k_r} = 1$, and the above rewriting exhibits x as a product of those elements of the form $a_i y_j a_k^{-1}$ which lie in H. Here, if $y_j = x_s^{-1}$, then $a_i y_j a_k^{-1} = (a_k x_s a_i^{-1})^{-1}$, so the elements of H of the form $a_i x_j a_k^{-1}$ generate H. Their number is dn. To reduce this number further, we choose the representatives a_i in a special way. Introduce some total order on the set $x_1, \dots, x_d, x_1^{-1}, \dots, x_d^{-1}$. Then order the set of all words in the x_i lexicographically, and in each coset Ha_i choose as the representative a_i, the first element in the lexicographic order that lies in this coset. If $a_i \ne 1$, let $a_i = b_i y_j$ be this expression of a_i as a first word in its coset, then the word b_i is also the representative of its coset, otherwise there exists a word c_i preceding b_i in Hb_i, and then $c_i y_j$ precedes a_i and lies in Ha_i.

Now if $y_j = x_j$, then $b_i x_j a_i^{-1} = 1$ is one of the special generators chosen above for H, and of course it is superfluous, while if $y_j = x_j^{-1}$, then $a_i x_j b_i^{-1}$ is eliminated in the same way. Thus we have eliminated $n - 1$ of the generators of H, and we are left with $dn - (n-1) = (d-1)n + 1$ generators, which is our claim.

(b). We connect R to H by a chain of subgroups, each maximal in its predecessor. Since subgroups of free pro-p groups are likewise free ([25], Theorem 5.4.6), it suffices to prove our claim when H is maximal in R. In that case H is normal and of index p. If K is another maximal subgroup of index p, then considering $R/\Phi(R)$, we see that we can find generators $x_1, \dots, x_d$ for R such that $H = \langle x_1, \dots, x_{d-1}, \Phi(R)\rangle$ and $K = \langle x_2, \dots, x_d, \Phi(R)\rangle$. It follows that an appropriate permutation of the x_i induces an automorphism of R interchanging H and K. Thus $d(H) = d(K)$, so it suffices to prove our claim for one particular maximal subgroup. Therefore it will suffice to exhibit a maximal subgroup H such that $d(H) \geq p(d-1)+1$. To that end it suffices to find a finite p-group G with d generators and a maximal subgroup H of G such that $d(H) = p(d-1)+1$, since then we can map R onto G. Let E be an elementary abelian group of order $p^{p(d-1)}$ and let G be the semidirect product of E by a cyclic group $C = \langle c \rangle$ of order p^2, acting in such a way that c permutes a basis of E as a product of $d-1$ cycles of length p. Then G and its subgroup $H = E\langle c^p \rangle$ satisfy our requirements.

Proof of Theorem 2. The number $a_p(R)$ of maximal subgroups is the number of maximal subspaces in the vector space $R/\Phi(R)$, and thus $p^{d-1} < a_p(R) = (p^d-1)/(p-1) < p^d$. The number $a_{p^k}(R)$ is bounded above by the number of maximal chains of length k starting at R, and by Lemma 2 we see that $a_{p^k}(R) < p^d.p^{(d-1)p+1}....p^{(d-1)p^{k-1}+1} = p^{k+(d-1)(p^k-1)/(p-1)}$, and $\limsup a_{p^k}(R)^{1/p^k} \leq p^{(d-1)/(p-1)}$.

To get a lower bound we start with one subgroup of index p^{k-1}. By Lemma 3 it contains at least $p^{(d-1)p^{k-1}}$ subgroups of index p^k in R. Each subgroup H of index p^k is obtained in this way from all the subgroups containing it with index p. These correspond to all the subgroups of order p in $N_R(H)/H$, whose number is at most $|N_R(H)/H| \leq p^k$, so in all we find that $a_{p^k}(R) \geq p^{(d-1)p^{k-1}} a_{p^{k-1}}(R)/p^k$. Iterating this yields $a_{p^k}(R) \geq p^{(d-1)(p^k-1)/(p-1)}/p^{\binom{k+1}{2}}$, which implies $\liminf(a_{p^k}(R))^{1/p^k} \geq p^{(d-1)/(p-1)}$ and proves the theorem.

It follows that given any finitely generated pro-p group G, the generating function $\sum a_n(G)z^n$ has a positive radius of convergence, and defines a holomorphic function inside its circle of convergence. Since $a_n(G) \neq 0$ only if n is a power of p, the circle of convergence is a natural boundary for this function ([24], 7.43). To my knowledge, these functions have not yet been investigated. The case when $G = R$ is free seems to be particularly interesting.

2 Normal subgroups

Our aim in this section is to find the normal subgroup growth of free pro-p groups, i.e., to consider the function $t_n(R)$ giving the number of normal subgroups of index n of a free pro-p group R. We are not able to give an exact asymptotic formula for $t_n(R)$, but will establish upper and lower bounds. The method differs considerably from that of the previous section. First, if $d(R) = d$, and G is any d-generator p-group of order n, there is an epimorphism of R onto G, whose kernel N is a normal subgroup of index n. Since different groups G determine in this way different normal subgroups, the number of normal subgroups of index n is at least the number of groups G. Each normal subgroup of index n is the kernel of some such epimorphism. Fixing G, the number of epimorphisms of R onto G is the number of generating d-tuples of G, and is bounded above by n^d. This is therefore also an upper bound for the number of kernels of epimorphisms onto G. Thus we obtain

Proposition 4. *Let $f(n, d)$ be the number of (isomorphism types of) p-groups of order n that can be generated by d elements. Then*

$$f(n, d) \leq t_n(R) \leq n^d f(n, d).$$

As we shall presently see, $f(n, d)$ is usually much bigger than n^d, so the main problem is to estimate f.

Theorem 5. *Assume that $d > 1$. There exist constants $c_1(d) > 0$ and $c_2(d)$, depending only on d, such that*

$$p^{c_1(d)k^2} \leq f(p^k, d) \leq p^{c_2(d)k^2}.$$

Conjecture. There exists a constant γ depending on d and p such that $f(p^k, d) = p^{\gamma k^2 + o(k^2)}$.

To prove the upper bound in Theorem 5, we first estimate the number of defining relations of p-groups.

Proposition 6. *Let G be a group of order p^k with d generators. Then G can be defined, in terms of these generators, by $(d + 1)k$ relations.*

Proof. Let z be a central element of order p in G, and write it as a word in the generators. By induction, the group $G/\langle z\rangle$ can be defined by $(d + 1)(k - 1)$ relations. Let $w = 1$ be one of these relations, where w is some word in the generators, then in G we have $w = z^i$, for some i. Replace the relation $w = 1$ by the relation $w = z^i$, where z is replaced by its expression in terms of the generators. Add the relations ensuring that z has order p and that it commutes with the generators. This adds $d + 1$ relations, so provides a presentation for G with $(d + 1)k$ relations.

Having proved the proposition, again let G be any group of order p^k with d generators, let z be as above and $N = \langle z\rangle$, choose some presentation for G/N with $(d + 1)(k - 1)$ relations, and construct a presentation for G as in the proposition. Each relation $w = 1$ of G/N is replaced by a relation $w = z^i$, and the element z^i can be any one of the p elements of N. Thus there are $p^{(d+1)(k-1)}$ possible presentations for G, and this is an upper bound for the number of possibilities for G, given G/N. An easy induction now shows that $f(p^k, d) \leq p^{(d+1)k(k-1)/2}$, proving one half of Theorem 5, and also showing that $c_2(d) \leq (d + 1)/2$ (this explicit estimate is from [18]).

To get lower bounds we use a method which was applied by G. Higman [8] to show the existence of many p-groups of a given order, but we have to modify his method so as to have our groups with a given number of generators. Here we only sketch the construction, referring to [15] for full details. We start by considering the lower p-central series of R. This is defined by $R_1 = R$, $R_{n+1} = [R_n, R]R_n^p$. The ranks of the elementary abelian groups R_n/R_{n+1} are known, so we also know how many subgroups lie between R_n and R_{n+1}. All these subgroups are normal in R. Moreover, the freeness of R implies that two such subgroups, say N and M, give rise to isomorphic factor groups only if they are conjugate under $\mathrm{Aut}(R)$. Now the automorphisms of R that are trivial on $R/\Phi(R)$ are also trivial on R_n/R_{n+1}, so the size of the orbit of N, say, under $\mathrm{Aut}(R)$, is at most the order of the group that $\mathrm{Aut}(R)$ induces on $R/\Phi(R)$, this induced group being $GL(d, p)$. Dividing the number of subgroups N by $|GL(d, p)|$ yields our lower bound. It is also not difficult to get an estimate for $c_1(d)$ by this method, but it seems to be unrealistically small.

We see that $t_n(R)$ is much smaller than $a_n(R)$. On the other hand, the calculation of the lower bound in the proof of Theorem 2 shows that the number of subnormal subgroups of index n and subnormality defect 2 of R is already exponential.

3 Boundedly generated subgroups and boundedly related groups[2]

Given a natural number k and a group G, we denote by $b_{n,k}(G)$ the number of subgroups H of index n in G that can be generated by k elements (if G is a pro-p group we mean of course topological generators), i.e., that satisfy $d(H) \leq k$. We denote by $a_{n,k}(G)$ the number of those subgroups of index n for which $d(H) = k$. In this section we prove the following

Theorem 7. *If G is a pro-p group, then as $n \to \infty$*

$$b_{n,k}(G) = o(n^k).$$

The constant implied by the o notation depends only k and p, not on G.

Before proceeding to the proof, we recall some easy probabilistic concepts. If G is a compact group, it has defined on it a so-called *Haar measure* μ, which is non-trivial, finite, defined on all Borel sets, σ-additive, and translation invariant, i.e., if $A \subseteq G$ is measurable, and $x \in G$, then $\mu(Ax) = \mu(xA) = \mu(A)$. Since $\mu(G)$ is finite, we can normalise it so that $\mu(G) = 1$, and G then becomes a probability space. Note that if H is a (closed, hence measurable) subgroup, then all its cosets have the same measure, so $\mu(H) \neq 0$ only if $|G : H|$ is finite, and then $\mu(H) = 1/|G : H|$. In particular, a profinite group is compact, so it has a Haar measure, which we always normalise as above.

Now suppose that G is finitely generated (as a profinite group). Then a subset $\{x_1, \ldots, x_d\}$ of G generates G if and only if there is no maximal subgroup M containing all the x_i. This means that the set of d-tuples generating G is the complement in G^d of the union of the products M^d, where M varies over all maximal subgroups of G. Since G is profinite, all its (closed) maximal subgroups have finite index, and for each positive integer there are only finitely many maximal subgroups of that index (because G is finitely generated), so there are at most countably many maximal subgroups in all. For each of these, M^d is an open subset of G^d. It follows that the set of generating d-tuples is closed in G^d, hence measurable. We denote its measure by $P(G, d)$, and refer to it as *the probability that d random elements generate G*. If $P(G, d) > 0$, for some

[2]The term "boundedly generated" has been used in recent years to describe groups that are products of finitely many cyclic groups. Such groups may be better termed "cyclic products" or "cyclically factored groups", or perhaps "multicyclic". My use of "boundedly generated" here has nothing to do with products, and will be self explanatory.

d, we say that G is *positively finitely generated*[3] (PFG for short, see [16]). Since G has only countably many finite index subgroups, it also follows that the set of elements that generate one of any given set of finite index subgroups is measurable.

If G is a pro-p group with minimally d generators, then $|G : \Phi(G)| = p^d$, and a subset $\{x_1, \ldots, x_k\}$ generates G if and only if it generates it mod $\Phi(G)$. It follows that if $k \geq d$, then $P(G, k) > 0$, so a f.g. pro-p group is PFG. Moreover, $P(G, k)$ depends only on p, d, and k, not on the particular group G. (*Exercise*: calculate this quantity!)

We now estimate the probability that a set of k random elements generates a subgroup of G of finite index. If H is any such subgroup, of index n, and $d(H) = r$, then our k elements lie in H with probability $1/n^k$, so the probability that they generate it is $P(H, k)/n^k$. This is 0 unless $k \geq r$, and if this inequality is met, then $P(H, k) \geq P(R, k)$, where R is a free pro-p group of rank k. Since generating distinct subgroups are disjoint events, the probability that we seek is

$$\sum_H P(H, k)/n^k \geq \sum_n P(R, k) a_{n,k}(G)/n^k.$$

Being a probability, the last sum is at most 1, so

$$\sum_n a_{n,k}(G)/n^k$$

converges, and is bounded by $1/P(R, k)$, which proves Theorem 7.

Since the growth of $a_{n,k}(G)$ is polynomial, we can define a Dirichlet series $\sum a_{n,k}(G)n^{-s}$, where s is a complex variable, which converges in the right half plane $\Re(s) \geq k$. Here $a_{n,k} \neq 0$ only if $n = p^r$ for some r, so the series becomes $\sum a_{p^r,k}(p^r)^{-s} = \sum a_{p^r,k}(p^{-s})^r$, a power series in p^{-s}. In the next section we will derive a functional equation involving these functions, which holds in a special class of pro-p groups. For more about these type of functions see Chapter 9.

We can apply similar considerations to estimate the number of normal subgroups of finite index that can be generated normally by a given number of elements. For a normal subgroup N of G, let $d(N, G)$ be the minimal number k such that N is the normal closure in G of k elements. Now N is the normal closure of a subset X if and only if X is not contained in any proper subgroup of N that is normal in G, and is maximal in N with respect to this property. A subgroup M is such a maximal subgroup if and only if $M \lhd G$, and $|N : M| = p$. This implies that $N/M \leq Z(G/M)$. Let $\Phi(N, G)$ denote the intersection of all these maximal subgroups. Then $\Phi(N, G) = N^p[N, G]$. It follows that X generates N normally if and only if it generates N modulo $\Phi(N, G)$, so $d(N, G) = d(N/\Phi(N, G))$. If N is finitely generated, in particular if it is of finite index, then certainly $d(N, G)$ is also finite. Moreover, the probability that k elements generate N normally is the same as the probability that k elements generate the elementary abelian group $N/\Phi(N, G)$, and this is equal to $P(R, k)$, where R is a free pro-p group of rank $d(N, G)$.

[3] In a PFG group the set of generating d-tuples is compact, carries a natural measure, inherited from the Haar measure of G^d, and the group G acts on it (by conjugation) by continuous measure preserving transformations. More generally, the automorphism group of G acts on this set in a similar manner (the uniqueness of the Haar measure ensures that all automorphisms are measure-preserving). We believe that this action is worth a further study.

Now let us denote by $t_{n,r}(G)$ the number of normal subgroups of G of index n that are the normal closure of r elements. Estimating, exactly as above, the probability that r random elements of G have normal closure of finite index, we obtain

Proposition 8. *Let G be a finitely generated pro-p group. Then, as $k \to \infty$,*

$$t_{p^k,r}(G) = o(p^{kr}).$$

In the previous section we used estimates on the number of groups of order p^k to obtain estimates for the number of normal subgroups. We can now reverse the procedure and apply Proposition 8 to estimate the number of groups of a given order and given number of defining relations. To say that a group G with generators $x_1, \dots, x_d$ has defining relations $w_1, \dots, w_r$, where the w_i are words in the generators, means that $G \cong F/N$, where F is the (abstract) free group with free generators $x_1, \dots, x_d$, and N is the normal closure in F of $w_1, \dots, w_r$. Thus the minimal number $r(G)$ of relations needed to define G is the minimum of $d(N, F)$, taken over all presentations $G \cong F/N$. It follows that the number of groups of order n that can be defined by r relations is not more than the number of normal subgroups N of F with r normal generators. Note that we are assuming that d is fixed, otherwise we have to take into account also the dependence of $r(G)$ on d. If G is finite, all the normal subgroups N occurring in presentations as above are conjugate under Aut(F), so $r(G)$ can be evaluated using any one of them. But we have considered above presentations $G \cong R/N$, where R is a free pro-p group, not an abstract free group. For such presentations $d(N, R)$ is the minimal number of relations needed to define G *as a pro-p group*, or, if G is finite, simply as a p-group. We denote this minimal number by $r_p(G)$. Since it is the minimal number of relations needed to define G, given the extra information that G is a pro-p group, it is clear that $r_p(G) \leq r(G)$. A more formal way of seeing this is to note that R may be taken to be the pro-p completion of F, and if $G \cong F/N$, then also $G \cong R/\bar{N}$, where $\bar{N}$ is the closure of N. It is not known yet if for a finite p-group G the two numbers $r(G)$ and $r_p(G)$ can indeed be different.

Note also that if R is generated freely by $x_1, \dots, x_d$, then the elements of R are profinite words in these generators, i.e., limits of sequences of ordinary (finite) words. But if N is a closed subgroup of finite index (hence open), which is the normal closure of $w_1, \dots, w_r$, then each of the cosets $w_i\Phi(N, R)$ is an open subset of R, therefore contains an element u_i of the dense subgroup F, and then the elements u_i, which are finite words, also generate N normally. This means that if $r_p(R/N) = r$, then $G = R/N$ has a set of r defining relations, as a p -group, consisting of ordinary words.

Now the relationship $r(G) \geq r_p(G) = d(N, R)$ and Proposition 8 yield

Theorem 9. *Let $g(n, r)$ be the number of groups of order n that can be defined by r relations. Then, as $k \to \infty$,*

$$g(p^k, r) = o(p^{kr}).$$

Note that we need not mention the number of generators d in Theorem 9, because if G is finite we have anyway $d \leq r$ (as can be seen, e.g., by considering G/G'. For p-groups a stronger inequality is provided by the Golod–Shafarevich Theorem $r_p(G) > d(G)^2/4$, see e.g. [2], Interlude D).

4 p-adic analytic groups and their relatives

For us, the term *p-adic analytic pro-p group* is synonymous with *pro-p group of finite rank*. Here, a group (abstract or profinite) is termed of finite rank, if all of its finitely generated subgroups are generated by a bounded number of elements (see [2]; naturally, for a profinite group we consider only closed subgroups). If G is such a group, its *rank* is defined to be the maximum value of $d(H)$ as H ranges over all finitely generated (closed) subgroups of G. These groups can be characterised by their subgroup growth.

Definition. A group G has *polynomial subgroup growth* (PSG) if there exists $s < \infty$ such that $a_n(G) \leq n^s$ for all n.

Theorem 10. *A pro-p group is a* PSG *group if and only if it is of finite rank.*

This result also holds if we replace "pro-p" by "finitely generated residually finite abstract", and then we can add to the conclusion "soluble-by-finite" [13]. Theorem 10 was an important step in the proof of the "abstract" theorem. There exists also a version for general profinite groups [23].

Let me make some remarks on the easy half of the proof of Theorem 10. That groups of finite rank are PSG is an immediate corollary of Theorem 7, but there is a shorter direct proof. Suppose that all subgroups of G can be generated by r elements. We saw in the proof of Theorem 2 that each subgroup has at most p^r subgroups of index p, so $a_{p^k}(G) \leq (p^r)^k = (p^k)^r$. This bound is better than the explicit bound $(1/P(R,r))(p^k)^r$ which is obtained by the proof of Theorem 7. On the other hand, Theorem 7 also states that $a_n(G) = o(n^r)$. For p odd, this is still proved easily directly, since for each subgroup H we have $a_p(H) \leq (p^r - 1)/(p-1) < (1/2)p^r$, so $a_{p^k}(G) < (1/2^k)(p^k)^r$. This inequality also establishes the extra claim that $\sum a_n(G)/n^r$ converges, but the whole argument fails for $p = 2$.

For the converse direction, that a pro-p PSG group has finite rank, we refer to [2], Chapter 3. The proof relies on properties of powerful groups. In Chapter 11 of that book you will find a proof of the following strengthening of Theorem 10, due to A. Shalev [21]:

Theorem 11. *Let the pro-p group G satisfy*

$$a_n(G) \leq n^{c \log_p n}$$

for some constant $c < 1/8$ and all large enough n. Then G is p-adic analytic.

The proof of that theorem applies in addition to the previous tools the theory of dimension subgroups. Shalev also gives examples of groups that are not p-adic analytic and that satisfy the inequality of Theorem 11 for $c > 2$. Very recently, Barnea and Guralnick have shown that for the group $G = SL^1(2, \mathbb{F}_p[[t]])$ (p odd), one has $a_{p^k}(G) \leq p^{k(k+1)/2+3k}$, so G satisfies the inequality for every $c > \frac{1}{2}$. Thus, for pro-p groups there is a "gap" in the possible subgroup growth. If the growth is not polynomial, it is at least $n^{c \log_p n}$, and the latter growth is actually possible. We have the following intriguing problem.

Problem. What is the supremum of the numbers c, such that if $a_n(G) \leq n^{c \log_p n}$ for all large n, then G is p-adic analytic?

In the other direction, Shalev [22] investigated groups of very small subgroup growth, proving

Theorem 12. *Let G be a profinite group. Then G satisfies $a_n(G) = o(n)$ if and only if G contains a procyclic central subgroup of finite index, and in that case the sequence $a_n(G)$ is actually bounded.*

The analogous "abstract" theorem follows easily from this. The proof of Theorem 12 is by reduction to the case when G is pro-p and powerful, and in that case the following remarkable "invariance of normalisers" result is crucial: *if G is a finitely generated powerful pro-p group and H a closed subgroup, then, denoting $G_i = G^{p^i}$ and $H_i = HG_i$, there exists a constant c such that $|N_G(H_i) : N_G(H)G_i| \leq c$, for all i.*

For groups of linear, rather than sublinear, subgroup growth, see Chapter 1.

Recall the notion of PFG groups, introduced in the previous section, and the probability $P(G, k)$ of G being generated by k elements. While this probability can be positive, and even very near to 1, it can never be equal to 1, because for any subgroup H of finite index, a random k-tuple of elements of G lies in H with positive probability, and thus cannot generate G. But it is possible that a random k-tuple generates a subgroup of finite index with probability 1. Let us denote by $Q(G, k)$ the probability that k random elements of G generate a subgroup of finite index. We now use this quantity to derive the functional equation referred to in the previous section.

Proposition 13. *Let G be a p-adic analytic pro-p group. Then $Q(G, k) = 1$, for some k.*

Proof. We need the following:

Borel–Cantelli Lemma (easy half). *Let A_i be a sequence of events in a probability space, where A_i has probability p_i. If $\sum p_i$ converges, then with probability 1 only finitely many of the events A_i happen.*

This follows easily from elementary properties of infinite series. To apply it to our situation, note that G has only countably many open subgroups, so, letting these subgroups be H_i, we let A_i be the event that k random elements lie in H_i. Then $p_i = 1/|G : H_i|^k$, and since for some k the sum $\sum a_n(G)/n^k$ converges, we see that for that k a random k-tuple lies almost surely in only finitely many subgroups of finite index. But the closed subgroup generated by a subset X is the intersection of all open subgroups containing X, so such k-tuples generate a closed subgroup of finite index.

Remark. The more difficult half of Borel–Cantelli states that if the events are pairwise independent and the sum of the probabilities diverges, then almost surely infinitely many events happen. That result is used in the characterisation of PFG groups [20].

Let G be a p-adic analytic pro-p group, of rank r. Then the previous proposition and Theorem 7 show that $Q(G, k) = 1$ for all $k \geq r$. Recall from the previous section the function $\zeta_{d,G}(s) := \sum a_{n,d}(G)n^{-s}$, which is defined and holomorphic for all s

such that $\Re(s) > r$. We have $Q(G,k) = \sum_H P(H,k)/|G:H|^k$, the sum taken over all finite index subgroups. Recall that if $d(H) = d$, then $P(H,k)$ depends only on d, k, and p. Let us denote this value by $P(d,k)$. Then the contribution of the subgroups such that $d(H) = d$ to the above sum is $P(d,k)\zeta_{d,G}(k)$, so we have

$$\sum_{d=1}^{r} P(d,k)\zeta_{d,G}(k) = 1 \text{ for all large k.}$$

We have noted that $\zeta_{d,G}(s)$ is a power series in p^{-s}. We also know from a previous exercise that $P(d,k) = (1-1/p^k)...(1-p^{d-1}/p^k)$. Define $P(d,s)$ by changing k to s in the last equality. Then $P(d,s)$ is a polynomial in p^{-s}. Now the equality

$$(1) \qquad \sum_{d=1}^{r} P(d,s)\zeta_{d,G}(s) = 1,$$

is an equality between two power series in $z = p^{-s}$, which holds for the infinitely many values $z = p^{-k}$, values which accumulate to 0. It follows that (1) holds for all $s > r$.

This functional equation takes a particularly simple form if all open subgroups H of G satisfy $d(H) = d$, a constant. The simplest groups with this property are the free abelian ones. Metaprocyclic groups that are not procyclic-by-finite also have this property. For such groups we obtain

$$(2) \qquad \zeta_G(s) = 1/P(d,s) = ((1-p^{-s})...(1-p^{-(s-d+1)}))^{-1},$$

where $\zeta_G(s) = \sum a_n(G)n^{-s}$.

Note the right hand side is the local factor of the product $\zeta(s)...\zeta(s-d+1)$, where $\zeta(s)$ is Riemann's ζ-function. Equation (2) for free abelian groups is classical. Four different proofs are given in [10] (one of which appears in §2 of Chapter 9). The generalisation to groups of constant rank is due to I. Ilani [9].

It was shown by Y. Barnea and M. Larsen [1] that there exist non-p-adic-analytic pro-p groups that satisfy $Q(G,k) = 1$. But such groups satisfy a weaker property: they have a finite *lower rank*, i.e., there exists a number r such that all open subgroups contain open subgroups that are generated by r elements. Indeed, the following proposition is slightly stronger.

Proposition 14. *Let G be a pro-p group satisfying $Q(G,k) = 1$, and let H be open in G. Then $Q(H,k) = 1$, and H is generated by its k-generator subgroups of finite index.*

Proof. Counting separately the k-tuples lying in H and the others yields: $Q(G.k) \leq (1/|G:H|^k)Q(H,k) + (1-1/|G:H|^k)$, so $Q(G,k) = 1$ implies $Q(H,k) = 1$. The proof of the remaining claim is left to the reader.

We will come back to these groups presently. For now let us turn to groups in which the number of normal subgroups, rather than the number of all subgroups, is restricted. Let us say that a group G has *polynomial normal subgroup growth* (PNSG) if $t_n(G) \leq n^s$, for some s. If G is a pro-p group in which all normal subgroups can be generated normally by r elements, then either Proposition 8 or a direct argument like the one following Theorem 10 establishes that G is a PNSG group. Unlike the case of PSG groups, the converse does not hold; see [19], and no characterisation is known of PNSG groups. But we can show that in a PNSG group the number of normal generators of subgroups grows very slowly, and sometimes also obtain some restrictions on the non-normal subgroups. Let $r_k(G) = \max\{d(N, G) \mid N \triangleleft G, \text{ and } |G : N| = p^k\}$.

Proposition 15. *Let G be a* PNSG *pro-p group. Then*

(i) $r_k(G) = O(\sqrt{k})$.

(ii) There exists a number r, such that each open subgroup H of G contains an open subgroup K, which is the normal closure in G of r elements.

If G is also finitely presented, then

(iii) If $|G : H| = n$, then $d(H) = O(\sqrt{n})$.

(iv) $a_n(G) \leq p^{c\sqrt{n}}$, for some constant c.

Proof. (ii) is proved by repeating the arguments of Propositions 13 and 14. We can derive (i) by imitating the proof in [20] of the corresponding statement for PSG groups. This applies also to PNSG groups that are not necessarily pro-p groups. But for pro-p groups we have the following simple and direct argument: let $N \triangleleft G$, $|G : N| = p^n$, and write $d(N, G) = r$. Then $N/\Phi(N, G)$ is an elementary abelian group of order p^r. Thus $N/\Phi(N, G)$ has at least $p^{r^2/4}$ subgroups of index $p^{[r/2]}$, and these correspond to normal subgroups of G of index $p^{n+[r/2]}$. By assumption, $t_n(G) \leq n^s$ for some s, hence $p^{r^2/4} \leq p^{s(n+[r/2])}$, which implies (i).

Now suppose that G is finitely presented, by t relations. Let $|G : H| = n$, and let K be an open normal subgroup, normally generated in G by r elements, which is contained in $\Phi(H)$. Then G/K is defined by $t + r$ relations, namely the relations of G supplemented by the normal generators of K. Then H/K, which has index n in G/K, is defined by $n(t + r)$ relations. By the Golod–Shafarevich theorem $d(H) = d(H/K) \leq 2\sqrt{n(t + r)}$. This proves (iii), and (iv) follows by standard arguments.

Claim (ii) may be formulated as saying that G has *finite lower normal rank.* Note that the proofs of (iii) and (iv) used only the fact that G has finite lower normal rank. Note also that if G has a finite lower rank, say, r, then each open normal subgroup contains an open normal subgroup generated by r elements, so it contains also the normal closure of these elements. Therefore G has finite lower normal rank. Thus (iii) and (iv) apply also to finitely presented groups of finite lower rank, and in particular, by Proposition 14, they apply to finitely presented pro-p groups satisfying $Q(G, k) = 1$.

5 Philip Hall's formula

'Rather curious' is Hall's typically understated and modest description of the following result [6]:

Theorem 16. *Let p be a prime. Then*

$$\sum \frac{1}{|G|} = \sum \frac{1}{|\mathrm{Aut}(G)|}$$

where the summation ranges over all finite abelian p-groups.

Hall proves this by first noting that the left hand side is equal to $\sum p(n)p^{-n}$, where $p(n)$ is the partition function (hence the sum converges), then enumerating partitions in a certain way, and applying an explicit expression for $|\mathrm{Aut}(G)|$. Another proof is given by I. G. Macdonald [17], which needs first the development of the formalism of Hall algebras. We now offer yet another proof, which is based on the equality (2), but we also need a partition identity due to Euler. This proof is taken from [14], to which we refer the reader for comparison with Hall's original proof, and some further remarks.

Let F be a free abelian pro-p group of rank n. We consider open subgroups H of F such that F/H satisfies $d(F/H) = n$. This is equivalent to $H \leq F^p$. Therefore the number of those subgroups H in our family which have index p^{n+k} in F is equal to the number of subgroups of index p^k in F^p, and this is equal to $a_{p^k}(F)$, because $F^p \cong F$.

We now count the subgroups H according to their factor groups $G = F/H$. Given an abelian p-group G, the number of subgroups $H \leq F$ such that $F/H \cong G$ is equal to $|\mathrm{Epi}(F, G)|/|\mathrm{Aut}(G)|$ ([5], 1.4). Here $\mathrm{Epi}(F, G)$ denotes the set of homomorphisms of F onto G, and $|\mathrm{Epi}(F, G)|$ is equal to the number of ordered n-tuples that generate G. Now an n-tuple generates G if and only if its images in G/G^p generate this factor group. Assuming that $d(G) = n$, we have that G/G^p is elementary abelian of order p^n, and the number of n-tuples generating it is $(p^n - 1)(p^n - p)...(p^n - p^{n-1})$. Since $|G| = p^{n+k}$, we have that $|G^p| = p^k$, so the number of n-tuples generating G is $(p^n - 1)(p^n - p)...(p^n - p^{n-1})p^{nk}$. We thus have

$$a_{p^k}(F)p^{-nk} = (p^n - 1)(p^n - p)...(p^n - p^{n-1}) \sum \frac{1}{|\mathrm{Aut}(G)|},$$

the summation being over all groups G of order p^{n+k} such that $d(G) = n$. Summing over all k, the left hand side becomes $\zeta_F(n)$, so using (2), we obtain

$$\frac{p^{-n^2}}{(1 - 1/p^n)^2(1 - 1/p^{n-1})^2...(1 - 1/p)^2} = \sum \frac{1}{|\mathrm{Aut}(G)|} \tag{3}$$

where this time the summation is over all abelian p-groups satisfying $d(G) = n$. Finally, we sum this expression over all n, and substitute the value $x = 1/p$ in the following identity of Euler ([7], Theorem 351)

$$\sum p(n)x^n = 1 + \sum \frac{x^{n^2}}{(1 - x)^2(1 - x^2)^2 \ldots (1 - x^n)^2}.$$

This yields $\sum p(n)p^{-n} = \sum 1/|\mathrm{Aut}(G)|$, proving Hall's formula.

Note that on the way we obtained, in equation (3), a closed formula for the sum of the reciprocals of the orders of the automorphism groups of all n-generator p-groups. Remarkably, this sum turns out to be a rational number of an easily described form. On the other hand, (3) can be used to show the following

Proposition 17. *The sum of the series occurring in Theorem 16 is an irrational number.*

Proof. Let A be this sum. Write the left hand side of (3) in the form $p^n/((p-1)...(p^n-1))^2$, so we have

$$A = 1 + \sum \frac{p^n}{((p-1)...(p^n-1))^2}. \tag{4}$$

If A is rational, then multiplying by some power, say p^e, of p, we obtain a rational number $B = p^e A = u/v$, where v is prime to p. Then for n large enough, say $n > N$, the denominator of the nth term in (4) is divisible by v. Therefore the difference between B and the Nth partial sum of (4), multiplied by p^e, is at least $1/((p-1)...(p^N-1))^2$. On the other hand, this difference is

$$\frac{1}{((p-1)...(p^N-1))^2} \sum_{n>N} \frac{p^{n+e}}{((p^{N+1}-1)...(p^n-1))^2},$$

and if N is large enough, the sum on the right is less than 1, a contradiction.

Remark. It is possible that, by starting from a different choice for the group F, or by considering a different family of quotients, similar identities may be obtained. See also [10], 2.2, for a different application of subgroup growth to partition identities.

References

[1] Y. Barnea and M. Larsen, Random generation in simple algebraic groups over local fields, submitted.

[2] J. Dixon, M. P. F. duSautoy, A. Mann, and D. Segal, *Analytic pro-p Groups*, 2nd edition, Cambridge University Press, 1999.

[3] M. P. F. duSautoy, Finitely generated groups, p-adic analytic groups and Poincaré series, *Ann. Math.*, **137** (1993), 639–670.

[4] M. P. F. duSautoy, p-groups, coclass and model theory: a proof of Conjecture P, submitted.

[5] P. Hall, The Eulerian functions of a group, *Quart. J. Math.*, **7** (1936), 134–151.

[6] P. Hall, A partition formula connected with Abelian groups, *Comm. Math. Helv.*, **11** (1938), 126–129.

[7] G. H. Hardy and E. M. Wright, *An Introduction to the Theory of Numbers*, 5th ed., Oxford University Press, 1979.

[8] G. Higman, Enumerating p-groups, I. Inequalities, *Proc. London Math. Soc.* (3), **10** (1960), 10–24.

[9] I. Ilani, Counting finite index subgroups and the P. Hall enumeration principle, *Israel J. Math.*, **68** (1989), 18–26.

[10] A. Lubotzky, *Subgroup growth*, Lecture Notes, University College, Galway, 1993.

[11] A. Lubotzky, Counting finite index subgroups, in *Groups '93 Galway/St Andrews*, Cambridge University Press 1995,

[12] A. Lubotzky, Subgroup growth and congruence subgroups, *Inv. Math.*, **119** (1995), 267–295.

[13] A. Lubotzky, A. Mann and D. Segal, Finitely generated groups of polynomial subgroup growth, *Israel J. Math.*, **82** (1993), 363–371.

[14] A. Mann, Philip Hall's 'rather curious' formula for abelian p-groups, *Israel J. Math.*, **96** (1996) (S. A. Amitsur Memorial volume), 445–448.

[15] A. Mann, Enumerating finite groups and their defining relations, *J. Group Theory*, **1** (1998), 59–64.

[16] A. Mann, Positively finitely generated groups, *Forum Math.*, **8** (1996), 429–459.

[17] I. G. Macdonald, The algebra of partitions, in *Group Theory: Essays for Philip Hall*, Academic Press, London 1984,

[18] A. McIver and P. M. Neumann, Enumerating finite groups, *Quart. J. Math.*, Oxford, **2** (1987), 473–488.

[19] A. Mann and D. Segal, Subgroup growth: some current developments, in *Infinite groups 94*, de Gruyter, Berlin 1995, 179–197.

[20] A. Mann and A. Shalev, Simple groups, maximal subgroups, and probabilistic aspects of profinite groups, *Israel J. Math.*, **96** (1996) (S. A. Amitsur Memorial volume), 449–468.

[21] A. Shalev, Growth functions, p-adic analytic groups and groups of finite coclass, *J. London Math. Soc.*, **46** (1992), 111–122.

[22] A. Shalev, Groups whose subgroup growth is less than linear, *Int. J. Alg. Comp.*, **7** (1997), 77–91.

[23] D. Segal and A. Shalev, Profinite groups with polynomial subgroup growth, *J. London Math. Soc.* (2), **55** (1997), 320–334.

[24] E. C. Titchmarsh, *The Theory of Functions*, 2nd ed., Oxford University Press, 1952.

[25] J. S. Wilson, *Profinite groups*, Oxford Univ. Press, 1998.

Institute of Mathematics
The Hebrew University
Jerusalem 91904
Israel
e-mail:mann@vms.huji.ac.il

9

Zeta Functions of Groups

Marcus du Sautoy and Dan Segal

1 Introduction

A zeta function is an analytic function whose analytic properties somehow encapsulate a tremendous amount of arithmetic information. These functions are an uncannily powerful tool in number theory; to mention just some celebrated examples, they are at the heart of the proofs of the Prime Number Theorem, Dirichlet's theorem on primes in arithmetic progressions, and the main theorems (in their original form) of class field theory, not to mention their more recent incarnation in the Weil conjectures.

While studying the classification of finitely generated nilpotent groups around twenty years ago [26], Fritz Grunewald and the second author came to feel that the arithmetical properties of such groups could be a rich field for investigation. Observing that the Dedekind zeta function of a number field can be defined as the generating function for the sequence (a_n), where a_n is the number of ideals of index n in the ring of algebraic integers, Grunewald and Segal associated to any finitely generated group G the sequence $(a_n(G))$, where $a_n(G)$ is the number of subgroups of index exactly n in G.

Any number of questions can then be asked: about the arithmetical behaviour of this sequence for a particular group G, and about how this varies when G ranges over a class of groups. The classical approach is to study the corresponding zeta function, in the hope that its analytical properties will shed some light on such questions; and the first thing to do is to find out what these zeta functions look like. As it turns out, these functions resemble the classical zeta functions in some ways, but differ from them in important and interesting respects.

We shall try in this chapter to give a brief but fairly comprehensive overview of the current state of knowledge on this topic. (The second author wishes to make clear that his role is strictly that of an expositor: the new results reported on in Sections 4 and 5 are due to Fritz Grunewald and the first author, while those in Sections 6 and 7 are due to the first author. The first author thanks the second author for doing most of the exposition.)

Following the model of the Dedekind zeta function of a number field, one defines the *zeta function* of a group G by

$$\zeta_G(s) = \sum_{n=1}^{\infty} a_n(G)n^{-s} = \sum_{H\leqslant G} |G:H|^{-s}.$$

This is in the first instance a formal Dirichlet series; but if it so happens that the sequence $a_n(G)$ grows at most polynomially with n, then the series has a finite abscissa of convergence α_G and defines an analytic function of the complex variable s, holomorphic in the right half-plane $\Re(s) > \alpha_G$. Thus α_G is finite precisely when the group G has polynomial subgroup growth (PSG). In that case, elementary analysis shows that

$$\alpha_G = \inf\left\{\alpha \mid \sum_{i=1}^{n} a_i(G) = O(n^{\alpha})\right\},$$

so α_G is an interesting group-theoretic invariant, the 'least exponent of polynomial subgroup growth'. The groups with PSG are now known ([38], [44]); it is very easy to see that they include all finitely generated nilpotent groups, and it is these that were considered in the first publication on the topic [27]. The precise value of α_G in general, even for groups G in this class, remained shrouded in mystery (and to a large extent still does).

If we are lucky, more advanced analysis will translate deeper analytic properties of the function ζ_G into more detailed information about the arithmetical properties of the sequence $a_i(G)$: until recently only a dream, this project is now beginning to produce substantial results [17], [18]. We report on these developments in Section 5 below.

When G is nilpotent, the arithmetical function $a_n(G)$ is weakly multiplicative; as Euler discovered, this fact is neatly expressed by the formal identity

$$\zeta_G(s) = \prod_p \zeta_{G,p}(s),$$

where the product is over all primes and $\zeta_{G,p}(s) = \sum_{k=0}^{\infty} a_{p^k}(G)p^{-ks}$. So far, the analogy with Dedekind's zeta function holds up rather well. What about the individual 'local factors' $\zeta_{G,p}$? In the classical situation these are exceedingly simple, for example taking the form $(1-p^{-fs})^{-g}$ when p is an unramified prime which splits into g factors in a Galois extension of degree fg over $\mathbb{Q}$. The remarkable fact, established in [27], is that for any torsion-free finitely generated nilpotent group G, the function $\zeta_{G,p}(s)$ is still a rational function of p^{-s}.

As it turns out, these rational functions, even for quite 'easy' groups, are very hard to compute. A small selection was worked out in [27]. For example, when G is the 3×3 unitriangular matrix group over the ring of integers of a quadratic number field

k, it was found that $\zeta_G^{\triangleleft}$ is a finite product of translates of ζ_k and of ζ (the Riemann zeta function); here $\zeta_G^{\triangleleft}$ denotes the function defined like ζ_G but counting only *normal* subgroups in G. On the basis of this and other examples, it seemed plausible that for any given finitely generated nilpotent group G, the rational functions $\zeta_{G,p}(s)$ would take only finitely many forms as p ranged over all primes, another close analogy with the Dedekind zeta function. More precisely, let us say that ζ_G is *uniform* if there exists a rational function $F(X, Y)$ of two variables, with rational coefficients, such that

$$\zeta_{G,p}(s) = F(p, p^{-s})$$

for almost all primes p, and that ζ_G is *finitely uniform* if there exist finitely many such rational functions $F_1(X, Y), \ldots, F_m(X, Y)$, such that for every prime p there exists i with $\zeta_{G,p}(s) = F_i(p, p^{-s})$. The question raised in [27] was whether ζ_G and $\zeta_G^{\triangleleft}$ are finitely uniform for every torsion-free finitely generated nilpotent group G; and it was conjectured that when G is a relatively free nilpotent group, ζ_G and $\zeta_G^{\triangleleft}$ are then uniform.

The answer to the first question is now known to be negative: the first author has constructed a group G of Hirsch length 9 for which ζ_G and $\zeta_G^{\triangleleft}$ are *not* finitely uniform (see §4 below). On the other hand, the conjecture about free-nilpotent groups is still open; it was proved for groups of class 2 in [27], and has recently been established for 2-generator groups in [20].

For any group G, it is clear that

$$\zeta_G(s) = \zeta_{\widehat{G}}(s)$$

where $\widehat{G}$ is the profinite completion of G, and by $a_n(\widehat{G})$ one understands the number of *open* subgroups of index n in $\widehat{G}$. If G is finitely generated and nilpotent, then the pro-p completion $\widehat{G}_p$ of G is a pro-p group of finite rank, so a compact p-adic analytic group. The rationality result above turned out to be a special case of a far more general fact, established by the first author in [10], and discussed in §3: *for every compact p-adic analytic group Γ, the function $\zeta_{\Gamma,p}(s)$ is a rational function of p^{-s}.*

The rationality of $\zeta_{G,p}(s)$ has a straightforward interpretation in terms of the sequence $(a_{p^n}(G))$: it means that the terms of this sequence, from some point on, satisfy a linear recurrence relation. Thus for a fixed prime p, the number of subgroups of p-power index in a group of the type discussed above grows in a very regular manner. However, the question of what these rational functions look like, in any particular case, remained highly obscure.

This has been greatly clarified in recent work of Grunewald and the first author [18], which we discuss in Sections 4 and 5. It now appears that the analogy with Dedekind zeta functions was too simplistic: a better analogy is with the Weil zeta functions of (the reductions modulo p) of an algebraic variety defined over $\mathbb{Z}$. In particular, it is now possible to say quite a lot about how the rational function $\zeta_{G,p}(s)$ varies as p ranges over all primes, when G is a finitely generated nilpotent group. Not only does this bring into sharper focus the question of uniformity; it also for the first time provides enough control over the global behaviour of the Euler product $\zeta_G(s)$ for this to be amenable to some classical analytic number theory. These methods lead to the results (a) that the abscissa α_G is always a *rational number*, and (b) that the global zeta function $\zeta_G(s)$ has a meromorphic continuation to the left of α_G, which in turn implies that the number of subgroups of index at most n in G grows asymptotically like $c \cdot n^{\alpha_G} \cdot (\log n)^{\beta}$ for some integer $\beta \geqslant 0$: a result that seems to lie remarkably deep.

A quite different and surprising outgrowth of these methods leads to new insights on some old and some new questions about *the number of isomorphism types of finite p-groups* [12], [14], [13]. One attaches a zeta function to a suitable weighted sum counting normal subgroups of finite index in certain pro-p groups. The rationality of this function can then be interpreted in terms of regular growth behaviour of the number of p-groups of order p^n, while an application of the analytic results from Section 5 gives an asymptotic formula for the number of d-generator nilpotent groups of class at most c and order at most n (for fixed c, d and $n \to \infty$). These matters are explained in Section 6. In the final section we discuss a variant of this method, whereby the rationality of a suitable zeta function is interpreted in terms of the *periodicity of the trees describing p-groups of a fixed coclass* (so establishing 'Conjecture P' of [42]).

A different kind of zeta function was also discussed in [27], associated to the p-adic points of a linear algebraic group. These zeta functions share many of the properties of those discussed here, and in some ways they are better behaved; for example when the algebraic group is the multiplicative group of a number field k (considered as an algebraic group over $\mathbb{Q}$), one gets the p-factor of the Dedekind zeta function ζ_k. A detailed study of these functions is given in [23]; we shall not deal with the topic here, as it really needs a whole chapter to itself. (We take this opportunity to mention that Lemma 3.3, and correspondingly Prop. 3.4, of [27] are wrongly stated, though Theorem 3.5 remains correct.)

2 Counting with p-adic integrals

Suppose we want to count the subgroups of finite index in $L = \mathbb{Z}_p^d$. Each such subgroup H has a basis of the form

$$\begin{array}{lcllll} h_1 & = & a_{11}\mathbf{e}_1+ & \cdots & \cdots & +a_{1d}\mathbf{e}_d \\ h_2 & = & & a_{22}\mathbf{e}_2+ & \cdots & +a_{2d}\mathbf{e}_d \\ \vdots & & & & & \vdots \\ h_d & = & & & & a_{dd}\mathbf{e}_d \end{array} \tag{1}$$

where $(\mathbf{e}_1, \ldots, \mathbf{e}_d)$ is the standard basis of L. Let $\mathcal{M}(H)$ denote the set of all upper-triangular $d \times d$ matrices (a_{ij}) over $\mathbb{Z}_p$ that arise in this way. For each such $A = (a_{ij})$ and $i = 1, \ldots, d$, put

$$\delta_i(A) = \prod_{j=1}^{i} a_{jj}.$$

Then the p-adic value $|\delta_i(A)|_p$ is equal to $|L : H + \langle \mathbf{e}_{i+1}, \ldots, \mathbf{e}_d \rangle|^{-1}$, so it is constant as A runs over $\mathcal{M}(H)$. In particular, we have $|L : H| = |\det A|_p^{-1} = |\delta_d(A)|_p^{-1}$.

Let $\mathcal{T}$ denote the $\mathbb{Z}_p$-module of all upper-triangular $d \times d$ matrices, and $\mathcal{U}$ the group of invertible matrices in $\mathcal{T}$. Choosing a representative $A_H \in \mathcal{M}(H)$, we see that $\mathcal{M}(H) = \mathcal{U}A_H$; writing μ to denote the normalised Haar measure on the additive group $\mathcal{T}$ with $\mu(\mathcal{T}) = 1$, we then find that

$$\mu(\mathcal{M}(H)) = (1 - p^{-1})^d |\delta_1(A_H) \dots \delta_{d-1}(A_H)|_p^{-1} |\delta_d(A_H)|_p^d. \quad (2)$$

We may therefore write

$$\begin{aligned}
\zeta_L(s) &= \sum_H |L : H|^{-s} = \sum_H |\delta_d(A_H)|_p^s \\
&= \sum_H \mu(\mathcal{M}(H))^{-1} \cdot \mu(\mathcal{M}(H)) |\delta_d(A_H)|_p^s \\
&= \sum_H (1 - p^{-1})^{-d} |\delta_1(A_H) \dots \delta_{d-1}(A_H)|_p |\delta_d(A_H)|_p^{-d} \int_{\mathcal{M}(H)} |\delta_d(A)|_p^s d\mu(A) \\
&= (1 - p^{-1})^{-d} \int_{A \in \mathcal{T}} |\delta_1(A) \dots \delta_{d-1}(A)|_p |\delta_d(A)|_p^{s-d} d\mu(A) \\
&= (1 - p^{-1})^{-d} \int_{A \in \mathcal{T}} |\delta_0(A)|_p |\delta_d(A)|_p^{s-d} d\mu(A), \quad (3)
\end{aligned}$$

where

$$\delta_0(A) = a_{11}^{d-1} a_{22}^{d-2} \dots a_{d-1.d-1};$$

for the penultimate equality, note that $\mathcal{T}$ is the disjoint union of the (countably many) open sets $\mathcal{M}(H)$ together with the set $\mathcal{T}_0$ of measure zero consisting of matrices with a zero on the diagonal (on which the integrand vanishes, anyway). To evaluate this integral, we partition $\mathcal{T} \setminus \mathcal{T}_0$ into the subsets

$$\mathcal{T}(n_1, \dots, n_d) = \left\{ A \in \mathcal{T} \mid |a_{ii}|_p = p^{-n_i} \text{ for } i = 1, \dots, d \right\}.$$

Noting that $\mu(\mathcal{T}(n_1, \dots, n_d)) = (1 - p^{-1})^d p^{-n_1 - \dots - n_d}$, we obtain

$$\begin{aligned}
\zeta_L(s) &= (1 - p^{-1})^{-d} \sum_{n_1, \dots, n_d = 0}^{\infty} \mu(\mathcal{T}(n_1, \dots, n_d)) p^{n_1 + 2n_2 + \dots + dn_d} p^{-(n_1 + \dots + n_d)s} \\
&= \sum_{n_1, \dots, n_d = 0}^{\infty} p^{-sn_1} p^{(2-s)n_2} \dots p^{(d-1-s)n_d} \\
&= (1 - p^{-s})^{-1} \left(1 - p^{1-s}\right)^{-1} \dots \left(1 - p^{d-1-s}\right)^{-1}.
\end{aligned}$$

The Euler product of these functions now gives

$$\zeta_{\mathbb{Z}^d}(s) = \zeta(s)\zeta(s-1)\dots\zeta(s-d+1),$$

a well-known formula easily obtained more directly; however one approaches the calculation, though it will necessarily involve the two key steps (a) finding the number of solutions of certain congruences in a given range, which corresponds to evaluating the measures $\mu(\mathcal{M}(H))$ and $\mu(\mathcal{T}(n_1, \dots, n_d))$, and (b) summing the resulting geometric progressions.

The above formalism has the advantage of being well adapted to dealing with more general situations. Suppose for example that the $\mathbb{Z}_p$-lattice L carries the structure of a $\mathbb{Z}_p$-algebra (which need not be associative), and that we wish to count *subalgebras* or *ideals* of finite index. The multiplication on L is given by a bilinear map $\beta : L \times L \to L$, and we write

$$\beta(\mathbf{e}_i, \mathbf{e}_j) = \sum c_{ij}(l)\mathbf{e}_l.$$

Let C_j be the matrix with (k, l)-entry $c_{kj}(l)$. Now suppose that $H \leqslant L$ has the basis (1); then H is an ideal of L if and only if $\beta(h_i, \mathbf{e}_j) \in H$ for each i and j. This holds only when the family of linear equations

$$(a_{i1}, \dots, a_{id})C_j = \mathbf{y}_{ij}A \quad (i, j = 1, \dots, d) \tag{4}$$

is soluble with $\mathbf{y}_{ij} \in \mathbb{Z}_p^d$. Putting

$$A^{\ddot{}} = A^{-1}\mathrm{diag}(\delta_1(A), \dots, \delta_d(A))$$

we rewrite (4) in the form

$$(a_{i1}, \dots, a_{id})C_jA^{\ddot{}} = (\delta_1(A)Y_{ij}^1, \dots, \delta_d(A)Y_{ij}^d).$$

Since $A = (a_{ij})$ is upper-triangular, the entries of $A^{\ddot{}}$ are homogeneous polynomials in the entries of A, with integer coefficients. So for each k, the kth component $g_{ijk}(A)$ of the vector $(a_{i1}, \dots, a_{id})C_jA^{\ddot{}}$ is equal to a homogeneous polynomial over $\mathbb{Z}$ in the entries of A; and we see that H is an ideal if and only if

$$\delta_k(A) \mid g_{ijk}(A) \quad (i, j, k = 1, \dots, d). \tag{5}$$

An entirely similar analysis shows that H is a subalgebra of L if and only if

$$\delta_k(A) \mid h_{ijk}(A) \quad (i, j, k = 1, \dots, d), \tag{6}$$

where $h_{ijk}(A)$ denotes the kth component of the vector $(a_{i1}, \dots, a_{id})(\sum_{l=j}^{d} a_{jl}C_l)A^{\ddot{}}$, again a homogeneous polynomial over $\mathbb{Z}$ in the entries of A.

Say L has b_n ideals of index n and a_n subalgebras of index n, and put

$$\zeta_L^{\triangleleft}(s) = \sum_{n=1}^{\infty} b_n n^{-s}, \quad \zeta_L(s) = \sum_{n=1}^{\infty} a_n n^{-s}.$$

Let $\mathcal{T}^{\triangleleft}, \mathcal{T}^s$ denote respectively the sets of matrices $A \in \mathcal{T}$ that satisfy (5), (6) respectively. Since H is an ideal (respectively, subalgebra) when $A \in \mathcal{T}^{\triangleleft}$ (respectively, $\mathcal{T}^s$) for every (or equivalently, any) $A \in \mathcal{M}(H)$, the above derivation (3) now gives

$$\begin{aligned} \zeta_L^{\triangleleft}(s) &= (1 - p^{-1})^{-d} \int_{A \in \mathcal{T}^{\triangleleft}} |\delta_0(A)|_p |\delta_d(A)|_p^{s-d} \, d\mu(A), \\ \zeta_L(s) &= (1 - p^{-1})^{-d} \int_{A \in \mathcal{T}^s} |\delta_0(A)|_p |\delta_d(A)|_p^{s-d} \, d\mu(A). \end{aligned} \tag{7}$$

These are examples of so-called *cone integrals*, introduced in [18]. In Section 4 we discuss an explicit finite procedure for evaluating such integrals, and some if its consequences.

Now we return to the original problem raised in the introduction, namely the counting of subgroups of finite index in groups. To begin with, let G be a torsion-free finitely generated nilpotent group, of Hirsch length d. The Mal'cev correspondence assigns to G a Lie algebra $\mathcal{L}(G)$ over $\mathbb{Q}$, of dimension equal to d, and a bijective mapping

$$\log : G^{\mathbb{Q}} \to \mathcal{L}(G)$$

(see for example Chapter 6 of [43]). The following is established in [27], Section 4: there is a positive integer f, depending only on d, such that (i) the subset $\log(G^f) = L$ is a full $\mathbb{Z}$-lattice and a Lie subring of $\mathcal{L}(G)$; (ii) for each prime p not dividing f, the correspondence $H \mapsto \log(H \cap G^f)$ gives an index-preserving bijection between the set of all subgroups of p-power index in G and the set of all Lie subrings of p-power index in the Lie ring L, under which normal subgroups of G correspond to ideals in L. It follows that for each prime $p \nmid f$, we have

$$\zeta_{G,p}(s) = \zeta_{L\otimes\mathbb{Z}_p}(s), \quad \zeta^{\triangleleft}_{G,p}(s) = \zeta^{\triangleleft}_{L\otimes\mathbb{Z}_p}(s).$$

Thus for almost all primes p, the local zeta functions of the group G are given by the cone integrals (7). We obtain the same conclusion if G is finitely generated and nilpotent, but not necessarily torsion-free, by taking $G/\tau(G)$ in place of G in the above discussion, and avoiding primes that divide the order of the torsion subgroup $\tau(G)$.

Consider next a uniform pro-p group G, of dimension d. Then G is a p-adic Lie group, and as such has an associated Lie algebra $\mathcal{L}(G) \cong \mathbb{Q}_p^d$. In this case, the image $L = \log(G)$ of G under the injective mapping $\log : G \to \mathcal{L}(G)$ is a $\mathbb{Z}_p$-Lie subalgebra and a full $\mathbb{Z}_p$-lattice in $\mathcal{L}(G)$; see [9], §7.2. Ishai Ilani [33] has proved the following: provided that $d \leqslant p$, the correspondence $H \mapsto \log(H)$ gives an index-preserving bijection between the set of all open subgroups in G and the set of all open $\mathbb{Z}_p$-Lie subalgebras L, under which normal subgroups of G correspond to ideals in L. It follows that as long as $d \leqslant p$, the zeta functions of G are again given by the cone integrals (7):

$$\zeta_G(s) = \zeta_L(s), \quad \zeta^{\triangleleft}_G(s) = \zeta^{\triangleleft}_L(s).$$

Can we obtain similar integral expressions for the local zeta functions at the 'bad' primes, those excluded in the preceding discussion? The answer is 'yes', but the procedure is more elaborate, and the resulting integrals are less tractable. Here we briefly explain how this is done in the case of finitely generated nilpotent groups, referring to [27], §2 for omitted details (but note that the bases are ordered in the opposite way). The case of uniform pro-p groups, and compact p-adic analytic groups in general, is discussed in the following section.

Let Γ be a torsion-free finitely generated nilpotent group, let $G = \widehat{\Gamma}_p$ be its pro-p completion, and let $(e_1, \ldots, e_d)$ be a Mal'cev basis for Γ: then the closed subgroups $G_i = \overline{\langle e_{i+1}, \ldots, e_d \rangle}$ in G form a central series $G = G_0 > G_1 > \cdots > G_d = 1$ with $G_i/G_{i+1} \cong \mathbb{Z}_p$ for each i. Each open subgroup H of G has a 'good basis'; this means an ordered generating set $(h_1, \ldots, h_d)$ such that for each i,

$$\overline{\langle h_{i+1}, \ldots, h_d\rangle} = H \cap G_i.$$

For each i we can write

$$\begin{aligned} h_i &= e_i^{a_{ii}} e_{i+1}^{a_{i,i+1}} \ldots e_d^{a_{id}} \\ &= \mathbf{e}^{\mathbf{a}_i}, \end{aligned}$$

with each $a_{ij} \in \mathbb{Z}_p$, and so associate to H the matrix $A = (a_{ij}) \in \mathcal{T}$. Let $\mathcal{M}(H)$ denote the set of all such matrices A corresponding to good bases for H. As in the abelian case, we have $|\delta_i(A)|_p = |G : HG_i|^{-1}$ and $|G : H| = |\det A|_p^{-1} = |\delta_d(A)|_p^{-1}$, and $\mu(\mathcal{M}(H))$ is still given by the formula (2). However, not every matrix in $\mathcal{T}$ arises in this way. Indeed, $(h_1, \ldots, h_d)$ is a good basis for some open subgroup H if and only if $|G : \overline{\langle h_1, \ldots, h_d\rangle}|$ is finite and

$$[h_i, h_j] \in \overline{\langle h_{j+1}, \ldots, h_d\rangle} \text{ for } 1 \leqslant i < j \leqslant d; \tag{8}$$

similarly it is a good basis for some open *normal* subgroup of G if and only if $|G : \overline{\langle h_1, \ldots, h_d\rangle}|$ is finite and

$$[h_i, e_j] \in \overline{\langle h_{i+1}, \ldots, h_d\rangle} \text{ for } 1 \leqslant i, j \leqslant d. \tag{9}$$

We have to translate these into conditions on the corresponding matrix A.

The multiplication law in Γ is expressed by certain integer-valued polynomial functions over $\mathbb{Q}$ (see [28], Chapter 6). This implies that there exist polynomials over $\mathbb{Q}$, depending only on Γ and the basis (e_i), defining functions

$$\begin{aligned} \lambda &: \mathbb{Z}_p^d \times \mathbb{Z}_p \to \mathbb{Z}_p^d \\ \mu^{(k)} &: (\mathbb{Z}_p^d)^k \to \mathbb{Z}_p^d, \quad (2 \leqslant k \leqslant d) \\ \kappa &: \mathbb{Z}_p^d \times \mathbb{Z}_p^d \to \mathbb{Z}_p^d, \end{aligned}$$

such that

$$\begin{aligned} (\mathbf{e}^{\mathbf{a}})^c &= \mathbf{e}^{\lambda(\mathbf{a},c)} \\ \mathbf{e}^{\mathbf{a}_1} \cdot \mathbf{e}^{\mathbf{a}_2} \cdot \ldots \cdot \mathbf{e}^{\mathbf{a}_k} &= \mathbf{e}^{\mu^{(k)}(\mathbf{a}_1, \ldots, \mathbf{a}_k)} \\ [\mathbf{e}^{\mathbf{a}}, \mathbf{e}^{\mathbf{b}}] &= \mathbf{e}^{\kappa(\mathbf{a},\mathbf{b})} \end{aligned}$$

for all $\mathbf{a}, \mathbf{a}_1, \mathbf{a}_2, \ldots, \mathbf{b} \in \mathbb{Z}_p^d$ and $c \in \mathbb{Z}_p$. Now consider the first-order formulas

$$\phi_{ij}(\mathbf{x}_1, \mathbf{x}_2, \ldots, \mathbf{x}_d) :$$
$$(\exists z_{j+1})\ldots(\exists z_d) \left(\kappa(\mathbf{x}_i, \mathbf{x}_j) = \mu^{(d-j)}(\lambda(\mathbf{x}_{j+1}, z_{j+1}), \ldots, \lambda(\mathbf{x}_d, z_d)\right),$$
$$\psi_{ij}(\mathbf{x}_1, \mathbf{x}_2, \ldots, \mathbf{x}_d) :$$
$$(\exists z_{i+1})\ldots(\exists z_d) \left(\kappa(\mathbf{x}_i, \mathbf{e}_j) = \mu^{(d-i)}(\lambda(\mathbf{x}_{i+1}, z_{i+1}), \ldots, \lambda(\mathbf{x}_d, z_d)\right),$$

where $\mathbf{x}_i = (x_{i1}, \ldots, x_{id})$ denotes a d-tuple of free variables. Define

$$\mathcal{T}^s = \{A \in \mathcal{T} \mid \phi_{ij}(\mathbf{a}_1, \ldots, \mathbf{a}_d) \text{ holds for } 1 \leqslant i < j \leqslant d\},$$
$$\mathcal{T}^{\lhd} = \{A \in \mathcal{T} \mid \psi_{ij}(\mathbf{a}_1, \ldots, \mathbf{a}_d) \text{ holds for } 1 \leqslant i, j \leqslant d\},$$

where $\mathbf{a}_i$ denotes the ith row of A, and the existential quantifiers range over $\mathbb{Z}_p$. Translating the conditions (8) and (9) gives the following: a matrix $A \in \mathcal{T}$ lies in $\mathcal{M}(H)$ for some open (respectively, open normal) subgroup H if and only if $\delta_d(A) \neq 0$ and $A \in \mathcal{T}^s$ (respectively, $A \in \mathcal{T}^{\lhd}$). It follows as before that $\zeta_G^{\lhd}(s)$ and $\zeta_G(s)$ are given by the formulas (7) (putting G in place of L).

The difference in the present case is that the domain of integration lacks such a simple definition: instead of the 'cone' conditions (5) and (6) we have to use the potentially more complicated first-order conditions ϕ_{ij} and ψ_{ij}. These make it very difficult to evaluate such integrals. Nevertheless, something very important can be said of them. A brilliant theorem due to Jan Denef [4], generalising pioneering ideas of Igusa [32], asserts that an integral of this kind is always rational as a function of p^{-s}. Indeed, what his proof shows is the following: let $\kappa(x_1, \ldots, x_n)$ be a formula in the first-order language of rings containing the n free variables x_i, let $V_{\kappa,p} = \left\{\mathbf{v} \in \mathbb{Z}_p^n \mid \kappa(\mathbf{v}) \text{ holds}\right\}$, and let f and g be polynomials in n variables over $\mathbb{Z}$. Then

$$I_p(\kappa, f, g; s) := \int_{V_{\kappa,p}} |f(\mathbf{v})|_p |g(\mathbf{v})|_p^s \, d\mu(\mathbf{v}) \tag{10}$$

is equal to a rational function over $\mathbb{Q}$ of p^{-s}. We shall see part of Denef's argument, applied to the special case of cone integrals, in Section 4. Applying Denef's theorem to the various integrals introduced above we deduce

Theorem 1. *Let $L_p \cong \mathbb{Z}_p^d$ be a $\mathbb{Z}_p$-algebra, let Γ be a torsion-free finitely generated nilpotent group, and let G be a uniform pro-p group of dimension at most p. Then each of the following functions is rational as a function of p^{-s}:*

$$\zeta_{L_p}, \quad \zeta_{L_p}^{\lhd}, \quad \zeta_{\Gamma,p}, \quad \zeta_{\Gamma,p}^{\lhd}, \quad \zeta_G, \quad \zeta_G^{\lhd}.$$

A refinement of Denef's theorem due to Macintyre [40] and Denef [5] states that as p varies over all primes, the rational functions $I_p(\kappa, f, g; s)$ of p^{-s} have numerators and denominators of bounded degrees. The same may therefore be said of the functions $\zeta_{\Gamma,p}$ and $\zeta_{\Gamma,p}^{\lhd}$ in the above theorem; this also holds for the functions ζ_{L_p} and $\zeta_{L_p}^{\lhd}$ if $L_p = L \otimes \mathbb{Z}_p$ for some fixed $\mathbb{Z}$-algebra $L \cong \mathbb{Z}^d$ (and these coincide with the functions $\zeta_{L,p}$ and $\zeta_{L,p}^{\lhd}$ that count subalgebras and ideals of p-power index in L). Macintyre's theorem, extending also results from [6], is actually even more general, giving a uniformity result for integrals including *parameters*. This was used in [11] to show that for a given natural number d, the rational functions corresponding to nilpotent groups of Hirsch length d, or to uniform pro-p groups or Lie algebras of dimension d, can be put over a 'universal denominator' of the form

$$\prod_{j=1}^{h} \left(1 - p^{-a_j s - b_j}\right);$$

here the a_j and b_j are integers that depend only on d (to deal with the case of uniform pro-p groups for the finitely many 'bad' primes $p < d$, the 'analytic' theory described in the following section has to be used.) The problem of 'uniformity' therefore comes down to understanding the *numerators* of these rational functions; this is discussed in Section 4 below.

3 Integrating on p-adic analytic groups

We begin by discussing a rationality theorem for certain integrals defined over a uniform pro-p group, generalising Denef's result for $\mathbb{Z}_p^d$. Let G be a uniform pro-p group of dimension d, with lower p-series $G = G_1 > G_2 > \dots$ Define $\omega(g) = n$ for $g \in G_n \setminus G_{n+1}$. A first-order language $\mathcal{L}_G$ is defined as follows. As well as the usual logical symbols (including =), this has two sorts of variables: those of sort x, to be interpreted as elements of G, and those of sort λ, to be interpreted as p-adic integers; certain function symbols: $x \cdot y, x^{-1}, x^{\lambda}, \phi_\alpha(x)$, interpreted respectively as the product, inverse, λth power, and certain automorphisms of G (to be specified); and a binary relation symbol to be interpreted as $\omega(x) \geqslant \omega(y)$. It also has constant symbols of the first sort, representing fixed elements of G. A subset $\mathcal{M}$ of $G^{(r)}$ is *definable* if there exists a formula κ in $\mathcal{L}_G$ with r free variables of sort x such that

$$\mathcal{M} = \mathcal{M}_\kappa = \{(g_1, \dots, g_r) \mid \kappa(g_1, \dots, g_r) \text{ is true}\};$$

a function $G^{(r)} \to G$ is definable if its graph is a definable subset of $G^{(r+1)}$. Finally, a function $f : G^{(r)} \to \mathbb{Z}$ is said to be *simple* if there exist definable functions $f_1, \dots, f_m : G^{(r)} \to G$ and integers $a_1, \dots, a_m$ such that

$$f(g_1, \dots, g_r) = \sum_{i=1}^{m} a_i \omega(f_i(g_1, \dots, g_r))$$

for all $g_1, \dots, g_r \in G$.

Let μ denote the normalised Haar measure on the compact group G (with $\mu(G) = 1$), and also the product measure on $G^{(r)}$. We can now state

Theorem 2 ([10] Theorem 1.17). *Let G be a uniform pro-p group. If $\mathcal{M}$ is a definable subset of $G^{(r)}$ and $f, k : G^{(r)} \to \mathbb{Z}$ are simple functions then the integral*

$$Z(f, k, \mathcal{M}; s) = \int_{\mathcal{M}} p^{-sf(\mathbf{g})} p^{-k(\mathbf{g})} d\mu(\mathbf{g})$$

is equal to a rational function over $\mathbb{Q}$ in p^{-s}.

This is proved by introducing p-adic co-ordinates (of the 'second kind') on G, and using the fact that the group operations (including taking p-adic powers) are expressed by analytic functions ([9], Chapters 4 and 9). The integral can then be interpreted as a 'definable integral' in the sense of [8]; in that paper, Denef and van den Dries generalise Denef's theorem, stated above, to the case where κ is a formula in a language $\mathcal{L}_{\text{an}}^D$ suitable for expressing the analytic theory of $\mathbb{Q}_p$.

Let us see how functions such as $\zeta_G(s)$ can be expressed as integrals of the above kind. Let H be an open subgroup of G, with $H \geqslant G_m$. A d-tuple $\mathbf{h} = (h_1, \ldots, h_d)$ is called a *good basis* for H if

(i) $\omega(h_i) \leqslant \omega(h_j)$ whenever $i \leqslant j$, and

(ii) for each $n \leqslant m$, the set

$$\left\{h_i^{p^{n-\omega(h_i)}} G_{n+1} \mid i = 1, \ldots, d;\ \omega(h_i) \leqslant n\right\}$$

is a basis for the $\mathbb{F}_p$-vector space $(H \cap G_n)G_{n+1}/G_{n+1}$.

It is easy to see that good bases exist, and that the set $\mathcal{M}(H)$ of all good bases for a given open subgroup H is an open subset of $G^{(d)}$. Writing $\mathbf{h}^\lambda = h_1^{\lambda_1} h_2^{\lambda_2} \ldots h_d^{\lambda_d}$ for $\lambda = (\lambda_1, \ldots, \lambda_d) \in \mathbb{Z}_p^d$, we have

Lemma 1 ([10] Lemma 2.4). *If* $\mathbf{h}$ *is a good basis for the open subgroup* H *of* G*, then*

$$H = \left\{\mathbf{h}^\lambda \mid \lambda \in \mathbb{Z}_p^d\right\},$$

and for each $\lambda \in \mathbb{Z}_p^d$

$$\omega(\mathbf{h}^\lambda) = \min\left\{\omega(h_i^{\lambda_i}) \mid i = 1, \ldots, d\right\}.$$

In order to express $\zeta_G(s)$ as a p-adic integral, we need (a) an expression for the index $|G : H|$ in terms of the good basis $\mathbf{h}$, (b) a formula for the measure of $\mathcal{M}(H)$, and (c) a suitable description of the set $\mathcal{M} = \bigcup_{H \leqslant_o G} \mathcal{M}(H)$ of all good bases for open subgroups of G.

The first step is to partition $\mathcal{M}$ into finitely many pieces, as follows. Let $P = (\lambda_1, \ldots, \lambda_r)$ be a partition of d, and for a d-tuple $\mathbf{h}$ in G and each $n \geqslant 1$ let $s_n(\mathbf{h}) = |\{i \mid \omega(h_i) = n\}|$. Let $\mathcal{M}_P$ denote the set of all $\mathbf{h} \in \mathcal{M}$ for which there exist distinct positive integers $n_1, \ldots, n_r$ such that

$$s_{n_i}(\mathbf{h}) = \lambda_i \text{ for } 1 \leqslant i \leqslant r,\ s_n(\mathbf{h}) = 0 \text{ for all other } n;$$

and put

$$q_P = \prod_{t \in P} \prod_{j=1}^{t} (1 - p^{-j}).$$

Note that if $\mathbf{h}$ is a good basis for H then $s_1(\mathbf{h}) + \cdots + s_n(\mathbf{h})$ is equal to the dimension over $\mathbb{F}_p$ of $(H \cap G_n)G_{n+1}/G_{n+1}$, for each n; so the numbers $s_i(\mathbf{h}) = s_i$ depend only on H, and $\mathcal{M}(H) \subseteq \mathcal{M}_P$ where $P = P_H$ consists of the positive numbers in the the set $\{s_n \mid n \in \mathbb{N}\}$, counted with their multiplicities. It also follows that

$$|G : H| = \prod_{n \geqslant 1} p^{d-s_1-\cdots-s_n}$$
$$= p^{-d} \cdot p^{\omega(h_1)+\cdots+\omega(h_d)}.$$

The measure of $\mathcal{M}(H)$ is given by

$$\mu(\mathcal{M}(H)) = q_P \cdot p^{d^2} \cdot p^{\sum_{i=1}^{d}(1-2i)\omega(h_i)};$$

the calculation is given in [10], Lemma 2.7 (apart from a small error in the formula for $\mu(\mathrm{GL}_{s_k}(\mathbb{Z}_p))$ on p. 656; this has a knock-on effect in a couple of places: the statement of Lemma 2.7 and formulas (2.1) and (2.5) are incorrect.).

Using Lemma 1 one may verify that a d-tuple $\mathbf{h} = (h_1, \ldots, h_d)$ is a good basis for some open subgroup if and only if the following hold:

(i) $h_i \neq 1$ for $1 \leqslant i \leqslant d$;

(ii) $\omega(h_i) \leqslant \omega(h_j)$ for $1 \leqslant i \leqslant j \leqslant d$;

(iii) for all $\lambda, \mu \in \mathbb{Z}_p^d$ there exists $\nu \in \mathbb{Z}_p^d$ such that

$$\mathbf{h}^{\nu} = \mathbf{h}^{\lambda}(\mathbf{h}^{\mu})^{-1};$$

(iv) for all $\lambda \in \mathbb{Z}_p^d$

$$\omega(\mathbf{h}^{\lambda}) = \min\left\{\omega(h_i^{\lambda_i}) \mid i = 1, \ldots, d\right\}.$$

This shows that the set $\mathcal{M}$ is definable. Writing the partition P as $(t_1, t_2 - t_1, \ldots, t_e - t_{e-1})$, we then have $\mathbf{h} \in \mathcal{M}_P$ if and only if

$$\omega(h_{i+1}) \leqslant \omega(h_i) \ \text{ if } t_{j-1} \leqslant i < t_j \text{ for some } j,$$
$$\omega(h_{t_j+1}) \nleqslant \omega(h_{t_j}) \ \text{ for each } j$$

(where $t_0 = 1$). Therefore each of the sets $\mathcal{M}_P$ is definable. The functions

$$f : \mathbf{h} \mapsto \sum_{i=1}^{d} \omega(h_i)$$
$$k : \mathbf{h} \mapsto \sum_{i=1}^{d} (1-2i)\omega(h_i)$$

are simple. So Theorem 2 applies to the integral $Z(f, k, \mathcal{M}_P; s)$, and we have

$$\begin{aligned}
\zeta_G(s) &= \sum_{H \leqslant_o G} |G : H|^{-s} \\
&= \sum_P \sum_{P_H = P} |G : H|^{-s} \cdot \mu(\mathcal{M}(H))^{-1} \int_{\mathcal{M}(H)} d\mu \\
&= p^{ds-d^2} \sum_P q_P^{-1} \int_{\mathcal{M}_P} p^{-sf(\mathbf{h})} p^{-k(\mathbf{h})} d\mu(\mathbf{h}) \\
&= p^{ds-d^2} \sum_P q_P^{-1} Z(f, k, \mathcal{M}_P; s).
\end{aligned}$$

More generally, suppose now that G is a compact p-adic analytic group. Then G has an open subgroup G_1 which is a uniform pro-p group ([9], Chapter 8). Fix a subgroup K lying between G_1 and G, and let

$$\begin{aligned}
\mathcal{H}(K) &= \{H \leqslant_o G \mid HG_1 = K\}, \\
\zeta_{G,p}^K(s) &= \sum_{H \in \mathcal{H}(K)} |K : H|^{-s} = \sum_{H \in \mathcal{H}(K)} |G_1 : H_1|^{-s},
\end{aligned}$$

where $H_1 = G_1 \cap H$. Clearly $\zeta_{G,p}(s)$ is a finite sum of terms like $|G : K|^{-s} \zeta_{G,p}^K(s)$, where $|G : K|$ is a power of p. Now fix a transversal $(1 = y_1, \ldots, y_n)$ to the cosets of G_1 in K. A $(d+n)$-tuple $(h_1, \ldots, h_d, t_1, \ldots, t_n)$ is called a *basis* for H if $(h_1, \ldots, h_d)$ is a good basis for H_1 and $(t_1 y_1, \ldots, t_n y_n)$ is a transversal to the cosets of H_1 in H.

For $H \in \mathcal{H}(K)$ let $\mathcal{K}(H) \subseteq G_1^{(d+n)}$ denote the set of all bases for H, and put $\mathcal{K} = \bigcup \mathcal{K}(H)$. It is easy to see that

$$\mu(\mathcal{K}(H)) = |G_1 : H_1|^{-n} \, \mu(\mathcal{M}(H_1))$$

(using μ to denote the product measures on both $G_1^{(d+n)}$ and $G_1^{(d)}$); so writing

$$\mathcal{K}_P = \{(\mathbf{h}, \mathbf{t}) \in \mathcal{K} \mid \mathbf{h} \in \mathcal{M}_P\}$$

for each partition P of d, we now have

$$\begin{aligned}
\zeta_{G,p}^K(s) &= \sum_P \sum_{P_H = P} |G_1 : H_1|^{-s} \cdot \mu(\mathcal{K}(H))^{-1} \int_{\mathcal{K}(H)} d\mu \\
&= p^{d(s-n)-d^2} \sum_P q_P^{-1} \int_{\mathcal{K}_P} p^{(n-s)f(\mathbf{h})} p^{-k(\mathbf{h})} d\mu(\mathbf{h}) \\
&= p^{d(s-n)-d^2} \sum_P q_P^{-1} Z(f, k - nf, \mathcal{K}_P; s).
\end{aligned}$$

Having seen already that the sets $\mathcal{M}_P$ are definable, to show that the $\mathcal{K}_P$ are definable we need only express the statement

$$H_1 t_1 y_1 \cup \ldots \cup H_1 t_n y_n \quad \text{is a subgroup of } G \tag{11}$$

in the language $\mathcal{L}_{G_1}$, where H_1 denotes the subgroup with a given good basis $(h_1, \dots, h_d)$. To do this, we interpret the symbol ϕ_i of $\mathcal{L}_{G_1}$ as the automorphism $g \mapsto y_i g y_i^{-1}$, for $i = 1, \dots, n$, and introduce constant symbols c_{ij}, for elements of G_1 such that

$$y_i y_j^{-1} = c_{ij} y_{\gamma(i,j)}.$$

Then (11) is equivalent to

$$h \in H_1 \& h' \in H_1 \Rightarrow$$
$$ht_i \cdot c_{ij} \cdot \phi_{\gamma(i,j)}(t_j^{-1} h'^{-1}) t_{\gamma(i,j)}^{-1} \in H_1 \quad (i, j = 1, \dots, n);$$

this is equivalent to a formula in $\mathcal{L}_{G_1}$ with free variables $h_1, \dots, h_d, t_1, \dots, t_n$ because

$$h \in H_1 \Leftrightarrow \exists \lambda \in \mathbb{Z}_p^d . h = h_1^{\lambda_1} \dots h_d^{\lambda_d}.$$

It follows that each $\mathcal{K}_P$ is a definable set.

As the functions f and $k - nf$ are simple, Theorem 2 is again applicable, showing that $Z(f, k - nf, \mathcal{K}_P; s)$ is rational in p^{-s}. Adding further definable conditions, in a similar spirit, to the set $\mathcal{K}_P$, we can specify additionally that the subgroup H with a given basis $(\mathbf{h}, \mathbf{t})$ be normal in G; we leave this as an exercise for the reader. The conclusion is

Theorem 3 [10]. *Let G be a compact p-adic analytic group. Then the functions $\zeta_{G,p}(s)$ and $\zeta^{\triangleleft}_{G,p}(s)$ are equal to rational functions over $\mathbb{Q}$ in p^{-s}.*

A more general version of this result will be discussed in Section 6 below. The converse of this theorem is also true, in the following sense: *if G is a pro-p group and $\zeta_G(s)$ is rational in p^{-s} then G is p-adic analytic.* The hypothesis implies that G has PSG, from which the conclusion follows by a theorem of Lubotzky and Mann (see [9], Interlude A). A (not quite obvious) application of Theorem 3, together with the characterisation of finitely generated abstract groups with PSG [38], gives

Corollary 1 [10]. *If Γ is a finitely generated group with PSG then $\zeta_{\Gamma,p}(s)$* is rational in p^{-s} *for every prime p.*

The method of combining a good basis with a transversal used in the above proof is not just of theoretical interest; in his thesis, John McDermott applied it practically to obtain explicit formulas for the zeta functions of the 17 'wallpaper' groups. These appear in [24].

4 Evaluating cone integrals

Denef [4] proved that integrals of the form (10) are rational in p^{-s} by first decomposing the domain of integration into nice pieces, and then making a change of variables so that on each piece, the integral can be evaluated directly by summing geometric progressions. The first stage of this decomposition depends on Macintyre's theory of quantifier elimination for the theory of $\mathbb{Q}_p$ [39]; the second stage depends on Hironaka's theorem on the resolution of singularities. When the defining formula κ takes a simple form like (5), the first – generally mysterious – stage can be dispensed with, and we are left with the second task, which can be carried out more or less explicitly. In this section we give an overview of the results; for the highly technical proofs the reader is referred to Sections 2 and 3 of [18].

Let f_i, g_i $(i = 0, 1, \dots, l)$ be polynomials in m variables over $\mathbb{Q}$. The statement

$$\psi(\mathbf{x}) : v(f_i(\mathbf{x})) \leqslant v(g_i(\mathbf{x})) \text{ for } i = 1, \dots, l \tag{12}$$

is called a *cone condition*; it is interpreted at $\mathbf{x} \in \mathbb{Q}_p^m$ by reading v as the p-adic valuation v_p. The collection $\mathcal{D} = \{f_0, g_0; f_1, g_1, \dots, f_l, g_l\}$ is called a set of *cone integral data*. For each prime p we write

$$V_p(\mathcal{D}) = \left\{\mathbf{x} \in \mathbb{Z}_p^m \mid \psi(\mathbf{x}) \text{ holds}\right\},$$

and to $\mathcal{D}$ we assign the so-called *cone integrals*

$$Z_{\mathcal{D}}(s, p) = \int_{V_p(\mathcal{D})} |f_0(\mathbf{x})|_p^s \, |g_0(\mathbf{x})|_p \, |d\mathbf{x}|,$$

where $|d\mathbf{x}|$ is the normalised additive Haar measure on $\mathbb{Z}_p^m$. Thus for example the local zeta functions (7) discussed in the preceding section take the form $(1-p^{-1})^{-d} Z_{\mathcal{D}}(s-d, p)$ for suitable choices of cone integral data $\mathcal{D}$.

Let $F = \prod_{i=0}^{l} f_i g_i \in \mathbb{Q}[X_1, \dots, X_m]$, and let $D = \operatorname{Spec}(\mathbb{Q}[X_1, \dots, X_m]/(F))$ be the hypersurface in affine space $\mathbb{A}^m$ defined by F. A *resolution* for F over $\mathbb{Q}$ consists of a smooth closed integral subscheme Y of projective space over X and a morphism $h : Y \to \mathbb{A}^m$ such that

- h restricts to an isomorphism $Y \setminus h^{-1}(D) \to \mathbb{A}^m \setminus D$, and
- the reduced scheme $(h^{-1}(D))_{\text{red}}$ associated to $h^{-1}(D)$ has only normal crossings as a subscheme of Y.

The precise meaning of the geometrical terms need not concern us unduly; for details see [5]. For present purposes, we need to know (1) that a resolution for F exists, and (2) that it has 'good reduction modulo p' for almost all primes p. Our attention will be focused on the *irreducible components* of the reduced scheme $(h^{-1}(D))_{\text{red}}$, which we label E_i $(i \in T)$; thus T is a finite indexing set and each E_i is a smooth, irreducible variety of dimension $m-1$ defined over $\mathbb{Q}$. 'Good reduction modulo p' means that the reductions $\overline{Y}$ and $\overline{E_i}$ modulo p are smooth varieties over $\mathbb{F}_p$, that $\bigcup_{i \in T} \overline{E_i}$ has only normal crossings as a subscheme of $\overline{Y}$, and that $\overline{E_i}$ and $\overline{E_j}$ have no common irreducible components when $i \neq j$.

We fix a resolution for F and keep the notation just introduced. Let us call a prime p *good* if the resolution has good reduction modulo p. For each good prime p and each subset I of the index set T, define

$$c_{p,I} = \left|\{a \in \overline{Y}(\mathbb{F}_p) \mid a \in \overline{E_i} \text{ if and only if } i \in I\}\right|$$
$$= \left|\bigcap_{i\in I} \overline{E_i}(\mathbb{F}_p) \setminus \bigcup_{j\in T\setminus I} \overline{E_j}(\mathbb{F}_p)\right|.$$

We can now state

Theorem 4 ([18] Theorem 3). *Let $\mathcal{D}$ be a set of cone integral data, and keep the notation introduced above. Then there exist rational functions $P_I(X, Y)$ over $\mathbb{Q}$, for $I \subseteq T$, such that for each good prime p,*

$$Z_{\mathcal{D}}(s, p) = \sum_{I\subseteq T} c_{p,I} P_I(p, p^{-s}).$$

One of the difficulties in working out integrals with such conditions as $v(f(x_1, \ldots, x_m)) \leqslant v(g(x_1, \ldots, x_m))$ is that the condition does not just depend on the valuations of $x_1, \ldots, x_m$, since in general $v(x_1 + x_2) \neq \min\{v(x_1), v(x_2)\}$. This problem does not arise if f and g happen to be *monomials*. Let us illustrate this by sketching the evaluation of $Z_{\mathcal{D}}(s, p)$ in the case where each f_i and g_i is a monomial. Say

$$f_i(\mathbf{X}) = u_i \prod_{j=1}^{m} X_j^{a_{ij}}, \quad g_i(\mathbf{X}) = v_i \prod_{j=1}^{m} X_j^{b_{ij}} \quad (i = 0, \ldots, l)$$

where u_i and v_i are units in $\mathbb{Z}_p$. Let

$$\Lambda = \left\{(n_1, \ldots, n_m) \in \mathbb{N}_0^m \mid \sum_{j=1}^{m} a_{ij} n_j \leqslant \sum_{j=1}^{m} b_{ij} n_j \text{ for } i = 1, \ldots, l\right\}.$$

Thus Λ is the set of integer lattice points in a certain *cone* C in $\mathbb{R}^m_{\geqslant 0}$; this motivates the term 'cone condition'. Writing $q_i = p^{-a_{0i}s - b_{0i} - 1}$ for each i, we then have

$$\begin{aligned} Z_{\mathcal{D}}(s, p) &= \int_{V_p(\mathcal{D})} |f_0(\mathbf{x})|^s \, |g_0(\mathbf{x})| \, |d\mathbf{x}| \\ &= \sum_{(n_1,\ldots,n_m)\in\Lambda} \prod_{j=1}^{m} (pq_j)^{n_j} \int_{v(x_j)=n_j (\forall j)} |d\mathbf{x}| \\ &= (1 - p^{-1})^m \sum_{(n_1,\ldots,n_m)\in\Lambda} q_1^{n_1} \cdots q_m^{n_m}. \end{aligned}$$

We can partition $C \setminus \{0\}$ into finitely many open simplicial cones of the form

$$C' = \bigoplus_{j=1}^{r} \mathbb{R}_{>0} v_j,$$

where $r \geqslant 1$, each v_j has non-negative integer co-ordinates, and a fundamental region of the lattice spanned by $v_1, \dots, v_r$ has no lattice-points of $\mathbb{Z}^m$ in its interior. To see this, note first that C is a union of finitely many rational simplicial cones, which may have some common boundary pieces of smaller dimension; each of these is the disjoint union of *open* simplicial cones of different dimensions, and C is the *disjoint* union of all these pieces. A typical such piece looks like C' above, for suitable $v_1, \dots, v_r$. If a point P of $\mathbb{Z}^m$ lies in the interior of a fundamental region of the lattice spanned by $v_1, \dots, v_r$, we draw the line from 0 through P and subdivide C' again so that this line appears as an edge of the new subcones. Since the volumes of the fundamental regions decrease, it is clear that this procedure terminates after finitely many steps.

Now let C' be as described. Then

$$\Lambda \cap C' = \bigoplus_{j=1}^{r} \mathbb{N} v_j,$$

and we have

$$\begin{aligned}\sum_{(n_1,\dots,n_m)\in\Lambda\cap C'} q_1^{n_1}\dots q_m^{n_m} &= \sum_{l_1=1}^{\infty}\dots\sum_{l_r=1}^{\infty}\prod_{j=1}^{m} q_j^{l_1 v_{1j}+\dots+l_r v_{rj}} \\ &= \prod_{k=1}^{r}\frac{q_1^{v_{k1}}\dots q_m^{v_{km}}}{1-q_1^{v_{k1}}\dots q_m^{v_{km}}} \\ &= \prod_{k=1}^{r}\frac{p^{-A_k s-B_k}}{1-p^{-A_k s-B_k}},\end{aligned}$$

where $A_k = \sum_{j=1}^{m} a_{0j} v_{kj}$ and $B_k = \sum_{j=1}^{m}(b_{0j}+1)v_{kj}$.

Adding up finitely many such expressions we conclude that

$$Z_{\mathcal{D}}(s,p) = \frac{H(p^{-1},p^{-s})}{\prod_{l=1}^{t}(1-p^{-\alpha_l s-\beta_l})} \tag{13}$$

where $H \in \mathbb{Z}[X,Y]$ and α_l, β_l are non-negative integers. In this case the expression is uniform in p, depending only on the combinatorics of the cone condition (12).

In general, what the resolution of singularities achieves is that the maps $f_j \circ h$ and $g_j \circ h$ all become monomial (in suitable coordinates), after which the preceding analysis can be applied. However these monomial expressions depend on where in $Y(\mathbb{Q}_p)$ one is located. One can define a map $\theta : H \to \overline{Y}(\mathbb{F}_p)$, 'reduction modulo p', where $H = \left\{b \in Y(\mathbb{Q}_p) : h(b) \in \mathbb{Z}_p^m\right\}$. Associated to the resolution are certain numerical data; these include integers $N_i(f_j)$ and $N_i(g_j)$ for each $i \in T$, and the following holds: if $a \in \bigcap_{i\in I}\overline{E_i}(\mathbb{F}_p)\setminus\bigcup_{j\in T\setminus I}\overline{E_j}(\mathbb{F}_p)$ then there exists a co-ordinate mapping $c : \theta^{-1}(a) \to (p\mathbb{Z}_p)^r$ such that

$$\begin{aligned}f_j \circ h \circ c^{-1}(y_1,\dots,y_r) &= u_{f_j}\cdot y_1^{N_{i_1}(f_j)}\dots y_r^{N_{i_r}(f_j)} \\ g_j \circ h \circ c^{-1}(y_1,\dots,y_r) &= u_{g_j}\cdot y_1^{N_{i_1}(g_j)}\dots y_r^{N_{i_r}(g_j)}\end{aligned}$$

where $I = \{i_1, \dots, i_r\}$ and u_{f_j} and u_{g_j} are p-adic units. In this way one can begin to see how to break the integral up according to where a sits on the configuration of varieties E_i: on each piece the integral is given by an expression like (13) above, which is uniform in p.

It should be emphasised that the proof, though intricate, constructs the rational functions P_I quite explicitly. This makes it possible in principle to determine just how the 'local factors' vary with the prime p (at least for almost all p), when one is dealing with a 'global' zeta function whose local factors (for almost all p) are given by a uniform set of cone conditions. This applies to the zeta functions discussed in Section 2. More precisely, it locates the question squarely where it belongs; that is, in the variation with p of the number of solutions modulo p of a system of equations over $\mathbb{Z}$.

How the number of points on a variety V mod p varies with p is in general wild. One does not expect the number of points to be finitely uniform; in other words there do not in general exist finitely many polynomials $H_1(X), \dots, H_m(X)$ such that for each prime p there exists some $i \in \{1, \dots, m\}$ with $|V(\mathbb{F}_p)| = H_i(p)$.

Example 1 ([34] §18.4). Let E be the elliptic curve $Y^2 = X^3 - X$. Put

$$N_p = |E(\mathbb{F}_p)| = \left|\left\{(x, y) \in \mathbb{F}_p^2 : y^2 = x^3 - x\right\}\right|.$$

If $p \equiv 3 \pmod 4$ then $N_p = p + 1$. However if $p \equiv 1 \pmod 4$, then

$$N_p = p + 1 - 2a,$$

where $p = a^2 + b^2$ and $a + ib \equiv 1 \pmod{2 + 2i}$.

The question now arises as to what possible varieties E_i actually occur in the present setting. Could it be, for example, that one can only realise affine spaces or flag varieties, where the number of points mod p is in fact uniform in p? The following example, discovered by the first author, dashes any such hope.

Example 2. Let G be the Hirsch length 9, class two nilpotent group given by the following presentation:

$$\left\langle \begin{array}{c} x_1, x_2, x_3, x_4, x_5, x_6, y_1, y_2, y_3 : [x_1, x_4] = y_3, [x_1, x_5] = y_1, [x_1, x_6] = y_2 \\ [x_2, x_4] = y_2, [x_2, x_6] = y_1, [x_3, x_4] = y_1, [x_3, x_5] = y_3 \end{array} \right\rangle$$

where all other commutators are defined to be 1. There exist rational functions $P_1(X, Y), P_2(X, Y) \in \mathbb{Q}(X, Y)$ such that for almost all primes p

$$\zeta^{\triangleleft}_{G,p}(s) = P_1(p, p^{-s}) + |E(\mathbb{F}_p)|\, P_2(p, p^{-s}),$$

where E is the elliptic curve $Y^2 = X^3 - X$.

To see where the elliptic curve is hidden in this presentation, take the determinant of the 3×3 matrix (a_{ij}) with entries $a_{ij} = [x_i, x_{j+3}]$ and you will get the projective version of E.

In general, then, the local zeta functions of an arbitrary nilpotent group will not be finitely uniform. However it is still believed that the local zeta functions of free nilpotent groups will be uniform. Using a more direct approach (without recourse to resolution of singularities), the first author and Grunewald [20] have proved this for the 'normal' zeta functions of two-generator free nilpotent groups of arbitrary class. If true in general, this will be relevant to Higman's PORC conjecture, as we shall see in Section 6.

In some sense the more conditions that are put on the presentation of a nilpotent group which take it away from being a free nilpotent group, the more complicated will be the resulting intersections of the components E_i. In the case of a free nilpotent group, the varieties are defined by relatively few equations in many variables, and one may hope that the corresponding intersections will resemble affine spaces.

Although the number of $\mathbb{F}_p$-rational points on the mod p reduction of a variety over $\mathbb{Z}$ in general varies wildly with p, the order of magnitude of this number is to a first approximation completely predictable: it is given by a famous theorem of Lang and Weil. In the following section, we explain how this can be used to understand the analytic properties of a 'global' zeta function whose Euler factors are cone integrals.

5 Euler products of cone integrals and subgroup growth

As in the preceding section, we let $\mathcal{D} = \{f_0, g_0; f_1, g_1, \ldots, f_l, g_l\}$ denote a set of cone integral data. We have seen that the corresponding cone integral $Z_\mathcal{D}(s, p)$ is equal to a rational function in p^{-s}; we denote by $a_{p,0} = Z_\mathcal{D}(\infty, p)$ the limit of this function as $s \to \infty$; thus $a_{p,0}$ is the constant term of the power series in p^{-s} that represents $Z_\mathcal{D}(s, p)$. In this section we consider the 'global zeta function' associated to $\mathcal{D}$, namely

$$Z_\mathcal{D}(s) = \prod_{p \text{ prime}} a_{p,0}^{-1} \cdot Z_\mathcal{D}(s, p).$$

Suppose that G is a finitely generated nilpotent group. We saw in §2 that for almost all primes p, the local zeta function $\zeta_{G,p}(s)$ takes the form $(1-p^{-1})^{-d} Z_\mathcal{D}(s-d, p)$ for suitable choice of cone integral data $\mathcal{D}$ (independent of p). Since these zeta functions have constant term equal to 1, it follows that $a_{p,0} = (1-p^{-1})^d$ for all such p, and hence that the global zeta function $\zeta_G(s)$ differs from $Z_\mathcal{D}(s - d)$ by at most finitely many Euler factors. Each of these factors is a rational function of p^{-s} (for the corresponding prime p), taking a form like (13), so it is a meromorphic function of s and has a rational abscissa of convergence. Thus from the analytic point of view, the function $\zeta_G(s)$ behaves in a similar way to $Z_\mathcal{D}(s - d)$; of course exactly the same holds for the function $\zeta_G^{\triangleleft}(s)$.

We now state the main theorem of [18]:

Theorem 5. *Let $\mathcal{D}$ be a set of cone integral data. Then*
(i) *the abscissa of convergence of $Z_\mathcal{D}(s)$ is a rational number, say, α;*

(ii) *the function $Z_{\mathcal{D}}(s)$ has a meromorphic continuation to the half plane $\Re(s) > \alpha - \delta$ for some $\delta > 0$.*

The result (ii) has an important application. The so-called *Tauberian Theorems* are used in analytic number theory to relate the analytic properties of a function represented by a Dirichlet series to the growth of the coefficients; we quote the following version from [41], p. 121:

Proposition 1. *Let the Dirichlet series $f(s) = \sum_{n=1}^{\infty} a_n n^{-s}$ with non-negative coefficients be convergent for $\Re(s) > \alpha > 0$. Suppose that for s in this half-plane, we have $f(s) = g(s)(s-\alpha)^{-\beta} + h(s)$, where $g(\alpha) \neq 0$, $\beta \in \mathbb{N}$ and $g(s), h(s)$ are holomorphic in the closed half-plane $\Re(s) \geqslant \alpha$. Then as $x \to \infty$,*

$$\sum_{n \leqslant x} a_n \sim \left(\frac{g(\alpha)}{\alpha \Gamma(\beta)}\right) x^{\alpha} (\log x)^{\beta - 1}.$$

In view of the preceding remarks, these results now give

Theorem 6. *Let G be a finitely generated nilpotent group and let $s_n(G)$, $s_n^{\triangleleft}(G)$ denote the number of subgroups, respectively normal subgroups of index at most n in G. Then there exist rational numbers α_G, $\alpha_G^{\triangleleft}$, positive real numbers c, c' and integers b, $b' \geqslant 0$ such that*

$$s_n(G) \sim c \cdot n^{\alpha_G} (\log n)^b ,$$
$$s_n^{\triangleleft}(G) \sim c' \cdot n^{\alpha_G^{\triangleleft}} (\log n)^{b'} .$$

Here α_G and $\alpha_G^{\triangleleft}$ are the abscissae of convergence of ζ_G and $\zeta_G^{\triangleleft}$ respectively.

Note that $b + 1$ is the order of the pole at $s = \alpha_G$. In [15] several examples are given where this order is greater than 1. For example, the zeta function of the discrete Heisenberg group G (the two-generator free nilpotent group of class 2) is given by

$$\zeta_G(s) = \zeta(s)\zeta(s-1)\zeta(2s-2)\zeta(2s-3) \cdot \zeta(3s-3)^{-1}. \tag{14}$$

This formula, obtained by Geoff Smith in his thesis [47], was the very first calculation of ζ_G for a non-abelian group G. As he observed, the double pole at $s = 2$ implies that the growth of subgroups is

$$s_n(G) \sim \frac{\zeta(2)^2}{2\zeta(3)} n^2 \log n.$$

We now indicate the main ideas in the proof of Theorem 5. The crucial facts in understanding the abscissa of convergence in these Euler products are the following:

(A) *An infinite product $\prod_{n \in J}(1 + a_n)$ converges absolutely if and only if the corresponding sum $\sum_{n \in J} |a_n|$ converges.*

(B) *$\sum_{p \text{ prime}} |p^{-s}|$ converges at $s \in \mathbb{C}$ if and only if $\Re(s) > 1$. The same holds for the sum $\sum_{p \in \mathcal{P}} |p^{-s}|$ whenever $\mathcal{P}$ is a dense set of primes.*

One can use these facts to prove the following:

Proposition 2. *Suppose that $P(X,Y) \in \mathbb{Q}[X,Y]$ has constant coefficient* 1, *and let $a_i > 0$ and b_i be integers. Then*

$$Z(s) = \prod_{p\ prime} \frac{P(p, p^{-s})}{(1-p^{-a_1 s+b_1})...(1-p^{-a_r s+b_r})} \tag{15}$$

has a rational abscissa of convergence.

Thus a zeta function whose local factors are uniform in p will have rational abscissa of convergence. To deal with our more general Euler products of cone integrals, we need some control over the numbers $c_{p,I}$ occurring in Theorem 4. This is provided by the Lang–Weil estimates for the number of points on a variety mod p:

Proposition 3 ([35]). *Let V be an absolutely irreducible variety of dimension d over a finite field $\mathbb{F}_q$, defined as the set of common zeros of polynomials $f_1, \ldots, f_r \in \mathbb{F}_q[X_1, \ldots, X_n]$ where each f_i has total degree $\leqslant e$. Then there is a positive constant C depending only on (e,n,r) such that*

$$\left| |V(\mathbb{F}_q)| - q^d \right| \leqslant Cq^{d-1/2}.$$

In the present setting one wants to count the number of points $c_{p,I}$ in sets like

$$\bigcap_{i\in I} \overline{E_i}(\mathbb{F}_p) \setminus \bigcup_{j\in T\setminus I} \overline{E_j}(\mathbb{F}_p),$$

where I ranges over subsets of an indexing set T. Put $E_I = \bigcap_{i\in I} E_i$ and let $F_{I,k}$ $(k \in C_I)$ be the irreducible components over $\mathbb{Q}$ of E_I with maximal dimension d_I. Now there is a complication if one wants to use Lang–Weil to count the number of points on $\overline{F_{I,k}}(\mathbb{F}_p)$. First, the reduction $\overline{F_{I,k}}$ of $F_{I,k}$ modulo p may no longer be irreducible; and secondly, not every irreducible component of $\overline{F_{I,k}}$ will necessarily be absolutely irreducible. However the helpful fact is that if a component is not absolutely irreducible, then it has no $\mathbb{F}_p$-rational points. Defining $l_p(F_{I,k})$ to be the number of irreducible components of $\overline{F_{I,k}}$ which are absolutely irreducible, one can deduce the following from Proposition 3:

Proposition 4. (i) *There exists $\gamma > 0$ such that for almost all primes p*

$$\left| c_{p,I} - \sum_{k\in C_I} l_p(F_{I,k}) p^{d_I} \right| \leqslant \gamma p^{d_I - 1/2}. \tag{16}$$

(ii) *$l_p(F_{I,k}) > 0$ for a dense set of primes p.*

There are a number of issues that one can brush under the carpet because one is taking estimates. For example, one can prove that $F_{I,k} \cap \bigcup_{j\in T\setminus I} E_j$ has dimension smaller than d_I, hence removing the points mod p in this set will not upset the estimate. Similarly, it is only necessary to consider the irreducible components $F_{I,k}$ of E_I of maximal dimension.

From Proposition 4 one derives

(C) *a sum of the form $\sum_p c_{p,I} r_p$ converges absolutely if and only if $\sum_p p^{d_I} r_p$ converges absolutely.*

Now there exists some subset T_0 of T such that the constant term $a_{p,0}$ actually appears as

$$a_{p,0} = \sum_{I \subseteq T_0} c_{p,I} P_I(p, p^{-s}).$$

Hence

$$a_{p,0}^{-1} \cdot Z_{\mathcal{D}}(s, p) = 1 + \sum_{I \subseteq T, I \nsubseteq T_0} \frac{c_{p,I}}{a_{p,0}} P_I(p, p^{-s}).$$

It can be shown that there exists $M \in \mathbb{N}$ such that for almost all p

$$1 \leqslant a_{p,0}^{-1} \leqslant 1 + M p^{-1/2}. \tag{17}$$

Combining this with (A) and (C) one sees that

$$Z_{\mathcal{D}}(s) = \prod_{p \text{ prime}} a_{p,0}^{-1} \cdot Z_{\mathcal{D}}(s, p)$$

has the same abscissa of convergence as

$$\prod_{p \text{ prime}} \left(1 + \sum_{I \subseteq T, I \nsubseteq T_0} p^{d_I} P_I(p, p^{-s}) \right).$$

One can then apply Proposition 2 to deduce that $Z_{\mathcal{D}}(s)$ has a rational abscissa of convergence, proving part (i) of Theorem 5.

However this approximation of $Z_{\mathcal{D}}(s, p)$ by a uniform expression in p does not help when it comes to extending $Z_{\mathcal{D}}(s)$ meromorphically beyond its abscissa of convergence. For example, the uniform expression

$$\prod_{p \text{ prime}} \left(1 + \frac{p^{-1-s}}{(1 - p^{-s})} \right) \tag{18}$$

converges for $\Re(s) > 0$ but has $\Re(s) = 0$ as a natural boundary. The convergence for $\Re(s) > 0$ follows from (A) and (B) above. The natural boundary is a result of the fact that $\left(1 - p^{-s}\right)^{-1}$ has a pole at $s = 2n\pi i / \log p$ for each $n \in \mathbb{Z}$. Hence the function is unbounded as one approaches any point on the line $\Re(s) = 0$.

In order to establish the meromorphic continuation, a more refined version of Theorem 4 is needed. Instead of a sum over the subsets of T (the indexing set of the irreducible components E_i), we express $Z_{\mathcal{D}}(s, p)$ as a sum over the open simplicial pieces of a certain polyhedral cone associated to the cone condition ψ. Write $t = |T|$.

Theorem 7. *There exist*

(a) *a closed polyhedral cone* $\overline{D}$ *in* $\mathbb{R}^t_{\geqslant 0}$, *which we write as the disjoint union of open simplicial pieces* R_k $(0 \leqslant k \leqslant w)$, *where* $R_0 = \{0\}$ *and* $R_1, \ldots, R_q$ *are the one-dimensional pieces;*

(b) *positive integers* A_j, B_j *for* $j = 1, \ldots, q$,

such that

$$Z_{\mathcal{D}}(s,p) = \sum_{k=0}^{w} (p-1)^{|I_k|} p^{-m} c_{p,I_k} \prod_{j \in M_k} \frac{p^{-(A_j s + B_j)}}{(1 - p^{-(A_j s + B_j)})}, \tag{19}$$

where for each k,

$M_k = \{i \in \{1, \ldots, q\} \mid R_i$ *is contained in the closure of* $R_k\}$,

I_k *is the subset of* T *defined by:* $i \in T \backslash I_k$ *if and only if the* i*th coordinate is zero for all points of* R_k.

We explained in the preceding section that on each of the pieces $\bigcap_{i \in I} \overline{E_i}(\mathbb{F}_p) \setminus \bigcup_{j \in T \setminus I} \overline{E_j}(\mathbb{F}_p)$, the original cone condition reduces to one involving only monomial f_i and g_i, and therefore specified by a certain cone that we denoted C. The cone in the statement of Theorem 7 is built out of the simplicial subcones of C introduced before, for each of the cones C associated to one of the pieces $\bigcap_{i \in I} \overline{E_i}(\mathbb{F}_p) \setminus \bigcup_{j \in T \setminus I} \overline{E_j}(\mathbb{F}_p)$. As the discussion in §4 suggests, one also gets an explicit expression for the integers A_j and B_j in terms of the numerical data of the resolution.

With this combinatorial expression to hand, it is possible to show that the pathologies of examples like (18) do not arise. In particular, the abscissa of convergence α of the global zeta function is determined by the terms in the expression (19) corresponding to the one-dimensional edges $R_1, \ldots, R_q$. This implies that the analytic behaviour near α is dictated by functions of the form

$$\prod_{p \text{ prime}} \left(1 + c_{p,I_k} p^{|I_k| - m} \frac{p^{-s}}{(1 - p^{-s})}\right). \tag{20}$$

Recall that d_I was defined to be the dimension of E_I. Because the E_i are hypersurfaces of dimension $m-1$ intersecting with normal crossings, the dimension of the intersection of E_i's goes down one for each E_i. Hence $d_I = m - |I_k|$. Combining this with the Lang–Weil estimate (16), one can show that the analytic behaviour of the function (20) near α will be the same as that of

$$\prod_{p \text{ prime}} \left(1 + l_p(F_{I,k}) \frac{p^{-s}}{(1 - p^{-s})}\right). \tag{21}$$

So far, we have replaced a function defined in terms of the delicate arithmetical data $c_{p,I}$ with one that depends only on the geometry of the configuration of varieties E_i; recall that $l_p(F_{I,k})$ was defined as the number of irreducible components of

$\overline{F_{I,k}}$ that are absolutely irreducible. Where do we go from here? In order to continue an analytic function past its abscissa of convergence, the idea is to find some known meromorphic function with which one can approximate the original function, up to a factor which converges a little better than the original. In the present context this role is taken by an *Artin L-function.*

The Artin L-function $L(s, \rho, K/\mathbb{Q})$ is associated to a linear representation ρ (in characteristic zero) of the Galois group $\mathrm{Gal}(K/\mathbb{Q})$ of a finite Galois extension K of $\mathbb{Q}$. It is defined as the Euler product over primes p of local factors which for unramified primes (i.e., almost all primes) take the form

$$\det(1 - \rho(\mathrm{Frob}_p) \cdot p^{-s})^{-1},$$

where Frob_p is a Frobenius automorphism of p. It is known that $L(s, \rho, K/\mathbb{Q})$ has meromorphic continuation to the whole of the complex plane ([3], Chapter VIII, §3). Indeed it is conjectured that it is an entire function except for a possible pole at $s = 1$; this is known as Artin's conjecture.

So where is such an L-function hiding in our setting? The extension of $\mathbb{Q}$ that one associates to each irreducible component $F_{I,k} = X$ is the field of definition K of the absolutely irreducible components $X_1, \ldots, X_n$ of X. The Galois group $\mathrm{Gal}(K/\mathbb{Q})$ permutes these absolutely irreducible components, so giving rise to a permutation representation ρ of $\mathrm{Gal}(K/\mathbb{Q})$.

The first task is to show that for almost all p, unramified in K,

$$l_p(X) = \mathrm{Tr}(\rho(\mathrm{Frob}_p)). \tag{22}$$

The trace is the number of fixed points in this permutation action. Let $\overline{X_i}$ denote the reduction of X_i mod $\mathfrak{p}$ where $\mathfrak{p}$ is a prime of K sitting over the unramified p. Then $\overline{X} = \overline{X_1} \cup \ldots \cup \overline{X_n}$ is the decomposition of $\overline{X}$ into absolutely irreducible components defined over the residue field of $\mathfrak{p}$. An absolutely irreducible component X_i is only fixed by a Frobenius automorphism of p if the corresponding component $\overline{X_i}$ is actually defined over $\mathbb{F}_p$. This proves (22).

To show that $L(s, \rho, K/\mathbb{Q})$ is a good approximation to our function (21), what is required is that

$$\prod_p \left(1 + l_p(X)\frac{p^{-s}}{(1-p^{-s})}\right)(\det(1 - \rho(\mathrm{Frob}_p) \cdot p^{-s}))$$

converges on $s > 1 - \delta$ for some $\delta > 0$ (the product being over almost all primes p). This follows from (A) and (B), (22) and the fact that

$$\det(1 - \rho(\mathrm{Frob}_p) \cdot p^{-s}) = 1 - \mathrm{Tr}(\rho(\mathrm{Frob}_p)) \cdot p^{-s} + \sum_{k=2}^{n} a_k p^{-ks}$$

for suitable a_k which can be bounded independently of p.

Retracing the steps we deduce that $Z_{\mathcal{D}}(s)$ has a meromorphic continuation to the left of its abscissa of convergence, as claimed in part (ii) of Theorem 5. For more complete details see [18].

We have already explained the connection between zeta functions of groups and the numbers of points on varieties. Perhaps the most striking feature of the above proof is the unexpected appearance of the Artin L-functions. Since Artin L-functions have a meromorphic continuation to the whole of the complex plane, is it possible that zeta functions of groups could similarly be meromorphically continued? After all, the example (14) of the Heisenberg group mentioned above has meromorphic continuation to the whole of $\mathbb{C}$ since it is built solely out of Riemann zeta functions. Similarly, it is shown in [27] that the normal zeta function of the Heisenberg group with entries from a quadratic number field can be expressed using the Dedekind zeta function of the number field, and hence has meromorphic continuation.

In [24] it is proved that crystallographic groups have zeta functions meromorphic on the whole of $\mathbb{C}$. More generally, the following is established:

Theorem 8. *Let G be a finite extension of a free abelian group. Then $\zeta_G(s)$ and $\zeta_G^{\triangleleft}(s)$ both have meromorphic continuation to the whole of $\mathbb{C}$.*

This is proved by relating such zeta functions to yet another sort of meromorphic zeta function, namely the zeta functions of central simple algebras studied by Hey, Solomon and more recently Bushnell and Reiner ([29], [1], [2]).

Although Theorem 5 establishes the possibility of some meromorphic continuation, in general zeta functions of groups possess a natural boundary. A general criterion for such natural boundaries has been established in [15]; this applies to all known examples of zeta functions of groups except for those with an obvious meromorphic continuation to $\mathbb{C}$. Here is a simple example:

Example 3. Let $F_{2,3}$ denote the free nilpotent group of class two on three generators. Then

$$\zeta_{F_{2,3}}^{\triangleleft}(s) = \zeta(s)\zeta(s-1)\zeta(s-2)\zeta(5s-8)\zeta(6s-9) \prod_{p \text{ prime}} V(p, p^{-s})$$

where

$$V(X,Y) = \left(1 + X^3Y^3 + X^4Y^3 + X^6Y^5 + X^7Y^5 + X^{10}Y^8\right).$$

$\zeta_{F_{2,3}}^{\triangleleft}(s)$ converges on $\Re(s) > 3$ and has meromorphic continuation to $\Re(s) > 7/5$. However $\Re(s) = 7/5$ is a natural boundary beyond which $\zeta_{F_{2,3}}^{\triangleleft}(s)$ cannot be meromorphically continued.

This zeta function was calculated in the paper [27]. Its analytic properties are established in [15]. That $\Re(s) = 7/5$ is a natural boundary follows from the fact that every point on this line is a limit point of zeros s (with $\Re(s) > 7/5$) of the functions $V(p, p^{-s})$ as p ranges over all primes. Now it turns out that as $p \to \infty$, the zeros of the polynomial $V(p, Y)$ tend to the zeros of a certain associated polynomial

$$\widetilde{V}(p,Y) = 1 + p^3Y^3 + p^7Y^5 + p^{10}Y^8 = (1 + p^3Y^3)(1 + p^7Y^5).$$

The corresponding Euler product

$$\prod_p \widetilde{V}(p, p^{-s}) = \frac{\zeta(3s-3)\zeta(5s-7)}{\zeta(6s-6)\zeta(10s-14)}$$

is meromorphic on $\mathbb{C}$. In a certain sense, the 'messy' function $\zeta^{\triangleleft}_{F_{2,3}}(s)$ is 'trying' to be a nice product of Riemann zeta functions.

This observation has been generalised in [21] and [22], where the first author and Grunewald associate to an arbitrary two-variable polynomial $V(X, Y)$ a so-called *ghost polynomial* $\widetilde{V}(X, Y)$; the procedure is quite definite and explicit. Replacing V by $\widetilde{V}$ in examples like the above one gets the corresponding 'ghost zeta function' of a group. The ghost is said to be *friendly* if the associated Euler product

$$\prod_{p \text{ prime}} \widetilde{V}(p, p^{-s})$$

has a meromorphic continuation to the whole of $\mathbb{C}$. The fascinating fact is that ghosts associated to groups are most often friendly, although not always. More on this topic may be found in the two papers cited above.

Another feature of the polynomials like $V(X, Y)$ that arise in zeta functions of groups is a certain palindromic symmetry that is called a *local functional equation.* For example, in the case of $F_{2,3}$ above we have

$$V(X^{-1}, Y^{-1}) = X^{-10}Y^{-8}V(X, Y).$$

It appears that in all known examples, both ζ_G and $\zeta^{\triangleleft}_G$ satisfy a similar local functional equation. For certain nilpotent groups G, a local functional equation was established in [23] and [16] for the zeta function $\zeta^{\wedge}_G$, counting subgroups $H \leqslant G$ for which $\widehat{H} \cong \widehat{G}$; this is proved by exploiting symmetries in the building of the associated algebraic automorphism group of G. To find an explanation of the phenomenon in general, however, is one of the most intriguing open problems in this area. It is possible that Theorem 7 may open up an avenue to understanding these local functional equations; for example, it may be that what lies behind them is the functional equation of the Weil zeta functions associated to the varieties in Theorem 7, and the functional equations associated to zeta functions of cones and their associated toric varieties. Some evidence for this is provided by Denef and Meuser's application [7] of the functional equation of the Weil zeta function in relation to the Igusa zeta function (the Igusa zeta function is a special case of a cone integral in which the cone condition is empty, i.e., it is an integral over the whole of $\mathbb{Z}_p^m$).

6 Counting p-groups. I: by size

The question of how many (isomorphism types of) groups of order p^n exist has a respectable history. Denote this number by $f(n, p)$. The following theorem was proved by Higman [30] (and improved by Sims [46]) nearly 40 years ago:

$$f(n, p) = p^{(2/27+o(1))n^3} \quad \text{as } n \to \infty.$$

Thus the asymptotic behaviour of this function as n grows is reasonably understood. On the other hand, there has been very little progress in understanding its arithmetical properties. In a second paper [31], Higman made the following conjecture: for each n there exist finitely many polynomials $P_{n,1}, \ldots, P_{n,N}$ over $\mathbb{Q}$ such that

$$p \equiv j \pmod N \Rightarrow f(n,p) = P_{n,j}(p);$$

in other words, that $f(n,p)$ is *PORC* ('polynomial on residue classes'). As we shall see below, this can be interpreted as a sort of finite uniformity for certain zeta functions, suggesting a promising new approach to this problem.

We turn first to the behaviour of $f(n,p)$ when p is fixed and n varies. More precisely, we consider the number

$$f(n,p,c,d)$$

of isomorphism types of *d-generator groups of order p^n and nilpotency class at most c*. The main result is

Theorem 9 [13]. *Let c, d be positive integers and let p be a prime. Then for sufficiently large n the function $n \mapsto f(n,p,c,d)$ satisfies a linear recurrence relation with constant coefficients.*

To prove this we introduce the generating function

$$\zeta_{c,d,p}(s) = \sum_{n=0}^{\infty} f(n,p,c,d)p^{-ns},$$

and observe that the theorem is a formal consequence of

Theorem 10. *$\zeta_{c,d,p}(s)$ is equal to a rational function over $\mathbb{Q}$ in p^{-s}.*

Let us fix the prime p and the integers c and d, and write $f(n) = f(n,p,c,d)$. Let F_p denote the pro-p completion of the free class-c nilpotent group on d generators. Then $f(n)$ is just the number of epimorphic images of F_p having order p^n. Now if N and N_1 are open normal subgroups of F_p, then $F/N_1 \cong F/N$ if and only if $N_1 = N^\gamma$ for some automorphism γ of F_p; this is a general property of free pro-$\mathcal{C}$ groups whenever $\mathcal{C}$ is a class of finite groups closed under subgroups, quotients and direct products ([25], Proposition 15.31). It follows that the number of quotients of F_p that are isomorphic to F_p/N is equal to $|A : \mathrm{N}_A(N)|$ where $A = \mathrm{Aut}(F_p)$, and we can write

$$\zeta_{c,d,p}(s) = \sum_{N \lhd_o F_p} |F_p : N|^{-s} \, |A : \mathrm{N}_A(N)|^{-1}.$$

This differs from the definition of $\zeta^{\lhd}_{F_p}(s)$ only in the presence of the extra factor $|A : \mathrm{N}_A(N)|^{-1}$, and its rationality will follow in a similar way. In fact we have the following generalisation of (the 'normal' part of) Theorem 3, which gives Theorem 10 on taking $G = F_p \rtimes \mathrm{Aut}(F_p)$ and $U = F_p$ (note that G in this case is a compact p-adic analytic group, by [9], Corollary 8.35 and Theorem 9.7):

Theorem 11. *Let G be a compact p-adic analytic group and U a closed normal subgroup. Let $f(n)$ denote the number of conjugacy classes in G of open normal subgroups of U having index p^n. Then the function*

$$\zeta_{G,U,p}(s) = \sum_{n=0}^{\infty} f(n)p^{-ns}$$

is a rational function over $\mathbb{Q}$ in p^{-s}.

The proof is similar to that of Theorem 3. We begin by observing that G has an open uniform pro-p subgroup G_1 such that both $G_1 \cap U = U_1$ and G_1/U_1 are uniform (this follows from elementary facts about uniform groups, see [9], Chapter 4). For each pair (K, L) with $U_1 \leqslant K \leqslant U$ and $G_1 \leqslant L \leqslant G$, write

$$\mathcal{H}(K, L) = \{H \lhd_o U \mid U_1 H = K \text{ and } G_1 \mathrm{N}_G(H) = L\}.$$

Then $\zeta_{G,U,p}(s)$ is a finite sum of terms of the form $|U : K|^{-s}\, |G : L|^{-1}\, Z_{K,L}(s)$ where

$$Z_{K,L}(s) = \sum_{H \in \mathcal{H}(K,L)} |U_1 : U_1 \cap H|^{-s}\, |G_1 : \mathrm{N}_{G_1}(H)|^{-1},$$

and $|U : K|$ is a power of p. We claim that for each such K and L, $Z_{K,L}(s)$ is equal to a finite linear combination, the coefficients being polynomials in p^{-s}, of integrals of the form $Z(f, k, \mathcal{M}; s)$, where f, k are simple functions and $\mathcal{M}$ is a definable set in the language $\mathcal{L}_{G_1}$ defined in Section 3. With Theorem 2 this will complete the proof of Theorem 11.

We make an important preliminary remark. In a uniform pro-p group M, the terms of the lower p-series are given by

$$M_n = \left\{a^{p^{n-1}} \mid a \in M\right\}$$

for $n \geqslant 1$ ([9], Chapter 3). Since G_1/U_1 is torsion-free, it follows that

$$G_n \cap U_1 = U_n$$

for each n. Therefore $\omega(a)$ for $a \in U_1$ is unambiguously defined, whether a is considered as an element of U_1 or of G_1. Moreover, the predicate '$x \in U_1$' is definable in $\mathcal{L}_{G_1}$, being equivalent to

$$\exists \lambda \in \mathbb{Z}_p^d \,.\, x = u_1^{\lambda_1} \ldots u_d^{\lambda_d},$$

where $u_1, \ldots, u_d$ are constant symbols representing the elements of some fixed minimal generating set for U_1. It follows that for subsets of $U_1^{(r)}$, being definable in $\mathcal{L}_{U_1}$ is equivalent to being definable in $\mathcal{L}_{G_1}$. We make tacit use of this principle in the following discussion.

Let us fix K and L as above, and fix a transversal $(1 = y_1, \ldots, y_n)$ to the cosets of U_1 in K. We consider pairs (H, X) where $H \in \mathcal{H}(K, L)$ and $X \leqslant_o G_1$. To each such pair we associate the set $\mathcal{K}(H, X) \subseteq U_1^{(d+n)} \times G_1^{(m)}$ consisting of triples $(\mathbf{h}, \mathbf{t}, \mathbf{x})$ where $(\mathbf{h}, \mathbf{t})$ is a basis for H and $\mathbf{x}$ is a good basis for X, as defined in Section 3 (here $d = \dim U_1$ and $m = \dim G_1$). Fix partitions P of d and Q of m, and put

$$\mathcal{H}_{P,Q} = \{H \in \mathcal{H}(K, L) \mid P_H = P \text{ and } P_{N(H)} = Q\},$$
$$\mathcal{K}_{P,Q} = \bigcup_{H \in \mathcal{H}_{P,Q}} \mathcal{K}(H, N(H))$$

where $N(H) = \mathrm{N}_{G_1}(H)$, and P_H, $P_{N(H)}$ are the partitions defined in §3.

Now let $H \in \mathcal{H}_{P,Q}$ and write $H_1 = U_1 \cap H, X = N(H)$. Let $(\mathbf{h}, \mathbf{t}, \mathbf{x}) \in \mathcal{K}(H, X)$. Adapting the notation from §3, we can write $\mathcal{K}(H, X) = \mathcal{K}(H) \times \mathcal{M}(X)$, and we have

$$\begin{aligned}\mu(\mathcal{K}(H, X)) &= \mu(\mathcal{K}(H)) \cdot \mu(\mathcal{M}(X)) \\ &= |U_1 : H_1|^{-n}\, q_P\, p^{d^2} p^{k(\mathbf{h})} \cdot q_Q\, p^{m^2} p^{k'(\mathbf{x})},\end{aligned}$$

where $k : G_1^{(d)} \to \mathbb{Z}$ and $k' : G_1^{(m)} \to \mathbb{Z}$ are simple functions as given in §3. Similarly we have

$$|U_1 : H_1| = p^{-d} p^{f(\mathbf{h})}, \quad |G_1 : X| = p^{-m} p^{f'(\mathbf{x})}$$

for simple functions f and f'.

It follows as before that

$$\begin{aligned}Z_{K,L}(s) &= \sum_{P,Q} \sum_{H \in \mathcal{H}_{P,Q}} |U_1 : H_1|^{-s}\, |G_1 : N(H)|^{-1} \\ &= \sum_{P,Q} q_P^{-1} q_Q^{-1} p^{-r} p^{-ds} Z(f^*, k^*, \mathcal{K}_{P,Q}; s),\end{aligned}$$

where $r = dn + d^2 + m^2 - m$ and f^* and k^* are simple functions on $G_1^{(d+m)}$. To complete the proof, therefore, it remains only to verify that the sets $\mathcal{K}_{P,Q}$ are definable.

Now the triple $(\mathbf{h}, \mathbf{t}, \mathbf{x}) \in U_1^{(d+n)} \times G_1^{(m)}$ belongs to $\mathcal{K}_{P,Q}$ if and only if

(i) $(\mathbf{h}, \mathbf{t}) \in \mathcal{K}_P$ and $\mathbf{x} \in \mathcal{M}_Q$,

(ii) $H \lhd U$,

(iii) $G_1 \mathrm{N}_G(H) = L$, and

(iv) $X = \mathrm{N}_{G_1}(H)$,

where $H = \langle h_i, t_j y_j \mid i = 1, \ldots, d,\ j = 1, \ldots, n\rangle$ and $X = \langle x_i \mid i = 1, \ldots, m\rangle$. We have seen in §3 that condition (i) is definable. The conjunction of (i) with either of '$g \in H_1$' or '$g \in X$' is then a definable predicate (in the free variables $g, \mathbf{h}, \mathbf{t}, \mathbf{x}$). We can write (iv) in the form

$$g \in X \Leftrightarrow \left(h \in H_1 \Rightarrow g^{-1} h t_j \phi_j(g) t_j^{-1} \in H_1\right) \; (j = 1, \ldots, n),$$

where $\phi_j(g) = y_j g y_j^{-1}$; this shows that (iv) is definable. For (ii) and (iii), we extend $(y_1, \dots, y_n)$ to a transversal $(y_1, \dots, y_{n+t})$ to U/U_1, and fix transversals $\{v_i \mid 1 \leqslant i \leqslant q\}$ and $\{w_i \mid 1 \leqslant i \leqslant r\}$ to L/G_1, G/L respectively, with $w_1 = 1$. Then (ii) is equivalent to

$$\text{for } i = 1, \dots, n, \ (g \in U_1 \Rightarrow g y_i H y_i^{-1} g^{-1} \subseteq H),$$

and (iii) is equivalent to

$$\text{for } i = 1, \dots, q, \ (\exists g) \left(g v_i H v_i^{-1} g^{-1} \subseteq H\right),$$
$$\text{for } i = 2, \dots, r, \ (\nexists g) \left(g w_i H w_i^{-1} g^{-1} \subseteq H\right).$$

These can similarly be expressed by formulas in $\mathcal{L}_{G_1}$, if we introduce function symbols representing the conjugation action of the y_i, v_i and w_i on G_1, and constant symbols representing elements of G_1 such as $y_i y_j y_i^{-1} y_l^{-1}$. This is left as a straightforward exercise for the reader. Thus the sets $\mathcal{K}_{P,Q}$ are definable, and Theorem 11 follows.

Of course, Theorem 11 is far more general than its original *raison d'être*, Theorem 9, and the gain in generality is paid for by a loss of information; for example, this approach tells us nothing about how $f(n, p, c, d)$ varies with the prime p. To do better, we can try to follow the method of §2. Still keeping c and d fixed, let F be the free-nilpotent group of class c on d generators, and F_p as above its pro-p completion. As explained in §2, there is a Lie ring $L = \log(F^f)$ (for a suitable integer f) whose ideals of index p^n correspond to normal subgroups of the same index in F, provided $p \nmid f$. Moreover, for such p we can identify the automorphisms of F_p with the Lie algebra automorphisms of $L_p = L \otimes \mathbb{Z}_p$. Now $\mathrm{Aut}(L_p) = \mathfrak{G}(\mathbb{Z}_p)$, where $\mathfrak{G}$ is a linear algebraic group defined over $\mathbb{Q}$; this group $\mathfrak{G}$ can be described quite explicitly, as an extension of a large unipotent radical by GL_d; and this makes it possible to express $\zeta_{c,d,p}(s) = \zeta_{F_p \rtimes \mathrm{Aut}(F_p), F_p}(s)$ as a *cone integral*, in the spirit of (7) (as long as $p \nmid f$). An important feature of this procedure is that the 'cone conditions' defining this integral are now *independent of the prime p*. We shall not prove these assertions, and refer the reader to [13]. We can however read off a number of consequences (all proved in [13]).

Theorem 4 gives the following: there exist finitely many rational functions $P_I(X, Y)$ over $\mathbb{Q}$ such that for almost all primes p,

$$\zeta_{c,d,p}(s) = \sum_I c_{p,I} P_I(p, p^{-s}),$$

where the coefficients $c_{p,I}$ represent the number of $\mathbb{F}_p$-points of certain (boolean combinations of) varieties defined over $\mathbb{Z}$. This implies that the rational functions in p^{-s} representing $\zeta_{c,d,p}(s)$ have numerators and denominators of bounded degrees as p varies over all primes p, and so yields the following refinement of Theorem 9:

Theorem 12. *Let c, d be positive integers. Then there exist positive integers k and l such that for each prime p, the function $n \mapsto f(n, p, c, d)$ satisfies for $n \geqslant k$ a linear recurrence relation of length l with constant coefficients.*

Since $f(n, p) = f(n, p, n-1, n)$, we can also deduce

Theorem 13. *Let $n \in \mathbb{N}$. There exist a variety Y_n defined over $\mathbb{Q}$, finitely many subvarieties $E_{i,n}$ $(i \in T(n))$ of Y_n, and for each $I \subseteq T(n)$ a polynomial $H_{n,I}(X)$, such that for almost all primes p*

$$f(n, p) = \sum_{I \subseteq T} e_{n,p,I} H_{n,I}(p)$$

where

$$e_{n,p,I} = \left|\{a \in \overline{Y_n}(\mathbb{F}_p) : a \in \overline{E_{i,n}} \text{ if and only if } i \in I\}\right|.$$

Thus Higman's PORC conjecture reduces to the question of how the 'solution numbers' $e_{n,p,I}$ vary with the prime p. Because the cone conditions that arise in this context are reasonably straightforward, there is some hope of answering this question by analysing the varieties obtained in the resolution of singularities. (It may be more practicable to do this rather for the function $\zeta_{F_p}^{\triangleleft}$, to show that this is uniform in p, as conjectured in [27] and discussed in the preceding section, and separately to show that the factor $\left|\mathfrak{G}(\mathbb{Z}_p) : \mathrm{N}_{\mathfrak{G}(\mathbb{Z}_p)}(N)\right|$, for normal subgroups N of index p^n in F_p, exhibits 'PORC' behaviour.)

The final application concerns the number of *nilpotent groups* of given order. Let $g(n, c, d)$ denote the number of isomorphism types of d-generator nilpotent groups of order n and class at most c. Since

$$g(n, c, d) = \prod f(n_p, p, c, d)$$

where $n = \prod p^{n_p}$, we have an Euler product

$$\sum_{n=1}^{\infty} g(n, c, d) n^{-s} = \prod_p \zeta_{c,d,p}(s).$$

So Theorem 5 with Proposition 1 (Section 5) gives

Theorem 14. *Let c, d be positive integers. Then there exist $\alpha \in \mathbb{Q}$, an integer $b \geqslant 0$ and $\gamma \in \mathbb{R}$ such that*

$$\sum_{i=1}^{n} g(i, c, d) \sim \gamma n^{\alpha} (\log n)^b.$$

7 Counting p-groups. II: Conjecture P

The classification theory of finite p-groups took a dramatically new direction in 1980, with the publication of Leedham-Green and Newman's *coclass conjectures* [37]. Their guiding insight was that while it is hopeless to classify p-groups of fixed nilpotency class c, because the classification rapidly gets more and more complicated as c grows, it is fruitful to classify the groups of fixed *coclass*: indeed, as it turns out, a similar pattern appears for each value of the coclass. The proof of all the conjectures in the succeeding 14 years (completed in [36] and [45]) has created a rich theory which in turn gives rise to many intriguing questions. Before explaining one of these, we refer the reader to Chapter 2 for the statement of the original conjectures, and to Chapter 10 of [9] for a self-contained proof of some of them.

Throughout this section, we keep fixed a prime p and a positive integer r. A group of order p^n and nilpotency class c is said to have *coclass* $r = n - c$; a pro-p group has coclass r if it is the inverse limit of an inverse system of finite p-groups of coclass r in which all the maps are surjective. The set of all isomorphism types of finite p-groups of coclass r is described by a directed graph $\mathcal{G} = \mathcal{G}(r, p)$, having a group at each vertex and a directed edge $Q \to P$ whenever $P \cong Q/\gamma_c(Q)$ where c is the nilpotency class of Q. One calls Q a *descendant* of P if there is a directed path from Q to P, which holds just when P is a quotient of Q. Now to each infinite pro-p group G of coclass r we associate the set $\mathcal{Q}_G$ consisting of those $P \in \mathcal{G}$ such that P is a quotient of G, but of no other infinite pro-p group of coclass r. The associated *family* $\mathcal{F}_G$ is then the subgraph of $\mathcal{G}$ consisting of $\mathcal{Q}_G$ and all descendants of members of $\mathcal{Q}_G$.

The main results of the theory imply the following:

(i) the graph $\mathcal{G}$ contains only finitely many families, and their union has finite complement in $\mathcal{G}$;

(ii) each family is a tree;

(iii) each tree $\mathcal{F} = \mathcal{F}_G$ contains a unique maximal infinite reverse-directed path (the *trunk*) $\mathcal{F}^* : P_1 \leftarrow P_2 \leftarrow \cdots \leftarrow P_n \leftarrow \cdots$, where $|P_n| = p^{n+e}$ for some fixed e. Here $\mathcal{Q}_G = \{P_1, P_2, \dots, \}$ and $G \cong \varprojlim P_i$; we shall denote this group $G(\mathcal{F})$.

The question then arises as to what is the shape of these trees? On the basis of extensive computer calculations, Newman and O'Brien [42] have made a new series of conjectures in answer to this question. Among these is

Conjecture P. *If $p = 2$, then each of the trees $\mathcal{F}$ is ultimately periodic, with period dividing 2^{r-1}.*

Denote by $\mathcal{F}_n$ the subtree consisting of all descendants of P_n, i.e., those $P \in \mathcal{F}$ such that the path from P to P_1 contains P_n; thus $\mathcal{F}_n$ is infinite while $\mathcal{F} \setminus \mathcal{F}_n$ is finite. The conjecture asserts that there exist $k \in \mathbb{N}$ and $l \mid 2^{r-1}$ such that $\mathcal{F}_k \cong \mathcal{F}_{k+l}$. This implies in particular that the *twig length*, that is the maximal distance from any $P \in \mathcal{F}$ to $\mathcal{F}^*$, is bounded over all n. This in fact is known to be true for $p = 2$ [36]; but it is false when p is odd. To save the conjecture for odd primes, we 'prune' the trees: for each positive integer M, let $\mathcal{F}(M)$ denote the subtree consisting of those $P \in \mathcal{F}$ such that the distance from P to the trunk $\mathcal{F}^*$ is at most M (so when $p = 2$, $\mathcal{F}(M) = \mathcal{F}$ for sufficiently large M). We can now state the main result of [14]:

Theorem 15. *Let $\mathcal{F}$ be a family in $\mathcal{G}(r, p)$ and let M be a positive integer. Then the tree $\mathcal{F}(M)$ is ultimately periodic.*

For $p = 2$ this implies the qualitative part of Conjecture P, without the bound on the period. In general, it says that if one looks at the graph $\mathcal{G}(r, p)$ through a set of blinkers, and ignores a finite piece at the top, one will see a repeating finite pattern. When p is odd, however, it says nothing about the total number $c(n, p, r)$ of p-groups of coclass r and order p^n: wider blinkers may reveal ever more groups. Some information on this question is provided by the following result from [14]:

Theorem 16. *Fix p and r. Then for sufficiently large n the function $n \mapsto c(n, p, r)$ satisfies a linear recurrence relation with constant coefficients.*

Both theorems are proved by establishing the rationality of a suitable zeta function, in the spirit of Theorem 11. The feasibility of such an approach depends on the main theorem of coclass theory, namely the truth of the original Conjecture A:

Proposition 5 ([36], [45]). *There exists a positive integer $h = h(p, r)$ such that every p-group of coclass r has a normal subgroup of class at most 2 (1, if $p = 2$) and index dividing p^h.*

Now it is easy to see that every p-group of coclass r can be generated by $r + 1$ elements. It follows that every such group is a homomorphic image of the group

$$U = F/\gamma_3\left(\gamma_h(F) \cdot F^{p^h}\right),$$

where F is the free pro-p group on $r + 1$ generators. On the other hand, not every image of this group will have coclass r. The possibility of singling out those that do depends on another theorem of Shalev:

Proposition 6 ([45], Cor. 4.4). *Let P be a finite p-group. Then P has coclass r if and only if $P/\gamma_f(P)$ has coclass r, where $f = 2p^r$ if p is odd, $f = 2^{r+3}$ if $p = 2$.*

Now let $\mathcal{H}$ denote the set of all open normal subgroups H in U such that U/H has coclass r, and put $\mathcal{H}_0 = \left\{H \in \mathcal{H} \mid H \geqslant \gamma_f(U)\right\}$. Then Proposition 6 ensures that

$$\mathcal{H} = \left\{H \lhd_o U \mid H\gamma_f(U) \in \mathcal{H}_0\right\}.$$

Since $F/(\gamma_h(F) \cdot F^{p^h})$ is finite, U is a pro-p group of finite rank, hence compact and p-adic analytic. It is also a free pro-$\mathcal{C}$ group for a certain variety of groups $\mathcal{C}$. Just as in Section 6, it follows that the number $c(n, p, r)$ is equal to the number of orbits of $\mathrm{Aut}(U)$ in the set $\{H \in \mathcal{H} \mid |U : H| = p^n\}$.

Put $G = U \rtimes \mathrm{Aut}(U)$, and recall the notation used in the proof of Theorem 11. We choose the uniform normal subgroup G_1 in G so that $G_1 \cap U = U_1$ satisfies

$$U_1 \leqslant \Phi(U^{p^{r+f-1}}),$$

and note the following simple lemma:

Lemma 2. *Let $H \lhd_o U$. Then $H \in \mathcal{H}$ if and only if $U_1 H \in \mathcal{H}$.*

Proof. Note that if $U_1H \in \mathcal{H}$, then $|U : U_1H\gamma_f(U)| \leqslant p^{r+f-1}$, which implies that $U_1 \leqslant \Phi(U_1H\gamma_f(U))$ and hence that $H\gamma_f(U) = U_1H\gamma_f(U) \in \mathcal{H}_0$. Therefore $H \in \mathcal{H}$. The converse is even simpler.

We deduce that $\sum_{n=0}^{\infty} c(n, p, r)p^{-ns} = \zeta_{\text{ccl}}^{r,p}(s)$ is equal to a finite linear combination of integrals of the form $Z(f^*, k^*, \mathcal{K}_{P,Q}; s)$, where f^*, k^* are simple functions and $\mathcal{K}_{P,Q}$ is a certain definable subset of $G_1^{(t)}$ for a suitable t; recall that these integrals were grouped into finitely many sets, labelled by the subgroups K lying between U_1 and U, and corresponding to open subgroups H of U with $U_1H = K$: in the present case we merely omit all those for which $K \notin \mathcal{H}$.

As before, we conclude that $\zeta_{\text{ccl}}^{r,p}(s)$ is a rational function of p^{-s}, and Theorem 16 follows.

Rather amazingly, an argument of similar nature suffices also to prove Theorem 15; of course, in order to encode the shape of a graph in a suitable zeta function the procedure will have to be more subtle. We outline the main steps, keeping the notation introduced above.

For each n and M put

$$\tau(n) = \mathcal{F}_n \setminus \mathcal{F}_{n+1}, \quad \tau(n, M) = \tau(n) \cap \mathcal{F}(M).$$

Thus $\tau(n)$ is the 'twig' of $\mathcal{F}$ attached to the trunk at P_n, and $\tau(n, M)$ the same twig after 'pruning'. We consider $\tau(n, M)$ as a rooted tree with root P_n. Now since the pro-p group U has finite rank g, say, each of the groups $H \in \mathcal{H}$ has at most $p^g - 1$ subgroups of index p; consequently each node of the tree $\mathcal{F}$ has valency at most $p^g = N$, say. Let $\mathcal{T}(M, N)$ denote the finite set of (isomorphism types of) rooted trees in which each node has valency at most N and distance at most M from the root. Then for each n, the rooted tree $\tau(n, M)$ is isomorphic to a member of $\mathcal{T}(M, N)$. Given $\tau \in \mathcal{T}(M, N)$, put

$$c_{\tau,n} = \begin{cases} 1 & \text{if } \tau(n, M) \cong \tau \\ \\ 0 & \text{if } \tau(n, M) \ncong \tau \end{cases}.$$

As $\mathcal{T}(M, N)$ is finite, we see that Theorem 15 is equivalent to

Proposition 7. *For each $\tau \in \mathcal{T}(M, N)$, there exists n_0 such that the sequence $(c_{\tau,n})_{n \geqslant n_0}$ is periodic.*

Now consider the function

$$\begin{aligned} \zeta_{\mathcal{F},M,\tau}(s) &= |P_1|^{-s} \sum_{n=1}^{\infty} c_{\tau,n} p^{-(n-1)s} \\ &= \sum_{\tau(n,M) \cong \tau} |P_n|^{-s}. \end{aligned}$$

It is a nice little exercise to show that a bounded sequence of integers that satisfies a linear recurrence relation must be eventually periodic. As a consequence, we see that Proposition 7 is equivalent to

Proposition 8. *The function* $\zeta_{\mathcal{F},M,\tau}(s)$ *is rational in* p^{-s}.

We claim that $\zeta_{\mathcal{F},M,\tau}(s)$ is a finite linear combination of integrals of the form $Z(f^*, k^*, \mathcal{K}^*_{P,Q}; s)$ as above. The set $\mathcal{K}^*_{P,Q}$ is defined by the previously stated conditions together with an additional condition. Let us call a group $H \in \mathcal{H}$ 'mainline' if $U/H \in \mathcal{F}^*$. This means that $U/H \cong P_k$ for some k, and in that case we write $\tau(H, M) = \tau(k, M)$. Then

$$\mathcal{K}^*_{P,Q} = \left\{(\mathbf{h}, \mathbf{t}, \mathbf{x}) \in \mathcal{K}_{P,Q} \mid H = \langle h_i, t_j y_j \rangle \text{ is mainline and } \tau(H, M) \cong \tau \right\}.$$

To show that this is a definable set, we fix an epimorphism $\pi : U \to G(\mathcal{F})$ and put $V = \ker \pi$, $V_1 = \pi^{-1}(P_1)$. Then $H \in \mathcal{H}$ is mainline if and only if there exists $\alpha \in G$ such that $V_1 \supseteq H^\alpha \supseteq V$, a definable condition. Next, we specify the tree τ by giving the valencies $1 + e_{i1}, \ldots, 1 + e_{ir(i)}$ of the nodes at distance i from the root, for $i = 0, \ldots, M - 1$ (note that nodes at distance M from the root have valency 1). Now suppose that H is mainline. Then $\tau(H, M) \cong \tau$ if and only if the following holds: there exist groups $L_{ij} \in \mathcal{H}$ $(1 \leqslant i \leqslant M,\ 1 \leqslant j \leqslant r(i))$ such that (putting $L_{01} = H$)

(i) L_{ij} is not mainline for $i \geqslant 1$;

(ii) $L_{i+1,l}$ is a maximal subgroup of L_{ij} for $e_{i1} + \cdots + e_{i,j-1} < l \leqslant e_{i1} + \cdots + e_{ij}$;

(iii) for $j \neq l$ there does not exist $\alpha \in G$ such that $L^\alpha_{ij} = L_{il}$;

(iv) if $B_m < B_{m-1} < \cdots < B_1 < B_0 = H$ is a chain of groups in $\mathcal{H}$, each maximal in the next, $m \leqslant M$, and B_1 is not mainline then there exists $\alpha \in G$ such that $B^\alpha_m = L_{mj}$ for some j.

The condition '$A \leqslant B$' is easily expressible in the language $\mathcal{L}_{U_1}$, in terms of bases for the open subgroups A and B of U . Note however that we have to consider subgroups A contained in H, but not necessarily satisfying $U_1 A = U_1 H$; so such an A will be specified by some *subset* $\{y_{i_j}\}$ of $\{y_1, \ldots, y_n\}$ and a basis of the form $(h_1, \ldots, h_d, t_{i_1}, \ldots, t_{i_k})$. It follows that (ii) and (iv) above are also definable in $\mathcal{L}_{G_1}$; hence $\mathcal{K}^*_{P,Q}$ is a definable set, and Proposition 8 follows as before from Theorem 2. This completes the proof of Theorem 15.

References

[1] C. J. Bushnell and I. Reiner, Zeta functions of arithmetic orders and Solomon's conjecture, *Math. Z.*, **173** (1980), 135–161.

[2] C. J. Bushnell and I. Reiner, L-functions of arithmetic orders and asymptotic distribution of ideals, *J. Reine Angew. Math.*, **327** (1981), 156–183.

[3] J. W. S. Cassels and A. Fröhlich, *Algebraic Number Theory*, Academic Press, London, 1967.

[4] J. Denef, The rationality of the Poincaré series associated to the p-adic points on a variety, *Invent. Math.*, **77** (1984), 1–23.

[5] J. Denef, On the degree of Igusa's local zeta function, *Amer. J. Math.*, **109** (1987), 991–1008.

[6] J. Denef, On the evaluation of certain p-adic integrals, *Séminaire de théorie des nombres, Paris 1983-84, Progress in Math.*, **59** (1985), 25–47.

[7] J. Denef and D. Meuser, A functional equation of Igusa's local zeta function, *Amer. J. Math.*, **113** (1991), 1135–1152.

[8] J. Denef and L. van den Dries, p-adic and real subanalytic sets, *Annals of Math.*, **128** (1988), 79–138.

[9] J. D. Dixon, M. P. F. du Sautoy, A. Mann and D. Segal, *Analytic pro-p groups*, 2nd edition, Cambridge studies in Advanced Math. 61, Cambridge Univ. Press, Cambridge, 1999.

[10] M. P. F. du Sautoy, Finitely generated groups, p-adic analytic groups and Poincaré series, *Annals of Math.*, **137** (1993), 639–670.

[11] M. P. F. du Sautoy, Zeta functions of groups and rings: uniformity, *Israel J. Math.*, **86** (1994), 1–23.

[12] M. P. F. du Sautoy, Zeta functions and counting finite p-groups, *Electronic research announcements, Amer. Math. Soc.*, **5** (1999), 112–122.

[13] M. P. F. du Sautoy, Counting p-groups and nilpotent groups, *preprint*.

[14] M. P. F. du Sautoy, p-groups, coclass and model theory: Conjecture P, *preprint*.

[15] M. P. F. du Sautoy, Natural boundaries for zeta functions of groups, *preprint*.

[16] M. P. F. du Sautoy, Functional equations and uniformity for local zeta functions of algebraic groups, (in preparation).

[17] M. P. F. du Sautoy and F. J. Grunewald, Analytic properties of Euler products of Igusa-type zeta functions and subgroup growth of nilpotent groups, *C.R. Acad. Sci. Paris, Série 1*, **329** (1999), 352–356.

[18] M. P. F. du Sautoy and F. J. Grunewald, Analytic properties of zeta functions and subgroup growth, *MPI preprint*, 99–87.

[19] M. P. F. du Sautoy and F. J. Grunewald, Counting subgroups of finite index in nilpotent groups of class two, (in preparation).

[20] M. P. F. du Sautoy and F. J. Grunewald, Uniformity for 2 generator free nilpotent groups, (in preparation).

[21] M. P. F. du Sautoy and F. J. Grunewald, Zeta functions of classical groups and their friendly ghosts, *C. R. Acad. Sci. Paris Série 1*, **327** (1998), 1–6.

[22] M. P. F. du Sautoy and F. J. Grunewald, Zeta functions of groups: zeros and friendly ghosts, *MPI preprint*, 99–110.

[23] M. P. F. du Sautoy and A. Lubotzky, Functional equations and uniformity for local zeta functions of nilpotent groups, *Amer. J. Math.*, **118** (1996), 39–90.

[24] M. P. F. du Sautoy, J. J. McDermott and G. C. Smith, Zeta functions of crystallographic groups and analytic continuation, *Proc. London Math. Soc.*, **79** (1999), 511–534.

[25] M. D. Fried and M. Jarden, *Field Arithmetic*, Springer-Verlag, Berlin–Heidelberg, 1986.

[26] F. J. Grunewald and D. Segal, Reflections on the classification of torsion-free nilpotent groups, In *Group theory: Essays for Philip Hall*, eds. K. W. Gruenberg and J. E. Roseblade, Academic Press, London, 1984.

[27] F. J. Grunewald, D. Segal and G. C. Smith, Subgroups of finite index in nilpotent groups, *Invent. Math.*, **93** (1988), 185–223.

[28] P. Hall, Nilpotent groups, In *Collected works of Philip Hall*, Clarendon Press, Oxford, 1988.

[29] K. Hey, Analytische Zahlentheorie in systemen hyperkomplexer Zahlen, Ph.D. thesis, Hamburg (1929).

[30] G. Higman, Enumerating p-groups, I, *Proc. London Math. Soc.*, **10** (1960), 24–30.

[31] G. Higman, Enumerating p-groups, II, *Proc. London Math. Soc.*, **10** (1960), 566–582.

[32] J.-I. Igusa, *Lectures on Forms of Higher Degree*, Tata Inst. Fund. Research, Bombay, 1978.

[33] I. Ilani, Analytic pro-p groups and their Lie algebras, *J. Alg.*, **176** (1995), 34–58.

[34] K. Ireland and M. Rosen, *A Classical Introduction to Modern Number Theory*, 2nd edition, Graduate Texts in Mathematics, 84, Springer-Verlag, New York, Berlin, Heidelberg, 1993.

[35] S. Lang and A. Weil, Numbers of points of varieties in finite fields, *Amer. J. Math.*, **76** (1954), 819–827.

[36] C. R. Leedham-Green, The structure of finite p-groups, *J. London Math. Soc.*, **50** (1994), 49–67.

[37] C. R. Leedham-Green and M. F. Newman, Space groups and groups of prime power order I, *Arch. Math.*, **35** (1980), 193–202.

[38] A. Lubotzky, A. Mann and D. Segal, Finitely generated groups of polynomial subgroup growth, *Israel J. Math.*, **82** (1993), 363–371.

[39] A. J. Macintyre, On definable subsets of p-adic fields, *J. Symbolic Logic*, **41** (1976), 605–610.

[40] A. J. Macintyre, Rationality of p-adic Poincaré series: uniformity in p, *Annals Pure Appl. Logic*, **49** (1990), 31–74.

[41] W. Narkiewicz, *Number Theory*, World Scientific Publishing Co, 1983.

[42] M. F. Newman and E. O'Brien, Classifying 2-groups by coclass, *Trans. Amer. Math. Soc.*, **351** (1999), 131–169.

[43] D. Segal, *Polycyclic Groups*, Cambridge Univ. Press, Cambridge, 1983.

[44] D. Segal and A. Shalev, Profinite groups with polynomial subgroup growth, *J. London Math. Soc.*, **55** (1997), 320–334.

[45] A. Shalev, The structure of finite p-groups: effective proof of the coclass conjectures, *Invent. Math.*, **115** (1994), 315–345.

[46] C. C. Sims, Enumerating p-groups, *Proc. London Math. Soc.*, **15** (1965), 151–166.

[47] G. C. Smith, Zeta functions of torsion free finitely generated nilpotent groups, Ph.D. thesis, (Manchester, UMIST 1983).

DPMMS, Centre for Mathematical Sciences
Wilberforce Road
Cambridge CB3 0WB
dusautoy@dpmms.cam.ac.uk
url: http://www.dpmms.cam.ac.uk/~dusautoy

All Souls College
Oxford OX1 4AL
UK
dan.segal@all-souls.ox.ac.uk

10

Where the Wild Things Are: Ramification Groups and the Nottingham Group

Marcus du Sautoy and Ivan Fesenko

1 Introduction

Group theorists have recently been interested in a pro-p group known as the Nottingham group. This group $\mathcal{N}(\mathbb{F}_p)$ is defined as the set of power series $t + a_2t^2 + \cdots \in t\mathbb{F}_p[[t]]$, the group operation being substitution of one power series in another. Interest in this group reached a peak with the proof by Rachel Camina of the following:

Theorem 1.1. ([3]). *$\mathcal{N}(\mathbb{F}_p)$ contains every countably-based pro-p group as a closed subgroup.*

Number theorists, on the other hand, have known this group for some time as the group of so-called *wild automorphisms* of the field $\mathbb{F}_p((t))$. It is important in the theory of ramification subgroups because every Galois group of a totally ramified local field extension with residue field $\mathbb{F}_p$ in which the ramification subgroups are open and can be embedded into $\mathcal{N}(\mathbb{F}_p)$ via a functor called the *field of norms*. Field extensions of this kind are called *arithmetically profinite*.

Clearly, in retrospect, there had to be some relationship between the two approaches. However the two camps—number theory and group theory—were not aware of their mutual interest in these issues, until an L.M.S. Research Symposium on 'Pro-p Groups and Related Topics' brought them together in 1997. One of its aims was to stress in particular the related topic of number theory, in the hope that the two camps might share such questions and answers. This chapter tells one of the stories of that meeting: how connections between the Nottingham Group and ramification theory came to light and generated answers on both sides of the divide.

It transpired that Camina in the course of her proof had almost established that every countably based pro-p group can be realised as the Galois group of an arithmetically profinite extension in characteristic p. Conversely, the field of norms construction provided an alternative way to complete Camina's proof, and this in turn yielded more sensitive information about the nature of the embeddings. We explain this new approach in Section 7 and also explain the key step in Camina's proof in the language of ramification subgroups.

Another theme that grew out of this meeting relates to an old theorem of Shankar Sen and a new solution to a problem of Coates and Greenberg. Both concern the question of which pro-p groups can occur as Galois groups of arithmetically profinite extensions in characteristic 0 (in contrast to Camina's setting of characteristic p). For many years the only result was the following:

Theorem 1.2. ([22]). *Let L/K be a totally ramified extension of local fields of characteristic zero whose Galois group is a p-adic analytic group. Then L/K is an arithmetically profinite extension.*

We present in Section 8 Sen's proof of this theorem, taking advantage of the new language for the theory of p-adic analytic groups introduced in [5]; we hope that this will make it accessible to the audiences—both group-theoretic and number-theoretic—that it deserves. It is an interesting theorem using a delicate interplay between number-theoretic ideas from local class field theory and group-theoretic ideas on the structure of p-adic analytic groups.

Coates and Greenberg [4] have recently extended ideas of Tate to show that arithmetically profinite extensions in characteristic zero have a good Kummer theory. In fact they established results for a more general class of field extensions in characteristic 0, the *deeply ramified* ones, where the ramification subgroups are non-trivial, while not necessarily open.

Prior to a meeting in Oberwolfach in the summer of 1994, the only known examples of arithmetically profinite or deeply ramified extensions in characteristic zero were those provided by Sen's theorem. Coates and Greenberg asked if there were other examples. In [6] (2.5) the first example was provided of an arithmetically profinite field extension whose Galois group is not p-adic analytic.

As we shall see, ramification subgroups are preserved on going to quotients. So by Sen's theorem, if a field extension has a Galois group with an infinite p-adic analytic quotient, it will be deeply ramified though not necessarily arithmetically profinite. The example in [6] (2.5) does in fact have p-adic analytic quotients, although it is not p-adic analytic itself. Coates and Greenberg were therefore interested in the following:

The Coates–Greenberg Problem. Does there exist a totally ramified and deeply ramified Galois extension L/F such that F is a finite extension of $\mathbb{Q}_p$ and the Galois group of L/F does not have any infinite p-adic analytic quotients?

Using an analysis of some natural closed subgroups of the Nottingham group it was possible to provide a positive answer to this question. The following theorem is proved in [7] and briefly discussed in Section 9:

Theorem 1.3. ([7]). *Let $q = p^r$ where $r > 1$ and define the following closed subgroup of the Nottingham group $\mathcal{N}(\mathbb{F}_p)$:*

$$T = \left\{ \sum_{i \geqslant 0} a_i X^{1+qi} : a_0 = 1, a_i \in \mathbb{F}_p \right\}.$$

Then (i) T has no infinite p-adic analytic quotients. (ii) Let F be an unramified extension of $\mathbb{Q}_p$ of degree $\geqslant q$. Then there exists a Galois extension L/F with Galois group isomorphic to T, and any such extension is arithmetically profinite.

Corollary 1.4. *The Coates-Greenberg Problem has a positive answer.*

The chapter begins by introducing the concepts of ramification subgroups and arithmetically profinite extensions. As soon as a pro-p group is the Galois group of some field extension of local fields, the filtration provided by the ramification subgroups becomes a very natural issue. When a finite group G is the Galois group of an extension L/F of local fields, a natural filtration on G is obtained by setting $G[n]$ to be the subgroup consisting of automorphisms that act trivially modulo the nth power of the maximal ideal. For group theorists this is a natural filtration very reminiscent of congruence subgroups in algebraic groups. These subgroups are called the *ramification subgroups in the lower numbering*, and their properties are explained in Section 3.

The curious fact is that this filtration does not behave well on going to quotients: this is why ramification subgroups are not as straightforward a subject as the above definition of $G[n]$ suggests. The problem is that if one ascends a tower of fields corresponding to an inverse system of finite p-groups, the effect of an automorphism modulo the maximal ideal at one point of the tower might be very different to its restriction at another point of the tower; this means that it is impossible, in general, to define a corresponding filtration for an infinite pro-p Galois group. We try to illustrate the problem at the beginning of Section 4 by considering the case of a tower of field extensions of formal power series $\mathbb{F}_p((T_L)) \geqslant \mathbb{F}_p((T_E)) \geqslant \mathbb{F}_p((T_K))$ (where T_L, T_E and T_K denote indeterminates, T_E being identified with an element of $\mathbb{F}_p((T_L))$, etc.).

All is not lost though. By taking the same filtration but reindexing it using a function $h_{L/F}$ attached to each field extension L/F—this is called the *Hasse–Herbrand function*—it is possible to define a new filtration $G(x) = G[h_{L/F}(x)]$ (where x now runs over rational as well as integer values) which does behave well on going to quotients. Essentially the Hasse–Herbrand function corresponding to each section of a tower tells us how the action of an automorphism changes with respect to the maximal ideals in the tower. The subgroups $G(x)$ with this new indexing are called *ramification subgroups in the upper numbering*. Note that the *groups* in the upper and lower numbering remain the same; it is just the *indexing* which changes.

This is all very well, but when we take the inverse limit of these groups to define the new filtration of ramification subgroups for our infinite pro-p group, it may be that these ramification subgroups are not of finite index, i.e., not open subgroups, because of the way we have had to reindex as we ascend the tower. Therefore we aren't getting a nice 'congruence' filtration for the infinite field. So it now becomes important to know whether in a particular infinite field extension, the ramification subgroups with this new indexing 'settle down' as we ascend a tower of finite subextensions. A similar issue was crucial in Camina's proof. As Camina built her tower of fields $\bigcup_{i\in\mathbb{N}} \mathbb{F}_p((T_i))$, it was important to ensure that the automorphisms had essentially the same effect modulo T_i as $i \to \infty$.

However there is an additional arithmetical significance of the Hasse–Herbrand function: it also relates how the filtrations of principal units behave as we ascend a tower of fields. This gives particular field-theoretic significance to the situation where the subgroups $G(x)$ are *open*, for every $x \in \mathbb{R}$. When this is the case we call the extension *arithmetically profinite*.

Camina's construction does not always produce arithmetically profinite extensions but something a bit weaker which was sufficient for her proof. However we shall see that it is possible to use Camina's argument to make the tower also arithmetically profinite.

These are all rather vague statements; precise details are given in Section 4.

The bridge between the ramification theory of arithmetically profinite extensions and the Nottingham group is provided by the *field of norms* functor, introduced in Section 5. To make this construction, one needs the extra information about the relationship between the filtrations of principal units provided by the condition that $G(x)$ is open for all x. What this functor does is to associate with an arithmetically profinite field extension L/F another field $N(L/F)$, called the *field of norms*, which essentially looks like $\mathbb{F}_p((T))$ if the residue field of L is $\mathbb{F}_p$. The functor naturally embeds the Galois group of the arithmetically profinite extension into the group of wild automorphisms of $N(L/F)$ – that is, into the Nottingham group when $N(L/F)$ is $\mathbb{F}_p((T))$. So although L is not a local field, the automorphisms of L/F can be made to act on a local field.

An intriguing feature of the construction is its close analogy to the intrinsic construction of the Lie algebra of a p-adic analytic group, as we shall see in Section 5.

This chapter plays the usual balancing act of any survey paper, hopefully providing enough details to get the reader up and running and whetting the appetite for further investigation, but not of course being able to provide proofs of all the statements that we quote.

Those frustrated by this or interested in more detainls are encouraged to consult the book by Fesenko and Vostokov [8] or the book by Serre [23], and the paper of Wintenberger [27], where a detailed and comprehensive introduction to the theory of ramification subgroups can be found, covering the material of Sections 2 to 5.

Lastly, a note on the nomenclature of $\mathcal{N}$: the group of wild automorphisms of a local field, the wild group, the Nottingham group, York's group, Snottingham group (the original name of the city of Nottingham according to some), first group of Cartan type (after its connections with the Witt Lie algebra in characteristic p, the first Lie algebra of Cartan type), the Paris or Limoges group (since Serre, Wintenberger and Laubie studied it well before group theorists got interested), the Richard group (after King Richard the Lionheart who had well known friends and enemies in Nottingham and whose life was ended near Limoges). Take your pick!

The first author would like to thank John Coates for a Part III course in Cambridge which first demystified Sen's theorem, Shankar Sen for his series of lectures in Durham and the second author for patiently explaining the subtleties of local fields. The second author would like to acknowledge the active role of the first author who wrote almost all the text of this chapter in eloquent English.

2 Local fields

In this section we recall some of the basic definitions relating to local fields.

Let F be a field and let $v : F^* \to \mathbb{Z}$ be a surjective homomorphism such that $v(\alpha + \beta) \geqslant \min(v(\alpha), v(\beta))$ for every $\alpha, \beta \in F^*$. Put $v(0) = +\infty$. Then v is called a *discrete valuation* on F and F is called a *discrete valuation field.*

Define the ring of integers $\mathcal{O} = \{\alpha \in F : v(\alpha) \geqslant 0\}$ and the maximal ideal $\mathcal{M} = \{\alpha \in F : v(\alpha) > 0\}$. The field F is a topological field with respect to the $\mathcal{M}$-adic topology which has powers of $\mathcal{M}$ as a fundamental system of open neighbourhoods of zero. F is called a *complete discrete valuation field* if it is complete with respect to the $\mathcal{M}$-adic topology.

It can be shown that the discrete valuation of a complete discrete valuation field is uniquely determined (see [8] Exercise 6 Section 2 Chapter II), so we denote by v_F, $\mathcal{O}_F$ and $\mathcal{M}_F$ the discrete valuation, ring of integers and maximal ideal of a complete discrete valuation field F. The quotient field $\mathcal{O}_F/\mathcal{M}_F = k_F$ is called the residue field of F. We always assume that the residue fields are perfect fields of characteristic p. A complete discrete valuation field with a finite residue field is called a *local field.* A generator $\pi = \pi_F$ of the maximal ideal $\mathcal{M}_F$ is called a *prime element* or *uniformiser* of F.

Let $v_p(n)$ denote the greatest power of p which divides an integer n. The completion of $\mathbb{Q}$ with respect to the discrete valuation $v_p(m/n) = v_p(m) - v_p(n)$ is the field of p-adic numbers $\mathbb{Q}_p$. The completion of the field of rational functions $\mathbb{F}(X)$ over a finite field $\mathbb{F}$ of characteristic $p \neq 0$ with respect to the discrete valuation $v(a_nX^n + \text{terms of higher order}) = n$ if $a_n \neq 0$ is the field of ('finite-tailed') Laurent series $\mathbb{F}((X))$. These two examples are the most important classes of local fields.

In fact they are essentially the only examples. In characteristic zero, a local field is a finite extension of $\mathbb{Q}_p$. While in characteristic p, a local field F is isomorphic to $\mathbb{F}((T))$ where $\mathbb{F}$ is the finite residue field. We shall generally use the notation that $T_F = T$, a uniformiser for the local field F in characteristic p.

If L is a finite extension of a complete discrete valuation field F, then the discrete valuation v_F can be uniquely extended to a discrete valuation v_L of L and L is complete with respect to v_L (see [23], [8] Section 2 Chapter II). The residue field k_F can be identified with a subfield of k_L. The number $e(L/F) = v_L(\pi_F)$ is called the *ramification index* of L over F. The degree $f(L/F)$ of the extension k_L/k_F is called the *inertia degree* or *residue degree*; we also have $f(L/F) = v_F(N_{L/F}(\pi_L))$ where $N_{L/F}$ is the norm map from L to F. Their product $e(L/F)f(L/F)$ is equal to the degree of L/F.

For a complete discrete valuation field F the finite or infinite number $e(F) = v_F(p)$ is called the *absolute ramification index* of F.

A finite extension L/F is called *unramified* if $e(L/F) = 1$; this means that a prime element π_F of F remains prime in L. At the other extreme if $f(L/F) = 1$ we say that the extension L/F is *totally ramified.* This happens if and only if $N_{L/F}(\pi_L)$ is a prime element of F.

Fix a separable closure F_s of F. The compositum $\tilde{F}$ of all unramified extensions of F inside F_s is called the maximal unramified extension of F. It is a discrete valuation field, with residue field isomorphic to the separable closure of k_F. For a separable extension L/F, the subfield $L \cap \tilde{F}$ is called the *inertia subfield* of L/F. A separable extension L/F is called *unramified* if its inertia subfield coincides with L and *totally ramified* if the inertia subfield coincides with F. If L/F is totally ramified, then $\mathcal{O}_L$ is generated as an $\mathcal{O}_F$-module by π_L, i.e., $\mathcal{O}_L = \mathcal{O}_F[\pi_L]$. The inertia subfield can also be defined as the splitting field of the polynomial $X^q - X$ over F where $q = p^{f(L/F)}$.

If L is an algebraic extension of the local field F, then it is the union of all its sub-extensions of finite degree over F. Since there are only finitely many of each fixed degree n, we can write $L = \bigcup_{n\in\mathbb{N}} L_i$ as an ascending chain of finite extensions L_i with $L_{i+1} \supset L_i$.

We denote by $U_F = U_{0,F}$ the group of units of $\mathcal{O}_F$ (this is also referred to as the group of units of F); it has a filtration of *higher principal unit subgroups* $U_{i,F} = 1 + \pi_F^i\mathcal{O}_F$ for $i \geqslant 1$. These form a base of neighbourhoods of the identity for the topology on $U_{0,F}$ induced by F. Using them we can see the units as an example of a finite extension of a pro-p group $U_{1,F}$:

Lemma 2.1. *(i)* $U_{0,F} = \varprojlim U_{0,F}/U_{i,F}$.

(ii) $U_{0,F}/U_{1,F} \cong k_F^*$.

(iii) $U_{i,F}/U_{i+1,F} \cong k_F^+$ *for* $i \geqslant 1$.

Recall the following definition from [5], Chapter 1:

Definition 2.1. The *lower p-series* for a finitely generated pro-p group G is defined by $P_1(G) = G$ and $P_n(G) = P_{n-1}(G)^p[P_{n-1}(G), G]$.

The next lemma identifies the lower p-series of the pro-p group $U_{N,F}$ in characteristic 0 and for N large enough. Since $U_{N,F}$ is commutative, to identify the lower p-series entails knowing about the action of the p-power map; it is described in the following lemma. Put $U_{i,F} = U_i$.

Lemma 2.2. *Let* $n \geqslant 1$.

(i) If $n \leqslant e(F)/(p-1)$, *or if* F *has characteristic* p, *then* $U_n^p \leqslant U_{np}$.

(ii) If $n > e(F)/(p-1)$ *and* F *has characteristic zero, then* $x \mapsto x^p$ *defines an isomorphism from* U_n *onto* $U_{n+e(F)}$.

(iii) If $N > e(F)/(p-1)$ *and* F *has characteristic zero, then* $P_{n+1}(U_{N,F}) = U_{N+ne(F),F}$.

Proof. Let $\pi = \pi_F$ be a prime element of F. Let $1 + z\pi^n \in U_n$; then

$$(1 + z\pi^n)^p = 1 + \sum_{i=1}^{p-1} \binom{p}{i} z^i \pi^{ni} + z^p \pi^{np}. \tag{1}$$

If F has characteristic p, then the lemma is clear since all the middle terms vanish.

Suppose that F has characteristic zero. Then $v_F\left(\binom{p}{i} z^i\pi^{ni}\right) \geqslant v_F(p) + ni$ and $v_F\left(z^p\pi^{np}\right) \geqslant np$.

(i) In this case $(p-1)n \leqslant e(F)$ and so $v_F\left(\binom{p}{i}z^i\pi^{ni}\right) \geqslant (p-1)n+n = np$. Hence $U_n^p \leqslant U_{np}$.

(ii) We want to prove that if $pn > n+e(F)$, then for each $y = 1+a\pi^{n+e(F)} \in U_{n+e(F)}$ there is a unique $x = 1+z\pi^n \in U_n$ with the property that $x^p = y$.

In this case $pz\pi^i$ is the term which determines the behaviour of $(1+z\pi^n)^p$. Since $\mathrm{char}(F) = 0$ we can write $p = \pi^{e(F)}b$ for some unit b. Hence the behaviour of $(1+z\pi^n)^p$ is determined by $zb\pi^{n+e(F)}$. So we can find an element $x_1 = 1+z_1\pi^n \in U_n$ such that $y_1 = yx_1^{-p} \in U_{n+e(F)+1}$. Continue for y_1 and use completeness of F.

Part (iii) then follows from (ii).

We recall another definition from [5], Chapter 4:

Definition 2.2. Let G be a finitely generated pro-p group.

(i) G is called *powerful* if $G^p \geqslant [G,G]$ and $p > 2$ or $G^4 \geqslant [G,G]$ and $p = 2$.

(ii) G is called *uniform* if G is powerful and the sections $P_i(G)/P_{i+1}(G)$ $(i \geqslant 1)$ of the lower p-series all have the same size p^d. The *dimension* of G is then defined to be d.

So in characteristic zero, for $N > e(F)/(p-1)$, U_N is a uniform pro-p group, and its dimension is

$$\log_p |U_N : U_{N+e(F)}| = \log_p |k_F|^{e(F)} = \deg(F/\mathbb{Q}_p).$$

Hence $U_N \cong \mathbb{Z}_p^{\deg(F/\mathbb{Q}_p)}$. On the other hand, in characteristic p the group $U_1 = 1+T\cdot\mathcal{O}_F$ is not finitely generated; e.g., when $F = \mathbb{F}_p((T))$ we have $U_1^p = 1+T^p\cdot\mathbb{F}_p[[T^p]]$.

3 Ramification groups: the lower numbering

Let L/F be a *finite* Galois extension of local fields with Galois group G. In this section we define a natural filtration on G that was introduced and studied by David Hilbert and his students almost ninety years ago. Let $\sigma \in G$. Then for each $n \geqslant 0$, the automorphism σ induces a ring automorphism

$$\sigma_n : \mathcal{O}_L/(\mathcal{M}_L)^{n+1} \to \mathcal{O}_L/(\mathcal{M}_L)^{n+1}.$$

So we have a homomorphism $\sigma \mapsto \sigma_n$ of G into the group of automorphisms of the ring $\mathcal{O}_L/(\mathcal{M}_L)^{n+1}$.

Definition 3.1. Let $G[n]$ denote the kernel of the map $\sigma \to \sigma_n$:

$$G[n] = \left\{\sigma \in G : \sigma(x) \equiv x \bmod (\mathcal{M}_L)^{n+1} \text{ for all } x \in \mathcal{O}_L\right\}.$$

The groups $G[n]$ $(n \geqslant -1)$ are called the *(higher) ramification groups in the lower numbering*. $G[1]$ is the *group of wild automorphisms* of L/F

We should stress at once that this filtration depends on the field L and not just on the group. It is possible for a group G to be the Galois group of two different field extensions, giving rise to two different filtrations.

Proposition 3.1. *(i) The sequence $G[n]$ is a chain of normal subgroups with $G[-1] = G$ and $G[m] = 1$ for m sufficiently large.*

(ii) $G[0] = \mathrm{Gal}(L/F_0)$ where F_0 is the inertia field of the extension L/F.

(iii) If $F \subset M \subset L$ and $H = \mathrm{Gal}(L/M)$ then $H[n] = H \cap G[n]$.

(iv) For each $n \geqslant 0$ the map $\sigma \mapsto (\sigma(\pi_L)/\pi_L)\, U_{n+1,L}$ induces an injective homomorphism

$$G[n]/G[n+1] \to U_{n,L}/U_{n+1,L}.$$

Proof. (i) If $\sigma \neq 1$, then there exists $x \in \mathcal{O}_L$ with $\sigma(x) \neq x$ and so $\sigma(x) \not\equiv x \bmod (\mathcal{M}_L)^{n+1}$ for n sufficiently large. Since G is finite, $G[m] = 1$ for all m sufficiently large.

(ii) Let k_0 denote the residue field of F_0. We have a commutative diagram:

$$\begin{array}{ccc} \mathrm{Gal}(L/F) & \xrightarrow{\alpha} & \mathrm{Gal}(k_L/k_F) \\ \beta \downarrow & & \downarrow \\ \mathrm{Gal}(F_0/F) & \longrightarrow & \mathrm{Gal}(k_0/k_F). \end{array}$$

The horizontal maps are induced by the map $\sigma \mapsto \sigma_0$ since σ_0 fixes k_F. The vertical maps are defined by restriction. Since L/F_0 is totally ramified, $k_L = k_0$. The alternative definition of F_0 as the splitting field of $X^q - X$ over F gives us an isomorphism

$$\mathrm{Gal}(F_0/F) \longrightarrow \mathrm{Gal}(k_0/k_F)$$

(see [10] Proposition 2.11). Hence

$$G[0] = \ker(\alpha) = \ker(\beta) = \mathrm{Gal}(L/F_0).$$

(iii) This is clear from the definition of $G[n]$.

(iv) Let π be any prime element of L. Then $\sigma \in G_n$ implies that $\sigma(\pi)\pi^{-1} \equiv 1 \bmod \mathcal{M}_L^n$, i.e., $\sigma(\pi)\pi^{-1} \in U_{n,L}$. If π' is another prime element then $\sigma(\pi)\pi^{-1} \equiv \sigma(\pi')\pi'^{-1} \bmod U_{n+1,L}$. Hence the map $G_n \to U_{n,L}/U_{n+1,L}$ given by $\sigma \mapsto \sigma(\pi)\pi^{-1} \bmod U_{n+1,L}$ is well defined.

It is easy to show that the latter is in fact a homomorphism with kernel G_{n+1}. See [10] Proposition 2.19 for details.

Combining Proposition 3.1 (iv) and Lemma 2.1 we can deduce some of the group-theoretic structure of $G = \mathrm{Gal}(L/F)$.

Corollary 3.2. *(i) $G/G[0]$ is cyclic of order $f(L/F)$.*

(ii) $G[0]/G[1]$ is a cyclic group of order dividing $p^{f(L/F)} - 1$.

(iii) For each $n \geqslant 1$, $G[n]/G[n+1]$ is an elementary abelian p-group of rank bounded by $f(L/F)$.

(iv) G is a soluble group.

4 Ramification groups: the upper numbering

In order to define ramification subgroups for an infinite Galois extension, we have to consider a new filtration (or rather the old filtration with a new numbering), defined in this section.

In Proposition 3.1(iii) we saw that the filtration $G[n]$ behaves well on restriction to subgroups. This is not the case when we consider quotients. In general if H is a normal subgroup of G, then $(G/H)[n] \neq G[n]H/H$.

Let us try to illustrate the problem. Suppose we have a Galois extension L/K with a normal intermediate field E such that G is the Galois group of L/K, H is the Galois group of L/E and G/H is the Galois group of E/K. We are going to consider the special case that the fields are $L = \mathbb{F}_p((T_L))$, $E = \mathbb{F}_p((T_E))$ and $K = \mathbb{F}_p((T_K))$ of characteristic p, since the analysis (which can be found in [3]) will be useful later when we consider the Nottingham group. Let $\varphi \in G$. Then

$$\varphi(T_L) = T_L + \sum_{i \geqslant i_L(\varphi)} c_i T_L^i$$

for some $i_L(\varphi) \geqslant 1$ defined so that $c_{i_L(\varphi)} \neq 0$. This number $i_L(\varphi)$ locates the automorphism φ in terms of the filtration: $\varphi \in G[i_L(\varphi) - 1]$ but $\varphi \notin G[i_L(\varphi) - 1 + \epsilon]$ for $\epsilon > 0$. But the induced action of φ on the uniformiser T_E of E may be different:

$$\varphi|_E(T_E) = T_E + \sum_{i \geqslant i_E(\varphi|_E)} d_i T_E^i.$$

Suppose now that L/E is cyclic separable of degree p. Then we can write the uniformiser T_E in terms of T_L as follows:

$$T_E = \sum_{i=p}^{\infty} a_i T_L^i. \tag{2}$$

We can get some understanding of the relationship between $i_L(\varphi)$ and $i_E(\varphi|_E)$ by comparing the expressions for $\varphi(T_E)$, one coming from the definition of $i_E(\varphi|_E)$ and the other coming from the action on the expression for T_E in terms of the uniformiser T_L. Let us start by looking at the special case where $\varphi = g$, the generator of the cyclic Galois group $H = \mathrm{Gal}(L/E)$, in which case $i_E(g|_E) = i_E(\mathrm{Id}_E) = \infty$. Here we make the following

Definition 4.1. Let L/E be a Galois extension with cyclic Galois group $H = \langle g \rangle$ of order p. Then $i_L(g) - 1$ is called *the ramification break* of the extension L/E, and is denoted $s(L/E)$.

The name derives from the fact that $s(L/E)$ is the point at which the ramification filtration 'breaks' from H to 1, i.e., $H[s(L/E)] = H$ but $H[s(L/E) + \epsilon] = 1$ for every $\epsilon > 0$. It is an important number.

Lemma 4.1. *Let a_u be the first non-zero coefficient in the expression (2) with $u \not\equiv 0 \bmod p$. Then $u = s(L/E)(p-1) + p$.*

Proof. Note that such a u exists since otherwise T_E would be a pth power in L, i.e., L/E would be inseparable. Since $g(T_E) = T_E$ we get from (2) that

$$\sum_{i=p}^{\infty} a_i T_L^i = g\left(\sum_{i=p}^{\infty} a_i T_L^i\right) = \sum_{i=p}^{\infty} a_i \left(T_L + \sum_{j \geqslant i_L(g)} c_j T_L^j\right)^i .$$

Delete $\sum_{i=p}^{\infty} a_i T_L^i$ from both sides. The remaining terms must cancel.
As we are in characteristic p, this means that

$$0 = a_p c_{i_L(g)} T_L^{p i_L(g)} + u a_u c_{i_L(g)} T_L^{u-1} T_L^{i_L(g)} + \text{(terms of higher degree)}.$$

Since $u a_u c_{i_L(g)} \neq 0$, the two displayed terms must have the same degree, and so

$$u = (i_L(g) - 1)(p-1) + p = s(L/E)(p-1) + p.$$

Using this relationship between u and $s(L/E)$ we can show how $i_L(\varphi)$ and $i_E(\varphi|_E)$ compare in general. To simplify the notation, write $s = s(L/E)$, $i_L = i_L(\varphi)$ and $i_E = i_E(\varphi|_E)$. Then we have:

Lemma 4.2. *i) If $i_E - 1 < s$, then $i_L = i_E$.*
ii) If $i_E - 1 \geqslant s$ and $i_L - 1 \neq s$, then

$$i_L - 1 = s + p(i_E - 1 - s).$$

Proof. As in the previous lemma consider the expression for $\varphi(T_E)$:

$$\begin{aligned}
\sum_{i=p}^{\infty} a_i T_L^i + \sum_{j \geqslant i_E} d_j \left(\sum_{i=p}^{\infty} a_i T_L^i\right)^j &= T_E + \sum_{j \geqslant i_E} d_j T_E^j \\
&= \varphi(T_E) \\
&= \varphi(\sum_{i=p}^{\infty} a_i T_L^i) \\
&= \sum_{i=p}^{\infty} a_i \left(T_L + \sum_{l \geqslant i_L} c_l T_L^l\right)^i .
\end{aligned}$$

Again we delete the initial $\sum_{i=p}^{\infty} a_i T_L^i$ and compare the resulting first terms. The first term on the left hand side is $d_{i_E} a_p T_L^{p i_E}$. The right hand side contributes two possible terms: $a_p c_{i_L} T_L^{p i_L}$ and $a_u T_L^{u-1} u c_{i_L} T_L^{i_L}$. So either $p i_E = \min\{p i_L, u - 1 + i_L\}$ or $p i_L = u - 1 + i_L \leqslant p i_E$. Noting that

$$u - 1 = (p-1)(s+1)$$

by the previous lemma, we separate cases.

Case 1 Where $pi_L < u - 1 + i_L$. In this case, $i_L - 1 < s$ and

$$i_E = i_L.$$

It follows that $i_E - 1 < s$ in this case.

Case 2 Where $pi_L > u - 1 + i_L$. Then $i_L - 1 > s$ and

$$\begin{aligned} pi_E = u - 1 + i_L &= (p-1)(s+1) + i_L \\ &> p(s+1); \end{aligned}$$

this implies that $i_L - 1 = s + p(i_E - 1 - s)$, and that $i_E - 1 > s$.

Case 3 Where $pi_L = u - 1 + i_L < pi_E$. Then $i_L - 1 = s$ and $i_E \geqslant i_L$; so $i_E - 1 \geqslant s$.

Case 4 Where $pi_L = u - 1 + i_L = pi_E$. In this case we have $i_E - 1 = i_L - 1 = s$.

The lemma follows, since the hypothesis (i) holds only in Case 1, while the hypothesis (ii) holds only in Case 2.

Note that if $s = i_L - 1$ and the two terms $a_p c_{i_L} T_L^{pi_L}$ and $a_u T_L^{u-1} u c_{i_L} T_L^{i_L}$ cancel (i.e., Case 3 of the above proof), then the conclusion will in fact be false, since i_E will have a larger value than predicted by the lemma. For example, when φ is the generator of the cyclic group $\mathrm{Gal}(L/E)$, we have $s(L/E) = i_L(\varphi) - 1$, by definition, while $\varphi|_E = 1$ so that $i_E(\varphi|_E) = \infty$.

The expression $i_L(\varphi) - 1$ tells us in which term of the filtration φ lies. So if we have a Galois extension E/K, we can define a function for a totally ramified cyclic extension of degree p of E which expresses the change in the filtration in the lower numbering as we go from E/K to L/K :

$$h_{L/E}(x) = \begin{cases} x \text{ if } x \leqslant s(L/E) \\ s(L/E) + p(x - s(L/E)) \text{ if } x > s(L/E). \end{cases}$$

We have shown that if G is the Galois group of L/K and G/H is the Galois group of E/K where H, of order p, is the Galois group of L/E, then

$$(G/H)[x] = G[h_{L/E}(x)]H/H.$$

We can generalize this to arbitrary local field extensions. First for a cyclic extension L/E of prime degree l with residue fields of characteristic p, define

$$h_{L/E}(x) = \begin{cases} x & \text{if } L/E \text{ is unramified, otherwise} \\ lx & \text{if } (l, p) = 1 \\ x & \text{if } x \leqslant s(L/E) \text{ and } l = p \\ s(L/E) + p(x - s(L/E)) & \text{if } x > s(L/E) \text{ and } l = p. \end{cases}$$

For any finite Galois extension L/K, we can take a tower of intermediate fields $L > E_{m-1} > ...E_1 > K$ with E_i/E_{i-1} a cyclic extension of prime degree and define a function

$$h_{L/K} = h_{L/E_{m-1}} \circ h_{E_{m-1}/E_{m-2}} \circ \ldots \circ h_{E_1/K}.$$

It can be shown that this function is independent of the choice of cyclic filtration. Then this function, called the *Hasse–Herbrand function* of the field extension, expresses the change in the filtration of ramification subgroups in the lower numbering on going to quotients. Namely, if $L \geqslant E \geqslant K$ with G the Galois group of L/K and H the Galois group of L/E, then

$$(G/H)\,[h_{E/K}(x)] = G[h_{L/K}(x)]H/H.$$

This statement is known as *Herbrand's Theorem*. The details of the above properties of the function $h_{L/K}$ may be found in [8].

Note that contained in this statement is the fact (which we have not proved) that in the case $i_L(\varphi) - 1 = s(L/E)$, which was excluded from the previous lemma, there exists $\psi \in G$ with $\psi|_E = \varphi|_E$ but $i_L(\psi) - 1 \neq s(L/E)$ and hence by the previous lemma $i_L(\psi) - 1 = h_{L/E}(i_E(\psi|_E) - 1)$. For example when φ is a generator for the cyclic extension L/E we can take $\psi = 1$.

Using this function we can change the numbering so that the index of filtration is preserved on going to quotients:

Definition 4.2. Let G be the Galois group of a field extension L/K. Define the *(higher) ramification subgroups in the upper numbering* to be

$$G(x) = G[h_{L/K}(x)].$$

Proposition 4.3. *Let G be a Galois group with normal subgroup H. Then $(G/H)(x) = G(x)H/H$.*

From the definition of the Hasse–Herbrand function one can deduce the following (see, for instance, (3.6) Chapter III [8]):

Corollary 4.4. *The left derivative of $h_{L/K}(x)$ is equal to $|G(0) : G(x)|$ and the right derivative of $h_{L/K}(x)$ is equal to $|G(0) : G(x)|$ if $h_{L/K}(x)$ is not an integer and to $|G[0] : G[h_{L/K}(x) + 1]|$ if $h_{L/K}(x)$ is an integer.*

There is an alternative definition of the Hasse–Herbrand function, the approach taken in [23]. First we define a piecewise linear function where the gradient of each piece is determined by the order of the ramification groups in the lower numbering:

Definition 4.3. For a finite field extension L/K define $\varphi_{L/K} : [-1, \infty) \to [-1, \infty)$ to be the unique function satisfying (i) $\varphi_{L/K}(w)$ is a continuous, piecewise-linear function in w with $\varphi_{L/K}(0) = 0$ and (ii) for $w \notin \mathbb{Z}$ the derivative is $\varphi'_{L/K}(w) = g_w/g_0$, where g_w is defined to be $|G[w]|$.

The following proposition gives two more explicit expressions for the function $\varphi_{L/K}$.

Proposition 4.5. *(i) If we define $|G[0] : G[w]| = 1$ for $w \in (-1, 0]$, then*

$$\varphi_{L/K}(w) = \int_0^w \frac{dt}{|G[0] : G[t]|}.$$

(ii) If n is a positive integer with $n \leqslant w \leqslant n+1$, then

$$\varphi_{L/K}(w) = \begin{cases} w \;\; \textit{for} \; -1 \leqslant w \leqslant 0 \\ \frac{1}{g_0}(g_1 + \cdots + g_n + (w-n)g_w) \;\; \textit{for} \; w > 0 \end{cases}.$$

The alternative definition of the Hasse–Herbrand function is now given by

Definition 4.4. $h_{L/K} : [-1, \infty) \to [-1, \infty)$ is the inverse function of $\varphi_{L/K}$.

This exists because $\varphi_{L/K}$ is a strictly increasing function. Note that since $\varphi_{L/K}$ is an inverse to $h_{L/K}$, we can use it to recover the lower numbering from the upper numbering: $G[w] = G(\varphi_{L/K}(w))$. We also have the following integral expression for the Hasse–Herbrand function:

$$h_{L/K}(v) = \int_0^v |G(0) : G(t)| dt. \tag{3}$$

It is sometimes helpful to picture the graph of $h_{L/K}$. If we go back to the cyclic extension K_{i+1}/K_i, then the value of $h_{K_{i+1}/K_i}(x)$ depends on whether x lies to the left or right of the ramification break s_i: to the left we get a line of gradient 1, to the right a line of steeper gradient, as in Figure 1. The graph of $h_{K_i/K}(x)$ consists of a series of lines starting with one of gradient 1 and the others getting increasingly steeper. In Figure 2, the points $\{b_1, ..., b_{i-1}\}$ mark the *breaks* of K_i/K; see Definition 4.6.

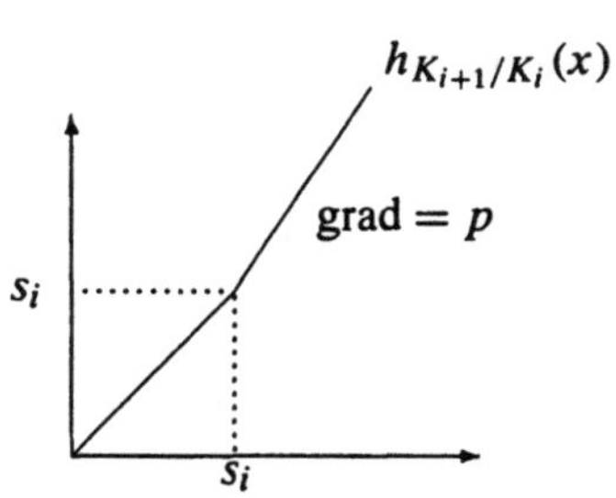

Figure 1

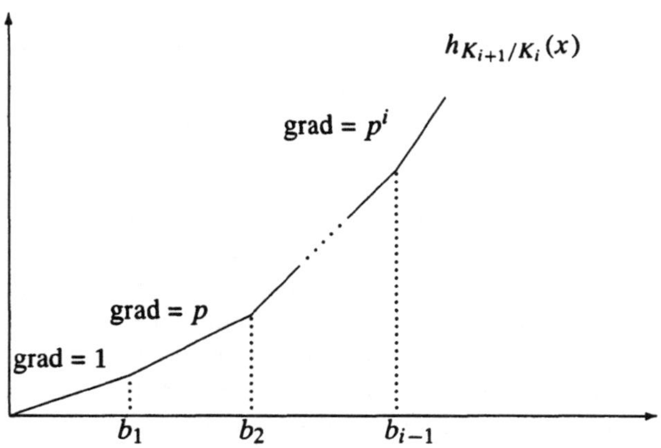

Figure 2

When we compose $h_{K_i/K}(x)$ with h_{K_{i+1}/K_i}, the portion of the graph below where the horizontal line at height s_i meets the graph will remain the same, while the rest will be bent ever steeper. We shall be concerned later with the limit of this operation as $i \to \infty$. The idea will be to focus on the situation when for every x, $h_{K_i/K}(x)$ eventually falls inside the piece which remains the same. These will be the arithmetically profinite extensions.

So far we have been concerned with finite field extensions. Consider now an infinite Galois extension L/K where L can be written as a union $\bigcup_{i\in I} K_i$ of finite Galois extensions K_i/K. The Galois group $\mathrm{Gal}(L/K)$ is defined to be the profinite group $\varprojlim \mathrm{Gal}(K_i/K)$ with the obvious connecting homomorphisms. If we want to define higher ramification subgroups for this infinite Galois group as the inverse limits of ramification subgroups of the finite subextensions, then we shall need that these ramification subgroups be preserved under going to quotients. In this way, therefore, we can only define higher ramification subgroups in the upper numbering.

Definition 4.5. Let L/K be an infinite Galois extension with Galois group G. We define for $w \in [-1, \infty)$ the ramification subgroups in the upper numbering by

$$G(w) = \varprojlim \mathrm{Gal}(K_i/K)(w)$$

where L is the union $\bigcup_{i\in I} K_i$ of finite Galois extensions K_i/K.

If we denote by $G_i(w)$ the preimage of $\mathrm{Gal}(K_i/K)(w)$ in G, then $G(w) = \bigcap_{i\in I} G_i(w)$.

Definition 4.6. The set $B(L/K)$ of *upper ramification breaks* of L/K is the set of those rational numbers x for which there is a finite Galois subextension M/K of L/K such that $\mathrm{Gal}(M/K)(x+\delta) \neq \mathrm{Gal}(M/K)(x)$ for every $\delta > 0$. Automorphisms in $\mathrm{Gal}(L/K)_1$ are called *wild automorphisms* of L/K.

Remark. It should be pointed out that this is different from the standard definition of *ramification breaks* as those x for which $\mathrm{Gal}(L/K)(x+\delta) \neq \mathrm{Gal}(L/K)(x)$ for every $\delta > 0$. It is proved in [13] that the set of ramification breaks is equal to the union of the upper ramification breaks together with the limits of decreasing sequences of the upper ramification breaks.

The ramification subgroups in the upper numbering give rather a mysterious filtration. It is not at all clear as we ascend a tower of fields which normal subgroups are going to be matched up under the Hasse–Herbrand function. So it may be that the inverse limit defines a subgroup which is not open. This is related to when the Hasse–Herbrand function has a limit. Take a tower of local fields of characteristic p, $L = \bigcup_{i\geqslant 1} \mathbb{F}_p((T_i))$ with each of degree p in the next, defining a totally ramified extension. If the ramification breaks $s_i = s(\mathbb{F}_p((T_{i+1}))/\mathbb{F}_p((T_i)))$ of the cyclic extensions are bounded, then we see that above this bound, the Hasse–Herbrand function for each cyclic extension keeps increasing without limit as we go up the tower. So the action of an automorphism may behave differently at each point of the tower. We are looking rather for a setting where the action of an automorphism settles down at some point of the tower, so that $i_{\mathbb{F}_p((T_i))}(\varphi)$ is the same for i large enough. So the action of the automorphism is determined in some finite quotient.

If s_i is an increasing sequence, as in Camina's proof, then, as we shall explain at the end of Section 7, it follows that $i_{\mathbb{F}_p((T_i))}(\varphi) = i_{\mathbb{F}_p((T_k))}(\varphi)$ where $i \geqslant k$ and $\varphi|_{\mathbb{F}_p((T_k))} \neq 1$. Hence for every automorphism φ of $L = \bigcup_{i\geqslant 1} \mathbb{F}_p((T_i))$, $i_L(\varphi) = \lim i_{\mathbb{F}_p((T_i))}(\varphi)$

is well defined. In fact as soon as $i_{\mathbb{F}_p((T_i))}(\varphi)$ is not infinite, i.e., as soon as φ acts non-trivially on some layer $\mathbb{F}_p((T_i))/\mathbb{F}_p((T_1))$, the action of the automorphism modulo the maximal ideal is determined.

As we shall see there is an added field theoretic significance to the Hasse–Herbrand function which means that it is not only important to know that *non-trivial* ramification subgroups are open, but that $G(x)$ is open for *all* values of x. Such extensions are called *arithmetically profinite field extensions*. This is equivalent to the Hasse–Herbrand function $h_{M/K}(x)$ having a limit for all x as we run over all finite subextensions M/K.

However as we shall see in Proposition 8.1, local class field theory relates the ramification subgroups in the upper numbering to the filtration of principal units. So the Hasse–Herbrand function is also giving information about the relationship between the filtrations of principal units when we restrict to a subfield. This is the approach taken in Chapter III of [8]. So if $h_{M/K}(x)$ has a limit for all x, we are capturing a more significant field theoretic setting. In particular, there is an interesting construction associated to such an extension, called the *field of norms*. The main purpose of the next section is to discuss this construction and in the subsequent section to use it to give an alternative proof that the Nottingham group contains all countably based pro-p groups as closed subgroups.

Let us illustrate why the Hasse–Herbrand function may not have a limit for all x even if $i_{\mathbb{F}_p((T_i))}(\varphi)$ has a limit for each automorphism φ. Consider our tower of local fields of characteristic p, $L = \bigcup_{i\geqslant 1} \mathbb{F}_p((T_i))$. If the Hasse–Herbrand function is going to have a limit, then it is certainly necessary that the breaks $B(L/\mathbb{F}_p((T_1)))$ of the extension be unbounded. Otherwise the graph of $h_{M/K}(x)$ keeps bending ever steeper above the supremum of the breaks. We shall see that when the ramification breaks $s_i = s(\mathbb{F}_p((T_{i+1}))/\mathbb{F}_p((T_i)))$ of each individual cyclic section are increasing then $i_{\mathbb{F}_p((T_i))}(\varphi)$ has a limit for each automorphism φ. However the following shows that this is not sufficient to imply that the breaks $B(L/\mathbb{F}_p((T_1)))$ of the tower L are unbounded:

Lemma 4.6. *Let $L = \bigcup_{i\geqslant 1} \mathbb{F}_p((T_i))$ be a tower of local fields of characteristic p, with each of degree p in the next, defining a totally ramified extension. If the ramification breaks $s_i = s(\mathbb{F}_p((T_{i+1}))/\mathbb{F}_p((T_i)))$ of the cyclic extensions are an increasing sequence then*

$$B(L/\mathbb{F}_p((T_1))) = \left\{s_1 + (s_2 - s_1)p^{-1} + \cdots + (s_i - s_{i-1})p^{-(i-1)} : i \geqslant 1\right\}.$$

Proof. To see that these are breaks, let $G_i = \mathrm{Gal}(\mathbb{F}_p((T_{i+1}))/\mathbb{F}_p((T_1)))$ and substitute $x = s_1 + (s_2 - s_1)p^{-1} + \cdots (s_i - s_{i-1})p^{-(i-1)}$ into

$$\begin{aligned} G_i(x)G_{i-1}/G_{i-1} &= G_i[h_{\mathbb{F}_p((T_{i+1}))/\mathbb{F}_p((T_1))}(x)]G_{i-1}/G_{i-1} \\ &= (G_i/G_{i-1})[h_{\mathbb{F}_p((T_i))/\mathbb{F}_p((T_1))}(x)] \\ &= (G_i/G_{i-1})[s_i]. \end{aligned}$$

Now s_i is the ramification break of $\mathbb{F}_p((T_{i+1}))/\mathbb{F}_p((T_i))$ whose Galois group is G_i/G_{i-1}. Hence $G_i(x)G_{i-1} = G_i$ but $G_i(x + \delta)G_{i-1} = G_{i-1}$. Alternatively one can understand this by composing the two graphs in Figures 1 and 2, where the breaks correspond to the kinks in the graph and the new break resulting from the composition will occur at the point that the horizontal line of height s_i meets the graph of Figure 2.

Therefore we must insist on the sequence s_i growing sufficiently fast if $B(L/K)$ is going to be unbounded. We shall explain later how Camina's proof depended only on constructing a tower as above with s_i increasing. She did not need the stronger condition that s_i grow fast enough to produce an arithmetically profinite tower. As we shall see, Camina's argument can be extended to make the s_i grow as fast as we want.

In the above lemma where the s_i are increasing, being arithmetically profinite is actually equivalent to the breaks being unbounded (see Proposition 5.3). However this is not always the case. In general the condition that the set of upper ramification breaks is unbounded is weaker than being arithmetically profinite. As we shall see later, in the case that L has finite residue field, if in addition to being unbounded the breaks are discrete, then the Hasse–Herbrand function has a limit for all x (see Proposition 5.7). Note however that if the breaks are unbounded, then it ensures still that the ramification subgroups $G(x)$ are non-trivial for all x, if not necessarily open. This is still an important field theoretic setting as explored in [4]. We therefore make the following:

Definition 4.7. A Galois extension K/F is called *deeply ramified* if its set of upper ramification breaks is unbounded (see [4], [6]).

Proposition 4.7. *A Galois extension K/F is deeply ramified if and only if the ramification subgroups $G(x)$ are non-trivial for all x where $G = \mathrm{Gal}(K/F)$.*

Proof. If $G(x)$ is trivial for some x, then the breaks are bounded by x. Conversely, suppose such a bound x exists. Then $G(y) = G(x)$ for all $y > x$. But for all normal subgroups N of finite index in G, $G(y)N = N$ for some y. Hence $G(x)N = N$ for every normal subgroup of finite index, i.e., $G(x) = 1$.

Such extensions were made famous by the work of many mathematicians including Iwasawa and Tate. Coates and Greenberg in [4] have shown that the deeply ramified field extensions are a good generalisation of $\mathbb{Z}_p$-extensions. In particular Tate's arguments generalise nicely to such field extensions and it is possible to develop Kummer theory for deeply ramified extensions.

We show in Section 8 that if the extension is a totally ramified p-adic analytic extension in characteristic zero, i.e., the Galois group is isomorphic to a p-adic analytic group, then the extension is deeply ramified. In fact it has the stronger property that the ramification subgroups are not only non-trivial but open, so it is arithmetically profinite.

In [6] (2.4) an example of a field extension is given which is deeply ramified but not arithmetically profinite. It is the compositum of a field extension whose set of breaks is bounded with an arithmetically profinite extension. Since the quotient of the ramification groups is non-trivial in the arithmetically profinite extension, the compositum must have non-trivial ramification subgroups. However the field itself cannot be arithmetically profinite since Galois sub-extensions of arithmetically profinite extensions are still arithmetically profinite, while the Galois group of the compositum maps onto a Galois group whose ramification breaks are bounded.

5 Arithmetically profinite extensions of local fields

Definition 5.1. A Galois extension L of a local field F is called *arithmetically profinite* if the upper ramification groups $\mathrm{Gal}(L/F)(x)$ are open in $\mathrm{Gal}(L/F)$ for every x.

The following proposition gives an equivalent condition for an extension to be arithmetically profinite (see [8] Section 5 Chapter III).

Proposition 5.1. *The following are equivalent:*

(1) L/F is arithmetically profinite;

(2) the inertia subfield of L/F is of finite degree over F and for every x the sequence $h_{M/F}(x)$ where M/F runs over all finite Galois subextensions in L/F has a real limit.

Proof. Note that the gradient of the graph of $h_{M/F}(x)$ is an integer for every finite extension M/F since it takes the value of the index of some ramification subgroup (see Corollary 4.4). Let L be the union of an increasing directed family of subfields L_i, which are finite extensions over F. Hence if $h_{L_i/F}(x)$ is going to have a limit, then for any $a > 0$ there exists an integer j, such that the derivative (left or right) of h_{L_i/L_j} for $x < h_{L_j/F}(a)$, $i > j$, is equal to 1. Hence the portion of the graph of $h_{L_i/F}$ below a remains the same. Hence $h_{L_i/F}(x) = h_{L_j/F}(x)$ for all $i > j$ and $x < a$. Let $G = \mathrm{Gal}(L/F)$ and let $G_i(x)$ be the preimage of $\mathrm{Gal}(L_i/F)(x)$. Then $G(x) = \bigcap G_i(x)$. But from Corollary 4.4, the index of $G_i(x)$ in G is constant (i.e., equal to the left derivative of $h_{L_j/F}$ at x) for $i > j$. So the index of $G(x)$ is finite, i.e., L/F is arithmetically profinite.

Conversely, if $G(a)$ is open, then for some j, $G_i(x) = G(x)$ for all $i > j$ and $x < a$. This implies then that h_{L_i/L_j} has derivative 1 for $x < h_{L_j/F}(a)$. Hence $h_{L_i/F}(x) = h_{L_j/F}(x)$ for all $i > j$ and $x < a$ and $h_{L_i/F}(x)$ has a limit.

Corollary 5.2. *If L/F is arithmetically profinite, then the breaks form a discrete subset.*

Proof. For each $a > 0$, the breaks below a are precisely the breaks of the finite extension L_j/F, since the graph of $h_{L_i/F}$ below a is the same as the graph of $h_{L_j/F}$. Hence the interval $(0, a)$ only contains finitely many breaks, i.e., the breaks form a discrete subset.

As we shall see at the end of this section, we can actually characterize arithmetically profinite extension as the deeply ramified extensions with discrete breaks.

When L/F is arithmetically profinite, if we define $h_{L/F}(x) = \lim h_{M/F}(x)$ as M runs over all finite extensions of F, then it is possible to also define the lower ramification subgroups for the infinite Galois group G by $G[i] = G(h_{L/F}^{-1}(i))$. The simplest example of an arithmetically profinite extension is a ramified $\mathbb{Z}_p$-extension of a finite extension of $\mathbb{Q}_p$, that is where the Galois group is isomorphic to $\mathbb{Z}_p$. As we shall see in Section 8 we can generalize this to show that all p-adic analytic extensions in characteristic zero are arithmetically profinite.

For an infinite arithmetically profinite Galois extension L/F, Fontaine and Wintenberger showed how to construct a field $N = N(L/F)$, called *the field of norms*, which was large enough to retain the faithful action of the Galois group, but small enough to be more manageable as a field extension. In particular, though L is not a local field, the Galois group of L/K can be made to act on $N = N(L/F)$ which is a local field of characteristic p; hence the Galois group embeds into the automorphism group of N, which is 'essentially' the Nottingham group (it contains the Nottingham group as an open subgroup). Therefore this is an ideal tool for an alternative proof of Camina's Theorem, as we shall explain in the next section.

Camina's own proof constructs a tower $\bigcup K_i$ which has cyclic extensions K_{i+1}/K_i with increasing breaks $s_i = s(K_{i+1}/K_i)$. This is sufficient by other means to complete her proof. However this will not in general be an arithmetically profinite extension:

Proposition 5.3. *Let $L = \bigcup K_i$ and suppose that the breaks $s_i = s(K_{i+1}/K_i)$ of each individual cyclic extension form an increasing sequence. Then the following are equivalent:*

(1) L/K is arithmetically profinite;

(2) L/K is deeply ramified;

(3) $s_1 + (s_2 - s_1)p^{-1} + \cdots + (s_i - s_{i-1})p^{-(i-1)} \to \infty$ as $i \to \infty$.

Proof. Recall from the end of section 4 that if the breaks $s_i = s(K_{i+1}/K_i)$ of each individual cyclic extension form an increasing sequence then the set of breaks $B(L/K)$ of the compositum $L = \bigcup K_i$ has the following description:

$$B(L/K) = \left\{ s_1 + (s_2 - s_1)p^{-1} + \cdots + (s_i - s_{i-1})p^{-(i-1)} : i \geqslant 1 \right\}.$$

For L/K to be deeply ramified means that this set must be unbounded. Hence (2) and (3) are equivalent. Now $h_{K_j/K}(s_1 + (s_2 - s_1)p^{-1} + \cdots + (s_j - s_{j-1})p^{-(j-1)}) = s_j \leqslant s_i$ and $h_{K_{i+1}/K_i}(x) = x$ for $x < s_i$. So $h_{K_i/K}(x) = h_{K_j/K}(x)$ for all $x < s_1 + (s_2 - s_1)p^{-1} + \cdots + (s_j - s_{j-1})p^{-(j-1)}$ and $i \geqslant j$, i.e., $h_{K_i/K}(x)$ has a limit for all $x < \sup B(L/K)$. Hence if $B(L/K)$ is unbounded, i.e., L/K is deeply ramified, then L/K must in fact be arithmetically profinite by Proposition 5.1.

Thus for L/K to be arithmetically profinite, the s_i must not only be increasing, but growing fast enough, that

$$s_1 + (s_2 - s_1)p^{-1} + \cdots + (s_i - s_{i-1})p^{-(i-1)} \to \infty \text{ as } i \to \infty.$$

We shall see in the next section that Camina's construction is flexible enough to allow such a choice for the breaks s_i.

We now sketch the construction of the field of norms. If $L \geqslant M_1 \geqslant M_2 \geqslant F$ where M_1 and M_2 are finite field extension, then the norm map N_{M_1/M_2} gives a homomorphism from M_1^* to M_2^*. The underlying set of the field of norms consists of zero together with

$$N(L/F)^* = \varprojlim M^*$$

where the inverse limit is over finite extensions M/F with connecting maps given by the norm functions.

Group theorists are used to taking inverse limits with respect to connecting maps that are group homomorphisms, which therefore guarantee a group structure on the inverse limit. Here the connecting maps are homomorphisms on the multiplicative structure, so the inverse limit inherits a multiplicative structure. The condition that the extension be arithmetically profinite comes into play in defining an *addition* on this inverse limit, making $N(L/F)$ into a field:

Definition 5.2. Let $a = (a_M)$ and $b = (b_M)$ with $a_M, b_M \in M^*$ be two norm-compatible sequences in $N(L/F)^*$.

(1) Define multiplication on $N(L/F)^*$ componentwise so that $a \cdot b = (a_M b_M)$.

(2) Define addition by $(a_M) + (b_M) = (c_M)$ where c_M is defined as $c_M = \lim_E N_{E/M}(a_E + b_E)$ and E runs over all finite subextension of M in L.

The definition of addition here is based on taking the addition further up the chain of fields and then bringing it down again using the norm maps. This is reminiscent of the addition one can define directly onto a uniform pro-p group G, where $g + h = \lim \left(g^{p^n} h^{p^n}\right)^{p^{-n}}$ for $g, h \in G$ (see Chapter 4 of [5]).

The remarkable thing, proved by Fontaine and Wintenberger (see [29]), is that provided L/F is arithmetically profinite, this is a well defined operation and makes $N(L/F)$ into a field, with the following properties:

Proposition 5.4. *(1) The field $N = N(L/F)$ is a complete discrete valuation field of characteristic p, with valuation $v_N((a_M)) = v_M(a_M)$. The residue field is isomorphic to the residue field of L, and a prime element is $X = (\pi_M)$ which is a sequence of norm-compatible prime elements of finite subextensions of L/F. The field N is therefore isomorphic to the field of formal power series over X with coefficients from the residue field of L.*
(2) Let L/F be in addition a totally ramified extension. Each F-automorphism τ of L induces a wild automorphism σ of the field of norms: $\sigma((\pi_M)) = (\tau\pi_M)$. The map $\tau \mapsto \sigma$ defines an embedding of the Galois group Gal(L/F) *into the group of wild automorphisms of the field N.*

At the heart of proving that the addition is well-defined is the fact that the functions $h_{L_i/F}$ determine how the filtrations of principal units relate as we ascend the tower $\bigcup L_i$ defining L. In particular:

Proposition 5.5. *Let $\varepsilon \in \mathcal{O}_{L_i}$. Then for $j < i$*

$$h_{L_i/L_j}\left(v_{L_j}\left(N_{L_i/L_j}(\varepsilon) - 1\right)\right) \geqslant v_{L_i}(\varepsilon - 1).$$

Also Proposition 8.1 in Section 8 shows how local class field theory relates the ramification subgroups to the filtration of principal units. For full details of a proof of Proposition 5.4 we refer the reader to Section 5.5, Chapter III of [8].

The field of norms functor W associates to an infinite arithmetically profinite Galois extension L/F its field of norms $N = N(L/F)$ and the image of Gal(L/F) in the group of continuous automorphisms of the local field N.

An important class of arithmetically profinite extensions is supplied by p-adic analytic extensions of local fields F of characteristic zero: the theorem of Sen ([22], the proof is given in Section 8) says that if L/F is an infinite totally ramified Galois extension with p-adic analytic Galois group $\mathrm{Gal}(L/F)$, then the filtration by upper ramification groups is cofinal with the lower p-series of $\mathrm{Gal}(L/F)$; in particular L/F is arithmetically profinite. For generalizations see [15].

The following proposition provides some other ways of getting arithmetically profinite extensions:

Proposition 5.6. *(1) An infinite Galois subextension of a Galois arithmetically profinite extension is arithmetically profinite.*

(2) If M/L is a finite separable extension and L/F is a Galois arithmetically profinite extension, then M/F is arithmetically profinite.

We can use (2) to define an associated object $N(K, L/F)$ to a Galois extension K of F containing an arithmetically profinite extension L/F :

Definition 5.3. If L/F is an arithmetically profinite extension and K is a Galois extension of L, one defines $N(K, L/F)$ to be the compositum of all $N(E/F)$ where E/F runs over Galois subextensions of K/F with $|E : L| < \infty$.

An infinite arithmetically profinite extension is deeply ramified since the ramification subgroups $G(x)$ being open certainly implies that they are non-trivial.

We can add the following equivalent reformulation of being arithmetically profinite, given in [6], (2.3):

Proposition 5.7. *A Galois extension L of a local field F with finite residue field is arithmetically profinite if and only if it is deeply ramified and the set $B(L/F)$ is discrete.*

Proof. If L/F is arithmetically profinite, then we know that it is deeply ramified and that the breaks are a discrete set (Corollary 5.2).

Conversely, assume that $B(L/F)$ is discrete and unbounded from above. Denote the breaks by $u_1 < u_2 < \dots$. If the cardinality of the residue field of L is p^f, we claim that $|\mathrm{Gal}(L/F)(u_j) : \mathrm{Gal}(L/F)(u_{j+1})| \le p^f$ for all $j \ge 1$. First, let M/F be a finite Galois extension with two successive breaks $u_1 < u_2$ with respect to the upper numbering. Let K_1, K_2 be the fixed fields of $\mathrm{Gal}(M/F)(u_1)$ and $\mathrm{Gal}(M/F)(u_2)$ respectively. Then, if $u_j = h_{M/F}(v_j)$ for appropriate integers v_j, the ramification group $\mathrm{Gal}(M/F)[v_2]$ coincides with $\mathrm{Gal}(M/F)[v_1+1]$ and $\mathrm{Gal}(M/F)[v_1]/\mathrm{Gal}(M/F)[v_1+1]$ injects into either the additive or the multiplicative group of the residue field of M by Proposition 3.1. We deduce that $|K_2 : K_1|$ is not greater than the cardinality of the residue field of M.

Now denote by E_j the fixed field in L of $\mathrm{Gal}(L/F)(u_j)$, put $E_0 = F$. We shall prove the inequality $|E_{j+1} : E_j| \leqslant p^f$ by induction on j. For $j = 0$ this is easy. We assume that the inequality holds for all integers smaller than j, and deduce it for j. From the inductive assumption E_j is of finite degree over E_0. Choose an infinite tower of finite Galois extensions M_n/F, $M_n \subset M_{n+1}$ such that $L = \bigcup M_n$. It is clear that for a sufficiently large n the field M_n contains E_j and u_j, u_{j+1} are two successive breaks with respect to the upper numbering of $\mathrm{Gal}(M_n/F)$. Then $E_j, E_{j+1} \cap M_n$ are the fixed fields of $\mathrm{Gal}(M_n/F)(u_j)$, $\mathrm{Gal}(M_n/F)(u_{j+1})$ respectively. Therefore, by the first part of the proof, we get $|E_{j+1} \cap M_n : E_j| \leqslant p^f$. Thus, $|E_{j+1} : E_j| \leqslant p^f$.

The fact that the Galois group of an arithmetically profinite extension embeds into the group of wild automorphisms of its field of norms makes it important to study the structure of this group; the following section is devoted to this.

6 The group of wild automorphisms of a local field

For a complete discrete valuation field F of positive characteristic p with residue field k we denote by $\mathcal{N}(k)$ the group of wild automorphisms of F, i.e., those continuous automorphisms σ of F which satisfy $(\sigma - 1)\mathcal{O}_F \subset \mathcal{M}_F$ where $\mathcal{M}_F$ is the maximal ideal of the ring of integers of $\mathcal{O}_F$. We shall usually denote this group by $\mathcal{N}$.

If we fix a prime element X of F, then F can be identified with the formal power series field $k((X))$ and $\mathcal{N}$ can be identified with the group of continuous automorphisms σ of F which satisfy $X^{-1}\sigma(X) \in U_{1,F}$. Associate to $\sigma \in \mathcal{N}$ the formal power series $\sigma(X) \in X + X^2k[[X]]$; then $\mathcal{N}$ can be interpreted as the group of formal power series $X + a_2X^2 + \dots$ with the group operation defined by substituting one power series in another. It is in this guise that for $k = \mathbb{F}_p$ group theorists know $\mathcal{N}$ as the *Nottingham group*.

The group was studied from a group theoretic angle by a number of people ([11], [12], [30], [3], [1], [2], [14]). Charles Leedham-Green gave the group the name Nottingham group in connection with Dave Johnson and Ian York in Nottingham. He writes: "The decision to call it the Nottingham group was taken (by myself) with great difficulty. I did not feel too happy about naming it after Dave Johnson, as other mathematicians (mainly French) had considered it; and since it is in principle just the automorphism group of a field there must be hundreds of mathematicians who could claim priority. However, Johnson first brought the group, with its exciting structure, to the notice of the group theorists, and the rest followed 'automatically', so he deserves primary credit among the algebraists. He had a student, York, who wrote a thesis on the Nottingham group. Having named the group in their honour, I can hardly back off unless a new name becomes popular."

However since $\mathcal{N}$ was known much earlier among number theorists as the group of wild automorphisms we have preferred to retain this name.

For a wild automorphism σ of F put $i(\sigma) = v_F(X^{-1}\sigma(X) - 1)$. Let $\mathcal{N}_i = \{\sigma \in \mathcal{N} : i(\sigma) \geqslant i\}$. The following proposition collects together properties of this group $\mathcal{N}$ and the filtration $(\mathcal{N}_i)$; these are all proved in Chapter 6.

Proposition 6.1. *(1) Each $\mathcal{N}_i$ is a normal subgroup of $\mathcal{N}$; $[\mathcal{N},\mathcal{N}] = [\mathcal{N},\mathcal{N}]\mathcal{N}^p = \mathcal{N}_3$; and $[\mathcal{N}_i,\mathcal{N}_j] \leqslant \mathcal{N}_{i+j}$.*

(2) $\mathcal{N}_i^p \leqslant \mathcal{N}_{ip}$.

(3) The group $\mathcal{N}$ is a hereditarily just-infinite group: every nontrivial normal closed subgroup of an open subgroup is open.

(4) The group $\mathcal{N}(\mathbb{F}_p)$ is a pro-p group, generated by the two elements $X/(1-X)$ (of order p) and $X + X^2$.

The field of norms functor W maps the Galois group of a totally ramified arithmetically profinite infinite extension L/F isomorphically onto a closed 'arithmetically profinite' subgroup G of the group $\mathcal{N} = \mathcal{N}(k)$ ($k = k_F$); this concept is defined below. It is interesting to know where the upper ramification subgroup $\mathrm{Gal}(L/F)(x)$ are mapped to under this isomorphism. The subgroups $\mathcal{N}_i$ are analogous to the ramification subgroups in the lower numbering. We can define for a closed subgroup of $\mathcal{N}$ groups analogous to the ramification subgroups in the upper numbering.

The group $\mathcal{N}$ is not p-adic analytic, since for every n relatively prime to p there is $\sigma \in \mathcal{N}_n \setminus \mathcal{N}_{n+1}$ such that $\sigma^p = e$ (it suffices to observe that given a natural number relatively prime to p there exists a cyclic totally ramified extension of degree p of a local field of characteristic p with the break equal to that number). If $\mathcal{N}$ were p-adic analytic, this would contradict the existence of an open uniform pro-p subgroup which is known to be torsion-free. Alternatively one may observe from Proposition 6.1 that $\mathcal{N}_n/\mathcal{N}_{2n}$ is an elementary abelian p-group of size $|k|^n$ and hence $\mathcal{N}$ does not have finite rank. (It is shown in Chapter 1, §5.1, that in fact $\mathcal{N}$ is not analytic over *any* pro-p ring.)

For a closed subgroup G of $\mathcal{N}$ put $G_i = G \cap \mathcal{N}_i$, $G_y = G_{\lceil y \rceil}$. Introduce the following function: $\varphi_G(x) = \int_0^x \frac{dy}{|G:G_y|}$. Then G is called an *arithmetically profinite subgroup* of $\mathcal{N}$ if $\lim_{x\to+\infty} \varphi_G(x) = +\infty$. If this is the case, define $\psi_G(x)$ as the inverse function to $\varphi_G(x)$ and put $G(x) = G_{\psi_G(x)}$. The points of discontinuity of the derivative of ψ_G are called *upper breaks* of G.

A theorem of Sen [21] says that the procyclic subgroup of $\mathcal{N}$ generated by an element σ of infinite order is arithmetically profinite and that $i(\sigma^{p^n}) \equiv i(\sigma^{p^{n-1}}) \bmod p^n$.

The following proposition describes where the ramification subgroups get mapped to under the field of norms functor.

Proposition 6.2. *Let L/F be an arithmetically profinite totally ramified extension and denote by G the image of* $\mathrm{Gal}(L/F)$ *in $\mathcal{N}$ under the functor W. Then for each x the upper ramification subgroup* $\mathrm{Gal}(L/F)(x)$ *is mapped by W onto the subgroup $G(x)$ of G.*

The question of which closed subgroups of $\mathcal{N}$ can be realised as images of the functor W has been considered by among others Wintenberger. In particular he has proved the following [28]:

Proposition 6.3. *For every abelian closed subgroup G of $\mathcal{N}(k)$ there exists a totally ramified arithmetically profinite extension L/F of local fields such that $W(L/F) = (k((X)), G)$.*

For example, the group topologically generated by an element σ of infinite order in $\mathcal{N}$ comes from an arithmetically profinite $\mathbb{Z}_p$-extension L/F. It is easy to deduce that the sequence $i(\sigma^{p^n})/p^n$ is increasing. Define e by $pe/(p-1) = \lim \frac{i(\sigma^{p^n})}{p^n}$. Then either $e = +\infty$ or $e \in \mathbb{N}$. In the first case F is of positive characteristic; in the second case F is of characteristic 0 and its absolute ramification index is e.

We can try to explain this as follows by using Proposition 8.1 below. This proposition shows how the local reciprocity map from local class field theory relates the filtration of principal units and the ramification subgroups in an abelian extension. Let a principal unit ϵ correspond to σ via the local reciprocity map. Then ϵ^{p^n} corresponds to σ^{p^n}, so we only need to study raising to the pth power in characteristics p and 0.

The point is that in characteristic p it moves the ith group of principal units to the pith group; while in characteristic 0, according to a celebrated result of Tate, for sufficiently large i it maps the ith group of principal units to $(i + e)$th group (see Lemma 2.2).

We also have the following characterization of e being infinite:

Lemma 6.4. *The following are equivalent:*
(1) $e = +\infty$;
(2) σ *belongs to the topological closure of the set of torsion elements of* $\mathcal{N}$;
(3) F *is of positive characteristic.*

Proof. The paragraph preceding the lemma gave some indication of the equivalence of (1) and (3) (for more details see [8]). We prove the equivalence of (1) and (2). Assume that the group G is topologically generated by σ and comes from a totally ramified arithmetically profinite $\mathbb{Z}_p$-extension L/F of fields of characteristic p; let τ be a topological generator for $\mathrm{Gal}(L/K)$. Let K_n be the subextension of L of degree p^n over K. Then K_n can be mapped isomorphically onto the field of norms $N(L/F)$: a prime element π_{K_n} belonging to a norm-compatible sequence of prime elements of finite subextensions in L/K goes to X. Let N_n be the image of K under this homomorphism, and let σ_n (of order p^n) be the image of τ. Then $i(\sigma\sigma_n^{-1})$ tends to $+\infty$ with n, so $\sigma = \lim_{n\to\infty} \sigma_n$ and (2) holds. Conversely, if σ is the limit of a sequence of automorphisms σ_n of finite order, then the upper breaks u_i of $G = \overline{\langle\sigma\rangle}$ satisfy $u_{i+1} \geqslant u_i^p$ (see [23]), therefore $e = +\infty$.

When $e \in \mathbb{N}$, F is then of characteristic 0 by the above and its absolute ramification index is e.

Thanks to work of Wintenberger [27] and Laubie [16] a lot is known on p-adic analytic subgroups in $\mathcal{N}$ which are in the image of the functor W.

7 Closed subgroups of the group of wild automorphisms

In this section we discuss the proof of Camina's theorem. We are given a countably based pro-p group G, and may suppose that $G = \varprojlim G_i$, where $(G_i)_{i\in\mathbb{N}}$ is an inverse system of finite p-groups in which all the maps are epimorphisms, and for each i the kernel of the epimorphism $G_{i+1} \to G_i$ is of order p. The aim is to embed G as a closed subgroup in $\mathcal{N}(\mathbb{F}_p)$, and we shall describe two ways of going about this, one based on the field of norms construction and the other on an extension of Camina's original approach.

Both approaches start by representing G as a Galois group. Let us write $F = \mathbb{F}_p((X))$. Suppose that for some i we find a totally ramified Galois extension K_i/F such that $G_i = \mathrm{Gal}(K_i/F)$. According to Iwasawa [9] one can realize the group G_{i+1} as the Galois group of a totally ramified Galois extension K_{i+1}/F where $F \subset K_i \subset K_{i+1}$, so that the restriction $\mathrm{Gal}(K_{i+1}/F) \to \mathrm{Gal}(K_i/F)$ coincides with the given map $G_{i+1} \to G_i$. Having constructed such a sequence (K_i) recursively, we take $L = \bigcup K_i$ and obtain $\mathrm{Gal}(L/F) = \varprojlim \mathrm{Gal}(K_i/F) = G$.

Denote the corresponding ramification breaks in the above sequence by

$$s_i = s(K_{i+1}/K_i).$$

The heart of the problem now is to show that we can make the field extensions in such a way that the sequence (s_i) grows very fast.

Given K_i, the extension K_{i+1} has the form

$$K_{i+1} = K_i(\beta)$$

where β is a solution of $\beta^p - \beta = \alpha$ for some $\alpha \in K_i$. Now suppose that $c \in F$ and $E = K_i(\beta_1)$ where β_1 satisfies $\beta_1^p - \beta_1 = \alpha + c$. Then E/F is still totally ramified, and we have a commutative diagram

$$\begin{array}{ccc} \mathrm{Gal}(E/F) & \stackrel{\simeq}{\to} & \mathrm{Gal}(K_{i+1}/F) \\ & \searrow \qquad \swarrow & \\ & \mathrm{Gal}(K_i/F) & \end{array},$$

so replacing K_{i+1} by $E = K_i(\beta_1)$ gives a solution of the same embedding problem. Thus we have the freedom to change the choice of α by adding elements c of the ground field $F = \mathbb{F}_p((X))$. However, different choices of c will give rise to different values for the ramification break s_i of the cyclic extension K_{i+1}/K_i. The following lemma shows how to read the ramification break from the valuation of α. Let T_i be a prime element of K_i and let $\mathcal{O}_i$ be its ring of integers and suppose that v_i is the valuation on K_i with $v_i(T_i) = 1$.

Lemma 7.1. *Suppose that $K_{i+1} = K_i(\beta_1)$ where β_1 is a solution of $\beta_1^p - \beta_1 = \alpha_1 \in K_i$. Provided that $v_i(\alpha_1) < 0$ is not divisible by p, the ramification break $s(K_{i+1}/K_i)$ is equal to $-v_i(\alpha_1)$.*

Proof. Since $\beta_1^p - \beta_1 = \alpha_1$ and $v_i(\alpha_1) = -n$, we have $v_i(\beta_1) = -n/p$ (extending the valuation v_i to K_{i+1}). Since p and n are coprime, there exist integers b and d with $bp - dn = 1$. Set $T_{i+1} = \beta_1^d T_i^b$. Then $v_i(T_{i+1}) = 1/p$ and $K_{i+1} = \mathbb{F}_p((T_{i+1}))$. We may suppose that $\mathrm{Gal}(K_{i+1}/K_i) = \langle g \rangle$ where $g(\beta_1) = \beta_1 + 1$, so that

$$\begin{aligned} g(T_{i+1}) &= g(\beta_1^d T_i^b) \\ &= (\beta_1 + 1)^d T_i^b \\ &= T_{i+1} + dT_{i+1}/\beta_1 + \cdots + T_{i+1}/\beta_1^d. \end{aligned}$$

Now $v_i(T_{i+1}/\beta_1^a) = v_i(T_{i+1}) + an/p = (an+1)/p$ for each a. Since $d \not\equiv 0 \bmod p$, this gives $v_i(g(T_{i+1}) - T_{i+1}) = (n+1)/p$. Hence $s(K_{i+1}/K_i) = n$.

If we want to make s_i large, it would seem sufficient now to choose c to have suitably large negative valuation. However as soon as $v_i(c) < v_i(\alpha)$, we have $v_i(\alpha_1) = v_i(\alpha + c) = v_i(c)$ which is divisible by the degree of K_i/F, since $c \in F$. In particular $v_i(\alpha_1)$ is divisible by p, and the preceding lemma is not applicable. Instead we have to be a bit more clever.

Let $\wp(X) = X^p - X$. According to Artin–Schreier theory, the kernel of $F/\wp(F) \to K_i/\wp(K_i)$ is finite-dimensional, hence so is the kernel of $F/\wp(F) \to K_i/(\wp(K_i) + T_i^{-k}\mathcal{O}_i)$ for each k. Since $F/\wp(F)$ is infinite-dimensional we can then choose $c \in F$ but not in $\wp(K_i) + T_i^{-k}\mathcal{O}_i$. Such an element can be split as a sum in K_i as follows:

Lemma 7.2. *Suppose that $c \in F$ but $c \notin \wp(K_i) + T_i^{-k}\mathcal{O}_i$. Then there exist $x, z \in K_i$ such that*

$$c = \wp(z) + x, \quad p \nmid v_i(x) < -k.$$

Proof. Since $v_i(c)$ is divisible by p, we can write $c = c_{-pn}T_i^{-pn} + c_{-pn+1}T_i^{-pn+1} + \cdots = \wp(c_{-pn}T_i^{-n}) + y$ with $v_i(y) > v_i(c)$. We can then continue in this way if $v_i(y)$ is negative and divisible by p. Eventually we deduce that $c = \wp(z) + x$ with either $v_i(x) \geqslant 0$ or $v_i(x)$ negative and prime to p. But we have chosen $c \notin \wp(K_i) + T_i^{-k}\mathcal{O}_i$; hence we must be in the second case with $v_i(x) < -k$ and prime to p.

Suppose now that we have reached the stage $K_{i+1} = K_i(\beta)$, where $\beta^p - \beta = \alpha \in K_i$. Let $k \geqslant -v_i(\alpha)$ be a large positive integer, choose $c \in F \setminus \left(\wp(K_i) + T_i^{-k}\mathcal{O}_i\right)$, and let β_1 be a solution of $\beta_1^p - \beta_1 = \alpha + c$. Let x, z be as in the last lemma and put $\beta_2 = \beta_1 - z$. Then β_2 satisfies $\beta_2^p - \beta_2 = \alpha + x$, where $v_i(\alpha + x) = v_i(x) < -k$, and $K_i(\beta_2) = K_i(\beta_1)$; and we find from Lemma 7.1 that

$$s(K_i(\beta_1)/K_i) = s(K_i(\beta_2)/K_i) = -v_i(\alpha + x) > k.$$

So replacing K_{i+1} by $K_i(\beta_1)$ we can make $s_i = s(K_{i+1}/K_i)$ as large as we like. In particular, we can construct the tower of fields (K_i) so that the sequence

$$s_1 + (s_2 - s_1)p^{-1} + \cdots + (s_i - s_{i-1})p^{-(i-1)}$$

is strictly increasing and unbounded. As we saw at the end of Section 4, this is enough to ensure that the extension L/F arithmetically profinite; so we have established

Proposition 7.3. *There exists an arithmetically profinite totally ramified Galois extension L of $\mathbb{F}_p((X))$ such that $G \cong \mathrm{Gal}(L/\mathbb{F}_p((X)))$.*

We can now deduce the main result:

Theorem 7.4. *Every countably based pro-p group is isomorphic to a subgroup of the closure of the torsion in $\mathcal{N}(\mathbb{F}_p)$.*

First proof. Apply the field of norms functor W to the arithmetically profinite extension L/F. By Proposition 5.4 this embeds G, the Galois group of the extension L/F, into $\mathcal{N} = \mathcal{N}(\mathbb{F}_p)$. (Note that L need not itself be a local field; so while Proposition 7.3 represents G as a group of automorphisms, this is not enough by itself to show that G embeds into $\mathcal{N}$. This, as we pointed out in the introduction, is one of the strengths of the field of norms construction: the Galois group of an arithmetically profinite extension acts naturally as a group of automorphisms of a local field $N(L/F)$. However, a direct argument, independent of the field of norms construction, is indicated below.)

Let $\tau \in \mathrm{Gal}(L/F)$ have infinite order, and fixed field L_τ. By Lemma 6.4 it will suffice to show that the image of τ in $\mathcal{N}$ comes from an arithmetically profinite $\mathbb{Z}_p$-extension in characteristic p.

It is an immediate consequence of the definition of the field of norms that if $|L_\tau : F|$ is finite, then $N(L/F) = N(L/L_\tau)$. In this case we are done, since L is an arithmetically profinite extension of L_τ, with Galois group $\mathbb{Z}_p$.

Suppose now that $|L_\tau : F|$ is infinite, and recall that we defined $N(L, L_\tau/F)$ to be the compositum of the fields $N(L'/F)$ as L' ranges over Galois subextensions of L/F with $|L' : L_\tau|$ finite.

Now the image of τ in $\mathcal{N} \leqslant \mathrm{Aut}(N(L/F))$ is determined by how τ acts on a fixed norm-compatible sequence of prime elements. The image of τ in the Galois group of $N(L, L_\tau/F)$ over $N(L_\tau/F)$ is determined by how it acts on a sequence of norm-compatible prime elements in finite extensions of $N(L_\tau/F)$ in $N(L, L_\tau/\mathbb{F}_p((X)))$. But prime elements there are themselves norm-compatible sequences of elements, so this image of τ is determined by how τ acts on a sequence of norm-compatible prime elements in finite extensions of F. Hence the image of τ in $\mathcal{N}$ is also the image of an element of $\mathrm{Gal}\,(N(L, L_\tau/F)/N(L_\tau/F))$, and again the result follows by Lemma 6.4. (For more formal arguments see Section 3.4 of [29].)

Remark. Recall that the field of norms functor relates the ramification breaks of the original Galois group with the filtration $(G(x))$ defined on closed subgroups G of $\mathcal{N}(\mathbb{F}_p)$. Since we can change the ramification breaks by making different choices of c in the construction of the tower defining L, this means that we can actually embed a pro-p group into $\mathcal{N}(\mathbb{F}_p)$ in infinitely many different ways.

Note that the closure of the torsion of $\mathcal{N}(\mathbb{F}_p)$ is distinct from $\mathcal{N}(\mathbb{F}_p)$: every automorphism with finite e (see the previous section) does not belong to it. However, similar arguments show the following, where the local field K may have characteristic p or zero:

Proposition 7.5. *Let M/K be a totally ramified arithmetically profinite extension. Then the image under W of $\mathrm{Gal}(M/K)$ in $\mathcal{N}(\mathbb{F}_p)$ is either virtually procyclic or else is contained in the closure of the torsion of $\mathcal{N}(\mathbb{F}_p)$.*

Proof. Let M/K be an arithmetically profinite extension (now possibly in characteristic 0) and suppose that $\mathrm{Gal}(M/K)$ is not virtually procyclic. Apply the functor W to the extension M/K. If $\tau \in \mathrm{Gal}(M/K)$ is of infinite order, then the fixed field M_τ of τ is an infinite extension of K, otherwise the group $\mathrm{Gal}(M/K)$ would be a finite extension of $\mathrm{Gal}(M/M_\tau)$ which is procyclic. Then the image of τ in $\mathcal{N}(\mathbb{F}_p)$ can be identified with the image of $\tau \in \mathrm{Gal}(N(M, M_\tau/K)/N(M_\tau/K))$ in $\mathcal{N}(\mathbb{F}_p)$, and hence as before belongs to the closure of the torsion of $\mathcal{N}(\mathbb{F}_p)$. Note how we have used the field of norms here to move from M/K, potentially characteristic zero, to $N(M, M_\tau/K)/N(M_\tau/K)$ which is of characteristic p so that we can apply Lemma 6.4.

We now sketch a **second proof of Theorem 7.4**; this is based on Rachel Camina's original argument for finitely generated groups and has been extended to the countably based case by Benjamin Klopsch. Instead of constructing a field isomorphic to $\mathbb{F}_p((X))$

on which G is made to act, the limiting process is applied directly to the sequence of group embeddings

$$G_n \hookrightarrow \mathrm{Aut}(K_n) = \mathrm{Aut}(\mathbb{F}_p((T_n))) \cong \mathrm{Aut}(\mathbb{F}_p((X))).$$

We assume that the tower (K_n) has been constructed so that the sequence of ramification breaks (s_n) is strictly increasing. Note that we shall not require for this proof the stronger property that (s_n) be growing fast enough to make the compositum $\bigcup K_n$ an arithmetically profinite extension of the base field. For each n let $H_n = \ker(G \to G_n)$, and choose an element $g_n \in H_{n-1} \setminus H_n$ (taking $H_0 = G$). Thus $\{H_n\}$ is a base for the neighbourhoods of 1 in G, and it is clear that the set $\{g_n \mid n \in \mathbb{N}\}$ generates G topologically, since $|H_{n-1}/H_n| = p$ for each n.

Identify $\mathrm{Aut}(K_n) = \mathrm{Aut}(\mathbb{F}_p((T_n)))$ with $\mathrm{Aut}(\mathbb{F}_p((X)))$ by identifying T_n with X, and let $\theta_n : G \to \mathcal{N}(\mathbb{F}_p) < \mathrm{Aut}(\mathbb{F}_p((X)))$ be the resulting homomorphism (composing $G \to G_n$ with $G_n \hookrightarrow \mathrm{Aut}(K_n)$); note that the image of G_n lies inside $\mathcal{N}(\mathbb{F}_p)$ since it is a p-group. Then for $g \in G$ we have

$$i(\theta_n(g)) = i_{K_n}(\theta_n(g)).$$

In particular, since $\theta_n(g_n)$ generates the cyclic group $\mathrm{Gal}(K_n/K_{n-1})$, we see from Definition 4.1 that

$$i(\theta_n(g_n)) = s_{n-1} + 1.$$

As the sequence (s_i) is strictly increasing, repeated applications of Lemma 4.2(i) then show that

$$i(\theta_m(g_n)) = i(\theta_n(g_n)) = s_{n-1} + 1 \quad \forall m > n. \tag{4}$$

Now put $h_n(m) = \theta_m(g_n) \in \mathcal{N} = \mathcal{N}(\mathbb{F}_p)$ for each m and n. We can pick strictly increasing sequences of natural numbers $\mathbf{m}_n = (m_{n1}, m_{n2}, \dots)$ such that

(a) for all j, $\mathbf{m}_{j+1}$ is a subsequence of $\mathbf{m}_j$, and

(b) for each n, the sequence $h_n(m_{nj})_{j \in \mathbb{N}}$ converges to an element $h_n \in \mathcal{N}$.

Note that $h_n(m) \in \mathcal{N}_{s_{n-1}+1}$ whenever $m \geqslant n$, by (4), and hence that $h_n \in \mathcal{N}_{s_{n-1}+1}$.

To complete the proof, we shall establish that the closed subgroup H of $\mathcal{N}$ generated by $\{h_n \mid n \in \mathbb{N}\}$ is isomorphic to G, and that H is contained in the closure of the torsion in $\mathcal{N}$.

Let Φ be the free pro-p group on a countably infinite set $\{x_n \mid n \in \mathbb{N}\}$. As $g_n \to 1$ in G and $h_n \to 1$ in $\mathcal{N}$, we have epimorphisms $\gamma : \Phi \to G$ and $\rho : \Phi \to H$ with $\gamma(x_n) = g_n$ and $\rho(x_n) = h_n$ for each n. We claim that $\ker\gamma = \ker\phi$. As these are both closed subgroups of Φ, this will follow once we show that for a finite word $w = w(x_1, \dots, x_n)$ in Φ, one has $w \in \ker\gamma \Leftrightarrow w \in \ker\phi$. To prove this, in turn, it suffices to show that for each $k > n$,

$$w(g_1, \dots, g_n) \in H_k \Leftrightarrow w(h_1, \dots, h_n) \in \mathcal{N}_{s_k+1}. \tag{5}$$

Now let $m > k$. Then

$$w(g_1, \dots, g_n) = g_1^{f_1} \dots g_m^{f_m} x$$

where $x \in H_m$ and $0 \leqslant f_j < p$ for each j. This gives

$$\begin{aligned} w(h_1(m), \dots, h_n(m)) &= \theta_m(w(g_1, \dots, g_n)) \\ &= h_1(m)^{f_1} \dots h_m(m)^{f_m}. \end{aligned}$$

Suppose now that $w(g_1, \dots, g_n) \in H_k$. Then $f_1 = \dots = f_k = 0$, so that

$$w(h_1(m), \dots, h_n(m)) = h_{k+1}(m)^{f_{k+1}} \dots h_m(m)^{f_m} \in \mathcal{N}_{s_k+1};$$

letting m tend to infinity along the sequence $\mathbf{m}_n$ we infer that $w(h_1, \dots, h_n) \in \mathcal{N}_{s_k+1}$.

Suppose on the other hand that $w(g_1, \dots, g_n) \notin H_k$, and let r be minimal such that $f_r \neq 0$; thus $r \leqslant k$. Arguing as above we now find that

$$w(h_1(m), \dots, h_n(m)) \equiv h_r(m)^{f_r} \pmod{\mathcal{N}_{s_r+1}},$$

and taking the limit as before gives

$$w(h_1, \dots, h_n) \equiv h_r^{f_r} \pmod{\mathcal{N}_{s_r+1}}.$$

Since $i(h_r^{f_r}) = i(h_r) = s_{r-1} + 1 < s_r + 1 \leqslant s_k + 1$ it follows that $w(h_1, \dots, h_n) \notin \mathcal{N}_{s_k+1}$.

This establishes (5), and shows that $\ker \gamma = \ker \phi$. The mapping $g_n \mapsto h_n$ ($n \in \mathbb{N}$) therefore extends to an isomorphism of G onto H, and we have successfully embedded G into $\mathcal{N}$.

It only remains to observe that H is contained in the closure of the torsion in $\mathcal{N}$. Since the finite words $w(h_1, \dots, h_n)$ are dense in H, it suffices to consider these; but $w(h_1, \dots, h_n) = \lim_{m \in \mathbf{m}_n} w(h_1(m), \dots, h_n(m))$, and for each m we have $w(h_1(m), \dots, h_n(m)) \in \theta_m(G)$, a finite subgroup of $\mathcal{N}$, so the result is clear.

8 Sen's Theorem on p-adic Lie extensions of local number fields

In characteristic zero the only progress for many years in identifying the ramification groups in the upper numbering for infinite non-abelian extensions was made in a paper by Shankar Sen in 1972 [22] and independently by Eckart Maus [19]. Sen proved that if L/K is a totally ramified Galois extension with a p-adic analytic Galois group G, then one can relate the ramification groups to the lower p-series of G. This confirmed a conjecture made by Serre [24].

We present here a proof of Sen's Theorem, making use of the language introduced since 1972 in [5] for p-adic analytic groups. Sen's proof is a cunning analysis of the abelian quotients of G combined with the knowledge local class field theory provides about ramification groups in abelian quotients.

Local class field theory gives a description of the abelian extensions of a local field K. It provides one with a monomorphism ρ_K from the multiplicative group K^* of the ground field K onto a dense subgroup of $\mathrm{Gal}(K_{ab}/K)$ where K_{ab} is the maximal abelian extension of K :

$$\Psi_K : K^* \to \mathrm{Gal}(K_{ab}/K).$$

This map is called the *reciprocity map* or the *norm residue map* or sometimes the *Artin map* of the local field K. If L/K is an abelian extension, then composing ρ_K with the restriction $\mathrm{Gal}(K_{ab}/K) \to \mathrm{Gal}(L/K)$, we get a reciprocity map

$$\Psi_{L/K} : K^* \to \mathrm{Gal}(L/K).$$

This reciprocity map provides a connection between the ramification filtration in the upper numbering on $\mathrm{Gal}(L/K)$ and the filtration detailed in Lemma 2.1 on the group of units of K. A proof of the following proposition can be found in [10] Theorem 7.12, [23] Chapter XV Section 2, or [8] Chapter IV, (7.4):

Proposition 8.1. *Let L/K be an abelian extension with Galois group G. For each $w \in (-1, \infty)$ we have*

$$G(w) = \Psi_{L/K}\left(U_{K,n} N_{L/K} L^*\right)$$

where $U_{K,n} = 1 + \mathcal{M}_K^n$ and $n = n(w)$, the least integer $\geqslant w$.

Combining Proposition 8.1 with Lemma 2.2 we see how the ramification groups $G(w)$ of an abelian extension behave under the p-power map.

Lemma 8.2. *Let L/K be an abelian extension with Galois group G.*
(i) If $x \leqslant e_K/(p-1)$, then $G(x)^p \leqslant G(px)$.
(ii) If $x > e_K/(p-1)$, then $G(x)^p = G(x + e_K)$.

Using knowledge of the structure of the abelian sections in a p-adic analytic group, we shall show how to deduce properties of the ramification groups of G. To do this we shall need to know how the ramification groups in the upper numbering of G relate to those of a subgroup H.

The readjustment of the numbering of our filtration to define the ramification groups in the upper numbering meant that the numbering was preserved under taking quotients. But we have had to pay for this by losing the property that the filtration is preserved on passing to subgroups. However the Hasse–Herbrand function keeps track of the change of the upper numbering when going to a subgroup:

Lemma 8.3. *Let L/K be a Galois extension with Galois group G, H an open normal subgroup and M the fixed field of H. Then*

$$G(x) \cap H = H(h_{M/K}(x))$$

for $x \geqslant 0$.

Proof. Herbrand's Theorem (see the discussion preceding Proposition 4.3) implies that $h_{L/M} \circ h_{M/K}(x) = h_{L/K}(x)$. So

$$\begin{aligned} G(x) \cap H &= G[h_{L/K}(x)] \cap H \\ &= H[h_{L/K}(x)] \\ &= H[h_{L/M} \circ h_{M/K}(x)] \\ &= H(h_{M/K}(x)). \end{aligned}$$

Suppose that $G = \mathrm{Gal}(L/K)$ is a p-adic analytic group, where L/K is a totally ramified extension of local fields of characteristic zero. We are going to prove that the ramification groups $G(x)$ are open subgroups of G. In view of Lemma 8.3, this will follow if we can show that the subgroups $H(y)$ are open in H for some open normal subgroup H of G. Similarly, if each of the subgroups $H(y)$ is different from 1, then each $G(x) \neq 1$; in other words, L/K is deeply ramified. Since G is p-adic analytic, G has an open normal subgroup H which is a *uniform pro-p group*. Replacing G by H, we may therefore assume that in fact G is a uniform pro-p group. Recall from Definition 2.2 and [5] that the p-power map then induces isomorphisms between the (elementary abelian) sections of the lower p-series: $P_n(G)/P_{n+1}(G) \longrightarrow P_{n+1}(G)/P_{n+2}(G)$. We write $G_n = P_n(G)$ for each n, and put $d = \dim(G)$.

We shall also use Lemma 8.3 to translate information we can glean about the abelian extension with Galois group G_n/G_{2n} to the full group G.

Definition 8.1. For a finite Galois extension E/K with Galois group H define

$$\begin{aligned} u_H &= \inf\{x : x \geqslant 0, H(x+\epsilon) = 1 \text{ for } \epsilon > 0\} \\ l_H &= \inf\{y : y \geqslant 0, H[y+\epsilon] = 1 \text{ for } \epsilon > 0\}. \end{aligned}$$

$u_H + 1$ is called the *conductor* of the field extension E/K.

Note that u_H and l_H are invariants of the field extension E/K and not of the abstract group H. The number u_H is the largest break in the finite extension E/K. So to show that the breaks of a tower of fields are unbounded, we will want to show that the conductors of the finite field extensions in the tower are unbounded.

We can get the following crude estimate from the integral expression (3) for the Hasse–Herbrand function

$$l_H = h_{L/K}(u_H) \leqslant |H| u_H.$$

Let us introduce some notation. Let $K_n \subseteq L$ be the fixed field of G_n. For each $n \geqslant 1$, put

$$\begin{aligned} x_n &= u_{G/G_n} \\ z_n &= l_{G/G_n} = h_{K_n/K}(x_n) \\ h_n &= h_{K_n/K} \\ e &= e_K \\ e_n &= e_{K_n} = |G : G_n| e = p^{(n-1)d} e. \end{aligned}$$

The last equality follows because we are assuming the field extension L/K is totally ramified. Lemma 8.3 tells us how to relate the ramification subgroups of G_n to those of G. In particular we have:

Lemma 8.4. *(i)* $G(x_n + te) = G_n(z_n + te_n)$ *for* $t > 0$.
(ii) For $n, r \geqslant 1$

$$u_{G_n/G_{n+r}} = z_n + (x_{n+r} - x_n)|G : G_n|.$$

Proof. (i) Corollary 4.4 and the definition of x_n show that the left derivative of $h_{K_n/K}(x)$ for $x > x_n$ is equal to $|G : G_n|$, so $h_{K_n/K}(x_n + te) = h_{K_n/K}(x_n) + te|G : G_n| = z_n + te_n$.

So the first part of the lemma follows now from Lemma 8.3.

(ii) Let $t = (x_{n+r} - x_n)/e$. To apply (i) we need to know that $t > 0$, i.e., that $x_{n+r} > x_n$. Set $x_{n+r} = x$. Then

$$(G/G_{n+1})\,(x) \neq 1 \text{ and } (G/G_{n+1})\,(x+\epsilon) = 1 \text{ for } \epsilon > 0.$$

The groups in the filtration in the upper numbering are exactly the same groups as in the lower numbering. Hence $(G/G_{n+1})\,(x)$ is just the last non-trivial section of the filtration in the lower numbering $(G/G_{n+1})\,[m]$. By Corollary 3.2, all these section are elementary abelian p-groups. Therefore $(G/G_{n+1})\,(x) = G(x)G_{n+1}/G_{n+1}$ is a group of exponent p. So if $\sigma \in G(x)$ then $\sigma^p \in G_{n+1}$, whence $\sigma \in G_n$ since G is a uniform pro-p group. Thus $G(x) \subseteq G_n$, and hence $x > x_n$.

We are now in good shape to apply part (i), and may infer that for each $\epsilon \geqslant 0$

$$G_n(z_n + (x_{n+r} - x_n)|G : G_n| + \epsilon) = G(x_{n+r} + \epsilon')$$

where $\epsilon' = \epsilon/|G : G_n|$. Thus we have $G(x_{n+r}) \nsubseteq G_{n+r}$ while if $\epsilon > 0$, then $G(x_{n+r} + \epsilon') \subseteq G_{n+r}$, by definition of x_{n+r}. Hence

$$u_{G_n/G_{n+r}} = z_n + (x_{n+r} - x_n)|G : G_n|.$$

Our first aim is to show that the breaks x_n are unbounded, i.e., that our p-adic analytic extension L/K is *deeply ramified*. The main step is to find abelian layers G_n/G_{n+r} with large conductor $u_{G_n/G_{n+r}}$; the previous lemma will then be used to give us information about the differences $x_{n+r} - x_n$.

Sen hit on the right idea for what we should call a large conductor in this context. It is based on the dichotomy exhibited in Lemma 8.2:

Definition 8.2. Let F_1/F_2 be an abelian extension with Galois group A. We say that the extension F_1/F_2 is *small* if

$$u_A \leqslant \frac{p}{p-1} e_{F_2}.$$

Small extensions can also be characterized by the action of the p-power map on the Galois group A :

Lemma 8.5. *F_1/F_2 is small if and only if $A(x)^p \leqslant A(px)$ for all $x \geqslant 0$.*

Proof. Suppose F_1/F_2 is small. Then

$$A\left(\frac{p}{p-1} e_{F_2} + \epsilon\right) = 1 \tag{6}$$

for $\epsilon > 0$. Hence for $x > \frac{p}{p-1}e_{F_2}$ we have $A(x)^p \leqslant A(px)$ since both groups are trivial. By Lemma 8.2(i), $A(x)^p \leqslant A(px)$ is always true for $x \leqslant e_{F_2}/(p-1)$. Now take x in the interval $\left(e_{F_2}/(p-1), \frac{p}{p-1}e_{F_2}\right]$. By Lemma 8.2 (ii),

$$\begin{aligned} A(x)^p &= A(x + e_{F_2}) \\ &\leqslant A\left(e_{F_2}/(p-1) + \epsilon + e_{F_2}\right). \end{aligned}$$

But $A\left(e_{F_2}/(p-1) + \epsilon + e_{F_2}\right) = 1$ by (6). Hence $A(x)^p \leqslant A(px)$ for all $x \geqslant 0$.

Conversely suppose that F_1/F_2 is not small, so $u_A - e_{F_2} > e_{F_2}/(p-1)$. Then by Lemma 8.2(ii),

$$A\left(u_A - e_{F_2}\right)^p = A(u_A) \neq 1.$$

But

$$A\left(p\left(u_A - e_{F_2}\right)\right) = A\left(u_A + \left((p-1)u_A - pe_{F_2}\right)\right) = 1$$

by definition of u_A. Hence $A(x)^p \nleqslant A(px)$ for $x = u_A - e_{F_2}$.

Our next goal is to show that our group has an abundance of abelian sections which are not small: specifically, that K_{n+r}/K_n is not small for all sufficiently large n and $1 \leqslant r \leqslant n$ (recall that G_n/G_{n+r} is always abelian for $r \leqslant n$). We start by showing that we have at least one layer that is not small; this will be deduced from the next two lemmas.

Lemma 8.6. *If $2 \leqslant r \leqslant n$ and K_{n+r}/K_n is small then*

$$z_{n+r}/e_{n+r} \geqslant (p-1)p^{r-2-d} \cdot z_{n+1}/e_{n+1}.$$

Proof. Let $A = G_n/G_{n+r}$ and $A_{i+1} = G_{n+i}/G_{n+r}$. We break the proof up into several pieces.

Claim 1 $z_{n+r} = l_A$ and $z_{n+1} = l_{A/A_2}$.

Proof of Claim 1. Clearly $l_A \leqslant z_{n+r}$ since $A[x] = G/G_{n+r}[x] \cap A$. By Corollary 3.2 (iii), the layers $(G/G_{n+r}[m]) \,/\, (G/G_{n+r}[m+1])$ have exponent p for $m \geqslant 1$. Now $m = z_{n+r}$ is the last non-trivial layer. Hence there exists a non-trivial element $g \in G/G_{n+r}[z_{n+r}]$ with $g^p = 1$. But $A_r \leqslant A$ is the group of elements of G/G_{n+r} of order dividing p, since G is a uniform pro-p group. Hence $g \in G/G_{n+r}[z_{n+r}] \cap A = A[z_{n+r}]$, i.e., $l_A \geqslant z_{n+r}$. Thus $l_A = z_{n+r}$. The same argument also gives $z_{n+1} = l_{A/A_2}$.

We focus now on the abelian group A.

We begin by trying to estimate $l_A = h_{K_{n+r}/K_n}(u_A)$. For any $\epsilon > 0$, by Corollary 4.4 we get

$$\begin{aligned} l_A &\geqslant h_{K_{n+r}/K_n}(u_A) - h_{K_{n+r}/K_n}(p^{-1}u_A + \epsilon) \\ &\geqslant (u_A - p^{-1}u_A - \epsilon)|A : A(p^{-1}u_A + \epsilon)|. \end{aligned} \tag{7}$$

Claim 2 $A(p^{-1}u_A + \epsilon) \leqslant A_r$.

Proof of Claim 2. Since A_r is the set of elements of order dividing p it suffices to show that $A(p^{-1}u_A+\epsilon)^p = 1$. But K_{n+r}/K_n is small. So by Lemma 8.5 $A(p^{-1}u_A+\epsilon)^p \leqslant A(u_A+p\epsilon) = 1$.

Claim 3 $u_A \geqslant p^{r-1}u_{A/A_2} \geqslant p^{r-1-d}l_{A/A_2}$.

Proof of Claim 3. By definition of u_{A/A_2} and the fact that upper ramification groups are preserved on taking quotients, there exists $g \in A(u_{A/A_2})\backslash A_2$. Since A_2 consists of all elements of order dividing p^{r-1} in the powerful group A, $g^{p^{r-1}} \neq 1$. But A is the Galois group of a small extension, so Lemma 8.5 implies that $g^{p^{r-1}} \in A(p^{r-1}u_{A/A_2})$. Hence $u_A \geqslant p^{r-1}u_{A/A_2}$. The second inequality in the claim follows from the estimate following Definition 8.1 which gives $u_H|H| \geqslant l_H$ for any finite group H.

Finally combining Claims 1, 2 and 3 with our estimate (7) for l_A, we get

$$\begin{aligned} z_{n+r}/e_{n+r} &= l_A/ep^{d(n+r-1)} \\ &\geqslant \left(\frac{p-1}{p}\right)p^{r-1-d}l_{A/A_2}\cdot\frac{|A:A_r|}{ep^{d(n+r-1)}} \\ &= (p-1)p^{r-2-d}\frac{z_{n+1}p^{d(r-1)}}{ep^{d(n+r-1)}} \\ &= (p-1)p^{r-2-d}\frac{z_{n+1}}{e_{n+1}}. \end{aligned}$$

Corollary 8.7. *If K_{n+d+3}/K_n is small for all $n \geqslant d+3$, then z_n/e_n is unbounded.*

Proof. For $i \geqslant 1$

$$\frac{z_{n+i(d+3)}}{e_{n+i(d+3)}} \geqslant (p-1)^i p^i \frac{z_{n+1}}{e_{n+1}}.$$

However, in our setting this sequence is not bounded:

Lemma 8.8. *The sequence z_n/e_n is bounded.*

Proof. Let $H = G/G_n$. Suppose there exists $s \in H[i]\backslash H[i+1]$ for $i \geqslant e_n+1$. We shall show that this contradicts Corollary 3.2 which shows that $H[i]$ is a finite p-group.

Let π denote a uniformiser for K_n and $\mathcal{M}$ the maximal ideal that it generates. Recall that $e_n = v_{K_n}(p)$. Since K_n/K is totally ramified, $\mathcal{O}_{K_n}$ is generated by π as an $\mathcal{O}_K$-module, i.e., $\mathcal{O}_{K_n} = \mathcal{O}_K[\pi]$. Hence $s(x) \equiv x \bmod \mathcal{M}^i$ if and only if $s(\pi) \equiv \pi \pmod{\mathcal{M}^i}$. Since $s \in H[i]\backslash H[i+1]$ there exists $a \in \mathcal{M}^i\backslash\mathcal{M}^{i+1}$ such that $s(\pi) = \pi(1+a)$. Now $a = \pi^i\alpha$ for some $\alpha \in \mathcal{O}_{K_n}$ and $s(\alpha) = \alpha \pmod{\mathcal{M}^i}$ so

$$\begin{aligned} s(a) &\equiv \pi^i(1+a)^i\alpha \pmod{\mathcal{M}^{2i}} \\ &\equiv a \pmod{\mathcal{M}^{2i}}. \end{aligned}$$

Hence

$$\begin{aligned} s^p(\pi) &\equiv \pi(1+a)^p \pmod{\mathcal{M}^{2i+1}} \\ &\equiv \pi(1+pa) \pmod{\mathcal{M}^{i+e_n+2}} \end{aligned}$$

since $i \geqslant e_n+1$. But $\pi pa \in \mathcal{M}^{i+e_n+1}\backslash\mathcal{M}^{i+e_n+2}$ so $s^p \in H[i+e_n]\backslash H[i+e_n+1]$. We can then repeat the above to show that for all $m > 0$, $s^{p^m} \in H[i+me_n]\backslash H[i+me_n+1]$.

But this contradicts s having order a power of p. Hence our original assumption that $H[i] \neq 1$ for $i \geqslant e_n + 1$ was false. So $z_n \leqslant e_n$, i.e., z_n/e_n is bounded, in fact bounded by 1. (It is possible to do better than this and get z_n/e_n bounded by $(p-1)^{-1}$ (see [23] IV Section 2 Exercise 3).

Combining Corollary 8.7 and Lemma 8.8 we reach our first goal:

Corollary 8.9. *There exist n_0 and n_1 for which K_{n_1}/K_{n_0} is a non-small abelian layer.*

Before establishing the existence of infinitely many such non-small layers, we show now why this would imply that the sequence (x_n) is unbounded:

Lemma 8.10. *Let $n > 1$ and $1 \leqslant r \leqslant n$ and suppose that the abelian extension K_{n+r}/K_n is not small. Then $x_{n+r} = x_{n+r-1} + e$.*

Proof. By Lemma 8.4(ii) it suffices to prove that

$$u_{G_n/G_{n+r}} = u_{G_n/G_{n+r-1}} + e_n;$$

recall that $e_n = e_{K_n} = |G : G_n|e$. Since K_{n+r}/K_n is not small, $u_{G_n/G_{n+r}} - e_n > e_n/(p-1)$. Hence, by Lemma 8.2(ii),

$$G_n/G_{n+r}(u_{G_n/G_{n+r}} - e_n + \epsilon)^p = G_n/G_{n+r}(u_{G_n/G_{n+r}} + \epsilon) \begin{cases} \neq 1 \text{ if } \epsilon = 0 \\ = 1 \text{ if } \epsilon > 0. \end{cases}$$

Hence

$$G_n/G_{n+r}(u_{G_n/G_{n+r}} - e_n + \epsilon) \begin{cases} \nleqslant G_{n+r-1}/G_{n+r} \text{ if } \epsilon = 0 \\ \leqslant G_{n+r-1}/G_{n+r} \text{ if } \epsilon > 0 \end{cases}$$

since the p-power map takes G_k bijectively onto G_{k+1} for all $k \geqslant 1$. Since ramification groups in the upper numbering are preserved on going to quotients we infer that

$$u_{G_n/G_{n+r-1}} = u_{G_n/G_{n+r}} - e_n$$

as required.

Having got one non-small abelian layer in Corollary 8.9, we go on to show that K_{n+2}/K_n is not small for all large n. The proof is not as direct as one might expect but involves a curious to and fro argument between the top and bottom of these non-small abelian layers. It relies on a sister lemma to Lemma 8.10. In Lemma 8.10 we showed that if K_{n+r}/K_n is not small then the difference between the conductors of K_{n+r} and K_{n+r-1} is given by $x_{n+r} - x_{n+r-1} = e$. In this next lemma we get instead a bound for the difference of the conductors of K_n and K_{n+1}.

Lemma 8.11. *If K_{n+r}/K_n is a small abelian layer, then*

$$x_{n+1} - x_n < \frac{e}{p^{r-2}(p-1)}.$$

Proof. By Lemma 8.4 (ii)

$$x_{n+1} - x_n < \frac{1}{|G : G_n|} u_{G_n/G_{n+1}}.$$

By Claim 3 of Lemma 8.6

$$p^{r-1} u_{G_n/G_{n+1}} \leqslant u_{G_n/G_{n+r}} \leqslant \frac{pe_n}{(p-1)}$$

because K_{n+r}/K_n is small. The result follows since $e_n = e|G : G_n|$.

We can now deduce

Proposition 8.12. *There exists n_1 such that for all $n \geqslant n_1$, $x_{n+1} = x_n + e$ and K_{n+2}/K_n is a non-small layer.*

Proof. By Corollary 8.9 K_{n_1}/K_{n_0} is a non-small layer for some n_0 and n_1. Hence $x_{n_1} = x_{n_1-1} + e$. But now we can apply Lemma 8.11 to deduce that K_{n_1+1}/K_{n_1-1} is not small. We apply Lemma 8.10 again to get $x_{n_1+1} = x_{n_1} + e$ which in turn implies that K_{n_1+2}/K_{n_1} is not small, by Lemma 8.11. Repeating the argument gives the result.

Corollary 8.13. *A totally ramified Galois extension of local fields of characteristic zero with p-adic analytic Galois group is deeply ramified.*

Proof. This follows from Proposition 8.12 by Lemma 8.3, since the p-adic analytic Galois group has an open normal uniform subgroup.

In fact the above argument can be applied in a more general setting, outside the context of p-adic analytic groups. We will use this in the last section.

Theorem 8.14. *Suppose that a pro-p group G has a filtration by a descending chain of open normal subgroups G_i which form a fundamental system of neighbourhoods of 1. Assume that $G_i = \{g \in G : g^p \in G_{i+1}\}$ for $i \geqslant 1$. Let G be the Galois group of a totally ramified extension L of a local number field K with the absolute ramification index e. Denote by K_i the fixed field of G_i. Suppose that G_i/G_{i+2} is abelian for every $i \geqslant i_1$ and let K_{i_1+1}/K_{i_0} be a non-small abelian extension for some $i_0 \leqslant i_1$. Then none of G_i/G_{i+2} is small and the sequence $x_i = u_{G/G_i}$ has the property $x_{i+1} = x_i + e$ for $i \geqslant i_1$.*

Proof. We argue as in Proposition 8.12. Since K_{i_1+1}/K_{i_0} is non-small, we get $x_{i_1} = x_{i_1-1} + e$. Apply Lemma 8.11 to deduce that K_{i_1+1}/K_{i_1-1} is not small. We apply Lemma 8.10 again to get $x_{i_1+1} = x_{i_1} + e$ which in turn implies K_{i_1+2}/K_{i_1} is not small by Lemma 8.11.

The main result is now a simple application of Herbrand's Theorem:

Theorem 8.15. *A p-adic analytic totally ramified extension L/K is arithmetically profinite.*

Proof. Let $G = \mathrm{Gal}(L/K)$, an infinite p-adic analytic group. By Corollary 8.13 we know that L/K is deeply ramified, i.e., $G(x)$ is a closed subgroup of G different from 1 for every x. Consider the quotient $G/G(x)$. It is a p-adic analytic group and the Galois group of some M over K. If it is infinite, then by Corollary 8.13 we should have $\mathrm{Gal}(M/K)(u) \neq 1$ for every u. However, $\mathrm{Gal}(M/K)(u) = G(u)G(x)/G(x) = G(x)/G(x) = 1$ for $u \geqslant x$, since the ramification groups are preserved on going to quotients. Hence $G/G(x)$ must be finite, i.e., $G(x)$ is open in G for every x and hence L/K is arithmetically profinite.

However Serre conjectured more than just the ramification subgroups being open. He conjectured that they be compatible with the lower p-series in the sense of the following theorem, the full force of Sen's result:

Theorem 8.16. *Let L/K be a totally ramified Galois extension of a local field of characteristic zero, with p-adic analytic Galois group G. Then there exists a constant c such that*

$$G(ne + c) \subset G_n \subset G(ne - c)$$

for all n.

So far we have exploited properties of the filtration G_n. The proof that a p-adic analytic extension is arithmetically profinite implies that each $G(x)$ contains some term of the lower p-series G_n. But to show the close relationship between the indexing by ramification subgroups and the lower p-series stated above, we need to use a little more of the Lie theory. Actually the previous work has already more or less established one half of this statement:

Lemma 8.17. *There exists a constant c such that for all $n \geqslant n_1$, $G(ne+c) \subseteq G_n$ and*

$$G(ne + c)^p G_{2n} = G((n + 1)e + c)G_{2n}.$$

Proof. By Proposition 8.12, with $c_1 = x_{n_1} - n_1 e$ for $n \geqslant n_1$ and $\varepsilon > 0$, $x_n = ne + c_1$, i.e., $G(ne + c_1) \nsubseteq G_n$ but $G(ne + c_1 + \varepsilon) \subseteq G_n$. Put $c = c_1 + e/(p - 1)$; then $G(ne + c) \subseteq G_n$. By Lemma 8.4 (i) we have $G(ne + c) = G(x_n + e/(p - 1)) = G_n(z_n + e_n/(p - 1))$. Since G_n/G_{2n} is abelian and $z_n + e_n/(p - 1) > e_n/(p - 1)$, Lemma 8.2 implies that

$$\begin{aligned} G_n(z_n + e_n/(p-1))^p G_{2n} &= G_n(z_n + e_n/(p-1) + e_n)G_{2n} \\ &= G(x_n + e/(p-1) + e)G_{2n} \\ &= G((n+1)e + c)G_{2n}. \end{aligned}$$

The following remarks are independent of the theory of ramification subgroups and apply to any sequence of normal subgroups $G\{n\} \subseteq G_n$ for $n \geqslant n_1$ satisfying $G\{n\}^p G_{2n} = G\{n + 1\}G_{2n}$. There is a natural way to define from such a sequence a submodule inside the Lie algebra $L(G)$ of G. Let us recall the implicit definition of the Lie algebra $L(G)$, referring to Chapter 4 of [5] for details.

The additive structure on $L(G)$ can be defined as the inverse limit of the abelian groups G_n/G_{2n} where the connecting maps are given by the homomorphisms $G_{n+1}/G_{2(n+1)} \to G_{n+1}/G_{2n+1} \overset{p^{-1}}{\to} G_n/G_{2n}$. We can define the Lie bracket on $L(G)$ by defining a Lie bracket on G_n/G_{2n} which is compatible with the inverse limit: for $g_n, h_n \in G_n$ define

$$(g_nG_{2n}, h_nG_{2n})_n = ([g_n, h_n]G_{3n})^{p^{-n}} = [g_n, h_n]^{p^{-n}}G_{2n}.$$

The condition $G\{n\}^pG_{2n} = G\{n+1\}G_{2n}$ means that the following is a well-defined additive subgroup of $L(G)$:

$$M = \varprojlim G\{n+1\}G_{2n}/G_{2n}.$$

The object M measures in some sense how far away the filtration $G\{n\}$ is from G_n. In particular:

Lemma 8.18. *If M has finite index as an additive subgroup in $L(G)$, then there exists n_1 such that $G_{n+n_1} \subseteq G\{n\}$ for n large enough.*

Proof. $p^{n_1}L(G) \subseteq M$ for some n_1 since M has finite index in $L(G)$. Therefore for n large enough

$$G_{n+n_1}/G_{2n} \subseteq G\{n+1\}G_{2n}/G_{2n}.$$

Therefore if $\{g_1, \ldots, g_d\}$ is a generating set for G_{n+n_1}, then there exist $\overline{g_1}, \ldots, \overline{g_d} \in G_{2n}$ such that $g_1\overline{g_1}, \ldots, g_d\overline{g_d} \in G\{n+1\}$. But G_{2n} is contained in the Frattini subgroup of G_{n+n_1} hence $G_{n+n_1} = \langle g_1, \ldots, g_d\rangle = \langle g_1\overline{g_1}, \ldots, g_d\overline{g_d}\rangle \subseteq G\{n+1\}$.

By identifying G_n/G_{2n} with G_1/G_{n+1} using the map $x \to x^{p^{-(n-1)}}$ we can identify the set $L(G)$ with G itself. We would like M to then define a normal subgroup inside G. This will be true if M satisfies $(L(G), M) \subseteq pM$ ($(L(G), M) \subseteq 4M$ if $p = 2$). This follows from the Baker–Campbell–Hausdorff formula. However this is not true in general for M. Instead we define the 'envelope of M':

$$N = \left\{g \in L(G) : p^ng \in M \text{ for some } n\right\} = \mathrm{env}(M).$$

N can also be defined as $\mathbb{Q}_pM \cap L(G)$, or the largest subgroup in which M is open.

The following is now true:

Lemma 8.19. *N satisfies $(L(G), N) \subseteq pN$ (resp. $(L(G), N) \subseteq 4N$ if $p = 2$) and hence defines a normal subgroup in G.*

Proof. Since $L(G)$ is powerful, $(L(G), L(G)) \subseteq pL(G)$ ($(L(G), L(G)) \subseteq 4L(G)$ if $p = 2$). Hence it suffices to check that $(L(G), N) \subseteq N$ since $p^nL(G) \cap N = p^nN$. Let $g \in L(G)$ and $h \in M$. The Lie bracket (g, h) when we identify $L(G)$ with G is then $(g, h) = \lim[g^{p^n}, h^{p^n}]^{p^{-2n}}$. Now N is closed. So it suffices to show that $[g^{p^n}, h^{p^n}]^{p^{-2n}} \in N$ for all n. This is the case if and only if $[g^{p^n}, h^{p^n}] \in N$, by the definition of N. So it suffices to show just that if $g \in L(G)$ and $h \in M$, then $[g, h] \in M$. Recall that we are identifying $L(G)$ with G to define $[g, h]$. Hence $M = \varprojlim G\{n+1\}^{p^{-(n-1)}} G_{n+1}/G_{n+1}$ under this identification. Since $G\{n\}$ is normal in G, this implies that M is stable under the action of g by conjugation (i.e., under the adjoint action). Therefore $[g, h] \in M$ and the lemma is established.

We now return to the case that $G\{n\} = G(ne + c)$, the ramification subgroups in the upper numbering.

Theorem 8.20. *There exists n_1 such that for large n, $G_{n+n_1} \leqslant G(ne + c)$.*

Proof. In view of Lemma 8.18, it will suffice to show that the additive subgroup M corresponding to the series of subgroups $G(ne + c)$ is open in $L(G)$. Since $L(G)$ is a finitely generated $\mathbb{Z}_p$-module, it suffices to prove that N is open as a normal subgroup of G.

Let us assume that N is of infinite index in G, and aim to derive a contradiction.

Since $\overline{G} = G/N$ is a powerful finitely generated pro-p group and is torsion-free, by the definition of N, $\overline{G}$ is a uniform pro-p group. Hence we can apply the same analysis as above to understand the ramification subgroups in the upper numbering $\overline{G}(x)$ of the field extension corresponding to $\overline{G}$. In particular, we define $\overline{x_n}$ to be the conductor of the field extension corresponding to the quotient $\overline{G}/\overline{G}_n$ where $\overline{G}_n = G_nN/N$ is the lower p-series of $\overline{G}$. Then since $\overline{G}$ is infinite, by assumption, there exists a constant $\overline{c_1}$ such that $\overline{x_n} = ne + \overline{c_1}$ for $n \geqslant \overline{n_1}$. Note that $\overline{e} = e$. But recall that ramification subgroups in the upper numbering are preserved on going to quotients. Hence for $n \geqslant n_1$

$$\begin{aligned} \overline{G}((n+1)e + c) &= G((n+1)e + c)N/N \\ &\leqslant G_{2n}N/N = \overline{G}_{2n}. \end{aligned}$$

This follows because $G((n+1)e + c)G_{2n} = MG_{2n} \leqslant NG_{2n}$ by definition of M and N. This means that for all $n \geqslant \max\{\overline{n_1}, n_1\}$ we have $(n+1)e + c > 2ne + \overline{c_1}$; as this is false for large n, we have the desired contradiction.

Theorem 8.16 follows from Theorem 8.20 and Lemma 8.17 by suitably adjusting our choice of c.

Since we have been interested in extensions whose Galois groups are p-adic analytic groups, we take the opportunity to record here a number of field theoretic ways to detect whether the extension is p-adic analytic.

Theorem 8.21. *Suppose that L is a Galois extension of a local field K with Galois group G. Then G is p-adic analytic if and only if the following equivalent conditions are satisfied:*

(1) Let $a_n(L/K)$ denote the number of field extensions of K contained in L of degree n; then $a_n(L/K)$ grows polynomially as a function of n;

(2) There is a bound N such that for all finite extensions $F' \geqslant F \geqslant K$ contained in L, $[F^\times : (F^\times)^p N(F'/F)] \leqslant N$.

Proof. Part (1) follows from the fact that finite extensions correspond to open subgroups and the number of open subgroups $a_n(G)$ of index n in a pro-p group G grows polynomially if and only if the pro-p group G is p-adic analytic.

Part (2) follows from local class field theory. In particular, let F_{ab} be the maximal abelian extension of F. Then $N(F'/F) = N((F' \cap F_{ab})/F)$ and the fundamental equality of local class field theory implies that $[F^\times : (F^\times)^p N((F' \cap F_{ab})/F)]$ is the size of the maximal abelian exponent-p quotient of $\mathrm{Gal}(L/F)$. (See Section 7.1 of [10].) Hence (2) is equivalent to a bound on the size of finite abelian exponent p-sections of finite index in G. This is the definition of the group G having finite rank which is equivalent to G being p-adic analytic.

Interlude A of [5] includes many other equivalent conditions on a pro-p group to be p-adic analytic which can be similarly translated. However these two above are nice in that they don't refer to the Galois group but just to the field extensions. For example the criterion that G contains an open normal powerful subgroup is equivalent to the existence of some finite Galois extension F/K, such that if F' is the fixed field of $\mathrm{Gal}(L/F)^p$ (or $\mathrm{Gal}(L/F)^4$ if $p = 2$) then $[F^\times : N(F'/F)] = [F' : F]$. This last equality only holds if the field extension is abelian. It would be nice if there were a purely field theoretic way to define the field F'.

9 Subgroups of $\mathcal{N}$ and the Coates–Greenberg problem

One corollary of Camina's work described in Section 7 is the fact that every finitely generated pro-p group can be realised as the Galois group of an arithmetically profinite extension in characteristic p. We have shown in the last section that the totally ramified field extensions of characteristic zero which have p-adic analytic Galois groups are always arithmetically profinite.

Until recently it was not known if there were any other examples of arithmetically profinite or deeply ramified extensions in characteristic zero which were not p-adic analytic or had infinite p-adic analytic quotients (which is sufficient to imply that the field is deeply ramified). In [6] an example of a non p-adic analytic arithmetically profinite extension is given. However this example does have infinite p-adic analytic quotients.

Since deeply ramified extensions have a nice Kummer theory as described by Coates and Greenberg, they were interested to know if there were other examples other than those coming from p-adic analytic groups. They therefore posed the following:

The Coates–Greenberg Problem Does there exist a totally ramified and deeply ramified Galois extension L/F such that F is a finite field extension of $\mathbb{Q}_p$ and the Galois group of L/F does not have any infinite p-adic analytic quotients?

We now describe a solution to this problem.

For $m \geqslant 2$ define the following closed subgroups in $\mathcal{N} = \mathcal{N}(\mathbb{F}_p)$:

$$S_m = \left\{ \sum_{i \geqslant 0} a_i X^{1+mi} : a_0 = 1, a_i \in \mathbb{F}_p \right\}.$$

For m relatively prime to p the group of principal units $1 + X^m\,\mathbb{F}_p[[X^m]]$ is uniquely m-divisible. Let $\sigma \in \mathcal{N}$, considered as a wild automorphism of $\mathbb{F}_p((X^m))$, where $\sigma(X^m) = X^m f(X^m)$ with $f(X) \in 1+X\mathbb{F}_p[[X]]$; to σ we associate the automorphism $\tau \in S_m$, considered as a wild automorphism of $\mathbb{F}_p((X))$, where $\tau(X) = X\sqrt[m]{f(X^m)}$. This gives an isomorphism of $\mathcal{N}$ with S_m and an isomorphism of S_{p^r} with S_{mp^r}.

Fix $r \geqslant 1$, write $q = p^r$, $T = S_q$ and put

$$T_i = \{f(X) \in T : f(X) \in X + X^{1+qi}\,\mathbb{F}_p[[X]]\}.$$

The subgroup T has the following properties: (1) If $\sigma \in T_i \setminus T_{i+1}$, then $\sigma^p \in T_{pi} \setminus T_{pi+1}$; (2) $[T_i, T_i] \leqslant T_{(q+1)i+1}$, the group $T_i/T_{(q+1)i}$ is abelian of exponent pq; (3) T is hereditarily just-infinite, (4) T has not more than q generators; (5) T is not p-adic analytic (see [7]).

For instance, if $\alpha \in F = \mathbb{F}_p((X))$ and $v_F(\alpha)$ is relatively prime to p, then $v_F(\alpha^{\sigma-1} - 1) = v_F(\alpha) + i(\sigma)$. Hence $i(\sigma^p) = v_F(t^{\sigma^p-1}) = v_F(t^{(\sigma-1)^p}) = pi(\sigma)$. It is more difficult to check that T is hereditarily just-infinite. If $i < j$ and j is relatively prime to p, then $[X + aX^{1+qj}, X + bX^{1+qi}] = X + jabX^{1+qj+q^2i} + \dots$. Therefore $T_l \leqslant T_{l+1}[T,T]$ for all $l \geqslant q+2$ relatively prime to p. Then $T_{pi} \leqslant T_{pi+1}T^p$. If T were a p-adic analytic group, then by [5], Theorem 9.34 it would contain an open subgroup G which is a uniformly powerful pro-p group, so in particular $G/\overline{G^p}$ is abelian, see [5], Section 3.1. However, the subgroup $[G,G]$ contains some elements $X + X^{1+qi} + \dots$ with i relatively prime to p which are obviously not in the group $\overline{G^p} = G^p = \{g^p : g \in G\}$ (see [5], Theorem 3.6).

Now let $q > p$ and let F be an unramified extension of $\mathbb{Q}_p$ of degree $\geqslant q$.

Let L/F be a Galois extension with the Galois group isomorphic to T (it exists, since according to Shafarevich's theorem [25] the pro-p part of the absolute Galois group of F is a free pro-p group with $|F : \mathbb{Q}_p| + 1$ generators).

Theorem 9.1. *(1) The extension L/F is deeply ramified.*

(2) The extension L/F is arithmetically profinite and $\mathrm{Gal}(K/M)$ *is not p-adic analytic for any intermediate subfields $F \subseteq M \subseteq K \subseteq L$ with $|M : F| < +\infty$ and K/M being an infinite Galois extension.*

Proof. (1) Denote by F_i the fixed field of the subgroup in $\mathrm{Gal}(L/F)$ corresponding to T_i. Let F_{unp} be the maximal unramified p-extension of F.

We are going to apply Theorem 8.14 to certain subgroups G_i of T.

Due to the description of T the quotient group T_1/T_q decomposes into the direct product of the cyclic group of order q generated by $t + t^{1+q}$, the cyclic group of order q generated by $t + t^{1+2q}$ and an abelian group. Therefore the Galois group of the extension $F_q/F_q \cap F_{unp}$ decomposes into the direct product of the cyclic group of order q and an abelian group. Since the field $F_q \cap F_{unp}$ is absolutely unramified, local class field theory implies that $F_q/F_q \cap F_{unp}$ is non-small. Put $I = \mathrm{Gal}(L/L \cap F_{unp})$. Then I as a normal closed subgroup of T is of finite index in T, since T is hereditarily just-infinite.

Define $G_i = T_{p^{i-1}} \cap I$, $G = G_1$. Since T_{p^i}/T_{qp^i} is abelian, G_i/G_{i+2} is abelian for every $i \geq 1$. The abelian field extension $F_q(L \cap F_{unp})/L \cap F_{unp}$ is non-small, since

$F_q/F_q \cap F_{unp}$ is non-small. From the description of the structure of the group T we deduce that G_i coincides with $\{\sigma \in G : \sigma^p \in G_{i+1}\}$. Thus, all conditions of Theorem 8.14 are satisfied and the extension L/F is deeply ramified.

(2) The extension L/F is arithmetically profinite, since $\mathrm{Gal}(L/F)(x)$ is a nontrivial normal closed subgroup of T, therefore it is open. Normal extensions of K in L are either finite over K or coincide with L; in the latter case they are not p-adic analytic, since T is not.

This theorem answers affirmatively the Coates–Greenberg problem as stated in [4].

References

[1] Y. Barnea and B. Klopsch, Index Subgroups of the Nottingham Group, preprint.

[2] Y. Barnea and A. Shalev, Hausdorff dimension, pro-p groups, and Kac–Moody algebras, *Trans. of the A.M.S.*, **349** (1997), 5073–5091.

[3] R. Camina, Subgroups of the Nottingham group, *J. Algebra*, **196** (1997), 101–113.

[4] J. Coates and R. Greenberg, Kummer theory for abelian varieties over local fields, *Invent. Math.*, **124** (1996), 129–174.

[5] J. D. Dixon, M. P. F. du Sautoy, A. Mann and D. Segal, *Analytic pro-p groups*, 2nd edition, Cambridge Studies in Advanced Math. 61, Cambridge Univ. Press, 1999.

[6] I. Fesenko, On deeply ramified extensions, *J. London Math. Soc.*, **57**(2) (1998), 325–335.

[7] I. Fesenko, On just infinite pro-p groups and arithmetically profinite extensions of local fields, *J. Reine Angew. Math.*, **517** (1999) 61–80.

[8] I. Fesenko and S. Vostokov, *Local Fields and Their Extensions*, A.M.S., Providence, R.I., 1993.

[9] K. Iwasawa, On solvable extensions of algebraic number fields, *Ann. Math.*, **58** (1953), 548–572.

[10] K. Iwasawa, *Local Class Field Theory*, Oxford Mathematical Monographs O.U.P., 1986.

[11] S. A. Jennings, Substitution groups of formal power series, *Canad. J. Math.*, **41** (1954), 325–340.

[12] D. L. Johnson, The group of formal power series under substitution, *J. Austral. Math. Soc. (Series A)*, **45** (1988), 298–302.

[13] Y. Kawada, On the ramification theory of infinite algebraic extensions, *Ann. Math.*, **58** (1953), 24–47.

[14] B. Klopsch, Normal subgroups and automorphisms of substitution groups of formal power series, Dissertation Oxford 1997.

[15] F. Laubie, Sur la ramification des extensions de Lie, *Comp. Math.*, **98** (1985), 253–262.

[16] F. Laubie, Extensions de Lie et groupes d'automorphismes de corps locaux, *Comp. Math.*, **67** (1988), 165–189.

[17] F. Laubie and M. Saine, Ramification of automorphisms of $k((t))$, *J. Number Theory*, **63** (1997), 143–145.

[18] A. Lubotzky and J. S. Wilson, An embedding theorem for profinite groups, *Arch. Math.*, **42** (1984), 397–399.

[19] E. Maus, Über die Verteilung der Grundverzweigungszahlen von wild verzweigten Erweiterungen p-adischer Zahlkörper, *J. Reine Angew. Math.*, **257** (1972), 47–79.

[20] E. Maus, Relationen in Verzweigungsgruppen, *J. Reine Angew. Math.*, **258** (1972), 23–50.

[21] Sh. Sen, On automorphisms of local fields, *Ann. Math.*, **90** (1969), 33–46.

[22] Sh. Sen, Ramification in p-adic Lie extensions, *Invent. Math.*, **17** (1972), 44–50.

[23] J.–P. Serre, *Corps Locaux*, 2nd ed., Hermann, Paris, 1968.

[24] J.–P. Serre, Sur les groupes de Galois attachés aux groupes p-divisibles, in *Proceedings of Conference on Local Fields Dreibergen, 1966*, Berlin-Heidelberg-New York, Springer, 1967.

[25] I.R. Shafarevich, On p-extensions (Russian), *Math. Sbornik*, **20** (1947), 351–363.

[26] A. Weil, *Basic Number Theory*, 2nd ed., Springer-Verlag, Berlin, 1973.

[27] J.–P. Wintenberger, Extensions de Lie et groupes d'automorphismes des corps locaux de caractéristique p, *C. R. Acad. Sci. série A*, **288** (1979), 477–479.

[28] J.–P. Wintenberger, Extensions abéliennes et groupes d'automorphismes des corps locaux, *C. R. Acad. Sc.*, **290**, série A (1980), 201–203.

[29] J.–P. Wintenberger, Le corps des normes de certaines extensions infinies des corps locaux; applications, *Ann. Sci. E.N.S., 4 série*, **16** (1983), 59–89.

[30] I. O. York, The group of formal power series under substitution, Ph.D. Thesis, Nottingham 1990.

DPMMS, Centre for Mathematical Sciences
Wilberforce Road
Cambridge CB3 0WB, UK
dusautoy@dpmms.cam.ac.uk
http://www.dpmms.cam.ac.uk/~dusautoy

Department of Mathematics
University of Nottingham
University Park
Nottingham NG7 2RD, UK
ibf@maths.nott.ac.uk

11

p-adic Galois Representations and pro-p Galois Groups

Nigel Boston

0 Introduction

We study the intimate interactions between the theory of p-adic Galois representations and the structure of pro-p Galois groups. In particular, information passes in both directions. Algebraic geometry, for instance in the guise of elliptic curves and modular forms, yields naturally occurring Galois representations, whereas on the other side, cohomological techniques and variants on class field theory tell us about the generators and relations of the pro-p Galois groups. In the case of pro-p extensions ramified at (primes above) p, this combination works together rather well to elucidate the structure of the set of Galois representations. In the case of pro-p extensions unramified at p, both sides are poorly understood, but there is the fundamental conjecture of Fontaine–Mazur claiming that such representations should have finite image (since algebraic geometry can produce no others). This has very interesting consequences for the corresponding pro-p Galois groups, possibly producing a new family of just-infinite pro-p groups.

Throughout this paper, since we are working with profinite groups, all subgroups and maps considered will be closed and continuous respectively. Generation will always refer to topological generation.

1 Introduction to p-adic Galois representations

It is an understatement to say that number theorists are interested in the structure of the profinite group $G_{\mathbb{Q}} := \mathrm{Gal}(\overline{\mathbb{Q}}/\mathbb{Q})$ (where $\overline{\mathbb{Q}}$ denotes an algebraic closure of $\mathbb{Q}$).

The author was partially supported by NSF grant DMS 96-22590.

A better understanding of this group would allow us to answer many more number-theoretic problems.

From [1], "... there are three separate but intertwined methods of studying the structure of finite groups. First, we have the oldest: *permutation groups*. One studies the G-sets for a group G, the sets that G acts on ... Second, there is the method of *representation theory*, where one deals with G-modules ... Third, there is the *local method*, that part of the subject that begins with the Sylow theorems."

We shall adopt the second approach through most of this paper. A challenge is to develop the third approach for profinite groups, in particular for $G_{\mathbb{Q}}$, which has some naturally occurring local subgroups, namely:

Definition. ([32]) Let $H = \mathrm{Gal}(K/\mathbb{Q})$ be any finite quotient of $G_{\mathbb{Q}}$ and $\wp$ a prime ideal of O_K (the ring of integers of K) above p, i.e., $\wp$ appears in the factorisation of pO_K (note that O_K is a Dedekind domain so has unique factorisation of ideals). The stabiliser of $\wp$ in the action of H on the prime ideals above p is the decomposition subgroup $D_\wp$ of H. Since the action is transitive, all such stabilisers are conjugate. The *decomposition subgroup* D_p of $G_{\mathbb{Q}}$ is defined (up to conjugacy) to be the inverse limit of the $D_\wp$, for compatibly chosen $\wp$, over all finite quotients H of $G_{\mathbb{Q}}$.

The kernel of the action of $D_\wp$ on $O_K/\wp$ is called the inertia subgroup $I_\wp$ of H. Taking inverse limits yields the *inertia subgroup* I_p of $G_{\mathbb{Q}}$, again defined up to conjugacy.

Indeed, I_p is a normal subgroup of D_p with quotient D_p/I_p, which is naturally isomorphic to the procyclic group $\mathrm{Gal}(\overline{\mathbb{F}_p}/\mathbb{F}_p)$ which has a distinguished generator, the Frobenius element Fr_p, which sends $x \mapsto x^p$. These are local subgroups because the first ramification subgroup (or wild inertia subgroup) P_p of I_p is its Sylow pro-p group [45], normal in both I_p and D_p. The group $D_p \cong G_{\mathbb{Q}_p}$ is a finitely generated prosoluble group, for which a presentation is known [30], at least for $p > 2$. The quotient D_p/P_p is easy to describe [45], namely $I_p/P_p \cong \prod_{q \neq p} \mathbb{Z}_q$ is procyclic, generated by τ, and $D_p/I_p \cong \hat{\mathbb{Z}}$ is procyclic, generated say by σ, and then there is the one relation $\tau^\sigma = \tau^p$. It is interesting to note [39] that D_p is embedded in $G_{\mathbb{Q}}$ like a Frobenius complement, i.e., $D_p \cap D_p^g = \{1\}\ \forall g \notin D_p$.

Complex conjugation fits into this picture by setting D_∞ to be the subgroup generated by it. Up to conjugation, this is the only torsion in $G_{\mathbb{Q}}$.

The theory of complex representations, $G_{\mathbb{Q}} \to GL_n(\mathbb{C})$, is important but of little interest to us here since the images are finite ($GL_n(\mathbb{C})$ has no infinite profinite subgroups). We will discuss the theory of p-adic representations, $G_{\mathbb{Q}} \to GL_n(\mathbb{Z}_p)$ (and over similar rings). We begin by recounting what is already known regarding the structure of $G_{\mathbb{Q}}$.

1.1 Properties of $G_{\mathbb{Q}}$

If $G_K (= \mathrm{Gal}(\overline{\mathbb{Q}}/K))$ is an open subgroup (so $K/\mathbb{Q}$ is a finite extension), then by class field theory [40] we know the abelianisation G_K^{ab} (often even as a $G_{\mathbb{Q}}/G_K$-module if $K/\mathbb{Q}$ is normal). For instance, $G_{\mathbb{Q}}^{ab} \cong \prod_p \mathbb{Z}_p^*$ since the maximal extension of $\mathbb{Q}$ with abelian Galois group, $\mathbb{Q}^{ab} = \mathbb{Q}(\mu_\infty)$, i.e., is the extension obtained by adding all roots of unity to $\mathbb{Q}$ [40]. Knowledge of G_K^{ab} will be repeatedly seen to be of use in this article.

In fact, we have the cyclotomic characters, $\chi_p : G_{\mathbb{Q}} \to GL_1(\mathbb{Z}_p) = \mathbb{Z}_p^*$, surjective maps defined by $\sigma(\zeta) = \zeta^{\chi_p(\sigma)} (\sigma \in G_{\mathbb{Q}}, \zeta \in \mu_{p^\infty}$, the group of p-power roots of 1 in $\overline{\mathbb{Q}}$). This completely describes the case of 1-dimensional p-adic representations since any such must be a (p-adic) power of χ_p (since it factors through $G_{\mathbb{Q}}^{ab}$ given explicitly above).

Further properties of $G_{\mathbb{Q}}$ include that it is complete [28], [56] (meaning that the natural map from it to its automorphism group is an isomorphism). Shafarevich has conjectured that its derived subgroup $G'_{\mathbb{Q}}$ is a free profinite group. Additionally there is the famous inverse problem of Galois theory, which asks whether every finite group occurs as a quotient of $G_{\mathbb{Q}}$. This has been proved for many finite groups [46] but not e.g., for the Mathieu group M_{23}. Birch has asked whether this can even be done just with tamely ramified quotients, i.e., ones that factor through the normal subgroup generated by all P_p.

Our naturally occurring p-adic representations of $G_{\mathbb{Q}}$ will actually factor through $G_{\mathbb{Q},S} := G_{\mathbb{Q}}/$(normal subgroup generated by all $I_p, p \notin S$), where S is some finite set of primes. We may consider Fr_ℓ as an element (at least defined up to conjugacy) of $G_{\mathbb{Q},S}$ if $\ell \notin S$. Since every finite nontrivial extension of $\mathbb{Q}$ is ramified at some prime but at most finitely many primes [32], it follows (i) that $G_{\mathbb{Q},\emptyset} = \{1\}$ and (ii) that given an open subgroup $G_K \trianglelefteq G_{\mathbb{Q}}$, there exists S such that $G_{\mathbb{Q}} \to G_{\mathbb{Q}}/G_K$ factors through $G_{\mathbb{Q},S}$.

Apart from Minkowski's result (i), very little is known in general regarding the structure of $G_{\mathbb{Q},S}$. It is even unknown if it is finitely generated. Harbater's article [25] gives a detailed discussion of what is known in the case $S = \{2\}$. In particular he asks which finite simple groups occur as quotients of $G_{\mathbb{Q},\{2\}}$. A broader question of Gross is whether there exists a quotient of $G_{\mathbb{Q},S}$ with finite insoluble Galois group and S consisting of a single prime $\leqslant 7$. The recent work of Gross on algebraic modular forms has produced evidence of this, such as a conjectural surjection from $G_{\mathbb{Q},\{5\}}$ onto $G_2(\mathbb{F}_5)$. We will see later that there are mod 11 representations associated to the modular form Δ and any of the elliptic curves of conductor 11 defined over $\mathbb{Q}$ (e.g. $y^2+y=x^3-x^2$) and in fact these provide examples of extensions of $\mathbb{Q}$ unramified outside 11, i.e., factoring through $G_{\mathbb{Q},\{11\}}$, with Galois group $GL_2(\mathbb{F}_{11})$. (Note that there are no elliptic curves over $\mathbb{Q}$ of conductor < 11.)

Questions involving linear Galois groups relate to the existence or not of Galois representations and so the key question to ask is what Galois representations exist. Can we classify them, e.g., by relating them to other kinds of objects? Before restricting ourselves to representations of the $G_{\mathbb{Q},S}$, let us note the recent result of Ramakrishna [42], showing that there do indeed exist irreducible representations $G_{\mathbb{Q}} \to GL_2(\mathbb{Z}_p)$ ramified at infinitely many primes (so not factoring through any $G_{\mathbb{Q},S}$). These cannot then come from algebraic geometry (see later) but, assuming the Generalised Riemann Hypothesis, they can be constructed to be "crystalline", introducing another kind of object to be understood. Reducible examples have been known for a long time.

1.2 p-adic representations associated to elliptic curves

The first method of producing p-adic representations employs *elliptic curves* over $\mathbb{Q}$, i.e., curves with the equation $E : y^2 = f(x)$, where $f \in \mathbb{Z}[x]$ is a cubic polynomial with distinct roots. We may sometimes give the curve differently, e.g., $E_{37} : y^2 + y = x^3 - x$ (which may be put in the other form by completing the square). For the theory of elliptic curves, see [53].

If R is any $\mathbb{Z}$-algebra, we will denote by $E(R)$ the set $\{(x, y) \in R^2 : y^2 = f(x)\} \cup \{\infty\}$, where ∞ is an extra point usually referred to as the point at infinity. In particular, $E(\mathbb{C})$ is complex analytically isomorphic to a torus, i.e., a product $S^1 \times S^1$ of two circles, via the Weierstrass $\wp$-function and its derivative. This gives $E(\mathbb{C})$ the structure of a group, with the group law being given formally in terms of the coefficients of f. This gives E the structure of a group scheme, which means that $E(K)$ is a group for any field K over which E is defined and an elliptic curve, with ∞ the identity of the group law.

For $K = \mathbb{R}$, this yields the description of addition as $P + Q + R = \infty$ if and only if P, Q, R are collinear. Since each line intersects the curve in exactly three points, this explains how to add two points on the curve. If two of the points are equal or one of them is ∞, this prescription must be modified. For instance, if r is a root of f, then to add $P = (r, 0)$ to itself, we take the tangent line at P. This apparently does not hit the curve again, but in fact it does at ∞. Thus $2P = \infty$ and P is a point of order 2. Likewise, $-Q$ is the reflection in the x-axis of Q and so to obtain $P + Q$, we draw a line through P and Q, see where it intersects the curve, and then reflect in the x-axis.

Next we consider $E[n] := \{P \in E(\mathbb{C}) : nP = \infty\} \cong \mathbb{Z}/n \times \mathbb{Z}/n$ (since $E(\mathbb{C}) \cong S^1 \times S^1$). Membership in $E[n]$ amounts to the coordinates satisfying polynomials in $\mathbb{Q}[x]$ and so $E[n] \subseteq E(\overline{\mathbb{Q}})$ and $G_{\mathbb{Q}}$ acts on $E[n]$. For instance, $E[2] = \{\infty\} \cup \{(r, 0) \mid f(r) = 0\}$.

Hence, we obtain representations $\rho_{E,n} : G_{\mathbb{Q}} \to GL_2(\mathbb{Z}/n)$. In the case $n = 2$, $\rho_{E,2} : G_{\mathbb{Q}} \to GL_2(\mathbb{Z}/2) \cong S_3$ is given by the action of $G_{\mathbb{Q}}$ on the roots of f. Letting n vary through powers of a prime p, we get a system of compatible maps yielding a representation $\rho_{E,p^\infty} : G_{\mathbb{Q}} \to GL_2(\mathbb{Z}_p)$ which when composed with the natural map $\mathbb{Z}_p \to \mathbb{Z}/p^k$ produces ρ_{E,p^k}.

The important fact, due to Faltings [17], is that ρ_{E,p^∞} and ρ_{E',p^∞} are isomorphic representations if and only if E and E' are what is known as isogenous curves [53]. This means that there is a surjective morphism from E to E' with finite kernel and is only slightly weaker than saying that E and E' are isomorphic. We see then that the isomorphism class of the associated p-adic representation encodes much of the data regarding the elliptic curve, whence their usefulness.

Properties of ρ_{E,p^∞} are as follows:

(1) It factors through $G_{\mathbb{Q},S}$ where $S = \Sigma_E \cup \{p\}$, where Σ_E is the finite set of primes of bad reduction for E (meaning primes dividing the discriminant of E, or equivalently for which the reduction of E mod p acquires singularities). For instance, $\Sigma_{E_{37}} = \{37\}$.

(2) det $\rho_{E,p^\infty} = \chi_p$ (so, in particular, det ρ_{E,p^∞}(complex conjugation) $= -1$ (this is what is termed an "odd" representation) and det $\rho_{E,p^\infty}(Fr_\ell) = \ell$ for $\ell \notin S$).

(3) tr $\rho_{E,p^\infty}(Fr_\ell) = \ell + 1 - \#E(\mathbb{F}_\ell)$ if $\ell \notin S$.

(4) If E has multiplicative reduction (i.e., a node) at $\ell \in S$, then we can even describe $\rho_{E,p^\infty}(D_\ell)$ [47].

(5) If E does not have complex multiplication (meaning that its endomorphism ring is no larger than $\mathbb{Z}$), then, for all but finitely many p, ρ_{E,p^∞} is surjective. (Indeed, for E_{37}, they are surjective for all p [47].)

Exercise. It is often important to see how large the image of a p-adic Galois representation is. One thing that helps us here is the following exercise.

Using an argument with the Frattini subgroup, show that if H is a subgroup of $SL_2(\mathbb{Z}_p)$ mapping onto $SL_2(\mathbb{Z}/p)$ ($p \geqslant 5$), then $H = SL_2(\mathbb{Z}_p)$. The same argument shows that a subgroup of $SL_n(R)$ (R a complete, noetherian local ring with maximal ideal m, R/m finite of characteristic > 2), mapping onto $SL_n(R/m^2)$, must be all of $SL_n(R)$. Is this true if SL_n is replaced by any semisimple group?

The above theory of representations associated to elliptic curves can be generalised in various directions, such as via abelian varieties, p-divisible group schemes, or modular (or automorphic) forms. Most generally, we say that a p-adic representation comes from algebraic geometry if it comes from the action of $G_{\mathbb{Q}}$ on a subquotient of an étale cohomology group of some variety over $\mathbb{Q}$, possibly tensored by a power of χ_p. Such representations always factor through some $G_{\mathbb{Q},S}$ and many such have been shown to be potentially semistable (a technical condition on their restriction to D_p) [19]. The conjecture of Fontaine and Mazur [19] says that these properties should exactly characterise those p-adic Galois representations that come from algebraic geometry.

1.3 p-adic representations associated to modular forms

We are interested in a particular kind of modular form, namely the following.

Definition. (See e.g., [49]) Let N, k be positive integers and $\epsilon : (\mathbb{Z}/N)^* \to \mathbb{C}^*$ be a homomorphism. A *cuspidal eigenform* of type (N, k, ϵ) is a holomorphic function on the upper half-plane Im $(z) > 0$

(i) that satisfies $f(\frac{az+b}{cz+d}) = \epsilon(d)(cz+d)^k f(z)$ for all matrices $\begin{pmatrix} a & b \\ c & d \end{pmatrix}$ in $\Gamma_0(N)$, which is the subgroup of $SL_2(\mathbb{Z})$ consisting of matrices with N dividing c, (taking $a = b = d = 1, c = 0$, it follows that f has a Fourier expansion $\sum a_n q^n$ where $q = e^{2\pi i z}$),

(ii) that vanishes at the cusps (i.e., the Fourier expansion of $z \mapsto (cz+d)^{-k} f(\frac{az+b}{cz+d})$ in powers of $exp(2\pi i z/N)$ only has positive powers, for any $\begin{pmatrix} a & b \\ c & d \end{pmatrix}$ in $SL_2(\mathbb{Z})$), (it follows that $a_n = 0$ for $n \leqslant 0$) and

(iii) that is an eigenvector for the Hecke operators
$T_\ell : \sum a_n q^n \mapsto \sum a_{\ell n} q^n + \epsilon(\ell)\ell^{k-1} \sum a_n q^{\ell n} (\ell \nmid N)$
$U_\ell : \sum a_n q^n \mapsto \sum a_{\ell n} q^n (\ell \mid N)$

Example. Let $\Delta = q \prod_{n=1}^{\infty}(1-q^n)^{24} = \sum_{n=1}^{\infty} \tau(n) q^n$, where $\tau(n)$ are Ramanujan's coefficients. Then Δ is the unique normalised ($a_1 = 1$) cuspidal eigenform of type $(1, 12, 1)$.

Serre realised [49] that all known congruences for $\tau(n)$ would follow if there exist Galois representations $\rho_{\Delta,n} : G_{\mathbb{Q}} \to GL_2(\mathbb{Z}/n)$ for each positive integer n with properties akin to the $\rho_{E,n}$ associated to elliptic curves. For instance, $\tau(\ell) \equiv 1 + \ell^{11} (\text{mod } 691)$ (ℓ prime $\neq 691$) follows if $\rho_{\Delta,691}(G_{\mathbb{Q}}) \subseteq \left\{ \begin{pmatrix} 1 & * \\ 0 & * \end{pmatrix} \right\} \subseteq GL_2(\mathbb{Z}/691)$ and the following properties are established:

(1) $\rho_{\Delta,p^\infty} : G_{\mathbb{Q}} \to GL_2(\mathbb{Z}_p)$ factors through $G_{\mathbb{Q},S}$ where $S = \Sigma_\Delta \cup \{p\}$ with $\Sigma_\Delta = \{\text{primes dividing} N\}$

(2) det $\rho_{\Delta,p^\infty} = \epsilon \chi_p^{k-1}$ (implying in particular that ρ_{Δ,p^∞} is an odd representation) (note that for Δ the character $\epsilon = 1$—the result is stated in the form that generalises to other cuspidal eigenforms)

(3) tr $\rho_{\Delta,p^\infty} = a_\ell (\ell \notin S)$

(4) we can usually describe $\rho_{\Delta,p^\infty}(D_\ell)(\ell \in S)$

(5) for all but finitely many p, $\rho_{\Delta,p^\infty}(G_{\mathbb{Q}})$ contains $SL_2(\mathbb{Z}_p)$ [54].

Deligne [14] established the existence of Galois representations $\rho_{\Delta,n}$ satisfying (1), (2), (3). In fact, he established the corresponding result for any cuspidal eigenform with the modification that $\mathbb{Z}_p$ be replaced by the (finite) extension of it generated by the a_n and values of ϵ.

For Δ, the finite set in (5) is $\{2, 3, 5, 7, 23, 691\}$. This result implies that these are the only primes p for which $\tau(\ell)$ depends on ℓ mod p. In fact, we have a complete congruence theory via the theory of p-adic representations, simply by computing the images of these representations. For this, results like the exercise at the end of the last section are helpful.

For example, the result for $\rho_{\Delta,691}$ now follows since by (2) $\rho_{\Delta,691}(Fr_\ell) = \begin{pmatrix} 1 & * \\ 0 & \ell^{11} \end{pmatrix}$ and so by (3) $\tau(\ell) = a_\ell \equiv 1 + \ell^{11} (\text{mod } 691)(\ell \neq 691)$.

Conjecture. (Serre, [50]) If $\overline{\rho} : G_{\mathbb{Q}} \to GL_2(k)$ (k a finite field) is an odd, absolutely irreducible representation, then it is associated to a cuspidal eigenform (weak conjecture) of a particular type (strong conjecture).

Note that the work of Ribet and others [43] shows that the weak conjecture implies the strong conjecture. There are similar ideas for more general $\overline{\rho}$. Reducible representations are known to have associated Eisenstein series and even representations (i.e., the determinant of the image of complex conjugation is 1) are expected to be associated to Maass wave-forms.

1.4 Interlude on Fermat's Last Theorem

Suppose $a^p + b^p + c^p = 0$, where a, b, c are nonzero integers and p is a prime $\geqslant 5$, i.e., that there is a counterexample to Fermat's Last Theorem (FLT). Then Frey's associated elliptic curve is defined to be $E : y^2 = x(x - a^p)(x + b^p)$, where we without loss of generality take b even and $a \equiv -1 (\text{mod } 4)$ (otherwise we rearrange, and possibly change the signs of, a, b, c).

By Ribet [44], if $\rho_{E,p}$ is associated to a cuspidal eigenform, then it is associated to one of type $(2, 2, 1)$, but there are none of this type! Thus $\rho_{E,p}$ presents a counterexample to Serre's conjecture above. In particular, we are done proving FLT if we can show that all such $\rho_{E,p}$ are associated to cuspidal eigenforms. This is contained in the conjecture that E is "modular". The conjecture of Taniyama-Shimura (whose proof has now been completed by C. Breuil, B. Conrad, F. Diamond and R. Taylor) says that all elliptic curves over $\mathbb{Q}$ are modular, i.e., we can consider three sets as follows.

Fix n. Let $\mathcal{E}$ denote the set of representations $\rho : G_{\mathbb{Q}} \to GL_2(\mathbb{Z}/n)$ that arise from elliptic curves. Let $\mathcal{M}$ denote the set of those representations that arise from modular forms. Since these representations share certain features, we can abstractly define a set, $\mathcal{A}$, of admissible representations $\rho : G_{\mathbb{Q}} \to GL_2(\mathbb{Z}/n)$ that contains both $\mathcal{E}$ and $\mathcal{M}$. (Technically [57] we restrict to semistable elliptic curves and abstractly define semistable representations—since Frey's curves are always semistable, this is permissible.) If we can show that $\mathcal{M}$ actually equals $\mathcal{A}$, then we have shown $\mathcal{E} \subseteq \mathcal{A} = \mathcal{M}$, which is enough to prove cases of the Taniyama-Shimura conjecture.

After breaking $\mathcal{M}$ and $\mathcal{A}$ up further into subsets, Wiles [57] proved his result by parametrising the two sets by rings $\mathcal{T}$ and $\mathcal{R}$, respectively. The inclusion $\mathcal{M} \subseteq \mathcal{A}$ yields a surjection $\mathcal{R} \to \mathcal{T}$ and one of Wiles' greatest contributions was in providing the clever commutative algebra that is enough to show this is an isomorphism. The parametrisation of sets of representations by rings is achieved as follows.

1.5 Deformations of Galois representations

We have seen that there is a conjectural description of the set of representations $\overline{\rho} : G_{\mathbb{Q},S} \to GL_2(k)$ (S any finite set of primes, k any finite field). In this section we generalise this to more general local rings A than finite fields. If the residue field of A is finite, then such a representation produces a representation into $GL_2(k)$ called its residual representation. It turns out that there is a nice theory if we lump together all those representations with the same residual representation.

Given representation $\overline{\rho} : G_{\mathbb{Q},S} \to GL_n(k)$ (k a finite field), we define a functor F from the category $\mathcal{C}$ of complete, noetherian local rings with residue field k (i.e., rings that are quotients of $W(k)[[T_1, ..., T_n]]$ where $W(k)$ is the ring of infinite Witt vectors over k (e.g., $W(\mathbb{F}_p) = \mathbb{Z}_p$), morphisms being local ring homomorphisms that induce the identity map on k) to the category of Sets. This functor is given by $F(R)$ being the set of lifts of $\overline{\rho}$ to R modulo an equivalence relation $\sim$, defined by $\rho_1 \sim \rho_2$ if they are conjugate by an element of $\ker(GL_n(R) \to GL_n(k))$.

Theorem. (Mazur [37]) *If $\overline{\rho}$ is absolutely irreducible (and in fact slightly more generally—if the centraliser of its image consists just of scalars), then F is representable. This means that there exists a universal deformation $\xi : G_{\mathbb{Q},S} \to GL_n(R(\overline{\rho}))$ (a lift defined up to $\sim$) such that $Hom(R(\overline{\rho}), R)$ produces (by composing with ξ) precisely $F(R)$.*

In other words, ξ parametrises all p-adic lifts of $\overline{\rho}$. For $n = 2$ and $\overline{\rho}$ odd, $R(\overline{\rho})$ is typically $W(k)[[T_1, T_2, T_3]]$. This means that, in calculated cases, for a fixed eigenform f (or elliptic curve over $\mathbb{Q}$, which Taniyama-Shimura indicates as a subcase), then "typically" above indicates that the set of primes for which the last statement holds has density 1 (e.g., [12]). We can impose conditions on the lifts to get representable subfunctors of F and to cut $R(\overline{\rho})$ down to more manageable size. For instance, insisting that $\rho(I_p) \subseteq \left\{ \begin{pmatrix} 1 & * \\ 0 & * \end{pmatrix} \right\}$ for our lifts ρ (we say ρ is "ordinary") ($p = \text{char } k$), assuming $\overline{\rho}$ is itself ordinary, produces a ring $R(\overline{\rho})^{\text{ord}}$, which is typically $W(k)[[T]]$. Work of Hida [26] produces representations $\tilde{\rho} : G_{\mathbb{Q}} \to GL_2(\mathbb{Z}_p[[T]])$ such that the specialisations $\mathbb{Z}_p[[T]] \to \mathbb{Z}_p$ given by $T \mapsto (1+p)^{k-1} - 1$ ($k = 2, 3, 4, \ldots$) yields ρ_{f_k, p^∞} for cuspidal eigenforms f_k of weight k. We can sometimes show that $\tilde{\rho}$ parametrises ordinary lifts of $\overline{\rho}$, i.e. is itself the universal ordinary lift, so all ordinary lifts are modular, as desired.

1.6 Relevance to pro-p groups

The natural question to ask is why this is relevant to pro-p group theory. The point is that $\ker(GL_n(R) \to GL_n(k))$ is a pro-p group [6]. Thus, if $\overline{\rho} : G_{\mathbb{Q},S} \to GL_n(k)$ is a given representation, all lifts of it will factor through $G_{\mathbb{Q},S}/N$, where $\ker \overline{\rho}/N$ is the maximal pro-p quotient of $\ker \overline{\rho}$.

To obtain $\xi : G_{\mathbb{Q},S} \to GL_n(R(\overline{\rho}))$, we therefore just need to find a presentation for the virtually pro-p group $G_{\mathbb{Q},S}/N$. This is what the next section is devoted to and the details of how to construct the universal deformation from the presentation will be given there.

1.7 A parting shot

I conjectured that for $n = 2$ and $\overline{\rho}$ odd, $p \in S$, $\xi : G_{\mathbb{Q},S} \to GL_n(R(\overline{\rho}))$ is never injective. As noted by Lubotzky and Shalev [35], this follows for $n = 2$ from $d(P) - r(P) \geqslant 2$, where $P = \ker\overline{\rho}/N)$ and from work of Zubkov [59] (now generalised for any n by Barnea and Larsen). The question then is to characterise $\ker \xi$ (or equivalently its fixed field). Indeed, with $G_{\mathbb{Q},S}$ replaced by any profinite group with the corresponding P finitely generated, this question makes sense—can we characterise the quotient of G cut out by such representations?

2 Applications to and from pro-p Galois theory

We continue the discussion begun at the end of the last section by investigating a specific example, namely $\overline{\rho} = \rho_{\Delta,23} : G_{\mathbb{Q},\{23\}} \to GL_2(\mathbb{Z}/23)$. By elementary means, we see that $\tau(n)$ satisfies congruences modulo 23, e.g., $\tau(\ell) \equiv 0 (\text{mod } 23)$ if ℓ is prime and not a square modulo 23. The only consistent way to have $\tau(\ell) \equiv \text{tr}\overline{\rho}(Fr_\ell)(\ell \neq 23)$ (to satisfy (3) of §1.3) is if the image of $\overline{\rho}$ is $\text{Gal}(K/\mathbb{Q}) \cong S_3 \leqslant GL_2(\mathbb{Z}/23)$, where K is the splitting field of $x^3 - x + 1$.

Suppose that we want to know all lifts of $\overline{\rho}$ to $GL_2(\mathbb{Z}_{23})$. What information does number theory provide for us?

(1) One lift, namely, $\rho_{\Delta,23^\infty}$, is already provided (thanks to Deligne [14]).

(2) Hida's work [26] provides a lift $\tilde{\rho} : G_{\mathbb{Q},\{23\}} \to GL_2(\mathbb{Z}_{23}[[T]])$, which provides a one-parameter family of lifts of $\overline{\rho}$ to $\mathbb{Z}_{23}$, by composing $\tilde{\rho}$ with $\mathbb{Z}_{23}[[T]] \to \mathbb{Z}_{23}$ defined by $T \mapsto (1+23)^{k-1} - 1 (k \in \mathbb{Z}_{23})$. (Recall that for $k = 2, 3, 4, \ldots$

this yields a representation associated to a cuspidal eigenform of weight k. In fact, with $k = 12$ we recover $\rho_{\Delta,23^\infty}$.)

We shall show that there exist many more representations and that we can find them all.

(3) Let $P = \ker(\overline{\rho})/N$ be the maximal pro-23 quotient of $\ker(\overline{\rho})$ and $G = G_{\mathbb{Q},\{23\}}/N$. We know the following [11] (since P is the Galois group over K of the maximal pro-23 extension unramified outside the set of primes above 23):

   (i) P is free pro-23 on 4 generators.

   (ii) Let $H = \mathrm{Gal}(K/\mathbb{Q})$. As an $\mathbb{F}_{23}[H]$-module, $P/\Phi(P)$ is a direct sum of the 3 non-isomorphic irreducible $\mathbb{F}_{23}[H]$-modules, each with multiplicity 1.

(We will see later where (i) and (ii) come from.)

Now group theory is used.

(4) By Schur-Zassenhaus (profinite version), $1 \to P \to G \to H \to 1$ splits.

(5) Given (by (ii) above) $H \to \mathrm{Aut}(P/\Phi(P))$, whatever lift $H \to \mathrm{Aut}(P)$ we pick, we get isomorphic semidirect products $P \rtimes H$ ([6], essentially since $\ker(\mathrm{Aut}(P) \to \mathrm{Aut}(P/\Phi(P)))$ is a pro-23 group).

By (5), we are free to pick the simplest form of lift that we can. Let $H (\cong S_3) = \langle \sigma, \tau \mid \sigma^2 = \tau^3 = 1, \tau^\sigma = \tau^{-1} \rangle$. We define an action of H on the generators $u_1, ..., u_4$ of P by $\tau(u_1) = u_2, \tau(u_2) = u_3, \tau(u_4) = u_4, \sigma(u_1) = u_1, \sigma(u_4) = u_4^{-1}$. By the above, $G \cong \langle \sigma, \tau, u, v \mid \sigma^2 = \tau^3 = 1, \tau^\sigma = \tau^{-1}, v^\tau = v, u^\sigma = u, v^\sigma = v^{-1} \rangle$ (where it is understood that $u, v, u^\tau, ...$ all have supernatural order p^∞).

Let $\rho : G_{\mathbb{Q},\{23\}} \to GL_2(R)$ be a lift of $\overline{\rho}$ to a complete, noetherian, local ring R (maximal ideal m). Then ρ factors through G.

By Schur-Zassenhaus, there is just one conjugacy class of lifts of $\langle \sigma, \tau \rangle \cong S_3$ to R and so there is a unique lift ρ such that $\sigma \mapsto \begin{pmatrix} 1 & 0 \\ 0 & -1 \end{pmatrix}, \tau \mapsto \begin{pmatrix} -\frac{1}{2} & \frac{1}{2} \\ -\frac{3}{2} & -\frac{1}{2} \end{pmatrix}$ say. We will normalise our lifts by always picking this one (and this exhausts the flexibility we have in picking equivalent lifts).

Then, from the relations for G, u maps to a matrix in $\ker(GL_2(R) \to GL_2(\mathbb{Z}/23))$ fixed under conjugation by the image of σ, i.e., to a matrix of the form $\begin{pmatrix} 1+t_1 & 0 \\ 0 & 1+t_2 \end{pmatrix}$ $(t_1, t_2 \in m)$ (and this is the only constraint on the image of u). Likewise, the image of v is fixed/inverted by conjugation by the image of τ/σ respectively, so τ maps to a matrix of the form $\begin{pmatrix} (1-3t_3^2)^{\frac{1}{2}} & t_3 \\ -3t_3 & (1-3t_3^2)^{\frac{1}{2}} \end{pmatrix}$ $(t_3 \in m)$. We deduce:

Theorem. ([11]) *$R(\overline{\rho}) \cong \mathbb{Z}_{23}[[T_1, T_2, T_3]]$ and the universal deformation is given explicitly by:* $\sigma \mapsto \begin{pmatrix} 1 & 0 \\ 0 & -1 \end{pmatrix}, \tau \mapsto \begin{pmatrix} -\frac{1}{2} & \frac{1}{2} \\ -\frac{3}{2} & -\frac{1}{2} \end{pmatrix},$

$$\begin{pmatrix} 1+T_1 & 0 \\ 0 & 1+T_2 \end{pmatrix}, v \mapsto \begin{pmatrix} (1-3T_3^2)^{\frac{1}{2}} & T_3 \\ -3T_3 & (1-3T_3^2)^{\frac{1}{2}} \end{pmatrix}.$$

Note that there is a major piece of information lacking here; we would like to know what Frobenius elements $Fr_\ell(\ell \neq 23) \in G$ map to, but the above does not help (Tchebotarev's density theorem [40] tells us that these elements are dense in G). A Challenge: Describe Fr_ℓ as an element of G in terms of σ, τ, u, v.

There are, however, advantages. We can easily describe interesting subsets of the set of all lifts to $\mathbb{Z}_p$, for instance those with image in the normaliser of a Cartan subgroup ($T_1 = T_2$).

By doing a similar analysis for $\overline{\rho}\mid_{D_p}$ and linking it to the above theorem, we can pick out lifts whose restriction to D_p or I_p is interesting, e.g., $\overline{\rho}(I_p) \subseteq \left\{\begin{pmatrix} 1 & * \\ 0 & * \end{pmatrix}\right\}$ (the ordinary lifts). The universal such is easily proved [11] to be Hida's $\tilde{\rho}$ and so all ordinary lifts are (pro-)modular (an inverse limit of modular lifts). This is the first instance of the type of result that Wiles produces.

Given $\overline{\rho} : G_{\mathbb{Q},S} \to GL_n(k)$ with $p \not| \#\overline{\rho}(G_{\mathbb{Q},S})$, a similar approach will always work so long as we know [6]

(i) the structure of the pro-p group $P = \ker(\overline{\rho})/N$ and

(ii) the $\mathbb{F}_p[\overline{\rho}(G_{\mathbb{Q},S})]$-module structure of $P/\Phi(P)$.

As regards (i), if $\overline{\rho}(G_{\mathbb{Q},S}) = \mathrm{Gal}(K/\mathbb{Q}), p \in S$, then $d(P) = r_2 + 1 + r(P)$, where $r(P) = (\sum_{v\in\tilde{S}} \delta(K_v)) - \delta(K) + \dim_{\mathbb{F}_p} B_{\tilde{S}}$ where r_2 is the number of pairs of complex places of K, $\tilde{S}$ is the set of primes of K above S, $\delta(F) = 1$ if F contains p pth roots of 1, 0 otherwise, and $B_{\tilde{S}} = \{x \in K^* \mid x\mathcal{O}_K = \mathcal{I}^p, x \in K_v^{*p}\ \forall v \in \tilde{S}\}/K^{*p}$.

As regards (ii), this is Class Field Theory. For example, if $r(P) = 0$, letting $H = \mathrm{Gal}(K/\mathbb{Q})$, $P/\Phi(P) \oplus \mathrm{Ind}_{<c>}^H \mathbb{F}_p \cong \mathbb{F}_p[H] \oplus \mathbb{F}_p$ as $\mathbb{F}_p[H]$-modules, where c is the image of complex conjugation.

2.1 Example

Let E be the elliptic curve over $\mathbb{Q}$ defined by $y^2 + xy = x^3 - x^2 - 2x - 1$. This has bad reduction only at 7. It also has complex multiplication so all the representations $\rho_{E,p^\infty}(p \neq 7)$ have image the normaliser of a Cartan subgroup. One finds [12] that if $p \geqslant 5$, $p \neq 7$, then $R(\rho_{E,p}) \cong \mathbb{Z}_p[[T_1, T_2, T_3]]$, whereas if $p = 3$, $R(\rho_{E,p}) \cong \mathbb{Z}_p[[T_1, T_2, T_3, T_4]]/((1+T_4)^3 - 1)$. This means that $\mathrm{Hom}(R(\rho_{E,3}), \mathbb{Z}_3[\omega])$ (ω a primitive cube root of 1) consists of three essentially three-dimensional sheets. The sheets are determined by $T_4 \mapsto \omega^i - 1$ and thus by the form of $\rho(I_7)$. The individual sheets can be given coordinates according to the form of $\rho(I_3)$.

Note that $R(\overline{\rho})$ ($\overline{\rho}$ odd, $n = 2$, $p \in S$) has always turned out to have the form $\mathbb{Z}_p[[T_1, ..., T_{m+3}]]/(r_1, ..., r_m)$ (or with $W(k)$ replacing $\mathbb{Z}_p$). For some partial results towards this, see [3]. These kind of rings might be called local complete intersection rings and this generalises the work of Taylor and Wiles [55], showing that specific quotients of the $R(\overline{\rho})$ are local complete intersections.

The basic challenge then is always, given $\overline{\rho} : G_{\mathbb{Q},S} \to GL_2(k)$ (not necessarily with $p \not| \#\overline{\rho}(G_{\mathbb{Q},S})$), to present the group $G_{\mathbb{Q},S}/N$. There are many results known for the cases of (i) $p \not| \#\overline{\rho}(G_{\mathbb{Q},S})$ and (ii) $\overline{\rho}$ reducible [38], but the most common case,

$SL_2(k) \subseteq \overline{\rho}(G_{\mathbb{Q},S})$, is poorly understood.

For example, recall that the image of $\rho_{\Delta,691}$ is contained in $\left\{\begin{pmatrix} 1 & * \\ 0 & * \end{pmatrix}\right\} \subseteq GL_2(\mathbb{Z}/691)$. Noting that the maximal pro-691 extension L unramified outside 691 of the fixed field K of ker $\rho_{\Delta,691}$ is also the maximal pro-691 extension of $\mathbb{Q}(\mu_{691})$ unramified outside 691, it follows that Gal$(L/\mathbb{Q})$ can be presented by the earlier methods.

Böckle [4] can also handle essentially all local Galois representations $\overline{\rho} : D_p \to GL_2(k)$ ($p =$ char k as usual) (the point is that D_p is prosoluble so there is never any problem handling insoluble images).

2.2 Local-global structure of deformation spaces

Let $\overline{\rho} : G_{\mathbb{Q},S} \to GL_n(k)$ be absolutely irreducible. At the joint AMS-LMS conference in Cambridge, 1992, I made various conjectures regarding the local-global structure of Hom$(R(\overline{\rho}), \mathbb{Z}_p)$. Liow [33] and Böckle [5] have established some of these.

In particular, if there is a lift to characteristic 0 (as for $n = 2$, odd $\overline{\rho}$, $p \in S$, is predicted by Serre's conjecture and has just been shown by Ramakrishna), then there should be a lift with "minimal" ramification (compare the work of Ribet and others [43]). Starting with Frey's elliptic curve E and setting $\overline{\rho} = \rho_{E,p}$, then we should get a lift to characteristic 0 with little ramification. If we could then use group theory to prove the non-existence of Galois representations with so little ramification, we would have an approach to Fermat's Last Theorem avoiding modular forms and sophisticated algebraic geometry.

The target conjecture to try first here is a special case of the following conjecture of Fontaine and Mazur [19]. With our current notation it is a little cumbersome; a more natural way of stating it will be given in the next section.

Conjecture. (Fontaine–Mazur) If $\rho : G_K \to GL_n(\mathbb{Z}_p)(K/\mathbb{Q}$ finite) is such that $I_p^g \cap G_K \subseteq \ker\rho (\forall g \in G_{\mathbb{Q}})$, then $\#\rho(G_K) < \infty$.

Another way of looking at this is to consider groups like the group P considered in this section, but now with $p \notin S$. For $p \in S$, we saw that our P are free or at least have large $d(P) - r(P)$ ($= r_2 + 1$, as noted just before Example 2.1). For $p \notin S$, the above conjecture says that P has no infinite analytic quotient. We will discuss methods of proving cases of this conjecture and consequences of it in the next section.

3 The Fontaine–Mazur conjecture

Number theory provides us with some naturally occurring families of pro-p groups. Namely, let S be a finite set of primes and K a finite extension of $\mathbb{Q}$. Let p be a rational prime. We consider the maximal pro-p extension of K unramified outside the primes of K above S and denote its Galois group over K by $P = P_{K,S,p}$. (Equivalently, P is the maximal pro-p quotient of G_K in which all $I_\ell^g (\ell \notin S, g \in G_{\mathbb{Q}})$ are killed.)

In Section 2, it became clear that the structure of P somewhat controls the theory of p-adic representations of $G_{\mathbb{Q}}$ and its open subgroups. The obvious question then is: what do we know about P?

The case of $p \in S$ was considered in detail in Section 2, where formulae for $d(P)$ and $r(P)$ were given. The most awkward term in those formulae is B_S. If $B_S = 0$, then all the relations of P come from local relations, i.e., relations of the corresponding decomposition groups and so are *Demuškin* (i.e., of the form $x_1^q(x_1, x_2)(x_3, x_4)...(x_{n-1}, x_n)$) for $v \mid p$ or *tame* (i.e., of the form $\tau^\sigma = \tau^q$) for $v \nmid p$.

The relations corresponding to B_S are problematic and indeed are called the "unknown relations".

The case of $p \notin S$ is completely different. In particular, we have

Conjecture. (Fontaine–Mazur, [19]) P has no infinite analytic quotient.

Since the property of having an infinite analytic quotient is inherited by open subgroups and finite extensions of a pro-p group, this is equivalent to the following:

Conjecture. P is uniform-deficient.

This means that no open subgroup of P has a uniform quotient. Note that P is at least $\mathbb{Z}_p$-deficient since the derived subgroups $P^{(n)}$ are of finite index by class field theory [40]. In terms of Galois representations, which was how this conjecture was originally stated (and it arose since all such representations coming from algebraic geometry satisfy this conjecture), this is equivalent to:

Conjecture. If K is a number field and S a finite set of primes of K not including any above p, then every representation $\rho : G_{K,S} \to GL_n(\mathbb{Z}_p)$ has finite image.

I have generalised this as follows [7]. This natural generalisation ensures that the deformation theory described in the last section is particularly simple and elegant in this case, namely it factors through a finite group (the deformation theory of such groups is explicitly worked out in [15]).

Conjecture. ([7]) If K is a number field and S a finite set of primes not containing any above p, then every representation $\rho : G_{K,S} \to GL_n(R)$, where R is any complete, noetherian, local ring with finite residue field of characteristic p, has finite image.

Suppose, to simplify the discussion for now, that $S = \emptyset$.

Early this century, Furtwängler asked if there exist "infinite Hilbert class towers". This follows if there exist K and p such that $P = P_{K,\emptyset,p}$ is infinite. The answer (yes) came in 1964 thanks to Golod and Shafarevich [22]:

(i) If G is a finite p-group, then $r(G) > \frac{1}{4}d(G)^2$, whereas

(ii) $0 \leqslant r(P) - d(P) \leqslant r_1 + r_2 + \delta(K) - 1$ where r_1 is the number of real places of K.

Note how the group-theoretic and number-theoretic ingredients are combined effectively.

Example. If K is an imaginary quadratic field, then
$0 \leqslant r(P) - d(P) \leqslant 0, 1$ (if p is odd, $= 2$ respectively).
Thus, if $d(P) \geqslant 4, 5$ (for p odd, $= 2$ respectively), then P is infinite.

Note that $d(P)$ equals, by class field theory, the p-rank of the ideal class group of K. For $p = 2$, genus theory provides us with arbitrarily large $d(P)$, whereas for any p Cohen-Lenstra heuristics [13] predict that any $d(P)$ can occur (with suitable choice of K).

There are refinements.

(i)′ If $G = F/R$ is a finite p-group, where F is free pro-p and $R \subseteq F_m$ (the index indicating the Zassenhaus filtration of F), then $r(G) > d(G)^m(m-1)^{m-1}/m^m$ ((i) is the case $m = 2$).

Example. For imaginary quadratic fields K and p odd, $R \subseteq F_3$ and so if $d(P) \geqslant 3$, then P is infinite [31]. For e.g., $K = \mathbb{Q}(\mu_{157})$ and $p = 157$, $R \subseteq F_5$ and so since $d(P) = r(P) = 2$, P is infinite [58].

In fact, the inequalities in (i) and (i)′ hold if G is an analytic pro-p group with $d(G) > 1$, except for $d(G) = 2, r(G) = 1, m = 2$ (Serre, Lubotzky [34]) (and indeed hold for many other groups; see e.g., Lubotzky–Shalev [35] for the case of $\mathbb{F}_p[[T]]$-analytic groups). The exceptions do not concern us (such a G could never be one of our P's since for us $r(P) \geqslant d(P)$). Thus if P fails the inequality in (i)′, then it is infinite and not analytic, proving several cases of the Fontaine–Mazur conjecture (see [23] for further discussion). It turns out though (see later) that the critical cases of the Fontaine–Mazur conjecture concern just-infinite pro-p Galois groups, all of which satisfy (i)′. Hajir [23] has asked whether every infinite $P = P_{K,\emptyset,p}$ has an open subgroup that fails the Golod–Shafarevich inequality. He gives examples in [24] of P that do not fail the inequality but turn out to have open subgroups that do.

Definition. ([7]) Let G be a pro-p group. We call G *self-similar* if it contains a filtration $G = G_1 \geqslant G_2 \geqslant \dots$ of open, characteristic subgroups G_i such that $\cap G_i = \{1\}$, G_i/G_{i+1} is abelian, and there are isomorphisms $\phi_i : G_{i-1}/G_i \to G_i/G_{i+1}$ that commute with every automorphism of G (note that since the factors G_i/G_{i+1} are abelian, we need only consider outer automorphisms).

Examples of this are fairly common, including uniform pro-p groups [16] (where each ϕ_i is of the form $x \mapsto x^p$) and $\mathbb{F}_p[[T]]$-perfect groups (D. Cartwright's recent Queen Mary College, London Ph.D. thesis) (where the ϕ_i are so-called T-maps).

Theorem. ([7, 8]) *Let F be a number field with class number not divisible by p. Let K be a cyclic extension of F of degree prime to p. Let L be a Galois extension of F containing K with L/K an unramified pro-p extension. Then $Gal(L/K)$ is not self-similar.*

The idea of the proof is the following. The number-theoretic ingredients are that (i) $\mathrm{Gal}(L/K)$ is insoluble by class field theory (which tells us that various abelian subquotients must be finite) and (ii) that $\mathrm{Gal}(L/F)$ has no normal subgroup of index p that contains all its inertia subgroups. Note that since L/K is unramified, these inertia subgroups map isomorphically to subgroups of $\mathrm{Gal}(K/F)$, so are of order prime to p.

Schur-Zassenhaus provides an automorphism σ of order $\#\mathrm{Gal}(K/F)$ of $U = \mathrm{Gal}(L/K)$ and so of U/U_n. By (i) and Shalev [51] (if U is uniform, we have that the U/U_n have bounded rank—otherwise the result needed is only conjectural [52]), there is a fixed point in some U/U_n. Taking n minimal, the fixed point is in some U_{n-1}/U_n. Using the ϕ_i, we obtain a fixed point in U/U_2, an abelian quotient of U. But this yields a C_p-quotient of Gal(L/F) which no inertia subgroup maps onto since their orders are prime to p. This contradicts (ii). This approach can be extended to S other than $\emptyset$ (still containing no primes above p) by using ray class groups in place of ideal class groups (and for the rest of the paper we work in this generality) but it is of limited use. If $p = 2$ and Gal$(K/\mathbb{Q}) \cong C_2$, then Gal$(L/\mathbb{Q})$ is generated by elements of order 2 and has an insoluble, self-similar pro-2 subgroup of index 2. This then yields no contradiction.

A related question (solved in [10]) was the question asked by Lemmermeyer. Does every finite 2-group embed with index 2 in some group generated by involutions? The answer (no) shows that Cohen-Lenstra heuristics [13] will not simply carry over to nonabelian extensions of quadratic fields.

This suggests that these infinite pro-p Galois groups should not have a particularly regular structure. We can investigate this subject further by using the classification (so far) of just-infinite pro-p groups. Since the class of groups linear over $\mathbb{Z}_p$ is closed under taking quotients, we have the following result.

Theorem. *The following are equivalent:*

(i) The Fontaine–Mazur conjecture.

(ii) The conjecture that if K is a number field and S a finite set of primes of K not including any above p, and if L/K is a pro-p extension unramified outside S with just-infinite Galois group, then Gal(L/K) is not analytic (i.e. linear over $\mathbb{Z}_p$).

The question that naturally arises then is the following. What kind of just-infinite pro-p group is Gal(L/K)? As explained in Chapter 2, the classification of just-infinite pro-p groups falls into four classes, I–IV. Class Field Theory says that our groups cannot be in class I. The Fontaine–Mazur conjecture says that they cannot be in class II. My generalisation says they cannot be in class III. There remain the class IV groups. It appears that the open subgroups of the Nottingham groups and of those of Fesenko [18] (see Chapter 10) are not suitable candidates—they may even be self-similar. It would be nice to settle this.

Of the presently known families of just-infinite pro-p groups, the most likely one for our Galois groups to be in is then the family of branch groups discovered and developed by Grigorchuk and others (see Chapter 4). Another possibility is that these Galois groups provide yet another family of just-infinite pro-p groups of a kind not previously known.

In [9] we investigate the possibility of these Galois groups being branch groups (we conjecture that they in fact are) by constructing representations of $G_{\mathbb{Q}}$ into wreath powers of C_2 that are unramified at 2 and are ramified at only finitely many primes. (Indeed [9] presents a first systematic study of the action of Galois groups on rooted trees, a natural theory that should be developed further.) As an explicit example, the Galois group of $x^{16} - 36x^{14} + 588x^{12} - 4974x^{10} + 24858x^8 - 74184x^6 + 136353x^4 - 161766x^2 + 105393$ is a subgroup of order 2^{13} of the 4th wreath power of C_2 (of order

2^{15}). Its splitting field is ramified only at {3, 19, 43} and so we have a quotient of the Galois group H of the maximal pro-2 extension of $\mathbb{Q}$ unramified outside {3, 19, 43}, which is known to have $d(H) = r(H) = 3$. (The number of generators and relations of the Galois group of the maximal pro-p extension of $\mathbb{Q}$ unramified outside S, plus presentations of its maximal quotient of nilpotency class 2, and the fact that the group has trivial Schur multiplicator can all be found in Fröhlich's book [21].)

Another approach here is the following. Suppose $I \leqslant G$ are groups. Define an equivalence relation on the set of subgroups of G by $H \sim N$ if and only if $I^g \cap H = I^g \cap N \forall g \in G$. With the choice of $G = G_{\mathbb{Q}}$ and $I = I_p$, the Fontaine–Mazur conjecture is equivalent to:

Conjecture. Suppose $N \trianglelefteq H$, H open, and $H \sim N$. Then H/N is not infinite, analytic pro-p (or equivalently not uniform pro-p).

We ask what choices of I, G satisfy the property that if $N \trianglelefteq H$ and H is a finite index subgroup of G such that $H \sim N$, then there are restrictions on the structure of H/N. One such example is the case of G being a Frobenius group and I a Frobenius complement. In this case H/N is a subquotient of the Frobenius kernel, so indeed has restrictions on its structure, namely it is nilpotent with bounds on its nilpotency class.

Maybe $G_{\mathbb{Q}}$ and I_p fit into some kind of generalisation of this. Some things are indeed similar. For instance, as noted earlier, $I_p \cap I_p^g = \{1\}$ if $g \notin D_p$. One might wonder whether our situation is a generalisation of the Frobenius group case.

3.1 Another tough test-case

Let $\mathbb{Q}(\sqrt{m}), \mathbb{Q}(\sqrt{n}), \mathbb{Q}(\sqrt{mn})$ be distinct quadratic fields, not all real, but all with class number exactly divisible by the odd prime p, and set $K = \mathbb{Q}(\sqrt{m}, \sqrt{n})$ (e.g. $p = 3, \mathbb{Q}(\sqrt{-104}, \sqrt{229})$). Then if $P = P_{K,\emptyset,p}$, $P^{ab} \cong C_p \times C_p \times C_p$ (where C_p denotes the cyclic group of order p), acted on by the augmentation representation of $\mathrm{Gal}(K/\mathbb{Q})$.

Note that Andozskii and Cvetkov [2] showed that if G is an infinite, finitely generated pro-p group such that (1) the Bockstein map $H^1(G, \mathbb{Z}/p) \to H^2(G, \mathbb{Z}/p)$ is an isomorphism, and (2) the cup-product $H^1(G, \mathbb{Z}/p) \wedge H^1(G, \mathbb{Z}/p) \to H^2(G, \mathbb{Z}/p)$ is an isomorphism, then $G \cong \ker(SL_2(\mathbb{Z}_p) \to SL_2(\mathbb{Z}/p))$. The Fontaine–Mazur conjecture says that P fails (1) or (2). The question is which one. Thus the conjecture has very down-to-earth consequences.

Remarks. It is known that every finite group occurs as the Galois group of some unramified extension of some number field [20]. In 1935, Magnus [36] asked whether there exists an infinite pro-p group P with all its derived subgroups $P^{(n)}$ of finite index. His idea was that if no such existed, then the answer to Furtwängler's question would be that $P_{K,\emptyset,p}$ is never infinite. Of course, all the just-infinite pro-p groups in classes II-IV answer Magnus's question (and in fact Itô [29] and Hobby [27] gave class II counterexamples in the early 1950s).

Several properties of $P_{K,\emptyset,p}$ are known. For instance, for K imaginary quadratic and p odd, $P = P_{K,\emptyset,p}$ is known to satisfy:

I. P is topologically finitely presented with generator rank $d(P)$ equal to its relation rank $r(P)$ (see earlier in this section).

II. If $1 \to R \to F \to P \to 1$ is a minimal presentation of P in the sense that F is free pro-p on $d(P)$ generators, then $R \subseteq F_3$ (F_i giving the Zassenhaus filtration on F). This implies that if $d(P) \geqslant 3$, then P is infinite, not analytic, and not $\mathbb{F}_p[[T]]$-analytic.

III. Every open (i.e., finite index) subgroup of P has finite abelianisation (finiteness of class numbers and class field theory).

IV. P has an automorphism of order 2 (namely complex conjugation) with no non-trivial fixed point on its abelianisation.

V. Let $Q = P/P''$. For every normal subgroup H of Q with cyclic Q/H, the order of Ker V is $[Q : H]$, where $V : Q/Q' \to H/H'$ is the transfer map (a result of K. Miyake).

VI. P has no uniform (more generally, self-similar) quotient stable under complex conjugation (the theorem of this section).

Conjecturally (i.e., if the Fontaine–Mazur conjecture is correct), P also has the property that every open subgroup of P has no infinite analytic quotients. This generalises III.

By Cohen–Lenstra heuristics [13], for any given odd prime p and positive integer d, there should exist examples of such groups with $d(P) = d$. While this seems a natural class of pro-p groups to attempt to classify, it is apparently not one familiar to pro-p group theorists. Most classes considered by group theorists satisfy the Golod–Shafarevich inequality.

Unfortunately the properties we can most readily derive are not the properties we are most interested in. One can play the game of attempting to take these properties and deduce others from them. My student, David Perry, has obtained a large number of pro-2 groups with 2 generators and 2 relations, with their derived subgroups of finite index (as far as the nilpotent quotient algorithm on MAGMA works), and with an open subgroup that fails the Golod–Shafarevich inequality (or rather Newman's modification [41] of it) so that the groups are indeed infinite, not analytic. It is interesting to explore what kind of group he produces (e.g., do they have torsion?—usually they do).

References

[1] J. L. Alperin, Finite groups viewed locally, *Bull. Amer. Math. Soc.* **83**(6) (1977).

[2] I. Andozskii and V. Cvetkov, A certain series of finite closed p-groups, *Izv. Akad. Nauk. SSSR Ser. Mat.* **38** (1974), 278–290.

[3] G. Böckle, The generic fiber of the universal deformation space associated to a tame Galois representation, *Manuscr. Math.* **96** (2) (1998), 231–246.

[4] G. Böckle, Demuškin groups with group actions and applications to deformations of local Galois representations, *Compositio Math. (to appear).*

[5] G. Böckle, A local-to-global principle for deformations of Galois representations, *J. Reine Angew. Math.* **509** (1999), 199–236.

[6] N. Boston, Explicit deformations of Galois representations, *Invent. Math.* **103** (1990), 181–196.

[7] N. Boston, Some cases of the Fontaine-Mazur conjecture, II, *J. Number Theory* **75**(2) (1999), 161–169.

[8] N. Boston, Some cases of the Fontaine-Mazur conjecture, *J. Number Theory* **42**(3) (1992), 285–291.

[9] N. Boston, Tree representations of Galois groups (preprint).

[10] N. Boston and C. R. Leedham-Green, Counterexamples to a conjecture of Lemmermeyer, *Arch. Math. (Basel)* **72**(3) (1999), 177–179.

[11] N. Boston and B. Mazur, Explicit universal deformations of Galois representations, *Adv. Studies in Pure Math.* **17** (1989), 1–21.

[12] N. Boston and S. V. Ullom, Representations related to CM elliptic curves, *Math. Proc. Camb. Phil. Soc.* **113** (1993), 71–85.

[13] H. Cohen and H. Lenstra, Heuristics on class groups of number fields, *Number Theory*, Noordwijkerhout 1983, Lecture Notes in Math. **1068** 33–62, Springer, Berlin, New York, 1984.

[14] P. Deligne, Formes modulaires et représentations ℓ-adiques, *Sem. Bourb.* **355** (1969).

[15] B. de Smit and H. Lenstra, Explicit construction of universal deformation rings, *Modular forms and Fermat's Last Theorem* (G. Cornell, J. Silverman, and G. Stevens, eds.) 313–326, Springer, Berlin, New York, 1997.

[16] J. D. Dixon, M. P. F. du Sautoy, A. Mann, and D. Segal, *Analytic pro-p Groups, 2nd ed.* Cambridge Studies in Advanced Maths. 61, Cambridge University Press, 1999.

[17] G. Faltings, Endlichkeitssätze für abelsche Varietäten über Zahlkörpern, *Invent. Math.* **73** (1983), 349–366.

[18] I. Fesenko, On just-infinite pro-p groups and arithmetically profinite extensions of local fields (preprint).

[19] J.-M. Fontaine and B. Mazur, Geometric Galois representations *Elliptic Curves and Modular Forms* (J. H. Coates and S. T. Yau, eds.), Proceedings of a conference held in Hong Kong, December 18–21, 1993, International Press, Cambridge, MA, and Hong Kong.

[20] A. Fröhlich, On non-ramified extensions with prescribed Galois group, *Mathematika* **9** (1962), 133–134.

[21] A. Fröhlich, *Central Extensions, Galois Groups, and Ideal Class Groups of Number Fields,* Contemporary Mathematics, Amer. Math. Soc., Providence, RI **24** (1983).

[22] E. Golod and I. Shafarevich, On class field towers, *Izv. Akad. Nauk. SSSR* **28** (1964), 261–272.

[23] F. Hajir, On the growth of p-class groups in p-class field towers, *J. Algebra* **188**(1) (1997), 256–271.

[24] F. Hajir, On a theorem of Koch, *Pacific J. Math.* **176**(1) (1996), 15–18.

[25] D. Harbater, Galois groups with prescribed ramification, *Arithmetic Geometry* (Tempe, AZ, 1993) 35–60, *Contemp. Math.* **174** Amer. Math. Soc., Providence, RI (1994).

[26] H. Hida, Galois representations into $GL_2(\mathbb{Z}_p[[X]])$ attached to ordinary cusp forms, *Invent. Math.* **85**(3) (1986), 545–613.

[27] C. Hobby, The derived series of a finite p-group, *Illinois J. Math.* **5** (1961), 228–233.

[28] M. Ikeda, Completeness of the absolute Galois group of the rational number field, *Arch. Math. (Basel)* **26**(6) (1975), 602–605.

[29] N. Itô, Note on p-groups, *Nagoya J. Math.* **1** (1950), 113–116.

[30] U. Jannsen and K. Wingberg, Die Struktur der absoluten Galoisgruppe $\wp$-adischer Zahlkörper, *Invent. Math.* **70** (1982), 71–98.

[31] H. Koch and Venkov, Über den p-Klassenkörperturm eines imaginarquadratischen Zahlkörpers, *Astérisque* **24** (1975), 57–67.

[32] S. Lang *Algebraic Number Theory,* Addison-Wesley, 1970.

[33] Y. Liow, *Internal structure of the deformation spaces of Galois representations,* University of Illinois at Urbana-Champaign, Ph.D. thesis (1997).

[34] A. Lubotzky, Group presentation, p-adic analytic groups and lattices in $SL_2(\mathbb{C})$, *Ann. Math.* **118** 115–130.

[35] A. Lubotzky and A. Shalev, On some Λ-analytic pro-p groups, *Israel J. Math* **85** (1994), 307–337.

[36] W. Magnus, Beziehung zwishen Gruppen und Idealen in einem speziellen Ring, *Math. Ann.* **111** (1935).

[37] B. Mazur, Deforming Galois representations, *Proceedings of the March 1987 Workshop on "Galois groups over* **Q**" held at MSRI, Berkeley, California.

[38] A. Mezard, Computation of a universal deformation ring in the Borel case, *Math. Proc. Cambridge Phil. Soc.* **126** (1999), 417–442.

[39] J. Neukirch, Kennzeichnung der endlich-algebraischen Zahlkörper durch die Galoisgruppe der maximalen auflosbaren Erweiterungen, *J. Reine Angew. Math.* **238** (1969), 135–147.

[40] J. Neukirch, *Class Field Theory* Springer-Verlag, Berlin, Heidelberg, New York, Tokyo, 1986.

[41] M. Newman, Proving a group infinite, *Arch. Math. (Basel)* **54**(3) (1990), 209–211.

[42] R. Ramakrishna, Infinitely ramified Galois representations *Annals of Math.* (to appear).

[43] K. Ribet, Report on mod ℓ representations of Gal($\overline{\mathbb{Q}}/\mathbb{Q}$), *Motives* (Seattle, WA, 1991) 639–676, Proc. Sympos. Pure Math. 55, Part 2, Amer. Math. Soc., Providence, RI (1994).

[44] K. Ribet, On modular representations of Gal($\overline{\mathbf{Q}}/\mathbb{Q}$) arising from modular forms, *Invent. Math.* **100** (1990), 431–476.

[45] J.-P. Serre, *Local Fields,* Springer-Verlag, New York, Heidelberg, Berlin, 1979.

[46] J.-P. Serre, *Topics in Galois Theory,* Research Notes in Mathematics 1, Jones and Bartlett Publishers, Boston, MA, 1992.

[47] J.-P. Serre, Propriétés galoisiennes des points d'ordre fini des courbes elliptiques, *Invent. Math.* **15** (1972), 259–331.

[48] J.-P. Serre, Modular forms of weight one and Galois representations, *Algebraic number fields: L-functions and Galois properties* (Proc. Sympos., Univ. Durham, Durham, 1975), pages 193–268 (1977).

[49] J.-P. Serre, Une interprétation des congruences relatives à la fonction τ de Ramanujan, (1969), *Séminaire Delange-Pisot-Poitou: 1967/68, Théorie des Nombres, Fasc. 1, Exp. 14*

[50] J.-P. Serre, Sur les représentations modulaires de degré 2 de Gal($\overline{\mathbb{Q}}/\mathbb{Q}$), *Duke Math. J.* **54** (1987), 179–230.

[51] A. Shalev, On almost fixed point free automorphisms, *J. Alg.* **157**(1) (1993), 271–282.

[52] A. Shalev, Finite p-groups, *Finite and locally Finite Groups* (Istanbul, 1994), 401–450, Kluwer Acad. Publ., Dordrecht, 1995.

[53] J. Silverman, *The Arithmetic of Elliptic Curves,* Springer-Verlag, New York, Berlin, Heidelberg, 1986.

[54] P. Swinnerton-Dyer, On ℓ-adic representations and congruences for coefficients of modular forms, *LNM* **350**, Springer-Verlag pp. 1–55, 1973.

[55] R. Taylor and A. Wiles, Ring-theoretic properties of certain Hecke algebras, *Ann. Math.* **141**(3) (1995), 553–572.

[56] K. Uchida, Isomorphisms of Galois groups, *J. Math. Soc. Japan* **28**(4) (1976), 617–620.

[57] A. Wiles, Modular elliptic curves and Fermat's Last Theorem, *Ann. Math.* **141**(3) (1995), 443–551.

[58] K. Wingberg, On the maximal unramified p-extension of an algebraic number field, *J. Reine Angew. Math.* **440** (1993), 129–156.

[59] A. Zubkov, Nonrepresentability of a free nonabelian pro-p group by second-order matrices (Russian), *Sibirsk. Mat. Zh.* **28**(5) (1987), 64–69.

Department of Mathematics
University of Illinois
Urbana, Illinois 61801, U.S.A.
boston@math.uiuc.edu

12

Cohomology of p-adic Analytic Groups

Peter Symonds and Thomas Weigel

1 Introduction

The purpose of this article is to give an exposition on the cohomology of compact p-adic analytic groups. The cohomology theory of profinite groups was initiated by J. Tate and developed by J-P. Serre [23] in the sixties, with applications to number theory. In his extraordinary work on p-adic analytic groups [17], M. Lazard also considered their cohomology and proved two striking theorems: Lazard's first theorem states that a compact p-adic analytic group G is a virtual Poincaré duality group; his second theorem states that the rational cohomology of G coincides with the G-stable cohomology of its associated $\mathbb{Q}_p$-Lie algebra $L(G)$. Our main goal is to discuss these results of Lazard in the spirit of the treatment of the structure of p-adic analytic groups in [10]. We also wish to emphasize the close parallels with the theory of discrete duality groups. In order to achieve this goal we need to set up the appropriate homological algebra.

For a profinite group G there are various choices of modules one can consider. They are all required to have a topology and G must act continuously. We denote by $\mathfrak{D}_p(G)$ the category of discrete G-modules whose objects are p-torsion modules. The dual category of compact G-modules whose objects are inverse limits of finite discrete p-torsion G-modules will be denoted by $\mathfrak{C}_p(G)$. Moreover, these categories are (anti-)equivalent by the Pontrjagin duality operation ([6], Prop. 2.3). The role played by the single category of left $\mathbb{Z}[G]$-modules for a discrete group G, is now played by

The second author gratefully acknowledges financial support of the Deutsche Forschungsgemeinschaft by a Heisenberg Stipendium.

the two categories $\mathfrak{D}_p(G)$ and $\mathfrak{C}_p(G)$.

It turns out that $\mathfrak{D}_p(G)$ has enough injectives and $\mathfrak{C}_p(G)$ has enough projectives. So one has a natural notion of cohomology $H^\bullet(G,\_)$ with coefficients in $ob(\mathfrak{D}_p(G))$ (as originally introduced by Tate) and also of homology $H_\bullet(G,\_)$ with coefficients in $ob(\mathfrak{C}_p(G))$ (as described in [6]). Moreover, these functors are linked by Pontrjagin duality, i.e., one has a natural isomorphism

$$H^\bullet(G,\_^*) \simeq H_\bullet(G,\_)^*, \tag{1.1}$$

where * denotes the Pontrjagin duality operation (3.4.6).

For certain profinite groups it is possible to enlarge the coefficient category for the cohomology functor. By analogy with the discrete case we call these groups of type p-$\mathbf{FP}_\infty$. These are the profinite groups G for which $\mathbb{Z}_p$ has a projective resolution $P_\bullet \to \mathbb{Z}_p$ in $\mathfrak{C}_p(G)$ with each P_k finitely generated as $\mathbb{Z}_p[\![G]\!]$-module. Here, $\mathbb{Z}_p[\![G]\!] = \varprojlim \mathbb{Z}_p[G/U]$ denotes the compact[1] $\mathbb{Z}_p$-group algebra of G ([10], §7). Although Lazard did not introduce this notion, he showed that compact p-adic analytic groups are of type p-$\mathbf{FP}_\infty$.

In order to have a cohomology and homology functor defined on the same category, we introduce the Pontrjagin category $\mathfrak{P}_p(G)$ which contains $\mathfrak{D}_p(G)$ and $\mathfrak{C}_p(G)$ as full subcategories. It is an exact category with the property that $\mathbf{Hom}_{\mathfrak{P}_p(G)}(A,B)$ is a topological $\mathbb{Z}_p$-module (with respect to the compact-open topology). Moreover, Pontrjagin duality provides a duality operation * on $\mathfrak{P}_p(G)$. For profinite groups of type p-$\mathbf{FP}_\infty$ it is possible to define homology and cohomology with coefficients in $\mathfrak{P}_p(G)$, and moreover the cohomology and homology groups naturally carry a topology.

One of the most striking results of Lazard's work on p-adic analytic groups is that torsion-free p-adic analytic pro-p groups satisfy *Poincaré duality*[2] (at the prime p), i.e., the cohomology $H^\bullet(G,\mathbb{F}_p)$ of G with coefficients in the trivial G-module $\mathbb{F}_p$ is a finite-dimensional $\mathbb{F}_p$-algebra under cup product, such that

(1) $dim_{\mathbb{F}_p}(H^n(G,\mathbb{F}_p)) = 1$, where n equals the dimension of G,

(2) $H^j(G,\mathbb{F}_p) = 0$ for $j > n$,

(3) the bilinear form

$$H^k(G,\mathbb{F}_p) \otimes H^{n-k}(G,\mathbb{F}_p) \longrightarrow H^n(G,\mathbb{F}_p) \tag{1.2}$$

induced by cup product is non-degenerate.

Discrete Poincaré duality groups were defined and studied by R. Bieri and B. Eckmann ([3]) some years later. A discrete group G of type FP_∞ of cohomological dimension n is called a *Poincaré duality group*, if

$$H^k(G,\mathbb{Z}[G]) = \begin{cases} \mathbb{Z} & \text{for } k = n, \\ 0 & \text{for } k \neq 0. \end{cases} \tag{1.3}$$

[1] In [10] this algebra is called the *completed group algebra*.

[2] This is the definition given by J-P. Serre ([23], Chap. I, §4.5).

For these groups one has a natural isomorphism between cohomology and homology with a twist by the dualizing module ([5], Chap.VIII, Thm. 10.1). One has a natural isomorphism

$$H^{n-k}(G, \_) \simeq H_k(G, D(G) \otimes \_), \tag{1.4}$$

where $D(G)$ denotes the dualizing module.

In order to have a notion of Poincaré duality for profinite groups (not just pro-p groups), we translate this definition to the profinite setting, i.e., a profinite group G of type p-$\mathbf{FP}_\infty$ of finite cohomological dimension n is called a *Poincaré duality group* in dimension n at the prime p, if

$$\mathbf{H}^k(G, \mathbb{Z}_p[\![G]\!]) = \begin{cases} \mathbb{Z}_p & \text{for } k = n, \\ 0 & \text{for } k \neq 0. \end{cases} \tag{1.5}$$

If G is a pro-p group this coincides with the definition above (1.2). Since it is possible to define the cohomology and homology functor on the same category, the theory is completely analogous to the theory for discrete Poincaré duality groups, i.e., (1.4) holds with $D(G)$ replaced by the compact dualizing module $D_p(G)$. This setup makes it easy to show Poincaré duality for p-torsion-free compact p-adic analytic groups using only results of Serre on the cohomological dimension [24] and the calculation of the cohomology of a uniformly powerful[3]pro-p group with coefficients in $\mathbb{F}_p$. This avoids referring to duality theorems due to Tate and J.L. Verdier as in the original proof given by Lazard [17], Prop. V.2.5.6.7. There are several possibilities for calculating the cohomology ring $H^\bullet(G, \mathbb{F}_p)$ for a uniformly powerful pro-p group ([17], Chap.V, (2.2), [26]], [27], Chap. II). However, the conceptual approach which provides a method to calculate (or at least to approximate) $H^\bullet(G, M)$ for a finitely generated pro-p group G and $M \in ob(\mathfrak{D}_p(G))$, $p.M = 0$, is given by the May spectral sequence (Thm. 5.1.12). For a uniformly powerful pro-p group P and $M = \mathbb{F}_p$ this spectral sequence collapses at the E_1-term and yields the desired result. Since we could not find this application anywhere in the literature,[4]we will discuss this technique in Section 5. The approach used by Lazard is certainly related to this technique ([9], Chap. 1, Thm. 1).

Finally, we discuss a result of Lazard which relates the rational cohomology $\mathbb{H}^\bullet(G, \_)$ of a compact p-adic analytic group G to the Lie algebra cohomology $H^\bullet(L(G), \_)$ of the $\mathbb{Q}_p$-Lie algebra $L(G)$ associated to G ([17], (V.2.4)). For a compact p-adic analytic group G it is possible to define cohomology with coefficients in $\mathfrak{fd}_p(G)$, the abelian category of finite-dimensional $\mathbb{Q}_p$-vector spaces with continuous left G-action. Every $\mathbb{Z}_p[\![G]\!]$-module $W \in ob(\mathfrak{fd}_p(G))$ can be considered as an $L(G)$-module for the $\mathbb{Q}_p$-Lie algebra $L(G)$ (Prop. 5.2.3). To distinguish between the different categories we use the symbol $L(W)$ if we consider W as an $L(G)$-module. The Lie algebra cohomology $H^\bullet(L(G), L(W))$ has a natural (right) G-action. Although not explicitly stated, Lazard establishes[5] a natural isomorphism

$$\mathbb{H}^\bullet(G, W) \simeq H^\bullet(L(G), L(W))^G \tag{1.6}$$

[3] This is the notion introduced in ([10]). Lazard called these groups *équi-p-valué*.

[4] For a finite powerful p-group this spectral sequence was considered in [9], (6.4).

(Thm. 5.2.4). This isomorphism can be used to study the action of a p-torsionfree compact p-adic analytic group G on its compact dualizing module $D_p(G)$. One obtains:

(1) The representation $\phi_{D_p(G)}: G \to \mathrm{End}_{\mathbb{Z}_p}(D_p(G)) = \mathbb{Z}_p^*$ is given by

$$\phi_{D_p(G)}(g) = det(Ad(g)), \tag{1.7}$$

where $Ad: G \to \mathrm{End}_{\mathbb{Z}_p}(L(G))$ denotes the (right) adjoint representation.

(2) The p-torsion-free compact p-adic analytic group G is virtually orientable if and only if $L(G)$ is a unimodular $\mathbb{Q}_p$-Lie algebra.

(Cor. 5.2.5, Cor. 5.2.6). Furthermore, a well known vanishing result for Lie algebra cohomology implies a vanishing result for rational cohomology of semi-simple compact p-adic analytic groups (Cor. 5.2.7).

2 Some homological algebra

Since our main subject of interest is the cohomology of profinite groups, we will have to use concepts and methods from homological algebra rather frequently. We assume that the reader is familiar with the standard notions such as categories, functors, natural transformations, additive and abelian categories, ([[8], [[18], [[14]). In Section §3 we describe the natural categories of representations for profinite groups. They are abelian categories, or exact categories, with the property that every kernel and cokernel of a morphism has a canonical representative. We call such an abelian (resp. exact) category an abelian (resp. exact) *module category*. We therefore start in §2.1 with a discussion of certain categories of topological $\mathbb{Z}_p$-modules and their natural duality operation, known as Pontrjagin duality. In §2.2 we give a short description of the construction of homological and cohomological functors and the connections between them. In §2.3 we discuss the cohomology of augmented algebras with a group action. Spectral sequences are considered in §2.4. Finally, we introduce a lemma of Serre, which is one of the main techniques for the construction of projective resolutions. For further information the reader is referred to the standard literature ([8], [18], [14]).

2.1 The categories $\mathfrak{F}_p$, $\mathfrak{C}_p$, $\mathfrak{D}_p$ and $\mathfrak{P}_p$

The category of left R-modules, ${}_R\mathrm{Mod}$, for a given ring R is an abelian module category. However, the category of topological R-modules (with morphisms defined to be continuous) for a topological ring R is in general not abelian. For $R = \mathbb{Z}_p$ we consider the categories $\mathfrak{F}_p$, $\mathfrak{C}_p$, $\mathfrak{D}_p$ and $\mathfrak{P}_p$.

By $\mathfrak{F}_p$ we denote the category whose objects are finite $\mathbb{Z}_p$-modules. For $A, B \in ob(\mathfrak{F}_p)$ we define

$$\mathrm{hom}_{\mathfrak{F}_p}(A, B) = \mathrm{Hom}_{\mathbb{Z}_p}(A, B). \tag{2.1.1}$$

[5] The proof of Lazard actually shows (1.6). However, he did not state this theorem in that form. In order to identify the Lie algebra cohomology he introduced the *stable rational cohomology*.

Thus $\mathfrak{F}_p$ is a full subcategory of ${}_{\mathbb{Z}_p}\mathrm{Mod}$, and it is easily seen that $\mathfrak{F}_p$ is an abelian module category. However, it has neither enough projectives nor enough injectives.

Let $\mathfrak{D}_p$ denote the category whose objects are p-torsion modules, i.e., $M \in {}_{\mathbb{Z}_p}\mathrm{Mod}$ is contained in $ob(\mathfrak{D}_p)$ if and only if for all $m \in M$ there exists some $k \in \mathbb{N}_0$ such that $p^k.m = 0$. As morphisms we define

$$\hom_{\mathfrak{D}_p}(A, B) = \mathrm{Hom}_{\mathbb{Z}_p}(A, B), \tag{2.1.2}$$

$A, B \in ob(\mathfrak{D}_p)$. So $\mathfrak{D}_p$ is again a full subcategory of ${}_{\mathbb{Z}_p}\mathrm{Mod}$, and it is an abelian module category. This category has enough injectives ([11]), but not enough projectives. Note that it is closed under direct limits and that $\mathfrak{F}_p$ is a full subcategory of $\mathfrak{D}_p$.

By $\mathfrak{C}_p$ we denote the category whose objects are the profinite $\mathbb{Z}_p$-modules. As morphisms we take

$$\hom_{\mathfrak{C}_p}(A, B) = \mathbf{Hom}_{\mathbb{Z}_p}(A, B) = \{\, \alpha \in \mathrm{Hom}_{\mathbb{Z}_p}(A, B) \mid \alpha \text{ continuous} \,\}, \tag{2.1.3}$$

$A, B \in \mathfrak{C}_p$. The category $\mathfrak{C}_p$ is a full subcategory of the category ${}_{\mathbb{Z}_p}\mathrm{TopMod}$ of topological $\mathbb{Z}_p$-modules. It is easy to check that it is an abelian module category. Note that a profinite $\mathbb{Z}_p$-module is just an abelian pro-p group. Since there exist free abelian pro-p groups, $\mathfrak{C}_p$ has enough projectives, but not enough injectives ([6], Lemma 1.6). The set $\mathbf{Hom}_{\mathbb{Z}_p}(A, B)$ carries naturally a topology—the compact-open topology—making $\mathbf{Hom}_{\mathbb{Z}_p}(A, B)$ a complete topological $\mathbb{Z}_p$-module. We therefore call $\mathfrak{C}_p$ a *topological abelian module category*. Note that $\mathfrak{C}_p$ is closed under inverse limits.

By $\mathfrak{P}_p$ we denote the category whose objects are either contained in $ob(\mathfrak{D}_p)$ or in $ob(\mathfrak{C}_p)$. As morphisms we take

$$\begin{aligned}\hom_{\mathfrak{P}_p}(A, B) &= \mathbf{Hom}_{\mathbb{Z}_p}(A, B)\\ &= \{\, \alpha \in \mathrm{Hom}_{\mathbb{Z}_p}(A, B) \mid \alpha \text{ continuous and strict} \,\}.\end{aligned} \tag{2.1.4}$$

Recall that a morphism of topological groups $\alpha\colon A \to B$ is called *strict*, if the canonical topologies (i.e., the quotient topology and induced topology) on $im(\alpha)$ coincide. Thus, for example, for $A \in ob(\mathfrak{D}_p)$, $B \in ob(\mathfrak{C}_p)$, $\phi \in \mathrm{Hom}_{\mathbb{Z}_p}(A, B)$ belongs to $\mathbf{Hom}_{\mathbb{Z}_p}(A, B)$ if and only if ϕ is *continuous* and has *finite image*. Again we consider $\mathbf{Hom}_{\mathbb{Z}_p}(A, B)$ as a topological space with the compact-open topology. This category satisfies all of the axioms for an abelian category except the existence of a direct product. Following D.A. Buchsbaum ([7]), we call such a category an *exact category*.

The category $\mathfrak{P}_p$ contains $\mathfrak{D}_p$ and $\mathfrak{C}_p$ as full subcategories. A short exact sequence in $\mathfrak{P}_p$ is either a short exact sequence in $\mathfrak{D}_p$ or a short exact sequence in $\mathfrak{C}_p$. The reason why we have introduced it here is because it has a duality operation analogous to taking the dual space of a vector space in the category of vector spaces. This operation is usually called the *Pontrjagin duality operation*.

Proposition 2.1.1. *Let* $I_p = \mathbb{Q}_p/\mathbb{Z}_p \in ob(\mathfrak{D}_p)$. *Then* $^*\colon \mathfrak{P}_p \to \mathfrak{P}_p$ *given by*

$$^* = \mathbf{Hom}_{\mathbb{Z}_p}(\_, I_p) \tag{2.1.5}$$

is a contravariant exact additive functor. The composition $^* \circ {}^*$ *is naturally isomorphic to the identity functor on* $\mathfrak{P}_p$. *Moreover, for* $A, B \in ob(\mathfrak{P}_p)$ *one has a natural* *isomorphism of topological $\mathbb{Z}_p$-modules*

$$\mathbf{Hom}_{\mathbb{Z}_p}(A, B) \simeq \mathbf{Hom}_{\mathbb{Z}_p}(B^*, A^*). \tag{2.1.6}$$

Proof. Note that for $A \in ob(\mathfrak{D}_p)$, $\mathbf{Hom}_{\mathbb{Z}_p}(A, I_p)$ carries the topology of pointwise convergence, while for $B \in ob(\mathfrak{C}_p)$, $\mathbf{Hom}_{\mathbb{Z}_p}(B, I_p)$ is a discrete topological $\mathbb{Z}_p$-module. So the first part of the proposition is well known ([6], Prop. 2.3, [13], p. 400). For the second part consider the right and left hand side as contravariant/covariant bi-functors to the category of abelian groups. The morphism function of the functor * yields the desired isomorphism as abelian groups. Then one can easily check, using the canonical neighbourhood basis of $0 \in \mathbf{Hom}_{\mathbb{Z}_p}(A, B)$, that the isomorphism is also a homeomorphism.

Pontrjagin duality can be used quite successfully to identify certain cohomology and homology groups of profinite groups. An immediate application of Pontrjagin duality is also the following:

Corollary 2.1.2. *If $M \in ob(\mathfrak{C}_p)$ is p-torsion-free, then it is a free abelian pro-p group and therefore free (and projective) in $\mathfrak{C}_p$.*

Proof. We have to find a free generating set in M. Let $W = M/p.M$ and $V = W^* \in ob(\mathfrak{D}_p)$. Then V is an $\mathbb{F}_p$-vector space. By Zorn's lemma, V has a basis $X \subset V$. For $x \in X$, let $x^* \in W$ be an element vanishing on $X \setminus \{x\}$ and satisfying $x^*(x) \neq 0$. Then $X^* = \{x^* \in W \mid x \in X\}$ is a free generating set of W as free elementary abelian pro-p group. Using the p-adic filtration on M then shows the desired result.

2.2 Homological and cohomological functors

Let $\mathfrak{A}$, $\mathfrak{B}$ be abelian or just exact categories and let $\Sigma(\mathfrak{A})$, $\Sigma(\mathfrak{B})$ denote the class of all short exact sequences in $\mathfrak{A}$, $\mathfrak{B}$. Let $\Sigma^* \subseteq \Sigma(\mathfrak{A})$ be a proper class of short exact sequences.[6] A Σ^*-*connected pair* (S, e, T) of covariant functors is a pair of covariant additive functors $S, T\colon \mathfrak{A} \to \mathfrak{B}$ together with a function e which assigns to each proper exact sequence $\mathbf{s}\colon 0 \to A \overset{\sigma}{\to} B \overset{\tau}{\to} C \to 0 \in \Sigma^*$ a morphism $e(\mathbf{s}) \in \hom_{\mathfrak{B}}(S(C), T(A))$. This $e(\mathbf{s})$ must have the property that for any short exact sequence $\mathbf{s}'\colon 0 \to A' \overset{\sigma'}{\to} B' \overset{\tau'}{\to} C' \to 0$ and any morphism $(\alpha, \beta, \gamma) \in \hom_{\Sigma^*}(\mathbf{s}, \mathbf{s}')$ of short exact sequences, the diagram

$$\begin{array}{ccc} S(C) & \xrightarrow{e(\mathbf{s})} & T(A) \\ {\scriptstyle S(\gamma)}\downarrow & & {\scriptstyle T(\alpha)}\downarrow \\ S(C') & \xrightarrow{e(\mathbf{s}')} & T(A') \end{array} \tag{2.2.1}$$

commutes. The function e is called the *connecting homomorphism* of the pair. The diagram (2.2.1) implies that e is a natural transformation if one considers S and T as functors from Σ^* to $\mathfrak{B}$. Moreover, if $\mathfrak{A}$ and $\mathfrak{B}$ are abelian then,

---

[6] For a definition see [18], XII. If $\mathfrak{A}$ is exact but not abelian, axiom (ii) must be omitted.

- if s is congruent to s′, then $e(\mathbf{s}) = e(\mathbf{s}')$,
- for $\gamma \in \hom_{\mathfrak{A}}(C', C)$ one has $e(\mathbf{s} \circ \gamma) = e(\mathbf{s}) \circ S(\gamma)$,
- for $\alpha \in \hom_{\mathfrak{A}}(A, A')$ one has $e(\alpha \circ \mathbf{s}) = T(\alpha) \circ e(\mathbf{s})$,

([18], XII.7.2(a)–(c)). Recall that the short exact sequences $\mathbf{s}\colon 0 \to A \xrightarrow{\sigma} B \xrightarrow{\tau} C \to 0$ and $\mathbf{s}'\colon 0 \to A \xrightarrow{\sigma'} B' \xrightarrow{\tau'} C \to 0$ are called *congruent*, if there exists a morphism $\boldsymbol{\beta} = (id_A, \beta, id_C) \in \hom_{\Sigma}(\mathbf{s}, \mathbf{s}')$. It is obvious that β is necessarily an isomorphism.

From this property we conclude that e induces a morphism of abelian groups

$$e\colon Ext^1_{\Sigma^*}(C, A) \longrightarrow \hom_{\mathfrak{B}}(S(C), T(A)) \tag{2.2.2}$$

and that

$$S(A) \xrightarrow{S(\sigma)} S(B) \xrightarrow{S(\tau)} S(C) \xrightarrow{e(\mathbf{s})} T(A) \xrightarrow{T(\sigma)} T(B) \xrightarrow{T(\tau)} T(C) \tag{2.2.3}$$

is a complex, i.e., the composition of any two successive maps is zero ([18], §XII.7).

A *morphism of* Σ^**-connected pairs* $(f, g)\colon (S, e, T) \to (S', e', T')$ of covariant additive functors is a pair of natural transformations $f\colon S \to S'$, $g\colon T \to T'$, such that the diagram

$$\begin{array}{ccc} S(C) & \xrightarrow{e(\mathbf{s})} & T(A) \\ {\scriptstyle f(C)}\downarrow & & \downarrow{\scriptstyle g(A)} \\ S'(C) & \xrightarrow{e'(\mathbf{s})} & T'(A) \end{array} \tag{2.2.4}$$

commutes for all $\mathbf{s} \in \Sigma^*$.

A Σ^**-connected sequence* $(T^\bullet, e^\bullet)$ of covariant additive functors is a sequence of covariant additve functors $T^k\colon \mathfrak{A} \to \mathfrak{B}$, $k \in \mathbb{Z}$, such that (T^k, e^k, T^{k+1}) is a Σ^*-connected pair for all $k \in \mathbb{Z}$. A *morphism of* Σ^**-connected sequences* $f^\bullet\colon (S^\bullet, e^\bullet_S) \to (T^\bullet, e^\bullet_T)$ is a sequence of natural transformations $f^\bullet = (f^k)_{k\in\mathbb{Z}}$ with the property that $(f^k, f^{k+1})\colon (S^k, e^k_S, S^{k+1}) \to (T^k, e^k_T, T^{k+1})$ is a morphism of Σ^*-connected pairs for all $k \in \mathbb{Z}$.

A Σ^*-connected sequence $(T^\bullet, e^\bullet)$ of covariant additve functors is called a Σ^*-*cohomological functor* if $T^k = 0$ for $k < 0$, and if for all short exact sequences $\mathbf{s}\colon 0 \to A \xrightarrow{\sigma} B \xrightarrow{\tau} C \to 0$ in Σ^*, the sequence

$$0 \longrightarrow T^0(A) \xrightarrow{T^0(\sigma)} T^0(B) \xrightarrow{T^0(\tau)} T^0(C) \xrightarrow{e^0(\mathbf{s})} T^1(A) \xrightarrow{T^1(\sigma)} T^1(B) \longrightarrow \cdots \tag{2.2.5}$$

is exact in $\mathfrak{B}$. If $\mathfrak{A}$ has enough Σ^*-injectives, we call the Σ-cohomological functor $(T^\bullet, e^\bullet)$ *co-effaceable* if $T^i(J) = 0$ for all Σ^*-injective objects $J \in ob(\mathfrak{A})$ and $i \geqslant 1$.

A Σ^*-connected sequence $(S^\bullet, e^\bullet)$ of covariant additive functors $S^k\colon \mathfrak{A} \to \mathfrak{B}$ is called a Σ^*-*homological functor* if $S^k = 0$ for $k > 0$, and if for all short exact sequences $\mathbf{s}\colon 0 \to A \xrightarrow{\sigma} B \xrightarrow{\tau} C \to 0$ in Σ^* the sequence

$$\cdots \longrightarrow S^{-1}(B) \xrightarrow{S^{-1}(\tau)} S^{-1}(C) \xrightarrow{e_{-1}(\mathbf{s})} S^0(A) \xrightarrow{S^0(\sigma)} S^0(B) \xrightarrow{S^0(\tau)} S^0(C) \longrightarrow 0 \tag{2.2.6}$$

is exact in $\mathfrak{B}$. If $\mathfrak{A}$ has enough Σ^*-projectives then the Σ^*-homological functor $(S^\bullet, e^\bullet)$ is called *effaceable* if $S^i(P) = 0$ for all Σ^*-projective objects $P \in ob(\mathfrak{A})$ and $i \leqslant -1$.

The following theorem summarizes one of the "general principles of homological algebra".

Theorem 2.2.1. *Let $\mathfrak{A}$, $\mathfrak{B}$ be abelian module categories and let $\Sigma^* \subseteq \Sigma(\mathfrak{A})$ be a proper class of short exact sequences.*
(a) Assume that $\mathfrak{A}$ has enough Σ^*-injectives and let $(T^\bullet, e_T^\bullet)$ be a co-effaceable Σ^*-cohomological functor, $T^k\colon \mathfrak{A} \to \mathfrak{B}$. Let $(U^\bullet, e_U^\bullet)$ be a non-negative Σ^*-connected sequence of covariant additive functors, i.e., $U^k\colon \mathfrak{A} \to \mathfrak{B}$ and $U^k = 0$ for $k < 0$. Then for every natural transformation $f\colon T^0 \to U^0$, there exists a unique morphism of connected sequences $f^\bullet\colon (T^\bullet, e_T^\bullet) \to (U^\bullet, e_U^\bullet)$ with $f^0 = f$. If f^0 is a natural isomorphism and $(U^\bullet, e_U^\bullet)$ is also a co-effaceable cohomological functor, then all f^i are natural isomorphisms.
(b) Assume that $\mathfrak{A}$ has enough Σ^*-projectives and let $(S^\bullet, e_S^\bullet)$ be an effaceable Σ^*-homological functor, $S^k\colon \mathfrak{A} \to \mathfrak{B}$. Let $(V^\bullet, e_V^\bullet)$ be a non-positive Σ^*-connected sequence of covariant additive functors, i.e., $V^k\colon \mathfrak{A} \to \mathfrak{B}$ and $V^k = 0$ for $k > 0$. Then for every natural transformation $g\colon V^0 \to S^0$, there exists a unique morphism of connected sequences $g^\bullet\colon (V^\bullet, e_V^\bullet) \to (S^\bullet, e_S^\bullet)$ with $g^0 = g$. If g^0 is a natural isomorphism and $(V^\bullet, e_V^\bullet)$ is also an effaceable homological functor, then all g^i are also natural isomorphisms.

Proof. In the notation of [18], $(T^\bullet, e_T^\bullet)$ is a right Σ^*-universal and $(S^\bullet, e_S^\bullet)$ a left Σ^*-couniversal connected sequence. Hence the claim follows from [18], Thm. XII.7.6, Thm. XII.7.2.

Our construction of homological and cohomological functors relies mainly on the exact homology sequence and the comparison theorem.

Let $\mathfrak{A}$ be an exact module category. A *chain complex* $(A_\bullet, \partial_\bullet)$ is a sequence of modules $A_k \in ob(\mathfrak{A})$, $k \in \mathbb{Z}$, and morphisms $\partial_k\colon A_k \to A_{k-1}$ such that $\partial_{k-1} \circ \partial_k = 0$. The sequence of mappings $\partial_\bullet = (\partial_k)_{k\in\mathbb{Z}}$ is called the *differential* of the chain complex. The *homology* of a chain complex is given by

$$H_k(A_\bullet, \partial_\bullet) = ker(\partial_k)/im(\partial_{k+1}). \tag{2.2.7}$$

A *mapping of chain complexes* $\sigma_\bullet\colon (A_\bullet, \partial_\bullet) \to (B_\bullet, \partial'_\bullet)$ is a sequence of mappings $\sigma_k\colon A_k \to B_k$ such that the diagram

$$\begin{array}{ccccccccc} \cdots & \longrightarrow & A_{k+1} & \xrightarrow{\partial_{k+1}} & A_k & \xrightarrow{\partial_k} & A_{k-1} & \longrightarrow & \cdots \\ & & \downarrow{\scriptstyle \sigma_{k+1}} & & \downarrow{\scriptstyle \sigma_k} & & \downarrow{\scriptstyle \sigma_{k-1}} & & \\ \cdots & \longrightarrow & B_{k+1} & \xrightarrow{\partial'_{k+1}} & B_k & \xrightarrow{\partial'_k} & B_{k-1} & \longrightarrow & \cdots \end{array} \tag{2.2.8}$$

commutes. The sequence of mappings of chain complexes

$$0 \longrightarrow (A_\bullet, \partial_\bullet) \xrightarrow{\sigma_\bullet} (B_\bullet, \partial'_\bullet) \xrightarrow{\tau_\bullet} (C_\bullet, \partial''_\bullet) \longrightarrow 0 \tag{2.2.9}$$

is called *short exact*, if $0 \to A_k \overset{\sigma_k}{\to} B_k \overset{\tau_k}{\to} C_k \to 0$ is short exact for all $k \in \mathbb{Z}$. The category of chain complexes $\mathbf{C}_\bullet(\mathfrak{A})$ with mappings of chain complexes as morphisms is also exact, and is abelian if $\mathfrak{A}$ is abelian. Moreover, the homology functors $H_n(\_)\colon \mathbf{C}_\bullet(\mathfrak{A}) \to \mathfrak{A}$ are covariant additive functors.

Dually, $(B^\bullet, \delta^\bullet)$ is called a *cochain complex* if B^k is a sequence of modules with $\delta^k\colon B^k \to B^{k+1}$, $\delta^{k+1} \circ \delta^k = 0$. The *cohomology* of a cochain complex is given by

$$H^k(B^\bullet, \delta^\bullet) = ker(\delta^k)/im(\delta^{k-1}). \tag{2.2.10}$$

Mappings of cochain complexes and short exact sequences are defined in the same way as for chain complexes. The category of cochain complexes will be denoted by $\mathbf{C}^\bullet(\mathfrak{A})$. It is also exact (abelian if $\mathfrak{A}$ is abelian). By reindexing, chain complexes can be made cochain complexes and vice versa.

Theorem 2.2.2. (Exact homology sequence) Let $\mathfrak{A}$ be an exact module category and $\mathbf{s}_\bullet\colon 0 \to (A_\bullet, \partial_\bullet) \overset{\sigma_\bullet}{\to} (B_\bullet, \partial'_\bullet) \overset{\tau_\bullet}{\to} (C_\bullet, \partial''_\bullet) \to 0$ a short exact sequence of chain complexes. Then there exists an exact sequence

$$\cdots \to H_n(A_\bullet) \overset{\sigma_n}{\to} H_n(B_\bullet) \overset{\tau_n}{\to} H_n(C_\bullet) \overset{e_k(\mathbf{s}_\bullet)}{\to} H_{n-1}(A_\bullet) \overset{\sigma_{n-1}}{\to} H_{n-1}(B_\bullet) \to \cdots, \tag{2.2.11}$$

where σ_n, τ_n denote the induced maps and $e_n(\mathbf{s}_\bullet)$ is given by the switchback rule. Moreover, $e_n(\_)$ is a natural transformation of the functors

$$H_n(\_{fin}), H_{n-1}(\_{in})\colon \mathbf{\Sigma}(\mathbf{C}_\bullet(\mathfrak{A})) \longrightarrow \mathfrak{A} \tag{2.2.12}$$

Proof. ([18], Thm. II.4.1, Prop. II.4.2). Here we have used the abbreviations $\_{fin}$ and $\_{in}$ for the final and initial term of a short exact sequence, and have used the same symbol for the induced mappings on the homology. For $\mathfrak{A} = {}_{\mathbb{Z}}\mathrm{Mod}$ the switchback rule is given by the following assignment. If $x \in C_n$ with $\partial''_n(x) = 0$, let $[x]$ denote the corresponding homology class in $H_n(C_\bullet, \partial''_\bullet)$. Choose $y \in B_n$ with $\tau_n(y) = x$. Then $\tau_{n-1}(\partial'_n(y)) = \partial''_n(\tau_n(y)) = 0$. Hence there exists $z \in A_{n-1}$ with $\sigma_{n-1}(z) = \partial'_n(y)$. Since $\sigma_{n-2}(\partial_{n-1}(z)) = \partial'_{n-1}(\sigma_{n-1}(z)) = 0$ and as σ_{n-2} is injective, one has $\partial_{n-1}(z) = 0$. Then $e_n(\mathbf{s}_\bullet)([x]) = [z]$ and this assignment is well defined and satisfies (2.2.11) and (2.2.12). Using this, it is now easy to establish a proof for general $\mathfrak{A}$ using only kernels and cokernels of mappings.

In order to simplify notations we omit the differential in the chain complex. Let $A_\bullet$, $B_\bullet$ be chain complexes and $f_\bullet, f'_\bullet\colon A_\bullet \to B_\bullet$ be chain mappings. One says that f and f' are *chain homotopic* if there exists a sequence of mappings $h_\bullet = (h_n)_{n\in\mathbb{Z}}$, $h_n\colon A_n \to B_{n+1}$, such that

$$f_n - f'_n = \partial_{n+1} \circ h_n + h_{n-1} \circ \partial_n. \tag{2.2.13}$$

For short we write $f_\bullet \simeq f'_\bullet$. It is an immediate consequence of the definition that chain homotopic maps induce the same mapping on the homology groups

$$f_\bullet = f'_\bullet\colon H_\bullet(A_\bullet) \longrightarrow H_\bullet(B_\bullet). \tag{2.2.14}$$

The complexes $A_\bullet$ and $B_\bullet$ are called *chain homotopy equivalent*, if there exist chain maps $g_\bullet\colon A_\bullet \to B_\bullet$ and $f_\bullet\colon B_\bullet \to A_\bullet$, such that $f_\bullet \circ g_\bullet \simeq 1_{A_\bullet}$, $g_\bullet \circ f_\bullet \simeq 1_{B_\bullet}$.

Let $M \in ob(\mathfrak{A})$. $P_\bullet \xrightarrow{\varepsilon} M$ is called a *projective resolution* of M, if $P_\bullet$ is a non-negative chain complex of projective modules $P_k \in ob(\mathfrak{A})$ and

$$\cdots \longrightarrow P_2 \xrightarrow{\partial_2} P_1 \xrightarrow{\partial_1} P_0 \xrightarrow{\varepsilon} M \tag{2.2.15}$$

is exact. Dually, $M \xrightarrow{\eta} J_\bullet$ is called an *injective resolution*, if $J_\bullet$ is a non-positive chain complex of injective modules $J_k \in ob(\mathfrak{A})$ and

$$M \xrightarrow{\eta} J_0 \xrightarrow{\partial_0} J_{-1} \xrightarrow{\partial_{-1}} J_{-2} \longrightarrow \cdots \tag{2.2.16}$$

is exact. The comparison theorem states that projective and injective resolutions are unique up to chain homotopy equivalence. More precisely, one has

Theorem 2.2.3. (Comparison theorem) *Let A, $B \in ob(\mathfrak{A})$ and $\alpha \in \hom_{\mathfrak{A}}(A, B)$.*

(1) Let $P_\bullet \xrightarrow{\varepsilon} A$ be a projective resolution of A and $Q_\bullet$ a chain complex, $\varepsilon\colon Q_0 \to B$ such that $\cdots \to Q_1 \xrightarrow{\partial_1} Q_0 \xrightarrow{\varepsilon} B \to 0$ is exact. Then there exists a chain mapping $f_\bullet\colon P_\bullet \to Q_\bullet$ making the diagram

$$\begin{array}{ccccccccc} \cdots & \longrightarrow & P_2 & \xrightarrow{\partial_2} & P_1 & \xrightarrow{\partial_1} & P_0 & \xrightarrow{\varepsilon} & A \\ & & {\scriptstyle f_2}\downarrow & & {\scriptstyle f_1}\downarrow & & {\scriptstyle f_0}\downarrow & & {\scriptstyle \alpha}\downarrow \\ \cdots & \longrightarrow & Q_2 & \xrightarrow{\partial_2} & Q_1 & \xrightarrow{\partial_1} & Q_0 & \xrightarrow{\varepsilon} & B \end{array} \tag{2.2.17}$$

commute. Moreover, any pair of chain mappings $f_\bullet$, $f'_\bullet$ satisfying (2.2.17) *are chain homotopic.*

(2) Let $A \xrightarrow{\eta} J_\bullet$, $B \xrightarrow{\eta} I_\bullet$ be injective resolutions. Then there exists a chain mapping $g_\bullet\colon J_\bullet \to I_\bullet$ making the diagram

$$\begin{array}{ccccccccc} A & \xrightarrow{\eta} & J_0 & \xrightarrow{\partial_0} & J_{-1} & \xrightarrow{\partial_{-1}} & J_{-2} & \longrightarrow & \cdots \\ {\scriptstyle \alpha}\downarrow & & {\scriptstyle g_0}\downarrow & & {\scriptstyle g_{-1}}\downarrow & & {\scriptstyle g_{-2}}\downarrow & & \\ B & \xrightarrow{\eta} & I_0 & \xrightarrow{\partial_0} & I_{-1} & \xrightarrow{\partial_{-1}} & I_{-2} & \longrightarrow & \cdots \end{array} \tag{2.2.18}$$

commute. Moreover, any pair of chain mappings $g_\bullet$, $g'_\bullet$ satisfying (2.2.18) *are chain homotopic.*

Proof. ([1], Thm. 2.4.2, [18], Thm. III.6.1, Thm. III.8.1) This yields:

Theorem 2.2.4. *Let $\mathfrak{A}$, $\mathfrak{B}$, $\mathfrak{Q}$ be exact module categories, $\mathfrak{A}_0 \subseteq \mathfrak{A}$ a full subcategory of $\mathfrak{A}$ and $M \in ob(\mathfrak{A}_0)$. Assume also that M has a projective resolution $P_\bullet \to M$ with $P_k \in ob(\mathfrak{A}_0)$.*

(a) *Let $T(\_,\_)\colon \mathfrak{A}_0 \times \mathfrak{B} \to \mathfrak{Q}$ be a biadditive left/left exact contravariant/covariant bifunctor such that $T(P,\_)\colon \mathfrak{B} \to \mathfrak{Q}$ is exact for all $P \in ob(\mathfrak{A}_0)$ projective in $\mathfrak{A}$. Then $(T^\bullet_M, e^\bullet)$ with*

$$T^\bullet_M(B) = H^\bullet(T(P_\bullet, B), \delta^\bullet), \tag{2.2.19}$$

and $e^\bullet$ given by the exact homology sequence, is a cohomological functor. Moreover, it does not depend on the particular choice of projective resolution of M in $\mathfrak{A}_0$.

If $\mathfrak{A}_0 = \mathfrak{A}$, $\mathfrak{B}$ has enough injectives and $T(\_, J)$ is exact for all $J \in ob(\mathfrak{B})$ injective, then $(T^\bullet_M, e^\bullet)$ is co-effaceable. In this case there is a (unique) natural isomorphism

$$(T^\bullet_M, e^\bullet) \simeq (\mathcal{R}^\bullet T_M, e^\bullet) \tag{2.2.20}$$

of cohomological functors, where $\mathcal{R}^\bullet T_M$ denote the right derived functors of $T_M(\_) = T(M,\_)$.

(b) *Let $S(\_,\_)\colon \mathfrak{A}_0 \times \mathfrak{B} \to \mathfrak{Q}$ be a biadditive right/right exact covariant/covariant bifunctor such that $S(P,\_)\colon \mathfrak{B} \to \mathfrak{Q}$ is exact for all $P \in ob(\mathfrak{A}_0)$ projective in $\mathfrak{A}$. Then $(S^M_\bullet, e_\bullet)$ with*

$$S^M_\bullet(B) = H_\bullet(S(P_\bullet, B), \partial_\bullet), \tag{2.2.21}$$

and $e_\bullet$ given by the exact homology sequence is a homological functor. Moreover, it does not depend on the particular choice of projective resolution of M in $\mathfrak{A}_0$.

If $\mathfrak{A}_0 = \mathfrak{A}$, $\mathfrak{B}$ has enough projectives and $S(\_, Q)$ is exact for all $Q \in ob(\mathfrak{B})$ projective, then $(S^M_\bullet, e_\bullet)$ is effaceable. In this case there is a (unique) natural isomorphism

$$(S^M_\bullet, e_\bullet) \simeq (\mathcal{L}_\bullet S^M, e_\bullet) \tag{2.2.22}$$

of homological functors where $\mathcal{L}_\bullet S^M$ denote the left derived functors of $S^M(\_) = S(M,\_)$.

Proof. (a) By hypothesis, if $s\colon 0 \to A \xrightarrow{\sigma} B \xrightarrow{\tau} C \to 0$ is a short exact sequence in $\mathfrak{B}$, then there is a short exact seqeunce of cochain complexes

$$0 \to T(P_\bullet, A) \xrightarrow{\sigma} T(P_\bullet, B) \xrightarrow{\tau} T(P_\bullet, C) \to 0. \tag{2.2.23}$$

The first part of (a) now follows from the exact homology sequence. For the second part, the hypothesis ensures that $(T^\bullet_M, e^\bullet)$ is co-effaceable. The cohomological functor $(\mathcal{R}^\bullet T_M, e^\bullet)$ is also co-effaceable ([18], §XII.9). Thus the claim follows from Theorem 2.2.1(a). Part (b) follows by similar arguments.

2.3 Homology and cohomology for augmented algebras

Let R be a commutative ring with unit 1_R and let (A, ε) be an augmented R-algebra which is also a free R-module, i.e., A has a unit 1_A and $\varepsilon\colon A \to R$ is a morphism of R-algebras with $\varepsilon(1_A) = 1_R$. By $A^+ = ker(\varepsilon)$ we denote the augmentation ideal. Then $(\mathbb{B}_\bullet(A), d_\bullet)$ with

$$\mathbb{B}_k(A) = A \otimes \overbrace{A^+ \otimes \cdots \otimes A^+}^{k \text{ times}}, \tag{2.3.1}$$

and $d_\bullet$ given by

$$\begin{aligned} d_k(x[x_1|..|x_k]) &= xx_1[x_2|..|x_k] \\ &\quad + \sum_{j=1}^{k-1}(-1)^j x[x_1|..|x_jx_{j+1}|..|x_k], \end{aligned} \tag{2.3.2}$$

is a complex of free left A-modules. Moreover, it is easy to check that $s_k\colon \mathbb{B}_k(A) \to \mathbb{B}_{k+1}(A)$ given by

$$s_k(x[x_1|..|x_k]) = 1.[x - \varepsilon(x).1|x_2|..|x_k] \tag{2.3.3}$$

is a homotopy contraction in positive degrees, i.e., for $k \geqslant 1$ one has

$$s_{k-1} \circ d_k + d_{k+1} \circ s_k = id_{\mathbb{B}_k(A)}. \tag{2.3.4}$$

Hence $\mathbb{B}_\bullet(A) \xrightarrow{\varepsilon} R$ is a projective resolution of the trivial A-module R, which is called the *(reduced) bar resolution*.

The bifunctors

$$\begin{aligned} \mathrm{Hom}_A(\_,\_)\colon {}_A\mathrm{Mod} \times {}_A\mathrm{Mod} &\longrightarrow {}_R\mathrm{Mod}, \\ \_\otimes_A\_\colon \mathrm{Mod}_A \times {}_A\mathrm{Mod} &\longrightarrow {}_R\mathrm{Mod}, \end{aligned} \tag{2.3.5}$$

satisfy the hypothesis of Theorem 2.2.4. We denote by

$$\begin{aligned} \mathrm{Ext}^\bullet_A(\_,\_)\colon {}_A\mathrm{Mod} \times {}_A\mathrm{Mod} &\to {}_R\mathrm{Mod}, \\ \mathrm{Tor}^A_\bullet(\_,\_)\colon \mathrm{Mod}_A \times {}_A\mathrm{Mod} &\to {}_R\mathrm{Mod}, \end{aligned} \tag{2.3.6}$$

the (derived) bifunctors obtained by Theorem 2.2.4. For $M \in {}_A\mathrm{Mod}$ we also put

$$\begin{aligned} H^\bullet(A, M) &= \mathrm{Ext}^\bullet_A(R, M) \\ H_\bullet(A, M) &= \mathrm{Tor}^A_\bullet(R, M). \end{aligned} \tag{2.3.7}$$

We also have to consider algebras A with a right action by some group G inducing R-algebra automorphisms and fixing the augmentation. The class of objects of the *twisted module category* ${}_A\mathrm{TMod}^G$ consists of all left A-modules M which are also right G-modules such that

$$a^g.m^g = (a.m)^g \tag{2.3.8}$$

for all $a \in A$, $m \in M$, $g \in G$. As morphisms one takes all A-module homomorphisms commuting with the G-action, i.e.,

$$\mathrm{Hom}_A^G(M, N) = \{\phi \in \mathrm{Hom}_A(M, N) \mid \phi(m^g) = \phi(m)^g\}. \tag{2.3.9}$$

Giving $\mathbb{B}_\bullet(A)$ the diagonal G-action, it is easy to see that $(\mathbb{B}_\bullet(A), d_\bullet)$ is actually contained in $ob(\mathbf{C}_\bullet({}_A\mathrm{TMod}^G))$. For $M, N \in {}_A\mathrm{TMod}^G$, $\mathrm{Hom}_A(M, N)$ has a canonical right G-action given by

$$\phi^g(m) = \phi(m^{g^{-1}})^g, \tag{2.3.10}$$

$\phi \in \mathrm{Hom}_A(M, N)$. It follows that $H^\bullet(A, M)$ has a right G-action for $M \in ob({}_A\mathrm{TMod}^G)$. In order to show that it is canonical and does not depend on the particular choice of projective resolution one has to work slightly harder.

Let A and B be two augmented R-algebras as described above, and let $M \in ob({}_A\mathrm{Mod})$, $N \in ob({}_B\mathrm{Mod})$. Let $\phi\colon A \to B$ be a morphism of augmented R-algebras. Then every left B-module can be also considered as left A-module via ϕ, so ϕ induces a covariant additive exact functor $\phi^*\colon {}_B\mathrm{Mod} \to {}_A\mathrm{Mod}$. Suppose that $\psi \in \mathrm{Hom}_R(N, M)$ is such that

$$a.\psi(n) = \psi(\phi(a).n). \tag{2.3.11}$$

Then ψ induces a natural transformation of contravariant additive functors

$$\psi_*\colon \mathrm{Hom}_B(\_, N) \longrightarrow \mathrm{Hom}_A(\phi^*(\_), M)) \tag{2.3.12}$$

given by $\psi^*(f)(q) = \psi(f(q))$ for $q \in Q$, $Q \in ob({}_B\mathrm{Mod})$. Let $P_\bullet \to R$ be a projective resolution of R as a left A-module, and let $Q_\bullet \to R$ be a projective resolution of R as a left B-module. By the Comparison Theorem (Thm. 2.2.3(1)), there exists a mapping of chain complexes $\tau_\bullet\colon P_\bullet \to \phi^*(Q_\bullet)$ with $\tau_k \in \mathrm{Hom}_A(P_k, \phi^*(Q_k))$ and which is uniquely determined up to chain homotopy. Hence the pair (ϕ, ψ) induces a mapping $(\phi, \psi)^\bullet\colon H^\bullet(B, N) \to H^\bullet(A, M)$ defined as the composite

$$H^\bullet(\mathrm{Hom}_B(Q_\bullet, N)) \overset{H^\bullet(\psi^*)}{\longrightarrow} H^\bullet(\mathrm{Hom}_A(\phi^*(Q_\bullet), M) \overset{H^\bullet(\tau_\bullet)}{\longrightarrow} H^\bullet(\mathrm{Hom}_A(P_\bullet, M)). \tag{2.3.13}$$

It is easily seen that $(\phi, \psi)^\bullet$ is independent of the particular choice of projective resolutions.

In our case the mappings ϕ_g and ψ_g are given by $\phi(a) = a^{g^{-1}}$, $\psi(m) = m^g$. By definition, (2.3.11) is satisfied. For $P_\bullet = Q_\bullet = \mathbb{B}_\bullet(A)$, $\tau_\bullet$ can be chosen to be given by the right action of the element $g^{-1} \in G$ on $\mathbb{B}_\bullet(A)$. For these mappings one can check easily that the action of $g \in G$ on $H^\bullet(A, M)$ is given by the map $(\phi_g, \psi_g)^\bullet$.

A typical example of an algebra with a group action is the group algebra $A = R[N]$ for a normal subgroup $N \lhd G$ and the action given by conjugation. Moreover, ${}_{R[G]}\mathrm{Mod}$ is a subcategory of ${}_{R[N]}\mathrm{TMod}^G$, where $M \in ob({}_{R[G]}\mathrm{Mod})$ is given the right action $m^g = g^{-1}.m$, $m \in M$, $g \in G$. If $G = N$, then (2.3.10) implies that the action of G on $H^\bullet(G, M)$ is trivial.

If $A \leqslant B$ is an augmented subalgebra of B, then the pair $(\iota_{A,B}, res_A^B) = (\phi, \psi)$ induces the restriction mapping

$$(\iota_{A,B}, res_A^B)^\bullet = res_A^B \colon H^\bullet(B, \_) \longrightarrow H^\bullet(A, \_). \tag{2.3.14}$$

If G acts on B by algebra automorphisms and if $A^g = A$ for all $g \in G$, then res_A^B is also a G-equivariant morphism. This can be seen easily from the above description using the reduced bar resolutions for A and B.

2.4 Spectral sequences

Throughout this section we assume that R is a commutative ring with 1. In order to compute the homology of a chain complex explicitly one often uses a filtration on the complex and considers the associated spectral sequence. A similar technique applies to the cohomology of a cochain complex.

Let $(A_\bullet, d_\bullet)$ be a non-negative chain complex of R-modules. A *filtration* $F_\bullet$ of $(A_\bullet, d_\bullet)$ is a series of sub chain complexes $(F_k A_\bullet, d_\bullet)_{k\in\mathbb{Z}}$, $F_k A_n \leqslant A_n$, such that

(1) $F_k A_n \leqslant F_{k+1} A_n$ for all $k \in \mathbb{Z}$, $n \in \mathbb{N}_0$,

(2) $d_n(F_k A_n) \leqslant F_k A_{n-1}$.

The filtration is called *exhaustive* if $\bigcup_{k\in\mathbb{Z}} F_k A_n = A_n$ for all $n \in \mathbb{N}_0$, and *separated* if $\bigcap_{k\in\mathbb{Z}} F_k A_n = 0$. The differential induces a differential d^0 on the associated (bi)graded object

$$E^0_{s,t} = F_s A_{s+t} / F_{s-1} A_{s+t} \tag{2.4.1}$$

of degree $(0, -1)$. A filtration $F_\bullet$ also induces a filtration $F_\bullet$ on $H_\bullet(A_\bullet, d_\bullet)$. Let

$$\begin{aligned} Z^r_{s,t} &= F_s A_{s+t} \cap d^{-1}_{s+t}(F_{s-r} A_{s+t-1}) \\ B^r_{s,t} &= F_s A_{s+t} \cap d_{s+t+1}(F_{s+r} A_{s+t+1}) \\ Z^\infty_{s,t} &= ker(d_{s+t}) \cap F_s A_{s+t} \\ B^\infty_{s,t} &= im(d_{s+t+1}) \cap F_s A_{s+t} \end{aligned} \tag{2.4.2}$$

Here we have set $d_n^{-1}(F_k A_{n-1})\colon = \{ x \in A_n \mid d_n(x) \in F_k A_{n-1} \}$.

Let $\eta_{s,t} \colon Z^\infty_{s,t} \to H_{s+t}(A_\bullet, d_\bullet)$ be the canonical map. Then we put

$$F_s H_{s+t}(A_\bullet, d_\bullet) = im(\eta_{s,t}) \tag{2.4.3}$$

and it is easy to check that this defines a filtration on $H_\bullet(A_\bullet, d_\bullet)$. Moreover, it is exhaustive if $F_\bullet$ is exhaustive on $(A_\bullet, d_\bullet)$. By definition, $F_\bullet$ is separated on $H_\bullet(A_\bullet, d_\bullet)$ if and only if

$$ker(d_{s+t}) \cap \bigcap_{s\in\mathbb{Z}} (im(d_{s+t+1}) + F_s A_{s+t}) = im(d_{s+t+1}). \tag{2.4.4}$$

If $F_\bullet$ is separated on $(A_\bullet, d_\bullet)$, then $ker(d_{s+t})$ is closed with respect to the topology $\mathfrak{T}_{F_\bullet}$ induced by the filtration $F_\bullet$. In this case (2.4.4) is equivalent to the fact that $im(d_{s+t+1})$

is closed with respect to $\mathfrak{T}_{F_\bullet}$. Therefore we call a filtration $F_\bullet$ on $(A_\bullet, d_\bullet)$ *convergent* if it is separated, exhaustive and $im(d_n)$ is closed with respect to $\mathfrak{T}_{F_\bullet}$.

A *spectral sequence of homological type* is a sequence of R-modules $(E^r_{s,t}, d^r)$, $r, t \in \mathbb{N}_0$, $s \in \mathbb{Z}$, and R-module mappings $d^r : E^r_{s,t} \to E^r_{s-r,t+r-1}$ such that:

(1) $(E^r_{\bullet,\bullet}, d^r)$ is a chain complex with respect to total degree, i.e., $d^r \circ d^r = 0$,

(2) $E^{r+1}_{\bullet,\bullet} = H_{\bullet,\bullet}(E^r_{\bullet,\bullet}, d^r)$.

A spectral sequence determines a bigraded R-module $E^\infty_{\bullet,\bullet}$, which can be constructed as follows. All the R-modules $E^{r+j}_{s,t}$ are subquotients of $E^r_{s,t}$, $j \in \mathbb{N}_0$. Let $(I^r_{s,t})_{r\in\mathbb{N}_0}$, $(K^r_{s,t})_{r\in\mathbb{N}_0}$ be the corresponding sequences of R-submodules of $E^0_{s,t}$ such that $E^r_{s,t} = K^r_{s,t}/I^r_{s,t}$. Then there are the inclusions

$$0 = K^0_{s,t} \leqslant K^1_{s,t} \leqslant K^2_{s,t} \leqslant \cdots \leqslant I^2_{s,t} \leqslant I^1_{s,t} \leqslant I^0_{s,t} = E^0_{s,t}. \tag{2.4.5}$$

Now define $I^\infty_{s,t} = \bigcap_{r\in\mathbb{N}_0} I^r_{s,t}$, $K^\infty_{s,t} = \bigcup_{r\in\mathbb{N}_0} K^r_{s,t}$ and

$$E^\infty_{s,t} = K^\infty_{s,t}/I^\infty_{s,t}, \tag{2.4.6}$$

and say that $(E^r_{\bullet,\bullet}, d^r)$ *converges* to $E^\infty_{\bullet,\bullet}$. The spectral sequence *collapses at* $r = N$ if $d^{N+j} = 0$ for all $j \geqslant 0$. In this case $E^\infty_{s,t} = E^N_{s,t}$.

Theorem 2.4.1. *Let $(A_\bullet, d_\bullet)$ be a non-negative chain complex and $F_\bullet$ a convergent filtration. Then there exists a first and second quadrant spectral sequence $(E^r_{s,t}, d^r)$ (i.e., $E^r_{s,t} = 0$ for $t < 0$,) of homological type with*

$$E^0_{s,t} = F_sA_{s+t}/F_{s-1}A_{s+t}, \tag{2.4.7}$$

converging to

$$E^\infty_{s,t} = F_sH_{s+t}(A_\bullet, d_\bullet)/F_{s-1}H_{s+t}(A_\bullet, d_\bullet). \tag{2.4.8}$$

Proof. If we set

$$E^r_{s,t} = Z^r_{s,t}/(B^{r-1}_{s,t} + Z^{r-1}_{s-1,t+1}), \tag{2.4.9}$$

then the proof follows from an elementary calculation ([18], Thm. XI.3.1).

One can define filtrations of cochain complexes in a similar way. Let $(A^\bullet, d^\bullet)$ be a cochain complex. A *filtration* $F^\bullet$ of $(A^\bullet, d^\bullet)$ is a series of sub cochain complexes $(F^kA^\bullet, d^\bullet)_{k\in\mathbb{Z}}$, $F^kA^n \leqslant A^n$ such that

(1) $F^{k+1}A^n \leqslant F^kA^n$ for all $k \in \mathbb{Z}$, $n \in \mathbb{N}_0$,

(2) $d^n(F^kA^n) \leqslant F^kA^{n+1}$.

Note that for cochain complexes the filtrations under consideration are decreasing, while for chain complexes we have considered increasing filtrations. The filtration is called *exhaustive* if $\bigcup_{k\in\mathbb{Z}} F^k A^n = A^n$ for all $n \in \mathbb{N}_0$, and *separated* if $\bigcap_{k\in\mathbb{Z}} F^k A^n = 0$. Let

$$E_0^{s,t} = F^s A^{s+t}/F^{s+1}A^{s+t} \tag{2.4.10}$$

be the associated graded object, and let $d_0\colon E_0^{s,t} \to E_0^{s,t+1}$ be the induced differential (of bidegree (0, 1)). In this case one defines

$$\begin{aligned} Z_r^{s,t} &= F^s A^{s+t} \cap (d^{s+t})^{-1}(F^{s+r}A^{s+t+1}),\\ B_r^{s,t} &= F^s A^{s+t} \cap d^{s+t-1}(F^{s-r}A^{s+t-1}),\\ Z_\infty^{s,t} &= F^s A^{s+t} \cap ker(d^{s+t}),\\ B_\infty^{s,t} &= F^s A^{s+t} \cap im(d^{s+t-1}). \end{aligned} \tag{2.4.11}$$

Let $\eta^{s,t}\colon Z_\infty^{s,t} \to H^{s+t}(A^\bullet, d^\bullet)$ denote the canonical projection and let

$$F^s H^{s+t}(A^\bullet, d^\bullet) = im(\eta^{s,t}). \tag{2.4.12}$$

The filtration $F^\bullet$ is called *convergent* if $F^\bullet$ is exhaustive, separated and $im(d^n)$ is closed with respect to the topology $\mathfrak{T}_{F^\bullet}$ for all $n \in \mathbb{N}_0$. As in the previous case one can show that the induced filtration $F^\bullet$ on $H^\bullet(A^\bullet, d^\bullet)$ is exhaustive and separated.

A *spectral sequence of cohomological type* is a sequence of R-modules $(E_r^{s,t}, d_r)$, $r, t \in \mathbb{N}_0$, $s \in \mathbb{Z}$, and R-module homomorphisms $d_r\colon E_r^{s,t} \to E_r^{s+r,t-r+1}$ such that:

(1) $(E_r^{\bullet,\bullet}, d_r)$ is a cochain complex (with respect to total degree),

(2) $E_{r+1}^{s,t} = H^{\bullet,\bullet}(E_r^{\bullet,\bullet}, d_r)$.

As before there is a bigraded R-module $E_\infty^{s,t}$ associated to the spectral sequence $(E_r^{\bullet,\bullet}, d_r)$. The dual version of Theorem 2.4.1 is

Theorem 2.4.2. *Let $(A^\bullet, d^\bullet)$ be a non-negative cochain complex and $F^\bullet$ a convergent filtration. Then there exists a first and second quadrant spectral sequence of cohomological type $(E_r^{s,t}, d_r)$ with*

$$E_0^{s,t} = F^s A^{s+t}/F^{s+1}A^{s+t}, \tag{2.4.13}$$

converging to

$$E_\infty^{s,t} = F^s H^{s+t}(A^\bullet, d^\bullet)/F^{s+1}H^{s+t}(A^\bullet, d^\bullet). \tag{2.4.14}$$

Proof. In this case one sets

$$E_r^{s,t} = Z_r^{s,t}/(Z_r^{s+1,t-1} + B_{r-1}^{s,t}) \tag{2.4.15}$$

and proves the required identities by calculation ([21], §2.2.2).

In general it is quite difficult to obtain information about the higher differentials of a spectral sequence, and in most cases one only knows the E_1 or E_2-term of the spectral sequence. However, in some cases the vanishing of the higher differentials d^r (resp. d_r) follows because of degree reasons (cf. (5.1.25.)).

Theorems 2.4.1 and 2.4.2 can be seen also in a more formal way. The construction of the spectral sequences shows that there are covariant functors

$$\begin{aligned} E^\bullet &\colon \mathbf{Fil}(\mathbf{C}_\bullet({}_R\mathrm{Mod})^+) \longrightarrow \mathbf{HomSpSeq}_R, \\ E_\bullet &\colon \mathbf{Fil}(\mathbf{C}^\bullet({}_R\mathrm{Mod})^+) \longrightarrow \mathbf{CohomSpSeq}_R, \end{aligned} \tag{2.4.16}$$

where $\mathbf{Fil}(\mathbf{C}_\bullet({}_R\mathrm{Mod})^+)$ (resp. $\mathbf{Fil}(\mathbf{C}^\bullet({}_R\mathrm{Mod})^+)$) denote the category of filtered non-negative chain (resp. cochain) complexes. We have also used the abbreviation $\mathbf{HomSpSeq}_R$ (resp. $\mathbf{CohomSpSeq}_R$) for the category of spectral sequences of homological (resp. cohomological) type. For example, this fact is also used to define a multiplicative structure on certain spectral sequences associated to the cohomology rings of cocommutative Hopf algebras.

2.5 Serre's lemma

The existence of certain projective resolutions is essential for the study of homological and cohomological functors. In this subsection we describe a method of lifting projective resolutions from the graded object $gr^\bullet(A)$ associated to a filtration of an algebra to the algebra A itself.

Let R be a commutative ring with a 1 and let (A, ε) be an augmented R-algebra with unit $\eta\colon R \to A$. We call $(P_\bullet, d_\bullet, \varepsilon_\bullet, s_\bullet)$ a *special projective resolution* if:

(a) $(P_0, \varepsilon_0) = (A, \eta \circ \varepsilon)$,

(b) $P_\bullet \xrightarrow{\varepsilon} R$ is a projective resolution over A of the trivial left A-module R,

(c) $s_\bullet$ is R-linear and a homotopy contraction, i.e.,

$$s_{k-1} \circ d_k + d_{k+1} \circ s_k = 1_{P_k} \tag{2.5.1}$$

for $k \geqslant 1$ and $d_1 \circ s_0 = 1_{P_0} - \varepsilon_0$.

Here we have considered $P_\bullet = \coprod_{k \in \mathbb{N}_0} P_k$ as $\mathbb{N}_0$-graded left A-module and $\varepsilon_\bullet = (\varepsilon_k)_{k \in \mathbb{N}_0}$, where $\varepsilon_k = 0$ for $k \geqslant 1$ and $\varepsilon_0\colon A \xrightarrow{\varepsilon} R \xrightarrow{\eta} A$. We have

$$\begin{aligned} d_\bullet &\in \mathrm{Hom}_A^{-1}(P_\bullet, P_\bullet), \\ \varepsilon_\bullet &\in \mathrm{Hom}_R^{0}(P_\bullet, P_\bullet), \\ s_\bullet &\in \mathrm{Hom}_R^{1}(P_\bullet, P_\bullet), \end{aligned} \tag{2.5.2}$$

where the upper index denotes the degree of a homomorphisms of graded modules. The fact that $(P_\bullet, d_\bullet, \varepsilon_\bullet, s_\bullet)$ is a special projective resolution can be expressed by the identities

$$\begin{aligned} d_\bullet \circ d_\bullet &= 0, \\ \varepsilon_\bullet \circ (1_\bullet - \varepsilon_\bullet) &= 0, \\ d_\bullet \circ s_\bullet + s_\bullet \circ d_\bullet &= id_{P_\bullet}, \\ \varepsilon_\bullet \circ s_\bullet = s_\bullet \circ \varepsilon_\bullet &= 0, \\ \varepsilon_\bullet \circ d_\bullet = d_\bullet \circ \varepsilon_\bullet &= 0. \end{aligned} \tag{2.5.3}$$

Note that if R is a field and $P_\bullet \xrightarrow{\varepsilon} R$ is a projective resolution of the trivial module with $P_0 = A$, then one can always find a chain homotopy contraction $s_\bullet$ such that $(P_\bullet, d_\bullet, \varepsilon_\bullet, s_\bullet)$ is special.

Let R be either $\mathbb{Z}_p$ or $\mathbb{Q}_p$ and let $F^\bullet$ denote the natural filtration associated to the p-adic valuation, i.e., $F^k\mathbb{Z}_p = p^k.\mathbb{Z}_p, k \in \mathbb{N}_0$, and $F^k\mathbb{Q}_p = p^k.\mathbb{Z}_p, k \in \mathbb{Z}$. We denote the graded object associated to a filtration of a module by $gr^\bullet(\_)$. Assume that (A, ε) is an augmented R-algebra with an integral (decreasing) filtration $F^\bullet = (F^k)_{k\in\mathbb{Z}}$ on A such that:

(1) A is a filtered augmented R-algebra, i.e.,

$$\begin{aligned} F^sR.F^tA &\leqslant F^{s+t}A, \\ F^sA.F^tA &\leqslant F^{s+t}A, \\ \eta(F^sR) &\leqslant F^sA, \\ \varepsilon(F^sA) &\leqslant F^sR; \end{aligned} \tag{2.5.4}$$

(2) the augmentation ideal $I(A)$ is complete with respect to the filtration $F^\bullet A$, and $gr^\bullet(I(A))$ is a free $gr^\bullet(R)$-module.

There is an important tool for constructing special resolutions ([17], (V.2.1)):

Lemma 2.5.1. (J-P. Serre) *Suppose that $(A, \varepsilon, F^\bullet)$ satisfies* (1) *and* (2) *as above. Let $(\bar{P}_\bullet, \bar{d}_\bullet, \bar{\varepsilon}_\bullet, \bar{s}_\bullet)$ be a special projective resolution of $gr^\bullet(A)$, where $\bar{d}_\bullet$ and $\bar{s}_\bullet$, $\bar{\varepsilon}_\bullet$ are mappings of graded left $gr^\bullet(A)$ and $gr^\bullet(R)$-modules, respectively. Assume further that each $\bar{P}_k$ is a free left $gr^\bullet(A)$-module of finite rank.*

Then there exist a special projective resolution $(P_\bullet, d_\bullet, \varepsilon_\bullet, s_\bullet)$ for A and a filtration $F^\bullet$ on every $P^\bullet$ such that:

(i) *P_k is a filtered-free left A-module of rank $rk_A(P_k) = rk_{gr^\bullet(A)}(\bar{P}_k)$,*

(ii) *$d_\bullet$ is a morphism of filtered left A-modules, and $\varepsilon_\bullet$, $s_\bullet$ are morphisms of filtered left R-modules,*

(iii) *$(\bar{P}_\bullet, \bar{d}_\bullet, \bar{\varepsilon}_\bullet, \bar{s}_\bullet) \simeq (gr^\bullet(P_\bullet), gr^\bullet(d_\bullet), gr^\bullet(\varepsilon_\bullet), gr^\bullet(s_\bullet))$ as special projective resolutions of $gr^\bullet(A)$.*

Here a filtered A-module Q of finite rank is called *filtered-free* if there exist a generating set $x_1, .., x_n$ and numbers $r_1, .., r_n$ such that

$$\begin{aligned} Q &= A.x_1 \oplus \cdots \oplus A.x_n, \\ \text{and } F^kQ &= F^{k-r_1}A.x_1 \oplus \cdots \oplus F^{k-r_n}A.x_n. \end{aligned} \tag{2.5.5}$$

Remark 2.5.2. Let R be a commutative ring with a 1 and let $\mathfrak{L}$ be an R-Lie algebra which is also a finitely generated free R-module. One knows that the *Koszul complex* $(\mathbf{Ko}_\bullet(\mathfrak{L}), d_\bullet)$, where

$$\mathbf{Ko}_\bullet(\mathfrak{L}) = \mathfrak{A}_R(\mathfrak{L}) \otimes \Lambda_\bullet(\mathfrak{L}) \tag{2.5.6}$$

and

$$d_n(x\,\langle y_1 \wedge \ldots \wedge y_n\rangle) = \sum_{1\leqslant i\leqslant n} (-1)^{i+1}\, x.y_i\,\langle \check{y}_i\rangle + \sum_{1\leqslant i<j\leqslant n} (-1)^{i+j}\, x\,\langle [y_i, y_j] \wedge \check{y}_{i,j}\rangle, \tag{2.5.7}$$

is a projective resolution of the trivial $\mathfrak{U}_R(\mathfrak{L})$-module ([16]). (Here we have used the abbreviation $\check{y}_i = y_1 \wedge .. \wedge y_{i-1} \wedge y_{i+1} \wedge .. \wedge y_n$.) In [17], (V.1.3.6) it is shown that there exists a homotopy contraction map $s_\bullet$ such that $(\mathbf{Ko}_\bullet(\mathfrak{L}), d_\bullet, \varepsilon_\bullet, s_\bullet)$ is a special projective resolution.

Assume that $R = R_\bullet$ is a $\mathbb{Z}$-graded ring and that $\mathfrak{L} = \mathfrak{L}_\bullet$ is a $\mathbb{Z}$-graded $R_\bullet$-Lie algebra such that $\mathfrak{L}_\bullet$ is a finitely generated free $R_\bullet$-module. Then $\mathfrak{U}_{R_\bullet}(\mathfrak{L}_\bullet)$ is a $\mathbb{Z}$-graded augmented $R_\bullet$-algebra. If $s^0_\bullet$ denotes the 0-component of the chain homotopy contraction described above, then $(\mathbf{Ko}_\bullet(\mathfrak{L}), d_\bullet, \varepsilon_\bullet, s^0_\bullet)$ is a special projective resolution with $d_\bullet, \varepsilon_\bullet, s^0_\bullet$ mappings of $\mathbb{Z}$-graded left $\mathfrak{U}_{R_\bullet}(\mathfrak{L}_\bullet)$-modules or $R_\bullet$-modules, as appropriate.

3 Cohomology and homology for profinite groups

For a profinite group G there are several natural categories of representations. In this section we will give a short description of these categories as well as the cohomological and homological functors they define. Unfortunately there seems to be no nice category of representations which contains them all. This is the reason why we must study different cohomology and homology theories and the relations between them.

When G is a profinite group we write $H \leqslant G$ for a closed subgroup H of G, and $K \leqslant_o G$ for an open subgroup K of G. Similarly we use the notation $H \triangleleft G$ for a closed normal subgroup H of G, and $K \triangleleft_o G$ for an open normal subgroup K of G.

3.1 Continuous discrete cohomology

Let G be a profinite group. A left G-module M is called *discrete* if M is a discrete abelian group with continuous G-action $\psi\colon G \times M \to M$. We denote the category of discrete left G-modules by $\mathfrak{D}(G)$[7] It is a full subcategory of ${}_{\mathbb{Z}[G]}\mathrm{Mod}$, where $\mathbb{Z}[G]$ denotes the group algebra of the abstract group G over $\mathbb{Z}$. A left $\mathbb{Z}[G]$-module M is contained in $ob(\mathfrak{D}(G))$ if and only if

$$\mathrm{Stab}_G(m) = \{\, g \in G \mid g.m = m \,\} \leqslant_o G \tag{3.1.1}$$

[7] In [23] this category was denoted by C_G.

is an open subgroup of G for all $m \in M$, i.e.,

$$M = \bigcup_{U \leqslant_o G} M^U, \tag{3.1.2}$$

where $M^U = \{ m \in M \mid u.m = m \text{ for all } u \in U \}$. For $M \in ob({}_{\mathbb{Z}[G]}\mathrm{Mod})$ we define

$$D(M) = \{ m \in M \mid \mathrm{Stab}_G(m) \leqslant_o G \} \in ob(\mathfrak{D}(G)). \tag{3.1.3}$$

For every morphism $\phi\colon N \to M \in mor({}_{\mathbb{Z}[G]}\mathrm{Mod})$ with $N \in ob(\mathfrak{D}(G))$ one has $im(\phi) \leqslant D(M)$. Hence

$$D\colon {}_{\mathbb{Z}[G]}\mathrm{Mod} \longrightarrow \mathfrak{D}(G) \tag{3.1.4}$$

with $D(\phi) = \phi|_{D(N)}$ for $\phi\colon N \to M \in mor({}_{\mathbb{Z}[G]}\mathrm{Mod})$ is a left exact covariant additive functor, and one can verify that it maps injectives to injectives. This yields the following result ([18], Chap. III, Thm. 7.4, Thm. 11.3).

Proposition 3.1.1. *$\mathfrak{D}(G)$ is an abelian module category with enough injectives. Moreover, every module $M \in ob(\mathfrak{D}(G))$ has a minimal injective envelope $i_M\colon M \to J(M)$, i.e., for every injective morphism $\phi\colon M \to J$ into an injective module $J \in ob(\mathfrak{D}(G))$ there exists an injective morphism $\phi^*\colon J(M) \to J$ such that $\phi = \phi^* \circ i_M$.*

For $M \in ob(\mathfrak{D}(G))$, $\mathrm{Hom}_{\mathfrak{D}(G)}(M, \_)\colon \mathfrak{D}(G) \to {}_{\mathbb{Z}}\mathrm{Mod}$ is a left exact covariant additive functor with right derived functors $\mathrm{Ext}^n_{\mathfrak{D}(G)}(M, \_)$ (Thm. 2.2.4(b), [18], §XII.9). The *continuous discrete cohomology* of G with coefficients in the discrete left G-module N is defined[8] by

$$H^\bullet(G, N) = \mathrm{Ext}^\bullet_{\mathfrak{D}(G)}(\mathbb{Z}, N). \tag{3.1.5}$$

Let $\mathfrak{D}_p(G)$ denote the full subcategory of $\mathfrak{D}(G)$ consisting of discrete left G-modules which are p-torsion modules. It has an important property ([6], Lemma 1.8, [23], Chap. I, Ann. 2, Lemme 3.3).

Proposition 3.1.2. *$\mathfrak{D}_p(G)$ is an abelian category with enough injectives. If $J \in ob(\mathfrak{D}_p(G))$ is injective, it is also injective in $\mathfrak{D}(G)$.*

Proposition 3.1.2 implies that the injective envelope $J(M)$ (Prop. 3.1.1) is contained in $ob(\mathfrak{D}_p(G))$ provided $M \in ob(\mathfrak{D}_p(G))$.

For $M \in ob(\mathfrak{D}_p(G))$ the canonical mapping $\mathrm{Hom}_{\mathfrak{D}(G)}(\mathbb{Z}_p, M) \to \mathrm{Hom}_{\mathfrak{D}(G)}(\mathbb{Z}, M)$ is an isomorphism. Hence

$$H^n(G, M) \simeq \mathrm{Ext}^n_{\mathfrak{D}(G)}(\mathbb{Z}_p, M). \tag{3.1.6}$$

3.2 The compact module category

From now on we fix a prime number p. By $\mathbb{Z}_p[\![G]\!]$ we denote the compact group algebra of G over $\mathbb{Z}_p$, i.e., $\mathbb{Z}_p[\![G]\!] = \varprojlim \mathbb{Z}_p[G/U]$, where U ranges over the inverse

[8] This definition coincides with the one given in [23] (cf. remark in [23], §2.2).

system of open normal subgroups of G (see [10], §7). Hence $\mathfrak{D}_p(G)$ coincides with the category of discrete left p-torsion $\mathbb{Z}_p[\![G]\!]$-modules ($\mathfrak{D}_p(G) = \mathfrak{D}_{\mathbb{Z}_p[\![G]\!]}$ in the notation of [6]).

Let $\mathfrak{F}_p(G) \leqslant \mathfrak{D}_p(G)$ denote the full subcategory of $\mathfrak{D}_p(G)$ whose objects are finite (discrete) left $\mathbb{Z}_p[\![G]\!]$-modules. By $\mathfrak{C}_p(G)$ we denote the category of topological $\mathbb{Z}_p[\![G]\!]$-modules whose objects consist of inverse limits of objects in $\mathfrak{F}_p(G)$ and whose morphisms are given by

$$\mathbf{Hom}_{\mathbb{Z}_p[\![G]\!]}(M, N) = \{\alpha \in \mathrm{Hom}_{\mathbb{Z}_p[\![G]\!]}(M, N) \mid \alpha \text{ continuous}\}, \tag{3.2.1}$$

$M, N \in ob(\mathfrak{C}_p(G))$. We consider $\mathbf{Hom}_{\mathbb{Z}_p[\![G]\!]}(M, N)$ as a topological space with the compact-open topology. This makes $\mathbf{Hom}_{\mathbb{Z}_p[\![G]\!]}(M, N)$ into a complete Hausdorff-topological $\mathbb{Z}_p$-module. Moreover, if M is a finitely generated topological $\mathbb{Z}_p[\![G]\!]$-module, then $\mathbf{Hom}_{\mathbb{Z}_p[\![G]\!]}(M, N)$ is compact. Note that every continuous morphism $\alpha: M \to N$ is also strict. One important property of $\mathfrak{C}_p(G)$ is ([6], Lemma 1.6):

Proposition 3.2.1. *$\mathfrak{C}_p(G)$ is a topological abelian module category with enough projectives.*

If $M \in ob(\mathfrak{C}_p(G))$ and $N \in ob(\mathfrak{D}_p(G))$, then $\mathbf{Hom}_{\mathbb{Z}_p[\![G]\!]}(M, N)$ is a discrete $\mathbb{Z}_p$-module. The bifunctor

$$\mathbf{Hom}_{\mathbb{Z}_p[\![G]\!]}(\_, \_): \mathfrak{C}_p(G) \times \mathfrak{D}_p(G) \longrightarrow \mathfrak{D}_p, \tag{3.2.2}$$

is contravariant/covariant and left/left exact. Furthermore, $\mathbf{Hom}_{\mathbb{Z}_p[\![G]\!]}(P, \_)$ is exact for $P \in ob(\mathfrak{C}_p(G))$ projective, and $\mathbf{Hom}_{\mathbb{Z}_p[\![G]\!]}(\_, J)$ is exact for $J \in ob(\mathfrak{D}_p(G))$ injective ([6], Lem. 2.2]). We denote the right-derived bifunctors of $\mathbf{Hom}_{\mathbb{Z}_p[\![G]\!]}(\_, \_)$ as in Theorem 2.2.4(a) by

$$\mathbf{Ext}^n_{\mathbb{Z}_p[\![G]\!]}(\_, \_): \mathfrak{C}_p(G) \times \mathfrak{D}_p(G) \longrightarrow \mathfrak{D}_p. \tag{3.2.3}$$

Then (3.1.6) implies that

$$H^\bullet(G, M) = \mathbf{Ext}^\bullet_{\mathbb{Z}_p[\![G]\!]}(\mathbb{Z}_p, M) \tag{3.2.4}$$

for all $M \in ob(\mathfrak{D}_p(G))$.

3.3 Continuous homology

If H is a closed subgroup of the profinite group G and $M \in ob(\mathfrak{C}_p(G))$ (resp. $ob(\mathfrak{D}_p(G))$) is a left $\mathbb{Z}_p[\![G]\!]$-module then we denote by $res^G_H(M) \in ob(\mathfrak{C}_p(H))$ (resp. $ob(\mathfrak{D}_p(H))$) the restriction of M to $\mathbb{Z}_p[\![H]\!]$. The next proposition allows us to define restriction maps in cohomology and homology.

Proposition 3.3.1. *Let H be a closed subgroup of the profinite group G and let $P \in ob(\mathfrak{C}_p(G))$ be projective. Then $res^G_H(P) \in ob(\mathfrak{C}_p(H))$ is projective.*

Proof. This follows from [6], Lemma 1.6 and Corollary 3.3.

On the category $\mathfrak{C}_p$ one has also a tensor product $\widehat{\otimes}_{\mathbb{Z}_p}$ ([6], §2). For $M, N \in ob(\mathfrak{C}_p)$ the tensor product is given by $M\widehat{\otimes}_{\mathbb{Z}_p}N = \varprojlim M/U \otimes_{\mathbb{Z}_p} N/V$ where the inverse system ranges over the pairs (U, V) of open submodules in M and N. Note that the abstract tensor product $M \otimes_{\mathbb{Z}_p} N$ is a dense submodule of $M\widehat{\otimes}_{\mathbb{Z}_p}N$. If $M, N \in ob(\mathfrak{C}_p(G))$ then $M\widehat{\otimes}_{\mathbb{Z}_p}N \in ob(\mathfrak{C}_p(G))$ has canonically the structure of a left (topological) $\mathbb{Z}_p[\![G]\!]$-module. The action of G on $M\widehat{\otimes}_{\mathbb{Z}_p}N$ is given by

$$g.(m\widehat{\otimes}n) = g.m\widehat{\otimes}g.n, \tag{3.3.1}$$

where $m \in M, n \in N, g \in G$. This tensor product has the following property which is well known for finite groups.

Proposition 3.3.2. *Let $M, P \in ob(\mathfrak{C}_p(G))$, and suppose that P is projective and M is p-torsion-free. Then $M\widehat{\otimes}_{\mathbb{Z}_p}P$ is projective.*

Proof. Note that by hypothesis $res_1^G(M) \in \mathfrak{C}_{\mathbb{Z}_p}$ is free (cf. Cor. 2.1.2). P is a direct summand of a free $\mathbb{Z}_p[\![G]\!]$-module $F \in ob(\mathfrak{C}_p(G))$ generated freely by a set I (cf. [6], Lemma 1.6). Assume that I is finite. The mapping $\phi\colon M\widehat{\otimes}_{\mathbb{Z}_p}F \to M\widehat{\otimes}_{\mathbb{Z}_p}F$, where

$$\phi(\sum_{i\in I} m_i\widehat{\otimes}g_i.i) = \sum_{i\in I} g_i^{-1}.m_i\widehat{\otimes}g_i.i, \tag{3.3.2}$$

$m_i \in M, g_i \in G$, is a continuous bijective mapping of $\mathbb{Z}_p[\![G]\!]$-modules. Note that the G-action on the module on the right hand side is given by $g.(m\widehat{\otimes}g_i.i) = m\widehat{\otimes}gg_i.i.$. Thus if J is a basis for M as profinite $\mathbb{Z}_p$-module, then $M\widehat{\otimes}_{\mathbb{Z}_p}F$ is a free $\mathbb{Z}_p[\![G]\!]$-module generated by $J \times I$. If I is infinite, then F is the inverse limit $F = \varprojlim F(X)$ where X runs over the finite subsets of I. By [6], Cor. 3.3, Lemma A.4, $M\widehat{\otimes}_{\mathbb{Z}_p}F$ is free, so $M\widehat{\otimes}_{\mathbb{Z}_p}P$ is projective.

Let $\mathfrak{C}_p(G)^\times$ denote the corresponding category of right $\mathbb{Z}_p[\![G]\!]$-modules. The bifunctor

$$\_\widehat{\otimes}_{\mathbb{Z}_p[\![G]\!]}\_ : \mathfrak{C}_p(G)^\times \times \mathfrak{C}_p(G) \longrightarrow \mathfrak{C}_p \tag{3.3.3}$$

is covariant/covariant, right/right exact and biadditive ([6], Lemma 2.1(i)). Let

$$\mathbf{Tor}_\bullet^{\mathbb{Z}_p[\![G]\!]}(\_,\_)\colon \mathfrak{C}_p(G)^\times \times \mathfrak{C}_p(G) \longrightarrow \mathfrak{C}_{\mathbb{Z}_p} \tag{3.3.4}$$

denote its left derived functors (Thm.2.2.4(b), [18], XII). We define the *continuous homology of G* with coefficients in $M \in ob(\mathfrak{C}_p(G))$ by

$$H_\bullet(G, M) = \mathbf{Tor}_\bullet^{\mathbb{Z}_p[\![G]\!]}(\mathbb{Z}_p, M). \tag{3.3.5}$$

Let $H \leqslant_o G$ be an open and $K \leqslant G$ be a closed subgroup of G. For $M \in ob(\mathfrak{C}_p(K))$, $N \in ob(\mathfrak{C}_p(H))$ we define

$$\begin{aligned} ind_G^K(M) &= \mathbb{Z}_p[\![G]\!]\widehat{\otimes}_{\mathbb{Z}_p[\![K]\!]}M, \\ coind_G^H(N) &= \mathbf{Hom}_{\mathbb{Z}_p[\![H]\!]}(\mathbb{Z}_p[\![G]\!], N) \end{aligned} \tag{3.3.6}$$

the *induced and coinduced* $\mathbb{Z}_p[\![G]\!]$-*modules* with the natural left $\mathbb{Z}_p[\![G]\!]$-action, i.e., for $f \in coind_G^H(M)$ the action is given by $(g.f)(x) = f(x.g)$. For an open subgroup $H \leqslant_o G$ one can prove easily that

$$ind_G^H(M) \simeq coind_G^H(M) \tag{3.3.7}$$

as $\mathbb{Z}_p[\![G]\!]$-modules (see [5], Chap. III, Prop. 5.9 for the discrete case).

Remark 3.3.3. The proof of Proposition 3.3.2 shows that that for $M \in ob(\mathfrak{C}_p(G))$ one has

$$M\widehat{\otimes}_{\mathbb{Z}_p}\mathbb{Z}_p[\![G]\!] \simeq ind_G^1(res_1^G(M)). \tag{3.3.8}$$

Hence, $M\widehat{\otimes}_{\mathbb{Z}_p}\mathbb{Z}_p[\![G]\!]$ is always induced from the trivial group.

If $H \leqslant G$ is a closed subgroup and $M \in ob(\mathfrak{C}_p(H))$, then the functor $\widehat{\otimes}_{\mathbb{Z}_p[\![H]\!]}M$ is naturally isomorphic to $\widehat{\otimes}_{\mathbb{Z}_p[\![G]\!]}\mathbb{Z}_p[\![G]\!]\widehat{\otimes}_{\mathbb{Z}_p[\![H]\!]}M$ considered as a functor from $\mathfrak{C}_p(G)$ to $\mathfrak{C}_p$. So Proposition 3.3.1 yields the following Eckmann-Shapiro type lemma:

Lemma 3.3.4. *Let $H \leqslant G$ be a closed subgroup of the profinite group G and let $M \in ob(\mathfrak{C}_p(H))$. Then*

$$H_\bullet(G, ind_G^H(M)) \simeq H_\bullet(H, M). \tag{3.3.9}$$

3.4 The Pontrjagin category

Our philosophy is that for "nice" profinite groups it should be possible to define cohomology and homology functors on a category containing $\mathfrak{D}_p(G)$ and $\mathfrak{C}_p(G)$ as full subcategories. This allows one to compare homology and cohomology. We therefore define $\mathfrak{P}_p(G)$ to be the category whose objects are the topological $\mathbb{Z}_p[\![G]\!]$-modules which belong to either $ob(\mathfrak{C}_p(G))$ or in $ob(\mathfrak{D}_p(G))$. The morphisms are given by

$$\mathbf{Hom}_{\mathbb{Z}_p[\![G]\!]}(M, N) = \{\alpha \in \mathrm{Hom}_{\mathbb{Z}_p[\![G]\!]}(M, N) \mid \alpha \text{ continuous and strict}\}. \tag{3.4.1}$$

Note that if $M \in ob(\mathfrak{D}_p(G))$, $N \in ob(\mathfrak{C}_p(G))$, then a continuous morphism $\alpha\colon M \to N$ is strict if and only if it has finite image.

Proposition 3.4.1. *$\mathfrak{P}_p(G)$ is a topological exact module category containing $\mathfrak{C}_p(G)$ and $\mathfrak{D}_p(G)$ as full subcategories. If $M \in ob(\mathfrak{C}_p(G))$ is projective in $\mathfrak{C}_p(G)$, then it is also projective in $\mathfrak{P}_p(G)$. If $N \in ob(\mathfrak{D}_p(G))$ is injective in $\mathfrak{D}_p(G)$, then it is also injective in $\mathfrak{P}_p(G)$.*

Proof. From the previous remark it follows easily that $\mathfrak{P}_p(G)$ is an exact category ([4], Chap. III, §2.8, remark after Prop. 24). If $P \in ob(\mathfrak{C}_p(G))$ is projective, then $\mathbf{Hom}_{\mathbb{Z}_p[\![G]\!]}(P, \_)\colon \mathfrak{D}_p(G) \to \mathfrak{D}_p$ is exact, while for $J \in ob(\mathfrak{D}_p(G))$ injective, the functor $\mathbf{Hom}_{\mathbb{Z}_p[\![G]\!]}(\_, J)\colon \mathfrak{C}_p(G) \to \mathfrak{D}_p$ is exact ([6], Lemma 2.2). This implies the assertion.

Note that every short exact sequence in $\mathfrak{P}_p(G)$ is either a short exact sequence in $\mathfrak{C}_p(G)$ or a short exact sequence in $\mathfrak{D}_p(G)$. Again we consider $\mathbf{Hom}_{\mathbb{Z}_p[\![G]\!]}(M, N)$,

$M, N \in ob(\mathfrak{P}_p(G))$, as a topological $\mathbb{Z}_p$-module where the topology is the compact-open topology. So for $A, B \in ob(\mathfrak{C}_p(G))$, $M, N \in \mathfrak{D}_p(G)$, $\mathbf{Hom}_{\mathbb{Z}_p[\![G]\!]}(A, M)$ is discrete, $\mathbf{Hom}_{\mathbb{Z}_p[\![G]\!]}(A, B)$ carries the topology of uniform convergence and $\mathbf{Hom}_{\mathbb{Z}_p[\![G]\!]}(M, N)$ carries the topology of pointwise convergence.

Set $\mathfrak{P}_p = \mathfrak{P}_p(1)$ and let $I_p = \mathbb{Q}_p/\mathbb{Z}_p \in ob(\mathfrak{P}_p)$. For $M \in \mathfrak{P}_p(G)$ we let

$$M^* = \mathbf{Hom}_{\mathbb{Z}_p}(M, I_p) \in ob(\mathfrak{P}_p(G)), \tag{3.4.2}$$

where the action is given by $(g.f)(x) = f(g^{-1}.x)$, $x \in M$, $f \in M^*$, $g \in G$. With the obvious assignment on morphisms we have (Prop. 2.1.1, [6], Prop. 2.3):

Proposition 3.4.2. *$*: \mathfrak{P}_p(G) \to \mathfrak{P}_p(G)$ is an exact contravariant functor such that $* \circ *$ is naturally equivalent to the identity functor. Moreover, the canonical mapping*

$$*: \mathbf{Hom}_{\mathbb{Z}_p[\![G]\!]}(M, N) \longrightarrow \mathbf{Hom}_{\mathbb{Z}_p[\![G]\!]}(N^*, M^*) \tag{3.4.3}$$

given by $\alpha^*(n^*)(m) = n^*(\alpha(m))$, $\alpha \in \mathbf{Hom}_{\mathbb{Z}_p[\![G]\!]}(M, N)$, $n^* \in N^*$, $m \in M$, is an isomorphism of topological $\mathbb{Z}_p$-modules.

Remark 3.4.3. (i) Since * must take projectives to injectives and injectives to projectives, it follows from Propositions 3.1.1 and 3.1.2 that every compact module $M \in ob(\mathfrak{C}_p(G))$ has a minimal projective cover $\gamma_M : P(M) \to M \in mor(\mathfrak{C}_p(G))$.
(ii) If $M \in ob(\mathfrak{C}_p(G))$, $N \in ob(\mathfrak{D}_p(G))$, then general principles of homological algebra ([18], Chap. XII, §9) show that (3.4.3) induces an isomorphism

$$\mathbf{Ext}^{\bullet}_{\mathbb{Z}_p[\![G]\!]}(N^*, M^*) \simeq \mathbf{Ext}^{\bullet}_{\mathbb{Z}_p[\![G]\!]}(M, N) \tag{3.4.4}$$

and that this isomorphism is natural in M and N.
(iii) Let M and N be as in (ii). For $N \in ob(\mathfrak{D}_p(G))$ let $N^{\times} \in ob(\mathfrak{C}_p(G)^{\times})$ be the module defined by (3.4.2) but with the canonical right $\mathbb{Z}_p[\![G]\!]$-module structure. Then there is also an isomorphism

$$\mathbf{Tor}^{\mathbb{Z}_p[\![G]\!]}_{\bullet}(N^{\times}, M) \simeq \mathbf{Ext}^{\bullet}_{\mathbb{Z}_p[\![G]\!]}(M, N)^* \tag{3.4.5}$$

and this isomorphism is natural in M and N ([6], Cor. 2.6).
(iv) From (3.4.4) and (3.4.5), it follows that for $M \in ob(\mathfrak{C}_p(G))$,

$$H_{\bullet}(G, M)^* \simeq H^{\bullet}(G, M^*) \tag{3.4.6}$$

and that this isomorphism is natural in M.
(v) Let $H \leqslant_o G$ be a closed subgroup and $J \in ob(\mathfrak{D}_p(G))$ injective. From Pontrjagin duality and Proposition 3.3.1 it follows that $res^H_G(J) \in ob(\mathfrak{D}_p(H))$ is injective. This can be used to show the following Eckmann-Shapiro type lemma: Let $H \leqslant_o G$ be a closed subgroup of the profinite group G, $M \in ob(\mathfrak{C}_p(H))$, $N \in ob(\mathfrak{D}_p(G))$. Then

$$\mathbf{Ext}^{\bullet}_{\mathbb{Z}_p[\![H]\!]}(M, N) \simeq \mathbf{Ext}^{\bullet}_{\mathbb{Z}_p[\![G]\!]}(ind^H_G(M), N). \tag{3.4.7}$$

3.5 Some relative homological algebra

When we have used the notion of a derived functor, it has always been with respect to the class Σ of all short exact sequences in $\mathfrak{D}_p(G)$ (resp. $\mathfrak{C}_p(G)$). If one uses the cup product to construct a morphism of connected sequences, then one has to deal with the problem that it behaves nicely only for short exact sequences which split as abelian groups. Therefore we define $\Sigma^{sp} \subset \Sigma$ to be the class of short exact sequences $\mathbf{s}: 0 \to A \xrightarrow{\sigma} B \xrightarrow{\tau} C \to 0$ in $\mathfrak{D}_p(G)$ (resp. $\mathfrak{C}_p(G)$) which split under restriction to $\mathfrak{D}_p$ (resp. $\mathfrak{C}_p$). It is easy to check that Σ^{sp} is a proper class of short exact sequences. With the obvious classes Σ^{sp}_{inj} and Σ^{sp}_{sur} of proper injective and surjective morphism, one can show easily:

Proposition 3.5.1.

(1) $\mathfrak{D}_p(G)$ contains enough Σ^{sp}-injectives, and every Σ-injective is also Σ^{sp}-injective.

(2) $\mathfrak{C}_p(G)$ contains enough Σ^{sp}-projectives, and every Σ-projective is also Σ^{sp}-projective.

Proof. (1) Let $M \in ob(\mathfrak{D}_p(G))$. The canonical mapping $i_M: M \to J_M = \mathbf{Hom}_{\mathbb{Z}_p}(\mathbb{Z}_p[\![G]\!], M)$ given by $i_M(m)(x) = x.m$, $m \in M$, $x \in \mathbb{Z}_p[\![G]\!]$, is a morphism of left $\mathbb{Z}_p[\![G]\!]$-modules, where the action on J_M is given by $(g.f)(x) = f(x.g)$. Moreover, $i_M \in \Sigma^{sp}_{inj}$, since evaluation at $1 \in \mathbb{Z}_p[\![G]\!]$ is an obvious section (in $\mathfrak{D}_p$). It is easy to check that J_M is injective relative to Σ^{sp} ([18], Lemma XII.10.1). Since $\Sigma^{sp}_{inj} \subset \Sigma_{inj}$, the second statement is obvious.
(2) follows by similar arguments using induced modules or also by Pontrjagin duality (Prop. 3.4.2).

For short we call a functor $T: \mathfrak{A} \to \mathfrak{B}$ an $\mathfrak{A}$-functor. Let $M \in ob(\mathfrak{C}_p(G))$. Proposition 3.5.1 implies that the right derived $\mathfrak{D}_p(G)$-functors $\mathbf{Ext}^k_{\mathbb{Z}_p[\![G]\!]}(M, \_)_{sp} = \mathbf{Ext}^k_{\mathbb{Z}_p[\![G]\!],\mathbb{Z}_p}(M, \_)$ and $\mathbf{Ext}^k_{\mathbb{Z}_p[\![G]\!]}(M, \_)$ are naturally isomorphic. However, considered together with the connecting homomorphisms, $(\mathbf{Ext}^\bullet_{\mathbb{Z}_p[\![G]\!]}(M, \_)_{sp}, e^\bullet_{sp})$ constitutes a right universal Σ^{sp}-connected sequence,[9] while $(\mathbf{Ext}^\bullet_{\mathbb{Z}_p[\![G]\!]}(M, \_), e^\bullet)$ constitutes a right universal Σ-connected sequence of $\mathfrak{D}_p(G)$-functors. Note that $e^\bullet|_{\Sigma^{sp}} = e^\bullet_{sp}$, so the two approaches differ only if one considers these functors as connected sequences. The following lemma shows that natural transformations of these functors are rather rigid with respect to the connecting homomorphisms.

Proposition 3.5.2. *Let $M \in ob(\mathfrak{C}_p(G))$, $T^k(\_) = \mathbf{Ext}^k_{\mathbb{Z}_p[\![G]\!]}(M, \_)$ and let $\mathbf{U}^\bullet = (U^n, e^n)_{n \in \mathbb{N}_0}$, $U^n: \mathfrak{D}_p(G) \to \mathfrak{D}_p$, be a Σ-connected sequence of covariant additive functors. Suppose that $\mathbf{f}^\bullet = (f^n)_{n \in \mathbb{N}_0}$, is a sequence of natural transformations such*

[9] For definitions see [18], Chap. XII, §8.

that

$$\begin{array}{ccccccccc} 0 & \longrightarrow & T^0(A) & \xrightarrow{T^0(\sigma)} & T^0(B) & \xrightarrow{T^0(\tau)} & T^0(C) & \xrightarrow{e^0(\mathbf{s})} & T^1(A)\cdots \\ \downarrow & & {\scriptstyle f^0(A)}\downarrow & & {\scriptstyle f^0(B)}\downarrow & & {\scriptstyle f^0(C)}\downarrow & & {\scriptstyle f^1(A)}\downarrow \\ 0 & \longrightarrow & U^0(A) & \xrightarrow{U^0(\sigma)} & U^0(B) & \xrightarrow{U^0(\tau)} & U^0(C) & \xrightarrow{e^0(\mathbf{s})} & U^1(A)\cdots \end{array} \tag{3.5.1}$$

commutes for all short exact sequences $\mathbf{s}\colon 0 \to A \xrightarrow{\sigma} B \xrightarrow{\tau} C \to 0$ *in* $\mathbf{\Sigma}^{sp}$. *Then (3.5.1) commutes for all short exact sequences in* $\mathbf{\Sigma}$.

Proof. Let $\mathbf{T}^\bullet = (T^n, e^n)_{n\in\mathbb{N}_0}$. The sequence $\mathbf{T}_\bullet|_{\mathbf{\Sigma}^{sp}}$ is a right universal $\mathbf{\Sigma}^{sp}$-connected sequence and $\mathbf{T}_\bullet$ is a right universal $\mathbf{\Sigma}$-connnected sequence ([18], Thm. XII.7.6). Hence there exists a unique mapping of $\mathbf{\Sigma}$-connected sequences $\mathbf{g}^\bullet\colon \mathbf{T}^\bullet \to \mathbf{U}^\bullet$ with $g^0 = f^0$. However $\mathbf{g}^\bullet$ is also a mapping of $\mathbf{\Sigma}^{sp}$-connected sequences, so $\mathbf{g}^\bullet = \mathbf{f}^\bullet$.

3.6 Continuous $\mathfrak{P}_p(G)$-functors

In this subsection we study functors between Pontrjagin categories and the connections between them. The necessity of this arises because $\mathfrak{P}_p(G)$ is an exact category which does not really satisfy all the usual requirements of homological algebra, because it does not have enough projectives or injectives. However, one can manage the resulting problems by introducing various continuity concepts. A covariant additive functor $T\colon \mathfrak{P}_p(G) \to \mathfrak{P}_p$ is called *continuous*, if

(1) $T(ob(\mathfrak{D}_p(G)) \subseteq \mathfrak{D}_p$,

(2) $T(ob(\mathfrak{C}_p(G)) \subseteq \mathfrak{C}_p$,

(3) the restriction $T|_{\mathfrak{D}_p(G)}$ commutes with direct limits, i.e.,

$$\varinjlim \circ T|_{\mathfrak{D}_p(G)} = T|_{\mathfrak{D}_p(G)} \circ \varinjlim, \tag{3.6.1}$$

(4) the restriction $T|_{\mathfrak{C}_p(G)}$ commutes with inverse limits, i.e.,

$$\varprojlim \circ T|_{\mathfrak{C}_p(G)} = T|_{\mathfrak{C}_p(G)} \circ \varprojlim . \tag{3.6.2}$$

Here we assume that all morphisms in the inductive system in the right-hand side of (3.6.1) are injective, and all morphisms in the inverse system in the right-hand side of (3.6.2) are surjective. In the dual way one defines a continuous contravariant additive $\mathfrak{P}_p(G)$-functor. The definition implies:

Proposition 3.6.1. *Let* S, $T\colon \mathfrak{P}_p(G) \to \mathfrak{P}_p$ *be continuous* $\mathfrak{P}_p(G)$*-functors and let* $f_\circ\colon S|_{\mathfrak{F}_p(G)} \to T|_{\mathfrak{F}_p(G)}$ *be a natural transformation of* $\mathfrak{F}_p(G)$*-functors. Then* $f_\circ$ *extends uniquely to a natural transformation of continuous* $\mathfrak{P}_p(G)$*-functors* $f\colon S \to T$.

Let $\mathbf{\Sigma} = \mathbf{\Sigma}(\mathfrak{P}_p(G))$ denote the category of all short exact sequences $\mathbf{s}\colon 0 \to A \xrightarrow{\sigma} B \xrightarrow{\tau} C \to 0$ in $\mathfrak{P}_p(G)$. Recall that such a sequence is either a short exact sequence in $\mathfrak{D}_p(G)$ or one in $\mathfrak{C}_p(G)$. Assume that $\mathbf{s}$ is a short exact sequence in $\mathfrak{C}_p(G)$. Then $B = \varprojlim_i B/U_i$ where U_i ranges over the open $\mathbb{Z}_p[\![G]\!]$-submodules. Let $B_i = B/U_i$, $C_i = C/\tau(U_i)$, $A_i = A/\sigma^{-1}(U_i)$. Then

$$\mathbf{s}_i : 0 \to A_i \xrightarrow{\sigma_i} B_i \xrightarrow{\tau_i} C_i \to 0, \tag{3.6.3}$$

where σ_i, τ_i are the canonical morphisms, is exact and there is a commutative diagram

$$\begin{array}{ccccccccc} 0 & \longrightarrow & A & \xrightarrow{\sigma} & B & \xrightarrow{\tau} & C & \longrightarrow & 0 \\ & & \downarrow & & \downarrow & & \downarrow & & \\ 0 & \longrightarrow & \varprojlim A_i & \xrightarrow{lim(\sigma_i)} & \varprojlim B_i & \xrightarrow{lim(\tau_i)} & \varprojlim C_i & \longrightarrow & 0 \end{array} \tag{3.6.4}$$

with exact rows and with all vertical arrows isomorphisms. A similar statement holds for $\mathbf{s}$ a short exact sequence in $\mathfrak{D}_p(G)$ and direct limits. Thus a $\boldsymbol{\Sigma}$-connected pair (S, e, T) of continuous covariant additive $\mathfrak{P}_p(G)$-functors has the following continuity properties:

(1) for all short exact sequences $\mathbf{s}: 0 \to A \xrightarrow{\sigma} B \xrightarrow{\tau} C \to 0$ in $\mathfrak{C}_p(G)$, the diagram (cf. (3.6.4))

$$\begin{array}{ccc} S(C) & \xrightarrow{e(\mathbf{s})} & T(A) \\ \downarrow & & \downarrow \\ \varprojlim S(C_i) & \xrightarrow{lim(e(\mathbf{s}_i))} & \varprojlim T(A_i) \end{array} \tag{3.6.5}$$

commutes,

(2) for all short exact sequences $\mathbf{s}: 0 \to A \xrightarrow{\sigma} B \xrightarrow{\tau} C \to 0$ in $\mathfrak{D}_p(G)$, the diagram

$$\begin{array}{ccc} \varinjlim S(C_i) & \xrightarrow{lim(e(\mathbf{s}_i))} & \varinjlim T(A_i) \\ \downarrow & & \downarrow \\ S(C) & \xrightarrow{e(\mathbf{s})} & T(A) \end{array} \tag{3.6.6}$$

commutes.

We call the $\boldsymbol{\Sigma}$-connected pair (S, e, T) of continuous covariant additive $\mathfrak{P}_p(G)$-functors *continuously right universal* if, for all $\boldsymbol{\Sigma}$-connected pairs (S', e', T') of continuous $\mathfrak{P}_p(G)$-functors and natural transformations $f: S \to S'$, there is a unique natural transformation $g: T \to T'$, such that $(f, g): (S, e, T) \to (S', e', T')$ is a morphism of $\boldsymbol{\Sigma}$-connected pairs. Dually, (S, e, T) is *continuously left couniversal* if, for all $\boldsymbol{\Sigma}$-connected pairs (S', e', T') of continuous $\mathfrak{P}_p(G)$-functors and natural transformation $g': T' \to T$, there is a unique natural transformations $f': S' \to S$, such that $(f', g'): (S', e', T') \to (S, e, T)$ is a morphism of $\boldsymbol{\Sigma}$-connected pairs.

Proposition 3.6.2. *Let (S, e, T) be a Σ-connected pair of continuous covariant additive $\mathfrak{P}_p(G)$-functors.*

(1) If

$$S(B) \xrightarrow{S(\tau)} S(C) \xrightarrow{e(\mathbf{s})} T(A) \longrightarrow 0 \tag{3.6.7}$$

is right exact for all short exact sequences $\mathbf{s}\colon 0 \to A \xrightarrow{\sigma} B \xrightarrow{\tau} C \to 0$ in $\mathfrak{D}_p(G)$ with B injective, then (S, e, T) is continuously right universal.

(2) If

$$0 \longrightarrow S(C) \xrightarrow{e(\mathbf{s})} T(A) \xrightarrow{T(\sigma)} T(B) \tag{3.6.8}$$

is left exact for all short exact sequences $\mathbf{s}\colon 0 \to A \xrightarrow{\sigma} B \xrightarrow{\tau} C \to 0$ in $\mathfrak{C}_p(G)$ with B projective, then (S, e, T) is continuously left couniversal.

Proof. (1) Let (S', e', T') be a Σ-connected pair of continuous $\mathfrak{P}_p(G)$-functors and $f\colon S \to S'$ a natural transformation. Since $(S|_{\mathfrak{D}_p(G)}, e, T|_{\mathfrak{D}(G)})$ is right universal ([18], Thm. XII.7.6), there exists a unique natural transformation $g|_{\mathfrak{D}_p(G)}$ such that

$$(f|_{\mathfrak{D}_p(G)}, g|_{\mathfrak{D}_p(G)})\colon (S|_{\mathfrak{D}_p(G)}, e, T|_{\mathfrak{D}_p(G)}) \longrightarrow (S'|_{\mathfrak{D}_p(G)}, e', T|'_{\mathfrak{D}_p(G)}) \tag{3.6.9}$$

is a morphism of connected pairs. By Proposition 3.6.1, $g|_{\mathfrak{D}_p(G)}$ can be uniquely extended to a natural transformation $g\colon T \to T'$. From (3.6.5) we conclude that (f, g) is a morphism of connected pairs and that g is unique. Part (2) follows by similar arguments using [18], Thm. XII.7.2 and (3.6.6).

A Σ-connected sequence $\mathbf{T}^\bullet = (T^n, e^n)$ of continuous $\mathfrak{P}_p(G)$-functors is called *continuous*. We call the continuous cohomological $\mathfrak{P}_p(G)$-functor $\mathbf{T}^\bullet = (T^n, e^n)$ *co-effaceable* if $T^i(J) = 0$ for all injective modules $J \in ob(\mathfrak{D}_p(G))$ and $i \geqslant 1$. Dually, we call the continuous homological $\mathfrak{P}_p(G)$-functor $\mathbf{S}^\bullet = (S^n, e^n)$ *effaceable* if $S^i(P) = 0$ for all projective modules $P \in ob(\mathfrak{C}_p(G))$ and $i \leqslant -1$. A direct consequence of Proposition 3.6.2 and Theorem 2.2.1 is the following:

Corollary 3.6.3. *Let $\mathbf{T}^\bullet$ be a co-effaceable continuous cohomological $\mathfrak{P}_p(G)$-functor, $\mathbf{S}^\bullet$ an effaceable continuous homological $\mathfrak{P}_p(G)$-functor, and $\mathbf{U}^\bullet$ a continuous connected sequence of covariant additive $\mathfrak{P}_p(G)$-functors.*

(1) Suppose that $U^j = 0$ for all $j < 0$. Then for every natural transformation $f\colon T^0 \to U^0$ there exists a unique mapping of connected sequences $\mathbf{f}^\bullet\colon \mathbf{T}^\bullet \to \mathbf{U}^\bullet$ such that $f^0 = f$. If f^0 is a natural isomorphism and $\mathbf{U}^\bullet$ is a co-effaceable continuous cohomological functor, then all the f^i are also natural isomorphisms.

(2) Suppose that $U^j = 0$ for all $j > 0$. Then for every natural transformation $g\colon U^0 \to S^0$ there exists a unique mapping of connected sequences $\mathbf{g}^\bullet\colon \mathbf{U}^\bullet \to \mathbf{S}^\bullet$ such that $g^0 = g$. If g^0 is a natural isomorphism and $\mathbf{U}^\bullet$ is an effaceable continuous homological functor, then all the g^i are also natural isomorphisms.

3.7 Profinite groups of type p-$\mathbf{FP}_\infty$

Let $\mathfrak{C}_p^0(G)$ denote the full subcategory of $\mathfrak{C}_p(G)$ whose objects consist of the topologically finitely generated $\mathbb{Z}_p[\![G]\!]$-modules in $ob(\mathfrak{C}_p(G))$. Note that if $M \in ob(\mathfrak{C}_p^0(G))$, then

(1) $\mathbf{Hom}_{\mathbb{Z}_p[\![G]\!]}(M, N) \in ob(\mathfrak{C}_p)$ is compact for all $N \in ob(\mathfrak{C}_p(G))$.

(2) $\mathbf{Hom}_{\mathbb{Z}_p[\![G]\!]}(M, \_)\colon \mathfrak{C}_p(G) \longrightarrow \mathfrak{C}_p$ commutes with inverse limits, i.e., if $N = \varprojlim N_i$, $N_i \in ob(\mathfrak{C}_p(G))$, and $\alpha_{ij}\colon N_i \to N_j$ are surjective, then the natural morphism

$$\phi\colon \mathbf{Hom}_{\mathbb{Z}_p[\![G]\!]}(M, N) \longrightarrow \varprojlim \mathbf{Hom}_{\mathbb{Z}_p[\![G]\!]}(M, N_i) \tag{3.7.1}$$

is an isomorphism of (topological) $\mathbb{Z}_p$-modules.

A compact module $M \in ob(\mathfrak{C}_p(G))$ is called of *type* $\mathbf{FP}_n$, if there exists a partial projective resolution

$$P_n \xrightarrow{\partial_n} \cdots \xrightarrow{\partial_3} P_2 \xrightarrow{\partial_2} P_1 \xrightarrow{\partial_1} P_0 \xrightarrow{\varepsilon} M \longrightarrow 0 \tag{3.7.2}$$

of M in $\mathfrak{C}_p(G)$ with each P_i finitely generated (as a topological $\mathbb{Z}_p[\![G]\!]$-module). Such a partial projective resolution will be called *of finite type*. By $\mathfrak{C}_p^n(G)$ we denote the full subcategory of $\mathfrak{C}_p(G)$ whose objects M are of type $\mathbf{FP}_n$. From the Horseshoe Lemma ([1], Lemma 2.5.1) one sees that $\mathfrak{C}_p^n(G)$ is closed under extensions, i.e., if $0 \to A \xrightarrow{\sigma} B \xrightarrow{\tau} C \to 0$ is a short exact sequence in $\mathfrak{C}_p(G)$ and $A, C \in ob(\mathfrak{C}_p^n(G))$, then $B \in ob(\mathfrak{C}_p^n(G))$. As a direct consequence of Schanuel's Lemma one obtains:

Proposition 3.7.1. *For $M \in \mathfrak{C}_p^0(G)$ and $n \geqslant 1$ the following conditions are equivalent:*

(1) There is a partial resolution $F_n \to \cdots \to F_0 \to M \to 0$ with each $F_i \in ob(\mathfrak{C}_p(G))$ free of finite rank.

(2) M is of type $\mathbf{FP}_n$.

(3) For every partial projective resolution $P_k \to \cdots \to P_0 \to M \to 0$ of finite type with $k < n$, $ker(P_k \to P_{k-1})$ is finitely generated.

Proof. The implications (1) $\Rightarrow$ (2) $\Rightarrow$ (3) $\Rightarrow$ (1) can be proved by the same arguments as in the discrete case ([5], Chap. VIII, Prop. 4.3.).

A module is called of *type* $\mathbf{FP}_\infty$ if it is of type $\mathbf{FP}_n$ for all $n \in \mathbb{N}_0$. The full subcategory of $\mathfrak{C}_p(G)$ whose objects consist of modules of type $\mathbf{FP}_\infty$ will be denoted by $\mathfrak{C}_p^\infty(G)$. By Proposition 3.7.1, a module $M \in ob(\mathfrak{C}_p^\infty(G))$ of type $\mathbf{FP}_\infty$ has a projective resolution $\cdots \xrightarrow{\partial_2} P_1 \xrightarrow{\partial_1} P_0 \xrightarrow{\varepsilon} M$ with all $P_i \in ob(\mathfrak{C}_p^0(G))$ finitely generated. Moreover each P_i can be chosen to be free. For $M \in ob(\mathfrak{C}_p^\infty(G))$, $N \in ob(\mathfrak{P}_p(G))$ we define

$$\mathbf{Ext}^n_{\mathbb{Z}_p[\![G]\!]}(M, N) = H^n(\mathbf{Hom}_{\mathbb{Z}_p[\![G]\!]}(P_\bullet, N), \delta^\bullet) \in ob(\mathfrak{P}_p), \tag{3.7.3}$$

where $\delta^\bullet$ denotes the induced differential.

Theorem 3.7.2. *Let $M \in ob(\mathfrak{C}_p^{\infty}(G))$. Then $\mathbf{T}^\bullet = (\mathbf{Ext}^\bullet_{\mathbb{Z}_p[\![G]\!]}(M,\_), e^\bullet)$ is a continuous co-effaceable cohomological $\mathfrak{P}_p(G)$-functor with $\mathbf{Ext}^0_{\mathbb{Z}_p[\![G]\!]}(M,\_) \simeq \mathbf{Hom}_{\mathbb{Z}_p[\![G]\!]}(M,\_)$.*

Proof. That $\mathbf{T}^\bullet$ is a cohomological $\mathfrak{P}_p(G)$-functor is a direct consequence of Theorem 2.2.4(a) for $\mathfrak{A}_0 = \mathfrak{C}_p^{\infty}(G)$, $\mathfrak{A} = \mathfrak{C}_p(G)$, $\mathfrak{B} = \mathfrak{P}_p(G)$, $\mathfrak{Q} = \mathfrak{P}_p$ and $T(\_,\_) = \mathbf{Hom}_{\mathbb{Z}_p[\![G]\!]}(\_,\_)$ (using also Prop. 3.4.1). For continuity we must check (3.6.1) and (3.6.2). Let $P_\bullet \xrightarrow{\varepsilon} M$ be a projective resolution of M with all the P_k finitely generated. Let $N = \varprojlim_i N_i \in ob(\mathfrak{C}_p(G))$ with all $\alpha_{ij}: N_i \to N_j$ surjective, and let $\pi_i : N \to N_i$ denote the canonical projections. Since P_k is projective, $\pi^*_{k,i}: \mathbf{Hom}_{\mathbb{Z}_p[\![G]\!]}(P_k, N) \to \mathbf{Hom}_{\mathbb{Z}_p[\![G]\!]}(P_k, N_i)$ is surjective. Let

$$\pi^*_k : \mathbf{Hom}_{\mathbb{Z}_p[\![G]\!]}(P_k, N) \to \varprojlim_i \mathbf{Hom}_{\mathbb{Z}_p[\![G]\!]}(P_k, N_i) \tag{3.7.4}$$

denote the canonical isomorphism (3.7.1) For the induced differentials

$$\begin{aligned} \delta^k_i &: \mathbf{Hom}_{\mathbb{Z}_p[\![G]\!]}(P_k, N_i) \longrightarrow \mathbf{Hom}_{\mathbb{Z}_p[\![G]\!]}(P_{k+1}, N_i), \\ \delta^k &: \mathbf{Hom}_{\mathbb{Z}_p[\![G]\!]}(P_k, N) \longrightarrow \mathbf{Hom}_{\mathbb{Z}_p[\![G]\!]}(P_{k+1}, N), \end{aligned} \tag{3.7.5}$$

one has $im(\delta^{k-1}_i) = \pi^*_{k,i}(im(\delta^{k-1}))$. The compactness of $\mathbf{Hom}_{\mathbb{Z}_p[\![G]\!]}(P_k, N)$ implies that $\pi^*_k(im(\delta^{k-1})) = \varprojlim_i im(\delta^{k-1}_i)$. Since $\pi^*_k(\ker(\delta^k)) = \varprojlim_i ker(\delta^k_i)$, the canonical mapping

$$\pi^*_k : ker(\delta^k)/im(\delta^{k-1}) \longrightarrow \varprojlim ker(\delta^k_i)/im(\delta^{k-1}_i) \tag{3.7.6}$$

is an isomorphism. This proves (3.6.2). A similar argument shows that (3.6.1) also holds. The co-effaceability follows from Proposition 3.4.1.

Remark 3.7.3. If $0 \to A \xrightarrow{\sigma} B \xrightarrow{\tau} C \to 0$ is a short exact sequence in $\mathfrak{C}_p^{\infty}(G)$, then one can use the Horseshoe Lemma and the exact homology sequence for chain complexes (Thm. 2.2.2) to show the existence of a long exact sequence

$$\begin{aligned} 0 \longrightarrow \mathbf{Hom}_{\mathbb{Z}_p[\![G]\!]}(C, N) &\xrightarrow{\tau^*} \mathbf{Hom}_{\mathbb{Z}_p[\![G]\!]}(B, N) \xrightarrow{\sigma^*} \mathbf{Hom}_{\mathbb{Z}_p[\![G]\!]}(A, N) \longrightarrow \\ \mathbf{Ext}^1_{\mathbb{Z}_p[\![G]\!]}(C, N) &\xrightarrow{\tau^*} \mathbf{Ext}^1_{\mathbb{Z}_p[\![G]\!]}(B, N) \xrightarrow{\sigma^*} \mathbf{Ext}^1_{\mathbb{Z}_p[\![G]\!]}(A, N) \longrightarrow \cdots \end{aligned} \tag{3.7.7}$$

for $N \in \mathfrak{P}_p(G)$.

For $M \in \mathfrak{C}^0_p(G)^\times$ the usual tensor product $M \otimes_{\mathbb{Z}_p[\![G]\!]} \_$ coincides with $M \widehat{\otimes}_{\mathbb{Z}_p[\![G]\!]} \_$ if they are considered as functors from $\mathfrak{C}_p(G)$ to $\mathfrak{C}_p$ ([6], Lemma 2.1(ii)). We extend the functor $M \widehat{\otimes}_{\mathbb{Z}_p[\![G]\!]} \_$ to $\mathfrak{P}_p(G)$ by setting

$$M \widehat{\otimes}_{\mathbb{Z}_p[\![G]\!]} N = M \otimes_{\mathbb{Z}_p[\![G]\!]} N \tag{3.7.8}$$

in the case that $N \in ob(\mathfrak{D}_p(G))$. Now $M \widehat{\otimes}_{\mathbb{Z}_p[\![G]\!]} \_ : \mathfrak{P}_p(G) \to \mathfrak{P}_p$ is a covariant additive right exact functor satisfying $M \widehat{\otimes}_{\mathbb{Z}_p[\![G]\!]} \_(\mathfrak{D}_p(G)) \subseteq \mathfrak{D}_p$,

$M\widehat{\otimes}_{\mathbb{Z}_p[\![G]\!]}\_(\mathfrak{C}_p(G)) \subseteq \mathfrak{C}_p$. By construction ([6], §2), $M\widehat{\otimes}_{\mathbb{Z}_p[\![G]\!]}\_$ satisfies (3.6.2). Since the functor $M\widehat{\otimes}_{\mathbb{Z}_p[\![G]\!]}\_|_{\mathfrak{D}_p(G)}$ coincides with the usual tensor product, it also satisfies (3.6.1). So $M\widehat{\otimes}_{\mathbb{Z}_p[\![G]\!]}\_$ is a continuous covariant right exact $\mathfrak{P}_p(G)$-functor. Moreover, if $P \in ob(\mathfrak{C}_p^0(G)^\times)$ is finitely generated and projective, then it is a direct summand of a finitely generated free (right) $\mathbb{Z}_p[\![G]\!]$-module ([6], Lemma 1.6) which is also free in $\mathrm{Mod}_{\mathbb{Z}_p[\![G]\!]}$. This shows that $P\widehat{\otimes}_{\mathbb{Z}_p[\![G]\!]}\_$ is a continuous covariant exact $\mathfrak{P}_p(G)$-functor. This allows us to define the functors $\mathbf{Tor}_\bullet^{\mathbb{Z}_p[\![G]\!]}(M,\_)\colon \mathfrak{P}_p(G) \to \mathfrak{P}_p$ as we did for $\mathbf{Ext}^\bullet_{\mathbb{Z}_p[\![G]\!]}$. Let $M \in ob(\mathfrak{C}_p^\infty(G)^\times)$ and let $\cdots \xrightarrow{\partial_2} P_1 \xrightarrow{\partial_1} P_0 \xrightarrow{\varepsilon} M$ be a projective resolution of finite type. For $N \in ob(\mathfrak{P}_p(G))$ we set

$$\mathbf{Tor}_\bullet^{\mathbb{Z}_p[\![G]\!]}(M,N) = H_n(P_\bullet\widehat{\otimes}_{\mathbb{Z}_p[\![G]\!]}N, \partial_\bullet), \tag{3.7.9}$$

where $\partial_\bullet$ also denotes the induced differential. From Theorem 2.2.4(b) one knows that $\mathbf{Tor}^{\mathbb{Z}_p[\![G]\!]}(M,\_)$ is a homological functor.

Theorem 3.7.4. *Let $M \in ob(\mathfrak{C}_p^\infty(G))$. Then $\mathbf{T}^{-\bullet} = (\mathbf{Tor}_\bullet^{\mathbb{Z}_p[\![G]\!]}(M,\_), e_\bullet)$ is a continuous effaceable homological $\mathfrak{P}_p(G)$-functor with $T^0 = M\widehat{\otimes}_{\mathbb{Z}_p[\![G]\!]}\_$.*

Proof. $T^{-n} = \mathbf{Tor}_n^{\mathbb{Z}_p[\![G]\!]}(M,\_)\colon \mathfrak{P}_p(G) \to \mathfrak{P}_p$ are covariant additive functors satisfying (1) and (2) of the definition of a continuous $\mathfrak{P}_p(G)$-functor. The previous remark has shown that for $N \in ob(\mathfrak{D}_p(G))$, $T^{-n}(N)$ coincides with $\mathrm{Tor}_n^{\mathbb{Z}_p[\![G]\!]}(M,N)$. Hence it satisfies (3.6.1) (cf. [25], 4.1.7, 5.1.9). (3.6.2) can be shown by a similar argument as in the proof of Theorem 3.7.2. The effaceability follows from [6], Lemma 2.1(iii).

By analogy with the case of discrete groups we say that the profinite group G is of *type p-*$\mathbf{FP}_n$, $n \in \mathbb{N}\cup\{\infty\}$, if the trivial $\mathbb{Z}_p[\![G]\!]$-module $\mathbb{Z}_p$ is contained in $ob(\mathfrak{C}_p^n(G))$. For a profinite group G of type p-$\mathbf{FP}_\infty$ it is possible to define continuous cohomological and homological $\mathfrak{P}_p(G)$-functors

$$\begin{aligned}\mathbf{H}^\bullet(G,\_) &= \mathbf{Ext}^\bullet_{\mathbb{Z}_p[\![G]\!]}(\mathbb{Z}_p,\_)\colon \mathfrak{P}_p(G) \longrightarrow \mathfrak{P}_p\\ \mathbf{H}_\bullet(G,\_) &= \mathbf{Tor}_\bullet^{\mathbb{Z}_p[\![G]\!]}(\mathbb{Z}_p,\_)\colon \mathfrak{P}_p(G) \longrightarrow \mathfrak{P}_p\end{aligned} \tag{3.7.10}$$

which extend the previous cohomology and homology functors and which are co-effaceable and effaceable respectively. Moreover Pontrjagin duality induces a natural isomorphism (3.4.6)

$$\mathbf{H}_\bullet(G,\_)^* \simeq \mathbf{H}^\bullet(G,\_^*). \tag{3.7.11}$$

3.8 Rational cohomology

Let $\mathfrak{fd}_p(G)$ denote the category whose objects are finite-dimensional $\mathbb{Q}_p$-vectorspaces which are also topological left $\mathbb{Z}_p[\![G]\!]$-modules and whose morphisms are given by

$$\mathbb{H}\mathrm{om}_{\mathbb{Z}_p[\![G]\!]}(M,N) = \mathrm{Hom}_{\mathbb{Z}_p[\![G]\!]}(M,N), \tag{3.8.1}$$

$M, N \in ob(\mathfrak{fd}_p(G))$.

Proposition 3.8.1.

(1) *$\mathfrak{fd}_p(G)$ is an abelian category.*

(2) *Every module $M \in ob(\mathfrak{fd}_p(G))$ contains a compact and open $\mathbb{Z}_p[\![G]\!]$-submodule $C \in ob(\mathfrak{C}_p(G))$ which is a finitely generated p-torsion-free $\mathbb{Z}_p$-module with $rk_{\mathbb{Z}_p}(C) = dim_{\mathbb{Q}_p}(M)$.*

(3) *Let $\phi\colon M \to N$ be a surjective morphism in $\mathfrak{fd}_p(G)$ and $C \leqslant N$ be a compact $\mathbb{Z}_p[\![G]\!]$-submodule of N. Then there exists a compact $\mathbb{Z}_p[\![G]\!]$-submodule C' of M such that $\phi(C') = C$.*

Proof. (1) is obvious. For (2) let $x_1,..,x_d$ be a $\mathbb{Q}_p$-basis of M. Then $C = \mathbb{Z}_p[\![G]\!].x_1 + \cdots + \mathbb{Z}_p[\![G]\!].x_d$ has the required properties. For (3) let $x_1, \ldots , x_k$ be a basis of C as a $\mathbb{Z}_p$-module. Let $y_1, \ldots , y_k \in M$ with $\phi(y_i) = x_i$. Then $C' = \mathbb{Z}_p[\![G]\!].y_1 + \cdots + \mathbb{Z}_p[\![G]\!].y_k$ satisfies $\phi(C') = C$ and is compact.

For $M \in ob(\mathfrak{C}_p(G))$, $N \in ob(\mathfrak{fd}_p(G))$ we set

$$\mathbb{H}\mathrm{om}_{\mathbb{Z}_p[\![G]\!]}(M, N) = \{\alpha \in \mathrm{Hom}_{\mathbb{Z}_p[\![G]\!]}(M, N) \mid \alpha \text{ continuous}\}, \tag{3.8.2}$$

where N is considered as a topological $\mathbb{Z}_p$-module. Hence

$$\mathbb{H}\mathrm{om}_{\mathbb{Z}_p[\![G]\!]}(\_,\_)\colon \mathfrak{C}_p(G) \times \mathfrak{fd}_p(G) \longrightarrow {}_{\mathbb{Q}_p}\mathrm{Mod} \tag{3.8.3}$$

is a left/left exact contravariant/covariant biadditive bifunctor. For $P \in ob(\mathfrak{C}_p(G))$ projective, Proposition 3.8.1(3) implies that

$$\mathbb{H}\mathrm{om}_{\mathbb{Z}_p[\![G]\!]}(P,\_)\colon \mathfrak{fd}_p(G) \longrightarrow {}_{\mathbb{Q}_p}\mathrm{Mod} \tag{3.8.4}$$

is exact. Thus we can apply Theorem 2.2.4(a) for $\mathfrak{A} = \mathfrak{A}_0 = \mathfrak{C}_p(G)$, $\mathfrak{B} = \mathfrak{fd}_p(G)$, $\mathfrak{Q} = {}_{\mathbb{Q}_p}\mathrm{Mod}$ and $T(\_,\_) = \mathbb{H}\mathrm{om}_{\mathbb{Z}_p[\![G]\!]}(\_,\_)$. For $M \in ob(\mathfrak{C}_p(G))$ we denote by $(\mathbb{E}\mathrm{xt}^\bullet_{\mathbb{Z}_p[\![G]\!]}(M,\_), e^\bullet)$ the corresponding cohomological functor. We define the *rational cohomology* of G with coefficients in the category $\mathfrak{fd}_p(G)$ by

$$\mathbb{H}^\bullet(G,\_) = \mathbb{E}\mathrm{xt}^\bullet_{\mathbb{Z}_p[\![G]\!]}(\mathbb{Z}_p,\_). \tag{3.8.5}$$

Theorem 3.8.2. *Let G be a profinite group of type p-$\mathbf{FP}_\infty$. Let $N \in ob(\mathfrak{fd}_p(G))$ and let $C \leqslant N$ be a compact and open $\mathbb{Z}_p[\![G]\!]$-submodule of N. Then the canonical mapping*

$$\mathbf{H}^\bullet(G, C) \otimes_{\mathbb{Z}_p} \mathbb{Q}_p \longrightarrow \mathbb{H}^\bullet(G, N) \tag{3.8.6}$$

is an isomorphism. In particular,

$$dim_{\mathbb{Q}_p}(\mathbb{H}^k(G, N)) < +\infty \tag{3.8.7}$$

for all $k \in \mathbb{N}_0$, $N \in ob(\mathfrak{fd}_p(G))$. Furthermore, the exact sequence $0 \to C \xrightarrow{\sigma} N \xrightarrow{\tau} N/C \to 0$ induces a long exact sequence

$$\begin{aligned} 0 \longrightarrow \mathbf{H}^0(G, C) &\xrightarrow{\sigma^*} \mathbb{H}^0(G, N) \xrightarrow{\tau^*} \mathbf{H}^0(G, N/C) \longrightarrow \\ \mathbf{H}^1(G, C) &\xrightarrow{\sigma^*} \mathbb{H}^1(G, N) \xrightarrow{\tau^*} \mathbf{H}^1(G, N/C) \longrightarrow \cdots \end{aligned} \tag{3.8.8}$$

of $\mathbb{Z}_p$-modules.

Proof. Let $P_\bullet \xrightarrow{\varepsilon} \mathbb{Z}_p$ be a projective resolution of finite type. In this case $\mathbf{Hom}_{\mathbb{Z}_p[\![G]\!]}(P_k, C)$ is a finitely generated $\mathbb{Z}_p$-module and the canonical morphism

$$\mathbf{Hom}_{\mathbb{Z}_p[\![G]\!]}(P_k, C) \otimes_{\mathbb{Z}_p} \mathbb{Q}_p \longrightarrow \mathbb{H}\mathrm{om}_{\mathbb{Z}_p[\![G]\!]}(P_k, N) \tag{3.8.9}$$

is an isomorphism. Also $\_\otimes_{\mathbb{Z}_p} \mathbb{Q}_p$ is an exact covariant additive functor on the abelian category of finitely generated $\mathbb{Z}_p$-modules. This proves (3.8.6) and (3.8.7). Note that $N/C \in ob(\mathfrak{D}_p(G))$, so for $\phi \in \mathbf{Hom}_{\mathbb{Z}_p[\![G]\!]}(P_k, N/C)$ the $\mathbb{Z}_p$-module $\phi(P_k)$ must be finite. This shows that

$$\begin{aligned} 0 \longrightarrow \mathbf{Hom}_{\mathbb{Z}_p[\![G]\!]}(P_k, C) &\xrightarrow{\sigma^*} \mathbf{Hom}_{\mathbb{Z}_p[\![G]\!]}(P_k, N) \\ &\xrightarrow{\tau^*} \mathbf{Hom}_{\mathbb{Z}_p[\![G]\!]}(P_k, N/C) \longrightarrow 0 \end{aligned} \tag{3.8.10}$$

is exact for $P_k \in ob(\mathfrak{C}_p^\infty(G))$ projective and finitely generated. Now (3.8.8) follows from the exact homology sequence (Thm. 2.2.2).

4 Finiteness conditions and Poincaré duality

Again let G be a profinite group and p a prime number. By $\mathfrak{S}_p(G) \subseteq ob(\mathfrak{D}_p(G))$ we denote the set of (isomorphism types of) simple discrete left G-modules which are p-torsion modules. In particular, every module $M \in \mathfrak{S}_p(G)$ is annihilated by $p\mathbb{Z}_p$. It is well-known that if G is a pro-p group, then $\mathfrak{S}_p(G)$ just consists of the trivial G-module $\mathbb{F}_p = \mathbb{Z}_p/p\mathbb{Z}_p$ ([23], Chap. I, Prop. 20). On the other hand if $|\mathfrak{S}_p(G)| = 1$, then G must be a pro-p group. More generally, one can show easily using Brauer character theory ([1], Cor. 5.3.5) that the set $\mathfrak{S}_p(G)$ is finite if and only if G is virtually pro-p.

4.1 The cohomological p-dimension

Recall that for a profinite group G the *cohomological p-dimension* $cd_p(G)$ is given by

$$cd_p(G) = sup\{\, n \in \mathbb{N}_0 \mid H^n(G, M) \neq 0 \text{ for some } M \in ob(\mathfrak{D}_p(G)) \,\}. \tag{4.1.1}$$

It has the following property.

Proposition 4.1.1. *The following are equivalent:*

(1) $cd_p(G) \leqslant n$,

(2) $H^{n+1}(G, F) = 0$ *for all* $F \in ob(\mathfrak{F}_p(G))$,

(3) $H^{n+1}(G, S) = 0$ *for all* $S \in \mathfrak{S}_p(G)$,

(4) There exists a projective resolution of $\mathbb{Z}_p$ *in* $\mathfrak{C}_p(G)$ *of length n.*

Proof. The implications (4) $\Rightarrow$ (1) $\Rightarrow$ (2) $\Rightarrow$ (3) are obvious. (3) $\Rightarrow$ (4) has been proved in [6], Cor. 3.2.

Let H be a closed subgroup of the profinite group G. Then it follows from Proposition 4.1.1 and Proposition 3.1.1 that

$$cd_p(H) \leqslant cd_p(G) \tag{4.1.2}$$

([[23], Prop. I.14). In the case that $|G : H|$ is coprime to p, or if H is open in G and $cd_p(G) < +\infty$, then equality holds in (4.1.2). This fact implies that profinite groups with finite cohomological p-dimension are necessarily p-torsion-free. If $N \triangleleft G$ is a closed normal subgroup of G, then the Lyndon-Hochschild-Serre spectral sequence for continuous discrete cohomology ([23], §I.2.6.(b)) implies that

$$cd_p(G) \leqslant cd_p(N) + cd_p(G/N). \tag{4.1.3}$$

4.2 Profinite groups of type p-$\mathbf{FP}_\infty$

If $H \leqslant_o G$ is an open subgroup of G, then $\mathbb{Z}_p[\![G]\!]$ is a free $\mathbb{Z}_p[\![H]\!]$-module of rank $|G : H|$. Now Proposition 3.7.1 implies (cf. proof of [5], Chap. VIII, Prop. 5.1):

Proposition 4.2.1. *Let $H \leqslant_o G$ be an open subgroup of G and $n \in \mathbb{N}_0 \cup \{\infty\}$. Then H is of type p-$\mathbf{FP}_n$ if and only if G is of type p-$\mathbf{FP}_n$.*

Let G be a pro-p group. Then $cd_p(G)$ is less or equal to 1 if and only if G is a free pro-p group [23], §I.4.2, Cor. 2). Hence a free pro-p group is of type p-$\mathbf{FP}_\infty$ if and only if it is finitely generated ([23], §I.4.2). This shows that a closed subgroup H of a profinite group G of type p-$\mathbf{FP}_\infty$ is not necessarily of type p-$\mathbf{FP}_\infty$. From the definition of the p-$\mathbf{FP}_\infty$ property one obtains also the following:

Proposition 4.2.2. *Let G be a profinite group of type p-$\mathbf{FP}_\infty$. Then for all $F \in ob(\mathfrak{F}_p(G))$, $n \in \mathbb{N}_0$,*

$$|H^n(G, F)| < +\infty. \tag{4.2.1}$$

One would like to have the p-$\mathbf{FP}_\infty$ condition as the conclusion of a statement rather than in the hypothesis. For virtual pro-p groups the p-$\mathbf{FP}_n$ condition can be characterized by (4.2.1).

Proposition 4.2.3. *Let G be a profinite group which is virtually pro-p and $n \in \mathbb{N}_0$. Then the following are equivalent:*

(1) G is of type p-$\mathbf{FP}_n$.

(2) $H^k(G, F)$ is finite for all $F \in ob(\mathfrak{F}_p(G))$, $k \leqslant n$,

(3) $\dim_{\mathbb{F}_p}(H^k(G, S)) < +\infty$ for all $S \in \mathfrak{S}_p(G)$, $k \leqslant n$.

Proof. The implications (1) $\Rightarrow$ (2) $\Rightarrow$ (3) are obvious. Assume that (3) holds and let $S \in \mathfrak{S}_p(G)$. Let $\cdots \xrightarrow{\partial_2} P_1 \xrightarrow{\partial_1} P_0 \xrightarrow{\varepsilon} \mathbb{Z}_p \longrightarrow 0$ denote the minimal projective resolution of $\mathbb{Z}_p$ in $\mathfrak{C}_p(G)$ (Remark 3.4.3(i)). Let $\boldsymbol{\Omega}^k(\mathbb{Z}_p) = ker(\partial_{k-1})$ ($\partial_0 = \varepsilon$), $\boldsymbol{\Omega}^0(\mathbb{Z}_p) = \mathbb{Z}_p$. Since $P_\bullet \xrightarrow{\varepsilon} \mathbb{Z}_p$ is the minimal projective resolution, the complex

$$\begin{aligned} 0 \longrightarrow \mathbf{Hom}_{\mathbb{Z}_p[\![G]\!]}(P_0, S) &\xrightarrow{\delta^0} \mathbf{Hom}_{\mathbb{Z}_p[\![G]\!]}(P_1, S) \\ &\xrightarrow{\delta^1} \mathbf{Hom}_{\mathbb{Z}_p[\![G]\!]}(P_2, S) \xrightarrow{\delta^2} \cdots \end{aligned} \tag{4.2.2}$$

has to have trivial differentials (cf. [1], Cor. 2.5.4). Thus

$$\mathbf{Ext}^k_{\mathbb{Z}_p[\![G]\!]}(\mathbb{Z}_p, S) \simeq \mathbf{Hom}_{\mathbb{Z}_p[\![G]\!]}(\boldsymbol{\Omega}^k(\mathbb{Z}_p), S). \tag{4.2.3}$$

Since P_k is finitely generated if and only if $\boldsymbol{\Omega}^k(\mathbb{Z}_p)$ is, the assertion follows from the following fact applied to $\boldsymbol{\Omega}^k(\mathbb{Z}_p)$.

Lemma 4.2.4. *Let G be a profinite group which is virtually pro-p and $M \in ob(\mathfrak{C}_p(G))$. Then the following conditions are equivalent:*

(1) M is finitely generated,

(2) $\dim_{\mathbb{F}_p}(\mathbf{Hom}_{\mathbb{Z}_p[\![G]\!]}(M, S)) < +\infty$ for all $S \in \mathfrak{S}_p(G)$.

Proof. The implication (1) ⇒ (2) is obvious. From (2) and the remark at the beginning of §4 one concludes that the number of maximal open submodules in M is finite. Hence

$$Rad(M) = \bigcap_{\substack{Q \leqslant_o M \\ Q \text{ maximal}}} Q \tag{4.2.4}$$

is open in M. So $M/Rad(M)$ is finite. This yields the assertion.

For pro-p groups one obtains:

Corollary 4.2.5. *A pro-p group G is of type p-$\mathbf{FP}_\infty$ if and only if $H^\bullet(G, \mathbb{F}_p)$ is an $\mathbb{F}_p$-algebra of finite type, i.e., $dim_{\mathbb{F}_p}(H^k(G, \mathbb{F}_p)) < +\infty$ for all $k \in \mathbb{N}_0$.*

If $\mathbb{Z}_p[\![G]\!]$ is noetherian[10], then G has to be of type p-$\mathbf{FP}_\infty$. Note that $\mathbb{Z}_p[\![G]\!]$ is noetherian if and only if $\mathfrak{C}_p^0(G) = \mathfrak{C}_p^\infty(G)$. In this case $\mathfrak{C}^0(G)$ is an abelian category with enough projectives.

For profinite groups of type p-$\mathbf{FP}_\infty$ one has also:

Theorem 4.2.6 (Lyndon–Hochschild–Serre spectral sequence). *Let G be a profinite group of type p-$\mathbf{FP}_\infty$, $N \lhd_o G$ an open normal subgroup.*

(a) *Let $M \in ob(\mathfrak{P}_p(G))$. Then there exists a first quadrant spectral sequence $(E_r^{s,t}, d_r)$ with*

$$E_2^{s,t} = \mathbf{H}^s(G/N, \mathbf{H}^t(N, M)) \Longrightarrow \mathbf{H}^{s+t}(G, M). \tag{4.2.5}$$

[10] i.e., every closed submodule of a (topologically) finitely generated $\mathbb{Z}_p[\![G]\!]$-module is finitely generated.

(b) *Let* $M \in ob(\mathfrak{fd}_p(G))$. *Then*

$$H^0(G/N,\_) \circ res_N^G : \mathbb{H}^\bullet(G,M) \longrightarrow \mathbb{H}^\bullet(N,M)^{G/N} \tag{4.2.6}$$

is an isomorphism.

Proof. (a) Let

$$\cdots \xrightarrow{\partial_3} P_2 \xrightarrow{\partial_2} P_1 \xrightarrow{\partial_1} P_0 \xrightarrow{\varepsilon} \mathbb{Z}_p \tag{4.2.7}$$

be a projective resolution of finite type over $\mathbb{Z}_p[G/N]$ and

$$\cdots \xrightarrow{\partial_3'} Q_2 \xrightarrow{\partial_2'} Q_1 \xrightarrow{\partial_1'} Q_0 \xrightarrow{\varepsilon'} \mathbb{Z}_p \tag{4.2.8}$$

a projective resolution of finite type over $\mathbb{Z}_p[\![G]\!]$. Consider the double complex

$$E_0^{s,t} = \mathbf{Hom}_{\mathbb{Z}_p[G/N]}(P_s, \mathbf{Hom}_{\mathbb{Z}_p[\![N]\!]}(Q_t, M)). \tag{4.2.9}$$

Since N is open in G, the restriction of $Q_\bullet \xrightarrow{\varepsilon'} \mathbb{Z}_p$ to N is also a projective resolution of finite type (Prop. 3.3.1). Note that $\mathbf{Hom}_{\mathbb{Z}_p[\![N]\!]}(Q_t, M)$ is a left $\mathbb{Z}_p[\![G]\!]$-module whose restriction to $\mathbb{Z}_p[\![N]\!]$ is trivial, so $\mathbf{Hom}_{\mathbb{Z}_p[\![N]\!]}(Q_t, M) \in ob(\mathfrak{P}_p(G/N))$. Let d_0 be the differential induced by $\partial_\bullet$ and d_1 be $(-1)^s$-times the differential induced by $\partial_\bullet'$. This sign convention ensures that

$$d_1 \circ d_0 + d_0 \circ d_1 = 0. \tag{4.2.10}$$

Now $(E_0^{s,t}, d_0, d_1)$ is a double complex ([2], §3.4). Since $\mathbf{Hom}_{\mathbb{Z}_p[G/N]}(P_s, \_)$ is exact, one obtains for spectral sequence associated to this double complex and applying d_1 first,

$$\begin{aligned} E_1^{s,t} &= \mathbf{Hom}_{\mathbb{Z}_p[G/N]}(P_s, \mathbf{H}^t(N,M)),\\ E_2^{s,t} &= \mathbf{H}^s(G/N, \mathbf{H}^t(N,M)). \end{aligned} \tag{4.2.11}$$

It remains to calculate $H^\bullet(\mathbf{tot}^\bullet(E_0^{\bullet,\bullet}), d_0 + d_1)$.

Lemma 4.2.7. *Let* $A \in ob(\mathfrak{C}_p^0(G/N))$, $B \in ob(\mathfrak{C}_p^0(G))$, $M \in ob(\mathfrak{P}_p(G))$. *Then*

$$\mathbf{Hom}_{\mathbb{Z}_p[G/N]}(A, \mathbf{Hom}_{\mathbb{Z}_p[\![N]\!]}(B,M)) \simeq \mathbf{Hom}_{\mathbb{Z}_p[\![G]\!]}(A \widehat{\otimes}_{\mathbb{Z}_p} B, M) \tag{4.2.12}$$

and the isomorphism is given by a natural transformation (of bifunctors), where A is considered as $\mathbb{Z}_p[\![G]\!]$*-module with* $\mathbb{Z}_p[\![N]\!]$ *acting trivially.*

Proof. ([2], Lemma 3.5.1) It follows that $(E_0^{\bullet,\bullet}, d_0, d_1)$ and $(\mathbf{Hom}_{\mathbb{Z}_p[\![G]\!]}(P_\bullet \widehat{\otimes}_{\mathbb{Z}_p} Q_\bullet, M)$, $\partial, (-1)^\bullet \partial')$ are isomorphic double complexes. Since P_s and Q_t are projective in $\mathfrak{C}_p$ (Prop. 3.3.1), the complex $\mathbf{tot}_\bullet(P_\bullet \widehat{\otimes}_{\mathbb{Z}_p} Q_\bullet, \partial + (-1)^\bullet \partial')$ is exact in positive degrees

([6], Lemma 2.1). By Proposition 3.3.2, $P_s \widehat{\otimes}_{\mathbb{Z}_p} Q_t \in ob(\mathfrak{C}_p(G))$ is projective. This yields

$$H^\bullet(\mathbf{tot}^\bullet(E_0^{\bullet,\bullet}), d_0 + d_1) = \mathbf{H}^\bullet(G, M)$$

and the proof is complete.
(b) The same argument shows the existence of a first quadrant spectral sequence with

$$E_2^{s,t} = H^s(G/N, \mathbb{H}^t(N, M)) \Longrightarrow \mathbb{H}^{s+t}(G, M). \tag{4.2.13}$$

As G/N is finite and $\mathbb{H}^\bullet(N, M)$ divisible, one has

$$E_\infty^{s,t} = E_2^{s,t} = \begin{cases} H^0(G/N, \mathbb{H}^t(N, M)) & \text{for } s = 0, \\ 0 & \text{for } s \neq 0. \end{cases} \tag{4.2.14}$$

Since the vertical edge homomorphism coincides with the stable restriction mapping, this yields the claim.

There is also the following Eckmann-Shapiro type lemma:

Lemma 4.2.8. *Let G be a profinite group of type p-$\mathbf{FP}_\infty$, $H \leqslant_o G$ an open subgroup of G and $M \in ob(\mathfrak{C}_p(H))$. Then*

$$\mathbf{H}^\bullet(H, M) \simeq \mathbf{H}^\bullet(G, ind_G^H(M)). \tag{4.2.15}$$

Proof. Let $Q \in ob(\mathfrak{C}_p^0(G))$. The mappings

$$\begin{aligned} \alpha\colon\ & \mathbf{Hom}_{\mathbb{Z}_p[\![H]\!]}(Q, M) \longrightarrow \mathbf{Hom}_{\mathbb{Z}_p[\![G]\!]}(Q, \mathbf{Hom}_{\mathbb{Z}_p[\![H]\!]}(\mathbb{Z}_p[\![G]\!], M)), \\ \beta\colon\ & \mathbf{Hom}_{\mathbb{Z}_p[\![G]\!]}(Q, \mathbf{Hom}_{\mathbb{Z}_p[\![H]\!]}(\mathbb{Z}_p[\![G]\!], M)) \longrightarrow \mathbf{Hom}_{\mathbb{Z}_p[\![H]\!]}(Q, M), \end{aligned} \tag{4.2.16}$$

given by

$$\begin{aligned} \alpha(f)(q)(x) &= f(x.q), \\ \beta(F)(q) &= F(q)(1), \end{aligned} \tag{4.2.17}$$

$q \in Q$, $x \in \mathbb{Z}_p[\![G]\!]$, are inverse natural transformations of covariant functors. Since the restriction of a projective resolution of $\mathbb{Z}_p$ in $\mathfrak{C}_p(G)$ of finite type is a projective resolution of $\mathbb{Z}_p$ in $\mathfrak{C}_p(H)$ of finite type, one obtains a natural isomorphism

$$\mathbf{H}^\bullet(H, M) \simeq \mathbf{H}^\bullet(G, coind_G^H(M)). \tag{4.2.18}$$

Now the assertion follows from (3.3.7).

Remark 4.2.9. Let G be of type p-$\mathbf{FP}_\infty$. Lemma 4.2.8 shows that the compact $\mathbb{Z}_p$-modules $\mathbf{H}^k(G, \mathbb{Z}_p[\![G]\!])$ are invariants of the commensurability class of G, i.e., one has

$$\mathbf{H}^\bullet(G, \mathbb{Z}_p[\![G]\!]) \simeq \mathbf{H}^\bullet(H, \mathbb{Z}_p[\![H]\!]) \tag{4.2.19}$$

for all open subgroups $H \leqslant_o G$.

4.3 Profinite groups of type p-FP

The profinite group G is called *of type p-***FP** if there exists a projective resolution of the trivial $\mathbb{Z}_p[\![G]\!]$-module $\mathbb{Z}_p$ which is of finite type and of finite length. If G is of type p-$\mathbf{FP}_\infty$ and of finite cohomological p-dimension, the minimal projective resolution of $\mathbb{Z}_p$ has to be of finite length (4.2.3) and of finite type (cf. Prop. 3.7.1). Thus a profinite group is of type p-**FP** if and only if it is of finite cohomological p-dimension and of type p-$\mathbf{FP}_\infty$. For groups of type p-**FP**, there is always equality in (4.1.3). More precisely:

Proposition 4.3.1. *Let G be a profinite group and $N \lhd G$ a closed normal subgroup of type p-***FP** *such that $cd_p(G/N) < +\infty$. Then*

$$cd_p(G) = cd_p(G/N) + cd_p(N). \tag{4.3.1}$$

Proof. Since $cd_p(G) = cd_p(H)$ for every closed subgroup $H \leqslant G$ which contains a pro-p-Sylow subgroup of G, we may assume that G/N is a pro-p group. Let $m = cd_p(N)$ and $n = cd_p(G/N)$. Then $H^m(N, S)$ is finite (Prop. 4.2.1) and non-trivial (Prop. 4.1.1) for some $S \in \mathfrak{S}_p(G)$. Hence the assertion follows from the following lemma and the Lyndon–Hochschild–Serre spectral sequence.

Lemma 4.3.2. *Let P be a pro-p group with $cd_p(P) = n$. Then $H^n(P, F) \neq 0$ for any $F \in ob(\mathfrak{F}_p(G))$, $F \neq 0$.*

Proof. Assume that the assertion is false and let $F \in ob(\mathfrak{F}_p(G))$, $F \neq 0$ and $H^n(P, F) = 0$. Assume further that F is minimal with respect to cardinality. Then $F \not\simeq \mathbb{F}_p$ (Prop. 4.1.1). Hence F has a non-trivial $\mathbb{Z}_p[\![P]\!]$-submodule F' with $F/F' \neq 0$. From the long exact sequence one concludes that $H^n(P, F/F') = 0$, a contradiction.

Remark 4.3.3. Let G be a pro-p group. Then every projective module in $\mathfrak{C}_p(G)$ is actually free. The minimal projective cover of the semi-simple module $\mathbb{F}_p^r$ coincides with the canonical map $\gamma\colon \mathbb{Z}_p[\![G]\!]^r \to \mathbb{F}_p^r$. Let G be a pro-p group of type p-$\mathbf{FP}_\infty$ and let $P_\bullet \xrightarrow{\varepsilon} \mathbb{Z}_p$ denote the minimal projective resolution of the trivial $\mathbb{Z}_p[\![G]\!]$-module $\mathbb{Z}_p$. Using (4.2.3) one concludes that $P_k \simeq \mathbb{Z}_p[\![G]\!]^{r_k}$, where $r_k = \dim_{\mathbb{F}_p} H^k(G, \mathbb{F}_p)$.

For profinite groups G of type p-**FP** with $cd_p(G) = n$ there are actually two ways to show the existence of the dualizing module $I_p(G)$ (or $D_p(G)$). A rather constructive way is to proceed as in the discrete case ([5], Chap. VIII, Prop. 6.8). Alternatively one can use the representability of the functor $H^n(G, \_)^*\colon \mathfrak{F}_p(G) \to \mathfrak{F}_p$ ([13], Chap. II, no. 4). One can prove ([23], Prop. I.17):

Proposition 4.3.4.

(1) *Let G be of type p-***FP** *with $cd_p(G) = n$. Then there exists a discrete $\mathbb{Z}_p[\![G]\!]$-module $I_p(G) \in ob(\mathfrak{D}_p(G))$ and a homomorphism $i\colon H^n(G, I_p(G)) \to I_p$ (of abelian groups) such that given any $M \in \mathfrak{F}_p(G)$, $f \in \mathrm{Hom}_{\mathbb{Z}_p[\![G]\!]}(M, I_p(G))$, the assignment*

$$T(f)\colon H^n(G, M) \xrightarrow{f} H^n(G, I_p(G)) \xrightarrow{i} I_p \tag{4.3.2}$$

defines a natural isomorphism of contravariant $\mathfrak{F}_p(G)$-functors

$$T : \mathrm{Hom}_{\mathbb{Z}_p[\![G]\!]}(\_, I_p(G)) \longrightarrow H^n(G, \_)^*. \tag{4.3.3}$$

(2) *The pair $(I_p(G), i)$ is uniquely defined up to isomorphism.*

(3) *If $H \leqslant_o G$ is an open subgroup of G, then*

$$I_p(H) = res_H^G(I_p(G)). \tag{4.3.4}$$

Pontrjagin duality (Prop. 3.4.2) implies that there is a natural isomorphism of contravariant $\mathfrak{F}_p(G)$-functors

$$T^* : \mathrm{Hom}_{\mathbb{Z}_p[\![G]\!]}(I_p(G)^*, \_^*) \longrightarrow H^n(G, \_)^*. \tag{4.3.5}$$

Moreover, both functors are right exact, so we may extend this natural isomorphism to a natural isomorphism of $\mathfrak{C}_p(G)$-functors

$$T^* : \mathbf{Hom}_{\mathbb{Z}_p[\![G]\!]}(I_p(G)^*, \_^*) \longrightarrow \mathbf{H}^n(G, \_)^*. \tag{4.3.6}$$

Note that (4.3.6) and Pontrjagin duality also imply also other identities, e.g., from (3.4.3) and (3.4.5) one can show that for any $M \in ob(\mathfrak{C}_p(G))$,

$$\begin{aligned}\mathbf{H}^n(G, M) &\simeq \mathbf{Hom}_{\mathbb{Z}_p[\![G]\!]}(I_p(G)^*, M^*)^* \\ &\simeq I_p(G)^\times \widehat{\otimes}_{\mathbb{Z}_p[\![G]\!]} M \simeq \mathbf{H}_0(G, I_p(G)^* \widehat{\otimes}_{\mathbb{Z}_p} M),\end{aligned} \tag{4.3.7}$$

and the isomorphism is natural in M ([6], Lem. 2.4). This is a well known identity for discrete groups of type FP ([5], (VIII.6.9)). Since $coind_G^1(I_p) = \mathbb{Z}_p[\![G]\!]^*$, we have

$$\mathbf{H}^n(G, \mathbb{Z}_p[\![G]\!])^* = \mathbf{Hom}_{\mathbb{Z}_p[\![G]\!]}(I_p(G)^*, coind_G^1(I_p)) = \mathbf{Hom}_{\mathbb{Z}_p}(I_p(G)^*, I_p). \tag{4.3.8}$$

Thus $\mathbf{H}^n(G, \mathbb{Z}_p[\![G]\!]) \simeq I_p(G)^*$ as compact $\mathbb{Z}_p$-modules. One can use (4.3.8) and the uniqueness of the dualizing module to show that $I_p(G)^\times \simeq \mathbf{H}^n(G, \mathbb{Z}_p[\![G]\!])$ as right $\mathbb{Z}_p[\![G]\!]$-modules ([5], Chap. VIII, §6). We will call $D_p(G) = I_p(G)^*$ the *compact dualizing module.*

Remark 4.3.5. Assume that G is of type p-**FP** and virtually pro-p. From (4.2.1) and (4.3.6) it follows that $\mathbf{Hom}_{\mathbb{Z}_p[\![G]\!]}(D_p(G), F)$ is finite for all $F \in ob(\mathfrak{F}_p(G))$. This implies that $D_p(G)$ is finitely generated (Lemma 4.2.4). One might expect this to be true without the assumption that G be virtually pro-p.

4.4 Duality groups

In accordance with the definition for discrete groups ([5], Chap. VIII, §10) we call G a *duality group of dimension n* (at the prime p), if G is of type p-**FP** with $cd_p(G) = n$, and

$$\mathbf{H}^k(G, \mathbb{Z}_p[\![G]\!]) = 0 \text{ for } k \neq n, \tag{4.4.1}$$

and $\mathbf{H}^n(G, \mathbb{Z}_p[\![G]\!])$ is p-torsionfree. If $\mathbf{H}^n(G, \mathbb{Z}_p[\![G]\!]) \simeq \mathbb{Z}_p$ (as abelian groups), G is called a *Poincaré duality group of dimension n.*

Proposition 4.4.1. *Let G be a profinite group with $cd_p(G) = n < +\infty$ and $H \leqslant_o G$ an open subgroup of G. Then G is a duality group (resp. Poincaré duality group), if and only if H is a duality group (resp. Poincaré duality group).*

Proof. G is of type p-**FP** if and only H is of type p-**FP** (4.1.2), Prop. 4.2.1). Now the assertion follows from (4.2.19).

Just as for discrete groups ([5], Chap. VIII, Prop. 5.2) we have:

Lemma 4.4.2. *Let G be a profinite group of type p-$\mathbf{FP}_\infty$. Let $k \in \mathbb{N}_0$ and let $F \in ob(\mathfrak{C}_p(G))$ be a free $\mathbb{Z}_p[\![G]\!]$-module such that $\mathbf{H}^k(G, F) = 0$. Then $\mathbf{H}^k(G, P) = 0$ for all projective $\mathbb{Z}_p[\![G]\!]$-modules $P \in ob(\mathfrak{C}_p(G))$.*

Proof. $\mathbb{Z}_p[\![G]\!]$ is a direct summand of F. Since $\mathbf{H}^k(G, \_)$ is additive, it follows that $\mathbf{H}^k(G, \mathbb{Z}_p[\![G]\!]) = 0$, and so $\mathbf{H}^k(G, M) = 0$ for all finitely generated free $\mathbb{Z}_p[\![G]\!]$-modules. As $\mathbf{H}^k(G, \_)$ commutes with inverse limits in $\mathfrak{C}_p(G)$ (Thm. 3.7.2), $\mathbf{H}^k(G, \_)$ vanishes on all free $\mathbb{Z}_p[\![G]\!]$-modules ([6], §1). Since any projective $\mathbb{Z}_p[\![G]\!]$-module is a direct summand of a free $\mathbb{Z}_p[\![G]\!]$-module ([6], Lemma 1.6), the additivity of $\mathbf{H}^k(G, \_)$ implies the assertion.

As in the discrete case, profinite duality groups can be characterized by the following properties ([5], Chap. VIII, Thm. 10.1).

Theorem 4.4.3. *Let G be a profinite group of type p-**FP** with $cd_p(G) = n$. Then the following are equivalent:*

(1) G is a duality group.

(2) There is a natural isomorphism of homological $\mathfrak{C}_p(G)$-functors

$$\mathbf{H}^{n-\bullet}(G, \_) \longrightarrow \mathbf{H}_\bullet(G, D_p(G)\widehat{\otimes}_{\mathbb{Z}_p}\_). \tag{4.4.2}$$

(3) There is a natural isomorphism of cohomological $\mathfrak{D}_p(G)$-functors

$$\mathbf{H}_{n-\bullet}(G, \_) \longrightarrow \mathbf{H}^\bullet(G, \mathbf{Hom}_{\mathbb{Z}_p}(D_p(G), \_)). \tag{4.4.3}$$

Proof. (1) $\Rightarrow$ (2). Assume that G is a duality group. Then $\mathbf{T}^\bullet = (T^\bullet, e^\bullet)$ with $T^{-k} = \mathbf{H}^{n-k}(G, \_)$ and the connecting homomorphisms given as in Theorem 3.7.2 is an effaceable homological $\mathfrak{C}_p(G)$-functor (Lemma 4.4.2). Since $D_p(G)$ is a free $\mathbb{Z}_p$-module (4.3.8), $D_p(G)\widehat{\otimes}_{\mathbb{Z}_p}\_ : \mathfrak{C}_p(G) \to \mathfrak{C}_p(G)$ is an exact covariant additive $\mathfrak{C}_p(G)$-functor taking projectives to projectives (Prop. 3.3.2). Hence $\mathbf{S}^\bullet = (S^\bullet, e^\bullet_S)$ with $S^{-k} = \mathbf{H}_k(G, D_p(G)\widehat{\otimes}_{\mathbb{Z}_p}\_)$ and the obvious connecting homomorphisms is also an effaceable homological functor. The claim follows from Theorem 2.2.1(b) and (4.3.7).

(2) $\Rightarrow$ (1). From (4.4.2) it follows that $\mathbf{H}^k(G, \mathbb{Z}_p[\![G]\!]) = 0$ for $k \neq n$. It remains to show that $D_p(G)$ is is p-torsion-free. By (3.3.8) and (3.3.9),

$$\begin{aligned} \mathbf{H}^{n-1}(G, \mathbb{F}_p\widehat{\otimes}_{\mathbb{Z}_p}\mathbb{Z}_p[\![G]\!]) &\simeq \mathbf{H}_1(G, \mathbb{F}_p\widehat{\otimes}_{\mathbb{Z}_p}D_p(G)\widehat{\otimes}_{\mathbb{Z}_p}\mathbb{Z}_p[\![G]\!]) \\ &\simeq \mathbf{H}_1(G, ind_G^1(res_1^G(\mathbb{F}_p\widehat{\otimes}_{\mathbb{Z}_p}D_p(G))) \\ &\simeq \mathbf{H}_1(1, res_1^G(\mathbb{F}_p\widehat{\otimes}_{\mathbb{Z}_p}D_p(G))) = 0. \end{aligned} \tag{4.4.4}$$

Hence

$$0 \longrightarrow \mathbf{H}^n(G, \mathbb{Z}_p[\![G]\!]) \xrightarrow{\cdot p} \mathbf{H}^n(G, \mathbb{Z}_p[\![G]\!]) \longrightarrow \mathbf{H}^n(G, \mathbb{F}_p \widehat{\otimes}_{\mathbb{Z}_p} \mathbb{Z}_p[\![G]\!]) \longrightarrow 0 \tag{4.4.5}$$

is exact. This shows that $D_p(G) \simeq \mathbf{H}^n(G, \mathbb{Z}_p[\![G]\!])$ is p-torsion-free. (4.4.2) and (4.4.3) are obviously equivalent by Pontrjagin duality (3.7.11).

Example 4.4.4. Let G be a free pro-p group over the finite set $Y = \{ y_1, \dots, y_r \}$. Then $cd_p(G) = 1$. If we put $x_i = y_i - 1$, then $\mathbb{Z}_p[\![G]\!]$ is a free $\mathbb{Z}_p$-algebra (in $\mathfrak{C}_{\mathbb{Z}_p}$) over the set $X = \{ x_1, .., x_r \}$ ([17], Prop. II.3.1.4). There is a projective resolution

$$0 \longrightarrow \oplus_{1 \leqslant i \leqslant r} \mathbb{Z}_p[\![G]\!].x_i \xrightarrow{\partial} \mathbb{Z}_p[\![G]\!] \xrightarrow{\varepsilon} \mathbb{Z}_p \longrightarrow 0 \tag{4.4.6}$$

with ∂ the obvious map. If we put $X^* = \mathrm{Hom}(X, \mathbb{Z}_p)$ and $x_i \in X^*, x_i^*(x_j) = \delta_{ij}$, we obtain

$$\begin{aligned} \mathbf{Hom}_{\mathbb{Z}_p[\![G]\!]}(\mathbb{Z}_p[\![G]\!], \mathbb{Z}_p[\![G]\!]) &= \mathbb{Z}_p[\![G]\!] \\ \mathbf{Hom}_{\mathbb{Z}_p[\![G]\!]}(\oplus_{1 \leqslant i \leqslant r} \mathbb{Z}_p[\![G]\!].x_i, \mathbb{Z}_p[\![G]\!]) &= X^* \widehat{\otimes}_{\mathbb{Z}_p} \mathbb{Z}_p[\![G]\!] \end{aligned} \tag{4.4.7}$$

with ∂^* given by

$$\partial^*(1) = \sum_{1 \leqslant i \leqslant r} x_i^* \widehat{\otimes} x_i =: \Delta(X). \tag{4.4.8}$$

Hence

$$\begin{aligned} \mathbf{H}^0(G, \mathbb{Z}_p[\![G]\!]) &= 0 \\ \mathbf{H}^1(G, \mathbb{Z}_p[\![G]\!]) &= X^* \widehat{\otimes}_{\mathbb{Z}_p} \mathbb{Z}_p[\![G]\!] / \Delta(X).\mathbb{Z}_p[\![G]\!] \end{aligned} \tag{4.4.9}$$

So G is a duality group, but a Poincaré duality group if and only if $|Y| = 1$.

4.5 Poincaré duality groups

If $res_1^G(D_p(G)) \simeq \mathbb{Z}_p$, then

$$\begin{aligned} D_p(G) \widehat{\otimes}_{\mathbb{Z}_p} \_ &: \mathfrak{P}_p(G) \longrightarrow \mathfrak{P}_p(G) \\ \mathbf{Hom}_{\mathbb{Z}_p}(D_p(G), \_) &: \mathfrak{P}_p(G) \longrightarrow \mathfrak{P}_p(G) \end{aligned} \tag{4.5.1}$$

are covariant exact functors. Furthermore, $D_p(G) \widehat{\otimes}_{\mathbb{Z}_p} \_$ maps projectives in $\mathfrak{C}_p(G)$ to projectives (Prop. 3.3.2) and $\mathbf{Hom}_{\mathbb{Z}_p}(D_p(G), \_)$ maps injectives in $\mathfrak{D}_p(G)$ to injectives. From Theorem 4.4.3 and Proposition 3.6.1 we can conclude

Proposition 4.5.1. *Let G be a profinite Poincaré duality group (at the prime p) of dimension n and $D_p(G)$ its compact dualizing module. Then there are natural isomorphisms*

$$\begin{aligned} \mathbf{H}^{n-k}(G, \_) &\simeq \mathbf{H}_k(G, D_p(G) \widehat{\otimes}_{\mathbb{Z}_p} \_), \\ \mathbf{H}_{n-k}(G, \_) &\simeq \mathbf{H}^k(G, \mathbf{Hom}_{\mathbb{Z}_p}(D_p(G), \_)), \end{aligned} \tag{4.5.2}$$

of continuous covariant additive $\mathfrak{P}_p(G)$-functors.

This yields

Corollary 4.5.2. *Let G be a profinite Poincaré duality group (at the prime p) of dimension n. Let M be a $\mathbb{Z}_p[\![G]\!]$-module which is isomorphic to $\mathbb{Z}_p$ as an abelian group. Then the following are equivalent:*

(1) $\mathbf{H}^n(G, M) \neq 0$.

(2) $D_p(G)\widehat{\otimes}_{\mathbb{Z}_p} M \simeq \mathbb{Z}_p$ *as $\mathbb{Z}_p[\![G]\!]$-modules, where $\mathbb{Z}_p$ denotes the trivial $\mathbb{Z}_p[\![G]\!]$-module.*

From the way that we have constructed the natural transformations in (4.5.2), we can see that they actually constitute a mapping of $\mathbf{\Sigma}$-connected sequences. These natural transformations can be described in a more explicit way using the cup product.

Let G be a profinite group. The cup product is a family of group homomorphisms

$$f^{s,t}\colon \mathbf{H}^s(G, M) \otimes_{\mathbb{Z}_p} \mathbf{H}^t(G, N) \longrightarrow \mathbf{H}^{s+t}(G, M \otimes_{\mathbb{Z}_p} N) \tag{4.5.3}$$

defined for all $s, t \in \mathbb{N}_0$ and $M, N \in \mathfrak{D}_p(G)$, such that

(1) $f^{0,0}\colon \mathbf{H}^0(G, M) \otimes_{\mathbb{Z}_p} \mathbf{H}^0(G, N) \to \mathbf{H}^0(G, M \otimes_{\mathbb{Z}_p} N)$ is the canonical map,

(2) $f^{s,t}$ is natural in M and N,

(3) $f^{s+1,t}(e^s_{sp}(\mathbf{s}) \otimes id) = e^{s+t}_{sp}(\mathbf{s} \otimes N) f^{s,t}$, for all proper short exact sequences $\mathbf{s}\colon 0 \to A \xrightarrow{\sigma} B \xrightarrow{\tau} C \to 0$ in $\mathbf{\Sigma}^{sp}(\mathfrak{D}_p(G))$,

(4) $f^{s,t+1}(id \otimes e^t_{sp}(\mathbf{s})) = e^{s+t}_{sp}(M \otimes \mathbf{s}) f^{s,t}$, for all proper short exact sequences $\mathbf{s} \in \mathbf{\Sigma}^{sp}(\mathfrak{D}_p(G))$.

Its construction and uniqueness can be proved as for discrete groups (see [18], Chap. XII, §10, [5], Chap. V.3). Note that the connecting homomorphism on the left hand side is defined with the usual sign convention ([18], Chap. XII, (10.1), [5], Chap. V, (3.3)).

Let G be a profinite group of type p-**FP** with $cd_p(G) = n$. For $M \in ob(\mathfrak{F}_p(G))$ we set

$$M^\vee = \mathbf{Hom}_{\mathbb{Z}_p}(M, I_p(G)) \in ob(\mathfrak{D}_p(G)). \tag{4.5.4}$$

Consider the mapping

$$\Psi^k_M\colon \mathbf{H}^k(G, M) \otimes \mathbf{H}^{n-k}(G, M^\vee) \longrightarrow I_p, \tag{4.5.5}$$

given by $\Psi^k_M = i \circ \mathbf{H}^n(ev) \circ f^{k,n-k}$, where i is given as in Proposition 4.3.4 and ev is the obvious evaluation mapping. This induces a map

$$\Phi^k_M\colon \mathbf{H}^k(G, M) \longrightarrow \mathbf{H}^{n-k}(G, M^\vee)^* \tag{4.5.6}$$

where $\Phi^k_M(x_k)(y_{n-k}) = \Psi^k_M(x_k \otimes y_{n-k})$, $x_k \in \mathbf{H}^k(G, M)$, $y_{n-k} \in \mathbf{H}^{n-k}(G, M^\vee)$.

Let $(\gamma\colon M \to N) \in mor(\mathfrak{F}_p(G))$ and let $\gamma^\vee\colon N^\vee \to M^\vee$ denote the induced mapping. The mappings

$$\alpha^k, \beta^k\colon \mathbf{H}^k(G, M) \otimes_{\mathbb{Z}_p} \mathbf{H}^{n-k}(G, N^\vee) \longrightarrow I_p \tag{4.5.7}$$

given by

$$\begin{aligned}\alpha^k &= i \circ \mathbf{H}^n(ev \circ (id \otimes \gamma^\vee)) \circ f^{k,n-k} \\ \beta^k &= i \circ \mathbf{H}^n(ev \circ (\gamma \otimes id)) \circ f^{k,n-k}\end{aligned} \tag{4.5.8}$$

coincide. From this and property (2) of the cup product it follows that

$$\Phi^k_{\_} : \mathbf{H}^k(G, \_)_{\mathfrak{F}} \longrightarrow \mathbf{H}^{n-k}(G, \_^\vee)^* \tag{4.5.9}$$

is a natural transformation of covariant $\mathfrak{F}_p(G)$-functors. One has to spend some more effort in order to show that it is a mapping of $\mathbf{\Sigma}^{sp}(\mathfrak{F}_p(G))$-connected sequences. Let $\mathbf{s}: 0 \to A \xrightarrow{\sigma} B \xrightarrow{\tau} C \to 0$ be a short exact sequence in $\mathbf{\Sigma}^{sp}(\mathfrak{F}_p(G))$, i.e., considered as a short exact sequence of abelian groups it splits. The dual sequence $\mathbf{s}^\vee : 0 \to C^\vee \xrightarrow{\tau^\vee} B^\vee \xrightarrow{\sigma^\vee} A^\vee \to 0$ is a short exact sequence in $\mathbf{\Sigma}^{sp}(\mathfrak{D}_p(G))$. Consider $\mathbf{s}$ and $\mathbf{s}^\vee$ as exact chain complexes with B and $B^\vee$ in degree 0. The total complex $(Q_\bullet, d_\bullet)$ of the associated double complex

$$\begin{array}{ccccc} A \otimes C^\vee & \xrightarrow{\sigma \otimes C^\vee} & B \otimes C^\vee & \xrightarrow{\tau \otimes C^\vee} & C \otimes C^\vee \\ {\scriptstyle -A\otimes\tau^\vee}\downarrow & & {\scriptstyle B\otimes\tau^\vee}\downarrow & & {\scriptstyle -C\otimes\tau^\vee}\downarrow \\ A \otimes B^\vee & \xrightarrow{\sigma \otimes B^\vee} & B \otimes B^\vee & \xrightarrow{\tau \otimes B^\vee} & C \otimes B^\vee \\ {\scriptstyle -A\otimes\sigma^\vee}\downarrow & & {\scriptstyle B\otimes\sigma^\vee}\downarrow & & {\scriptstyle -C\otimes\sigma^\vee}\downarrow \\ A \otimes A^\vee & \xrightarrow{\sigma \otimes A^\vee} & B \otimes A^\vee & \xrightarrow{\tau \otimes A^\vee} & C \otimes A^\vee \end{array} \tag{4.5.10}$$

is also exact. Note that $ker(d_{-1})$ is the pull-back of the diagram

$$\begin{array}{ccc} & & C \otimes B^\vee \\ & & \downarrow{\scriptstyle C\otimes\sigma^\vee} \\ B \otimes A^\vee & \xrightarrow{\tau \otimes A^\vee} & C \otimes A^\vee \end{array} \tag{4.5.11}$$

Now the short exact sequence $\mathbf{t} = (\mathbf{s} \otimes A^\vee \oplus (-1)C \otimes \mathbf{s}^\vee) \circ \Delta_{C \otimes A^\vee}$ is given by

$$\mathbf{t}:\ 0 \longrightarrow A \otimes A^\vee \oplus C \otimes C^\vee \xrightarrow{\lambda} ker(d_{-1}) \xrightarrow{\eta} C \otimes A^\vee \longrightarrow 0, \tag{4.5.12}$$

where η is given by the pull-back diagram and $\lambda = \sigma \otimes A^\vee \oplus (-1)C \otimes \tau^\vee$. The sequence $\mathbf{t}$ is congruent to

$$\mathbf{t}':\ A \otimes A^\vee \oplus C \otimes C^\vee \xrightarrow{\gamma} A \otimes A^\vee \oplus B \otimes B^\vee \oplus C \otimes C^\vee / im(d_1) \xrightarrow{\eta \circ \bar{d}_0} C \otimes A^\vee \tag{4.5.13}$$

where γ denotes the canonical mapping and $\bar{d}_0$ the induced mapping. This shows that there is a commutative diagram

$$\begin{array}{ccccccccc} 0 & \longrightarrow & A \otimes A^\vee \oplus C \otimes C^\vee & \xrightarrow{\lambda} & ker(d_{-1}) & \xrightarrow{\eta} & C \otimes A^\vee & \longrightarrow & 0 \\ & & \big\downarrow{\scriptstyle ev_A + ev_C} & & \big\downarrow{\scriptstyle \chi} & & & & \\ 0 & \longrightarrow & I_p(G) & = & I_p(G) & \longrightarrow & 0. & & \end{array} \tag{4.5.14}$$

Lemma 4.5.3. *Let G be a profinite group and $\mathbf{s} \in \Sigma^{sp}(\mathfrak{F}_p(G))$, $\mathbf{s}^\vee \in \Sigma^{sp}(\mathfrak{D}_p(G))$ as above. Let (S, e, T) be a Σ^{sp}-connected pair of $(\mathfrak{D}_p(G))$-functors. Then the diagram*

$$\begin{array}{ccc} S(C \otimes A^\vee) & \xrightarrow{e(\mathbf{s} \otimes A^\vee)} & T(A \otimes A^\vee) \\ \big\downarrow{\scriptstyle e(C \otimes \mathbf{s}^\vee)} & & \big\downarrow{\scriptstyle T(ev)} \\ T(C \otimes C^\vee) & \xrightarrow{T(ev)} & T(I_p(G)) \end{array} \tag{4.5.15}$$

commutes.

Proof. From (4.5.14) we see that the composition

$$S(C \otimes A^\vee) \xrightarrow{e(\mathbf{t})} T(A \otimes A^\vee \oplus C \otimes C^\vee) \xrightarrow{T(ev_A + ev_C)} T(I_p(G)) \tag{4.5.16}$$

is trivial. Hence the additivity and functoriality of the connecting homomorphism ([18], Chap. XII, (7.2a–c)) yield the desired result.

Assume that $D_p(G)$ is a finitely generated free $\mathbb{Z}_p$-module. Then Φ^k can be extended to a natural transformation of continuous covariant $\mathfrak{P}_p(G)$-functors (Prop. 3.6.1), which we will denote by the same letter. Moreover, it has the following property.

Proposition 4.5.4. *Let G be a profinite group of type p-**FP**, $cd_p(G) = n$. Suppose that $D_p(G)$ is a finitely generated free $\mathbb{Z}_p$-module. Then*

$$\Phi^\bullet : (\mathbf{H}^\bullet(G, \_), e^\bullet) \longrightarrow (\mathbf{H}^{n-\bullet}(G, \_^\vee)^*, (-1)^{\bullet+1}(e^{n-\bullet} \circ {}^\vee)^*) \tag{4.5.17}$$

is a mapping of continuous Σ-connected sequences of covariant $\mathfrak{P}_p(G)$-functors. Furthermore, Φ^n coincides with the Pontrjagin dual of the mapping T (4.3.3) and hence is a natural isomorphism. If G is a Poincaré duality group, then Φ^k is a natural isomorphism for all $k = 0, \ldots, n$.

Proof. By hypothesis, $^\vee$ is an exact contravariant additive $\mathfrak{P}_p(G)$-functor. So the right hand side of (4.5.17) is a continuous Σ-connected sequence of covariant $\mathfrak{P}_p(G)$-functors.

Let $\mathbf{s}\colon 0 \to A \xrightarrow{\sigma} B \xrightarrow{\tau} C \to 0$ be a short exact sequence in $\Sigma^{sp}(\mathfrak{F}_p(G))$ and let $\mathbf{s}^\vee$ denote the dual sequence. Using property (3) and (4) of the cup product and Lemma

4.5.3, one can check easily that the diagram

$$\begin{array}{ccc} \mathbf{H}^k(G,C) & \xrightarrow{\Phi^k} & \mathbf{H}^{n-k}(G,C^\vee)^* \\ {\scriptstyle e_k(s)}\downarrow & & \downarrow{\scriptstyle (-1)^{k+1}e^{n-k-1}(s^\vee)^*} \\ \mathbf{H}^{k+1}(G,A) & \xrightarrow{\Phi^{k+1}} & \mathbf{H}^{n-k-1}(G,A^\vee)^* \end{array} \tag{4.5.18}$$

is commutative. Hence $\Phi^\bullet$ is a morphism of $\Sigma^{sp}(\mathfrak{F}_p(G))$-connected sequences of $\mathfrak{F}_p(G)$-functors. From (3.6.6) one concludes that $\Phi^\bullet$ is also a morphism of $\Sigma^{sp}(\mathfrak{D}_p(G))$-connected sequences of $\mathfrak{D}_p(G)$-functors. Then Proposition 3.5.2 implies that $\Phi^\bullet$ is a morphism of $\Sigma(\mathfrak{D}_p(G))$-connected sequence of $\mathfrak{D}_p(G)$-functors. Now (3.6.5) implies that $\Phi^\bullet$ is a morphism of $\Sigma(\mathfrak{P}_p(G))$-connected sequences of $\mathfrak{P}_p(G)$-functors.

Let $M \in ob(\mathfrak{F}_p(G))$. The map

$$\mathbf{H}^n(ev) \circ f^{n,0}\colon \mathbf{H}^n(G,M) \otimes_{\mathbb{Z}_p} \mathbf{H}^0(G,M^\vee) \longrightarrow \mathbf{H}^n(G,I_p(G)) \tag{4.5.19}$$

coincides with the coefficient morphism, i.e., $\mathbf{H}^n(ev) \circ f^{n,0}(x_n)(y_0) = y_0(x_n)$ where $x_n \in \mathbf{H}^n(G,M)$, $y_0 \in \mathbf{H}^0(G,M^\vee)$ ([5], Chap. 3, Exer. 1). This shows that $\Phi^n = T^*$. If G is a Poincaré duality group, then the final remark follows from Corollary 3.6.3.

Remark 4.5.5. Proposition 4.5.4 allows us to make the natural isomorphisms occuring in Proposition 4.5.1 much more precise. A more elegant way would have been to use the cap product which we do not introduce in this article. The Pontrjagin dual $i^*\colon \mathbb{Z}_p \to \mathbf{H}^n(G,I_p(G))$ of the dualizing morphism i (Prop. 4.3.4) defines a cohomology class $z = i^*(1)$. For discrete Poincaré duality groups the cap product with the equivalent of this cohomology class z yields the first of the natural isomorphisms occuring in (4.5.2) (cf. [5], Thm. 10.1, second proof).

The cup product can be used to prove the following Poincaré duality criterion for pro-p groups which is due to J-P. Serre.[11]

Theorem 4.5.6. *Let G be a pro-p group of type p-**FP** with $cd_p(G) = n$. Then the following are equivalent:*

(1) G is a Poincaré duality group.

(2) $dim_{\mathbb{F}_p}(H^n(G,\mathbb{F}_p)) = 1$ and the canonical pairing

$$H^k(G,\mathbb{F}_p) \otimes H^{n-k}(G,\mathbb{F}_p) \longrightarrow H^n(G,\mathbb{F}_p) \tag{4.5.20}$$

induced by the cup product is non-degenerate.

Proof.[12] If G is a Poincaré duality group, then

$$H^n(G,\mathbb{F}_p) = H^n(G,\mathrm{Hom}_{\mathbb{Z}_p}(\mathbb{F}_p,I_p(G))) = H^0(G,\mathbb{F}_p) = \mathbb{F}_p. \tag{4.5.21}$$

[11] In [23], §I.4.5 Poincaré duality groups were defined by the property (2). In [23], Prop. I.30 it was shown that they satisfy property (1).

[12] The proof of the proposition is basically the same as in [23].

The non-degeneracy of (4.5.20) follows from Proposition 4.5.3. Assume that G satisfies (2). Let $\mathbb{F}_p\mathfrak{P}_p(G)$ denote the proper subcategory of $\mathfrak{P}_p(G)$ consisting of all modules in $ob(\mathfrak{P}_p(G))$ which are annihilated by $p \in \mathbb{Z}_p$ and let

$$\alpha_k \colon \mathbf{H}^k(G, \_) \longrightarrow \mathbf{H}^{n-k}(G, \_^*)^* \tag{4.5.22}$$

be the natural transformations of $\mathbb{F}_p\mathfrak{P}_p(G)$-functors induced by the cup product (Prop. 4.5.4). Since $\mathbf{H}^k(G, \_)$ and $\mathbf{H}^{n-k}(G, \_^*)^*$ are cohomological $\mathbb{F}_p\mathfrak{P}_p(G)$-functors, it follows from the Five Lemma and induction that the α_k are natural isomorphisms of $\mathbb{F}_p\mathfrak{P}_p(G)$-functors. In particular, by Lemma 3.3.4,

$$\begin{aligned} \mathbf{H}^{n-1}(G, \mathbb{F}_p[\![G]\!]) &\simeq \mathbf{H}^1(G, \mathbb{F}_p[\![G]\!]^*)^* \\ &\simeq \mathbf{H}_1(G, \mathbb{F}_p[\![G]\!]) \simeq \mathbf{H}_1(G, ind_G^1(\mathbb{F}_p)) = 0. \end{aligned} \tag{4.5.23}$$

The long exact sequence implies that $0 \to D_p(G) \xrightarrow{\cdot p} D_p(G)$ is exact. Thus $D_p(G)$ is p-torsion-free and therefore a free $\mathbb{Z}_p$-module (cf. Cor. 2.5.2). On the other hand the uniqueness of the dualizing module for the category $\mathbb{F}_p\mathfrak{P}_p(G)$ implies that $D_p(G)/p.D_p(G) \simeq \mathbb{F}_p$. Hence $D_p(G)$ is isomorphic to $\mathbb{Z}_p$ as a $\mathbb{Z}_p$-module and $\mathbf{H}^\bullet(G, \mathbf{Hom}_{\mathbb{Z}_p}(\_, I_p(G)))^* \simeq \mathbf{H}_\bullet(G, D_p(G)\hat{\otimes}_{\mathbb{Z}_p}\_)$ is a continuous homological $\mathfrak{P}_p(G)$-functor (3.7.11). By the Five Lemma, induction and continuity, one concludes that Φ^k (4.5.9) is a natural isomorphisms of $\mathfrak{P}_p(G)$-functors for $k = 0, \ldots, n$. Hence G is a Poincaré duality group (Thm. 4.4.3).

Let G be a Poincaré duality group and let

$$\phi_{D_p(G)} \colon G \longrightarrow \mathbb{Z}_p^* = GL_{\mathbb{Z}_p}(D_p(G)) \tag{4.5.24}$$

denote its representation on its compact dualizing module. One can distinguish three types of Poincaré duality groups.

(1) $D_p(G) \simeq \mathbb{Z}_p$ is the trivial $\mathbb{Z}_p[\![G]\!]$-module. In this case G is called *orientable.*

(2) $im(\phi_{D_p(G)})$ is finite. In this case we call G *virtually orientable.*

(3) $im(\phi_{D_p(G)})$ is infinite.

We will see later that, for a p-torsion-free compact p-adic analytic group virtual orientability is related to the unimodularity of its associated $\mathbb{Q}_p$-Lie algebra (Cor. 5.2.6).

4.6 The mod p cohomology ring

Let $\mathfrak{X}$ be a group theoretic property for profinite groups which is closed with respect to open subgroups, i.e., if G has the property $\mathfrak{X}$ and $H \leqslant_o G$, then H has also the property $\mathfrak{X}$. We say that G has *virtually* the property $\mathfrak{X}$ (or is virtually $\mathfrak{X}$), if G contains an open subgroup with property $\mathfrak{X}$. So one has the notions of profinite virtual Poincaré duality groups, profinite groups which are virtually of type p-**FP**, profinite groups with finite virtual cohomological p-dimension, etc.

Some of the group theoretic properties $\mathfrak{X}$ we have studied in this section have strong implications on the structure of the mod p cohomology ring $H^\bullet(G, \mathbb{F}_p)$, where

G has property $\mathfrak{X}$. For example, this applies to groups of finite cohomological p-dimension, groups of type p-$\mathbf{FP}_\infty$ and Poincaré duality groups (Prop. 4.1.1, Prop. 4.2.2, Prop. 4.5.1). Moreover, the class of pro-p groups with one of these properties is characterized by the structure of the mod p cohomology ring (Prop. 4.1.1(3), Cor. 4.2.5, Thm. 4.5.6). For profinite groups of type p-$\mathbf{VFP}$ (i.e., profinite groups which are virtually of type p-$\mathbf{FP}$) one has the following:

Proposition 4.6.1. *Let G be profinite group of type p-$\mathbf{VFP}$. Then $H^\bullet(G, \mathbb{F}_p)$ (with cup product) is a finitely generated $\mathbb{F}_p$-algebra.*

Proof. See [22], Prop. 13.5. To us it is not known whether this particular structure of the mod p cohomology ring characterizes pro-p groups with this property.

For groups of type p-$\mathbf{FP}_\infty$ it is possible to define a cup product on $H^\bullet(G, \mathbb{Z}_p)$. Using similar arguments as in [12], §7.4, one can also prove that for profinite groups of type p-$\mathbf{VFP}$, $H^\bullet(G, \mathbb{Z}_p)$ is a finitely generated $\mathbb{Z}_p$-algebra.

5 Cohomology of p-adic analytic groups

5.1 Cohomological finiteness conditions

One of the central results in the theory of p-adic analytic groups is that they contain always an open subgroup which is uniformly powerful ([17, Chap. V, 2.2.7.1]). More precisely, one can show

Theorem 5.1.1 (M. Lazard). *Let G be a compact p-adic analytic group. Then G contains an open characteristic uniformly powerful pro-p group.*

Proof. ([10, Cor. 8.34]). This result is actually the key to the proofs of many theorems concerning compact p-adic analytic groups. Uniformly powerful subgroups of compact p-adic analytic groups play the role of the connected component for real analytic groups. One consequence of Theorem 5.1.1 is also the following:

Theorem 5.1.2 (M. Lazard). *Let G be a compact p-adic analytic group. Then $\mathbb{Z}_p[\![G]\!]$ is noetherian. In particular, compact p-adic analytic groups are of type p-$\mathbf{FP}_\infty$.*

Proof. ([17, Prop. V.2.2.4], [10, Cor. 7.25], and §4.2).

We do not know whether there exist virtual pro-p- groups G for which $\mathbb{Z}_p[\![G]\!]$ is noetherian and which are not p-adic analytic. We call a profinite group G *noetherian* if every closed subgroup $H \leqslant G$ is finitely generated. For solvable groups one has the following characterization:

Theorem 5.1.3. *Let G be a solvable virtual pro-p group. Then the following are equivalent.*

(1) $\mathbb{Z}_p[\![G]\!]$ is noetherian.

(2) G is noetherian.

(3) G is p-adic analytic.

The proof of Theorem 5.1.3 makes use of the following

Lemma 5.1.4. *Let G be a virtual pro-p group such that $\mathbb{Z}_p[\![G]\!]$ is noetherian. Then every closed subgroup $A \leqslant G$ is of type p-$\mathbf{FP}_\infty$. In particular, G is noetherian.*

Proof. In order to prove the lemma we may assume that G a pro-p group (Prop. 4.2.1). Let $A \leqslant G$ be a closed subgroup. By Corollary 4.2.5, it suffices to show that $H^k(A, \mathbb{F}_p)$ is finite for all $K \in \mathbb{N}_0$. From the Eckmann-Shapiro lemma (3.4.7) one obtains

$$H^k(A, \mathbb{F}_p) = \mathbf{Ext}^k_{\mathbb{Z}_p[\![G]\!]}(ind_G^A(\mathbb{Z}_p), \mathbb{F}_p). \tag{5.1.1}$$

As $\mathbb{Z}_p[\![G]\!]$ is noetherian, one has $\mathfrak{C}^0(G) = \mathfrak{C}^\infty(G)$. Since $ind_G^A(\mathbb{Z}_p)$ is a finitely generated $\mathbb{Z}_p[\![G]\!]$-module, it follows that it is a $\mathbb{Z}_p[\![G]\!]$-module of type p-$\mathbf{FP}_\infty$. Thus $H^k(A, \mathbb{F}_p)$ is finite.

Proof of Theorem 5.1.3. It remains to prove the implication (2) $\Rightarrow$ (3) for G a pro-p group. We proceed by induction on the derived length $dl(G)$ of G. Every finitely generated abelian pro-p group is p-adic analytic. Hence the assertion holds for $dl(G) = 1$. Let G be a noetherian solvable pro-p group with $dl(G) = n$. Then $A = G^{(n-1)}$, the $(n-1)^{th}$ closed derived subgroup of G, is a closed abelian subgroup, and by assumption, A is finitely generated. Thus A is p-adic analytic. Moreover, G/A is a noetherian with $dl(G/A) = n-1$. So the induction hypothesis yields that G/A is p-adic analytic. By a theorem of Serre, extensions of p-adic analytic groups are p-adic analytic ([17, Thm. III.3.4.5], [10, Thm. 9.7]). This completes the proof.

For uniformly powerful pro-p groups the mod p cohomology ring can be calculated explicitly. One obtains the following striking result:

Theorem 5.1.5 (Lazard). *Let G be a uniformly powerful pro-p group and let $P^{el.ab.} = P/P^p$. Then*

$$H^\bullet(P, \mathbb{F}_p) \simeq \Lambda_\bullet((P^{el.ab.})^*), \tag{5.1.2}$$

where $\Lambda_\bullet(\_)$ denotes the exterior $\mathbb{F}_p$-algebra. In particular, P is a Poincaré duality group in dimension $n = rk_p(P)$ (at the prime p).

The proof is given at the end of this subsection. For a different proof close to Lazard's original proof see [27, Chap. II].

For p odd powerful pro-p groups can be characterized by their mod p cohomology ring.

Theorem 5.1.6. *Let p be odd and let P be a finitely generated pro-p group. Then the following are equivalent.*

(1) The canonical mapping

$$\gamma_P\colon H^1(P, \mathbb{F}_p) \wedge H^1(P, \mathbb{F}_p) \to H^2(P, \mathbb{F}_p) \tag{5.1.3}$$

is injective.

(2) P is powerful.

Proof. Let $Q = P/\Phi(P)$ where $\Phi(P) := \overline{P^p[P,P]}$ denotes the Frattini group, and let $\pi : P \to Q$ denote the canonical projection. One has a commutative diagram

$$\begin{array}{ccc} H^1(P,\mathbb{F}_p) \wedge H^1(P,\mathbb{F}_p) & \xrightarrow{\gamma_P} & H^2(P,\mathbb{F}_p) \\ {\scriptstyle \pi_1^*\wedge\pi_1^*}\uparrow & & {\scriptstyle \pi_2^*}\uparrow \\ H^1(Q,\mathbb{F}_p) \wedge H^1(Q,\mathbb{F}_p) & \xrightarrow{\gamma_Q} & H^2(Q,\mathbb{F}_p) \end{array} \tag{5.1.4}$$

where the horizontal arrows are induced by cup product, and the vertical arrows are given by inflation. By construction,

$$\pi_1^* : H^1(Q,\mathbb{F}_p) \to H^1(P,\mathbb{F}_p) \tag{5.1.5}$$

is an isomorphism [23, I.4.2]. So $\pi_1^* \wedge \pi_1^*$ is an isomorphism.

For $d \in H^2(Q,\mathbb{F}_p)$ let

$$1 \longrightarrow \mathbb{F}_p \longrightarrow Q_d \longrightarrow Q \longrightarrow 1 \tag{5.1.6}$$

denote the equivalence class of group extensions classified by the element d. As $H^2(Q,\mathbb{F}_p) = im(\gamma_Q) \oplus im(\beta)$, where β denotes the first Bockstein operator ([12, §3.4]), one has

Fact 5.1.6.1. *$Q_d^p = 1$ if and only if $d \in im(\gamma_Q)$.*

For short we set $Q_c = Q_{\gamma_Q(c)}$ and denote by $\pi_c : Q_c \to Q$ the canonical projection. Let

$$P_c = Q_c \times_{(\pi_c,\pi)} P := \{ (q,p) \in Q_c \times P \mid \pi_c(q) = \pi(p) \} \tag{5.1.7}$$

Now $\gamma_P(c_P) = 0$ if and only if the extension $\mathbb{F}_p \to P_c \to P$ splits. So $\gamma_P(c_P) = 0$, if and only if there exists a group homomorphism τ which makes the diagram

$$\begin{array}{ccc} P & = & P \\ {\scriptstyle \tau}\downarrow & & {\scriptstyle \pi}\downarrow \\ Q_c & \xrightarrow{\pi_C} & Q \end{array} \tag{5.1.8}$$

commute. If $c \neq 0$ such a map τ induces a surjection $\tau^* : P/\Phi(P) \to Q_c/\Phi(Q_c)$, and thus τ must be surjective.

Let P be powerful and $c \in H^1(Q,\mathbb{F}_p) \wedge H^1(Q,\mathbb{F}_p)$, $c \neq 0$. As $P/\overline{P^p}$ is elementary abelian and $Q_c^p = 1$, there exist no surjective group homomorphisms $\tau : P \to Q_c$. Thus $\gamma_P(c_P) \neq 0$ for all $c \neq 0$ and γ_P is injective.

Let γ_P be injective and assume that P is not powerful. Then $\overline{P^p} \lneq \Phi(P)$. Hence there exists $N \triangleleft_o P$ such that $\overline{P^p} \leqslant N \lneq \Phi(P)$ and $|\Phi(P)/N| = p$. Then $P^* := P/N$ is a non-abelian p-group of exponent p and $P^* \simeq Q_d$ for some $d \in H^2(Q,\mathbb{F}_p)$, $d \neq 0$. As P^* is of exponent p, $d \in im(\gamma_Q)$ (Fact 5.1.6.1). Let $d = \gamma_Q(c)$, $c \in H^1(Q,\mathbb{F}_p) \wedge H^1(Q,\mathbb{F}_p)$. In particular, $c \neq 0$. Let $c_P := \pi_1^* \wedge \pi_1^*(c)$. By hypothesis, one has $\gamma_P(c_P) \neq 0$. However, from the previous remark (5.1.8) one concludes that $\gamma_P(c_P) = 0$, a contradiction, and the theorem is proved.

Profinite groups of finite cohomological p-dimension are necessarily p-torsionfree. Note that for a finitely generated powerful pro-p group, being uniformly powerful is equivalent to being torsionfree ([10, Chap. 4]). Hence one obtains the following straight forward consequence of Theorem 5.1.6.

Corollary 5.1.7. *Let p be odd and let G be a finitely generated pro-p group with*

$$H^\bullet(P, \mathbb{F}_p) \simeq \Lambda_\bullet((P^{el.ab.})^*). \tag{5.1.9}$$

Then P is uniformly powerful.

In order to prove Poincaré duality for p-torsion-free compact p-adic analytic groups we have to make use of the following result on the cohomological p-dimension due to Serre ([24]):

Theorem 5.1.8 (Serre). *Let G be a p-torsionfree profinite group and $H \leqslant_o G$ an open subgroup of G. Then $cd_p(H) = cd_p(G)$.*

From Proposition 4.4.1, Theorem 5.1.5 and Theorem 5.1.8 we conclude:

Theorem 5.1.9 (Lazard). *Let G be a compact p-adic analytic group. Then G is a virtual Poincaré duality group in dimension $n = dim(G)$ at the prime p, where $dim(G)$ denotes the dimension of G as p-adic analytic manifold.*

Note that Theorem 5.1.9 also implies that a compact p-adic analytic group G is a Poincaré duality group provided G is p-torsion-free.

For profinite groups which are virtually of type p-**FP** it is possible to define the p-Euler-Poincaré charcteristic $\chi_p(\_)$ as in the discrete case. If G is of type p-**FP** the p-Euler-Poincaré charcteristic is given by

$$\chi(G) = \sum_{i=0}^{cd_p(G)} (-1)^i dim_{\mathbb{F}_p}(H^i(G, \mathbb{F}_p)). \tag{5.1.10}$$

If G is virtually of type p-**FP** one defines $\chi(G) = \frac{1}{|G/U|} \cdot \chi(U)$, where $U \leqslant_o G$ is an open subgroup of G of finite cohomological p-dimension. It turns out that this definition is independent of the choice of the subgroup $U \leqslant_o G$ ([23, §I.4.1]).

Corollary 5.1.10. *Let G be an infinite compact p-adic analytic group. Then one has*

$$\chi_p(G) = 0. \tag{5.1.11}$$

Remark 5.1.11. The Demuškin pro-p groups furnish another class of Poincaré duality groups ([23, Chap. I, §4.5, Ex. 2]). Their p-Euler-Poincaré characteristic can differ from 0. However, if G is a Demuškin pro-p group with $\chi_p(G) = 0$, then G is p-adic analytic.

Lazard's proof of Theorem 5.1.5 ([17, Chap. V, §2]) can be seen as a constructive version of a more general principle which was introduced by May ([19], [20]) and which is known as the *May spectral sequence*. The advantage of this technique is that it yields approximations for the cohomology and homology of a finitely generated pro-p group with not necessarily trivial coefficients.

Let G be a finitely generated pro-p group and let

$$\mathbb{F}_p[\![G]\!] = \varprojlim_{U \lhd_o G} \mathbb{F}_p[G/U], \tag{5.1.12}$$

be the compact $\mathbb{F}_p$-group algebra. It has an obvious augmentation $\varepsilon\colon \mathbb{F}_p[\![G]\!] \to \mathbb{F}_p$. We set $I(G) := ker(\varepsilon)$. We denote by $gr_\bullet(G)$ the restricted $\mathbb{F}_p$-Lie algebra associated to the Jennings filtration of G and which we assume to be graded by negative[13] integers in the obvious way. Let $\Delta_\bullet(G) = \mathfrak{U}^{[p]}(gr_\bullet(G))$ denote its $-\mathbb{N}_0$-graded restricted universal enveloping algebra which coincides with the $-\mathbb{N}_0$-graded $\mathbb{F}_p$-group algebra associated to the augmentation filtration of $\mathbb{F}_p[\![G]\!]$ by the Jennings–Quillen theorem ([10, Chap. 12]).

For $M \in ob(\mathfrak{C}^0_{\mathbb{F}_p[\![G]\!]})$ there is an increasing filtration given by

$$F_{-k}M = I(\mathbb{F}_p[\![G]\!])^k.M \tag{5.1.13}$$

for $k \in \mathbb{N}_0$ and $F_kM = M$ for $k \in \mathbb{N}$. Since G is assumed to be finitely generated, it is obvious that $F_\bullet$ is exhaustive and separated. Let $gr_\bullet(M)$ denote the associated $\mathbb{Z}$-graded $\Delta_\bullet(G)$-module. Since we assumed M to be finitely generated as an $\mathbb{F}_p[\![G]\!]$-module, $gr_\bullet(M)$ is discrete as abelian group.

A discrete module $N \in ob(\mathfrak{D}_{\mathbb{F}_p[\![G]\!]})$ is filtered in the opposite way. There is an increasing filtration $F_\bullet$ given by $F_kN = 0$ for $k < 0$ and

$$F_kN = \{ n \in N \mid I(\mathbb{F}_p[\![G]\!])^k.n = 0 \} \tag{5.1.14}$$

Again, it is easy to check that $F_\bullet$ is exhaustive and separated. We denote by $gr_\bullet(N)$ the associated $\mathbb{Z}$-graded left $\Delta_\bullet(G)$-module.

Theorem 5.1.12 (May spectral sequence). *Let G be a finitely generated pro-p group.*

(1) Let $M \in ob(\mathfrak{C}^0_{\mathbb{F}_p[\![G]\!]})$. Then there exists a second quadrant spectral sequence $(E^r_{\bullet,\bullet}, d^r)$, $d^r : E^r_{s,t} \to E^r_{s-r,t+r-1}$ of homological type with

$$E^1_{s,t} = \mathrm{Tor}^{\Delta_\bullet(G)}_{s,t}(\mathbb{F}_p, gr_\bullet(M)), \tag{5.1.15}$$

(where s denotes the filtration degree and t denotes the complementary[14] degree) converging to

$$E^\infty_{s,t} = F_sH_{s+t}(G, M)/F_{s-1}H_{s+t}(G, M). \tag{5.1.16}$$

For $M = \mathbb{F}_p$ the spectral sequence converges as a spectral sequence of $\mathbb{F}_p$-coalgebras. Moreover, the induced filtration $F_\bullet$ on $H_\bullet(G, M)$ is exhaustive and separated.

(2) Let $N \in ob(\mathfrak{D}_{\mathbb{F}_p[\![G]\!]})$. Then there exists a second quadrant spectral sequence $(E_r^{\bullet,\bullet}, d_r)$, $d_r : E_r^{s,t} \to E_r^{s+r,t-r+1}$ of cohomological type with

$$E_1^{s,t} = \mathrm{Ext}^{s,t}_{\Delta_\bullet(G)}(\mathbb{F}_p, gr_\bullet(N)), \tag{5.1.17}$$

[13] This is equivalent to considering the Jennings filtration as an increasing filtration.

[14] This means that $s + t$ equals the homological degree.

(where s denotes the filtration degree and t the complentary degree) converging to

$$E_{\infty}^{s,t} = F^s H^{s+t}(G,N)/F^{s+1}H^{s+t}(G,N). \tag{5.1.18}$$

For $N = \mathbb{F}_p$ the spectral sequence converges as a spectral sequence of $\mathbb{F}_p$-algebras. Moreover, the induced filtration $F^\bullet$ on $H^\bullet(G,N)$ is exhaustive and separated.

Proof. For finite p-groups this theorem is well-known and was discussed in great detail in [9]. In order to handle the convergency problems we have to recall and extend the proof given in [9].

(1) Let $(\hat{\mathbb{B}}_\bullet, d_\bullet)$ denote the topological unreduced bar resolution of the trivial right $\mathbb{F}_p[\![G]\!]$-module $\mathbb{F}_p$, i.e.,

$$\hat{\mathbb{B}}_k = \overbrace{\mathbb{F}_p[\![G]\!]\widehat{\otimes}_{\mathbb{F}_p}\cdots\widehat{\otimes}_{\mathbb{F}_p}\mathbb{F}_p[\![G]\!]}^{k\text{-times}}\widehat{\otimes}_{\mathbb{F}_p}\mathbb{F}_p[\![G]\!] \tag{5.1.19}$$

and d_k is given by the formula

$$\begin{aligned} d_k(\langle x_1|\cdots|x_k\rangle x) = {} & \langle x_2|\cdots|x_k\rangle.x \\ & + \sum_{1\leqslant j\leqslant k-1}(-1)^i\langle x_1|\cdots|x_jx_{j+1}|\cdots|x_k\rangle.x \\ & + (-1)^k\langle x_1|\cdots|x_{k-1}\rangle x_kx. \end{aligned} \tag{5.1.20}$$

It is well known that $(\hat{\mathbb{B}}_\bullet, d_\bullet)$ is a projective resolution of $\mathbb{F}_p$ as right $\mathbb{F}_p[\![G]\!]$-module ([6, Cor. 3.3]). Let $(C_\bullet, \partial_\bullet) = (\hat{\mathbb{B}}_\bullet\widehat{\otimes}_{\mathbb{F}_p[\![G]\!]}M, d_\bullet\widehat{\otimes}M)$, and let $F_\bullet$ denote the canonical increasing filtration on $C_\bullet$, i.e.,

$$F_sC_k = \sum_{s=s_1+\cdots+s_k+s_{k+1}} F_{s_1}\mathbb{F}_p[\![G]\!]\widehat{\otimes}_{\mathbb{F}_p}\cdots\widehat{\otimes}_{\mathbb{F}_p}F_{s_k}\mathbb{F}_p[\![G]\!]\widehat{\otimes}_{\mathbb{F}_p[\![G]\!]}F_{s_{k+1}}M, \tag{5.1.21}$$

where $F_\bullet\mathbb{F}_p[\![G]\!]$ is given by (5.1.13). Then $F_\bullet$ is an exhaustive and separated filtration on C_k for all $k \in \mathbb{N}_0$. Moreover, the topology induced by the filtration $F_\bullet$ coincides with the natural topology on C_k. From (5.1.20) it follows that ∂_k is continuous. Hence $F_\bullet$ is convergent (§2.4). It is easy to see that the E^1-term of the associated spectral sequence coincides with (5.1.15). The final remark follows since the Alexander-Whitney map ([1, p. 59]) is a mapping of filtered chain complexes ((2.4.16)).

(2) Let $(\hat{\mathbb{B}}_\bullet, d_\bullet)$ denote the unreduced bar resolution of the trivial left $\mathbb{F}_p$-module and let $F_\bullet$ be the canonical increasing filtration on $\hat{\mathbb{B}}_\bullet$ (5.1.21). For $N \in ob(\mathfrak{D}_{\mathbb{F}_p[\![G]\!]})$, let $F^\bullet$ be the decreasing filtration on

$$(C^\bullet, \delta^\bullet) = (\mathbf{Hom}_{\mathbb{F}_p[\![G]\!]}(\hat{\mathbb{B}}_\bullet, N), d_\bullet^*) \tag{5.1.22}$$

given by

$$F^sC^k = \{\phi \in \mathbf{Hom}_{\mathbb{F}_p[\![G]\!]}(\hat{\mathbb{B}}_k, N) \mid \phi(F_j\hat{\mathbb{B}}_k) \leqslant F_{j-s}N \text{ for all } j \in \mathbb{Z}\}. \tag{5.1.23}$$

Then $F^1C^k = 0$. A morphism $\psi \in \mathbf{Hom}_{\mathbb{F}_p[\![G]\!]}(\mathbb{B}_k, N)$ must have finite image. This implies that $\psi \in \bigcup_{s\in -\mathbb{N}_0} F^sC^k$ and that the filtration $F^\bullet$ is exhaustive and separated on $C^\bullet$. For the spectral sequence associated to the filtered cochain complex $(C^\bullet, \delta^\bullet, F^\bullet)$ one obtains

$$\begin{aligned} E_0^{s,t} &= \mathrm{Hom}^s_{\Delta_\bullet(G)}(gr_\bullet(\mathbb{B}_{s+t}), gr_\bullet(N)) \\ E_1^{s,t} &= \mathrm{Ext}^{s,t}_{\Delta_\bullet(G)}(\mathbb{F}_p, gr_\bullet(N)) \end{aligned} \tag{5.1.24}$$

([9, §2.3]). The topology induced by the filtration coincides with the discrete topology. So $F^\bullet$ is convergent. The final remark can be proved as in (1).

Unfortunately there is nothing known about the higher differentials in these spectral sequences. Nevertheless, Theorem 5.1.12(2) can be used to prove Lazard's theorem as follows.

Proof of Theorem 5.1.5. Let P be a uniformly powerful group and $P^{el.ab.} = \mathbb{F}_p.x_1 \oplus \cdots \oplus \mathbb{F}_p.x_n$. Then $\Delta_\bullet(P) = S_\bullet[x_1, \ldots, x_n]$ is a polynomial $\mathbb{F}_p$-algebra generated in degree -1. Using the Koszul resolution (Rem. 2.5.2]) one concludes that

$$\mathrm{Ext}^{s,t}_{\Delta_\bullet(G)}(\mathbb{F}_p, \mathbb{F}_p) = \Lambda_{\bullet,\bullet}[y_1, \ldots, y_n], \tag{5.1.25}$$

where each generator y_i is of bidegree $(-1, 2)$. Since $(E_r^{\bullet,\bullet}, d_r)$ is a spectral sequence of differential graded algebras ([9, Lemma 14]), one concludes that $E_\infty^{\bullet,\bullet} \simeq E_1^{\bullet,\bullet}$ as bigraded $\mathbb{F}_p$-algebras. From the filtration on $H^n(G, \mathbb{F}_p)$, it is obvious that $H^\bullet(G, \mathbb{F}_p)$ is isomorphic to $\mathbf{tot}_\bullet(\Lambda_{\bullet,\bullet}[y_1, \ldots, y_n])$ as an $\mathbb{N}_0$-graded $\mathbb{F}_p$-algebra. The final claim follows from Serre's Poincaré duality criterion (Thm. 4.5.6).

5.2 Rational cohomology

In this subsection we consider the rational cohomology of a p-adic analytic group G with coefficients in $\mathfrak{fd}_p(G)$. For simplicity let $\mathbf{p} = p$ if p is odd and $\mathbf{p} = 4$ if $p = 2$.

In [17] Lazard defined several classes of finitely generated pro-p groups by the existence of a certain filtration on the group and its properties, e.g., the class of p-valuable groups and p-saturable groups. We will restrict our considerations to the class of uniformly powerful groups (=u.p.f. groups) and denote the associated category by $\mathfrak{P}f^{t.f.}$. It is well known that every finitely generated uniformly powerful group is p-saturable and therefore p-valuable. As corresponding p-filtration ([17, Chap. III, (2.1)]) one can take the lower p-central series (suitably re-indexed for $p = 2$). For $P \in ob(\mathfrak{P}f^{t.f.})$ we put

$$\begin{aligned} F^kP &= P^{p^{k-1}}, \qquad p \text{ odd} \\ F^{k+1}P &= P^{p^{k-1}}, \qquad p = 2. \end{aligned} \tag{5.2.1}$$

We define the associated valuation $\omega_P \colon P \to \mathbb{N}_0 \cup \{\infty\}$ by

$$\omega_P(g) = \max\{\, k \in \mathbb{N}_0 \cup \{\infty\} \mid g \in F^kP \,\}. \tag{5.2.2}$$

The associated graded $\mathbb{F}_p$-module will be denoted by $gr^\bullet(P)$. Note that $gr^\bullet(P)$ is an $\mathbb{F}_p[t]$-Lie algebra, $deg(t) = 1$, with operations given by

$$= (g,h).F^{k+l+1}P, \qquad t.(g.F^{k+1}P) = g^p, \tag{5.2.3}$$

where $g \in F^kP$, $h \in F^lP$ and $(g,h) = g^{-1}h^{-1}gh$ denotes the group commutator. As P is a finitely generated uniformly powerful group, $gr^\bullet(P)$ is a finitely generated free $\mathbb{F}_p[t]$-module. The compact group algebra $\mathbb{Z}_p[\![P]\!]$ has a canonical (decreasing) filtration given by

$$\begin{aligned} F^k\mathbb{Z}_p[\![P]\!] &= J^k, \ J = p.\mathbb{Z}_p.1 + I(\mathbb{Z}_p[\![P]\!]), && p \text{ odd} \\ F^k\mathbb{Z}_p[\![P]\!] &= 2^k.\mathbb{Z}_p + 2^{k-2}I(\mathbb{Z}_p[\![P]\!]) \\ &\quad +2^{k-4}I(\mathbb{Z}_p[\![P]\!])^2 + \cdots + I(\mathbb{Z}_p[\![P]\!])^{\lceil \frac{k}{2}\rceil} && p = 2, \end{aligned} \tag{5.2.4}$$

where $I(\,)$ denotes the augmentation ideal and $\lceil x \rceil$ stands for the smallest integer larger that x. Again we set

$$\omega(x) = \max\{\, k \in \mathbb{N}_0 \cup \{\infty\} \mid x \in F^k\mathbb{Z}_p[\![P]\!]\,\}. \tag{5.2.5}$$

By $gr^\bullet(\mathbb{Z}_p[\![P]\!])$ we denote the associated graded $\mathbb{F}_p$-module, which is canonically and $\mathbb{F}_p[t]$-algebra, where $t = p + F^2\mathbb{Z}_p[\![P]\!]$ is of degree 1.

Theorem 5.2.1. *Let $P \in ob(\mathfrak{P}f^{t.f.})$ and let $F^\bullet$ be given as in (5.2.1), (5.2.4). Then:*

(1) There is a canonical isomorphism

$$\mathfrak{U}_{\mathbb{F}_p[t]}(gr^\bullet(P)) \simeq gr^\bullet(\mathbb{Z}_p[\![P]\!]), \tag{5.2.6}$$

$$\begin{aligned} \omega(1) &= 0, \\ \omega(x+y) &\geqslant \min\{\omega(x), \omega(y)\}, \\ \omega(x \cdot y) &\geqslant \omega(x) + \omega(y), \\ \omega(\lambda \cdot x) &= v(\lambda) + \omega(x) \end{aligned} \tag{5.2.7}$$

(3) For $g \in P$ one has

$$\omega_P(g) = \omega(g-1). \tag{5.2.8}$$

In particular,

$$P \cap (1 + F^k\mathbb{Z}_p[\![P]\!]) = F^kP. \tag{5.2.9}$$

Proof. The theorem is just a special case of [17, Thm. III.2.3.3]. Let $i\colon P \to \mathbb{Z}_p[\![P]\!]$ denote the canonical mapping given by $i(g) = g - 1$. For $g - 1 \in F^k\mathbb{Z}_p[\![P]\!]$ ($k \geqslant 2$

for $p = 2$) one has $g^p - 1 \equiv p.(g-1) \mod F^{k+2}\mathbb{Z}_p[\![P]\!]$. Hence one proves easily by induction that $\omega_P(g) \leqslant \omega(g-1)$ for all $g \in P$. This implies that ω coincides with the canonical filtration ω on $\mathbb{Z}_p[\![P]\!]$ induced by ω_P ([17, Def. III.2.3.1.2]). Hence the induced morphism

$$gr^\bullet(i)\colon gr^\bullet(P) \longrightarrow gr^\bullet(\mathbb{Z}_p[\![P]\!]) \tag{5.2.10}$$

is an injective morphism of $\mathbb{F}_p[t]$-Lie algebras and

$$gr^\bullet(i)^*\colon \mathfrak{U}_{\mathbb{F}_p[t]}(gr^\bullet(P)) \longrightarrow gr^\bullet(\mathbb{Z}_p[\![P]\!]) \tag{5.2.11}$$

is an isomorphism of associative $\mathbb{F}_p[t]$-algebras ([17, Thm. III.2.3.3]). This proves (1). Since $gr^\bullet(\mathbb{Z}_p[\![P]\!])$ is a free $\mathbb{F}_p[t]$-module, the filtration is actually a valuation and (5.2.7) is satisfied. The filtration is separated and all ideals $F^k\mathbb{Z}_p[\![P]\!]$ are open and of finite index in $\mathbb{Z}_p[\![P]\!]$. Since $\mathbb{Z}_p[\![P]\!]$ is compact with respect to the topology induced by the filtration $F^\bullet$, this topology has to coincide with the natural topology. This shows (2). (3) is a direct conseqeunce of (1).

The valuation ω can be extended to $\mathbb{Q}_p \otimes_{\mathbb{Z}_p} \mathbb{Z}_p[\![P]\!]$ by the assignment

$$\omega(q \otimes x) = v(q) + \omega(x), \tag{5.2.12}$$

where $v\colon \mathbb{Q}_p \to \mathbb{Z} \cup \{\infty\}$ denotes the p-adic valuation with $v(p) = 1$. Obviously, all statements of (5.2.7) are also satisfied for $(\mathbb{Q}_p \otimes_{\mathbb{Z}_p} \mathbb{Z}_p[\![P]\!], \omega)$. We denote by $\hat{A}[\![P]\!]$ the completion of $\mathbb{Q}_p \otimes_{\mathbb{Z}_p} \mathbb{Z}_p[\![P]\!]$ and denote by ω again also the extension of ω to $\hat{A}[\![P]\!]$. The associated filtration on $\hat{A}[\![P]\!]$ is given by

$$F^k\hat{A}[\![P]\!] = \{\, x \in \hat{A}[\![P]\!] \mid \omega(x) \geqslant k \,\}. \tag{5.2.13}$$

We call $\hat{A}[\![P]\!]$ the *completed group algebra*[15] of P. From the construction it is clear that every morphism $\phi\colon P \to Q$ in $\mathfrak{P}f^{t.f.}$ extends uniquely to a continuous morphism

$$\phi^*\colon \hat{A}[\![P]\!] \longrightarrow \hat{A}[\![Q]\!]. \tag{5.2.14}$$

So $\hat{A}[\![\_]\!]$ can be considered as a covariant functor from $\mathfrak{P}f^{t.f.}$ to the category of associative $\mathbb{Q}_p$-algebras (with valuation). The $\mathbb{Z}_p$-Lie algebra of P is defined by

$$\mathrm{Log}(P) = \{\, log(g) \in \hat{A}[\![P]\!] \mid g \in P \,\}, \tag{5.2.15}$$

where the bracket is given by $[x, y] = x.y - y.x$, and log is given by the power series $log(x) = \sum_{k\in\mathbb{N}} \frac{-1^k}{k}.(x-1)^k$. Note that for $g \in F^kP$,

$$log(g) \equiv (g-1) \mod F^k\hat{A}[\![P]\!]. \tag{5.2.16}$$

The filtration $F^\bullet$ on $\hat{A}[\![P]\!]$ induces a filtration on $\mathrm{Log}(P)$, which we also denote by $F^\bullet$.

---

[15] This definition coincides with the definition in [10, §7] only for p odd. For $p = 2$ see [10, Exer. 7.10].

Proposition 5.2.2.

(1) *Let* $\mathrm{Lie}_{\mathbb{Z}_p}^{p.f.}$ *denote the category of* $\mathbb{Z}_p$*-Lie algebras whose objects are* $\mathbb{Z}_p$*-Lie algebras* $\mathfrak{L}$ *which are finitely generated free* $\mathbb{Z}_p$*-modules satisfying* $[\mathfrak{L}, \mathfrak{L}] \leqslant \mathbf{p}.\mathfrak{L}$. *Then* $\mathrm{Log}\colon \mathfrak{P}f^{t.f.} \to \mathrm{Lie}_{\mathbb{Z}_p}^{p.f.}$ *is a covariant (exact) functor which induces an equivalence of categories.*

(2) *For all* $P \in ob(\mathfrak{P}f^{t.f.})$ *there exists a canonical isomorphism*

$$gr^{\bullet}(P) \simeq gr^{\bullet}(\mathrm{Log}(P)) \tag{5.2.17}$$

of graded $\mathbb{F}_p[t]$*-Lie algebras.*

(3) *The associative* $\mathbb{Q}_p$*-algebra generated by* $\mathrm{Log}(P) \leqslant \hat{A}[\![P]\!]$ *is isomorphic to* $\mathbb{Q}_p \otimes_{\mathbb{Z}_p} \mathfrak{A}_{\mathbb{Z}_p}(\mathrm{Log}(P))$ *and it is dense in* $\hat{A}[\![P]\!]$.

Proof. (1) The fact that Log(_) is a covariant functor follows directly from the construction. The fact that it induces an equivalence of categories follows from the existence of the functor $Exp\colon \mathrm{Lie}_{\mathbb{Z}_p}^{p.f.} \to \mathfrak{P}f^{t.f.}$ ([10, Thm. 9.10], [17, Chap. IV]). (2) follows from (5.2.16) and the fact that $gr^{\bullet}(i)$ (5.2.10) is an injective homomorphism of graded $\mathbb{F}_p[t]$-Lie algebras. (3) follows from (2) and Theorem 5.2.1(1).

Let G be a compact p-adic analytic group and $P \lhd_o G$ an open normal uniformly powerful subgroup of G. Then P has a continuous right G-action by conjugation, $h^g = g^{-1}hg$, $g \in G$, $h \in P$. This implies that there are also a canonical right G-actions on $\hat{A}[\![P]\!]$ and $\mathrm{Log}(P)$ where G acts by $\mathbb{Q}_p$-algebra and $\mathbb{Z}_p$-Lie algebra homomorphisms respectively.

If G is a compact p-adic analytic group then the $\mathbb{Q}_p$-Lie algebra $L(G)$ associated to G is defined by

$$L(G) = \varinjlim_{\substack{U \lhd_o G \\ U \text{ u.p.f.}}} \mathbb{Q}_p \otimes \mathrm{Log}(U). \tag{5.2.18}$$

All the morphisms are given by inclusion and are therefore compatible with the right G-action, so $L(G)$ has a canonical right G-action which we will denote by

$$Ad\colon G \longrightarrow Aut(L(G)). \tag{5.2.19}$$

It is easily seen that $L(G) \simeq \mathbb{Q}_p \otimes \mathrm{Log}(U)$ for every open normal uniformly powerful subgroup $U \lhd_o G$ ([17, Chap. V, (2.4.2.4)]) and that the isomorphism is also an isomorphism of right G-modules.

It is easy to check that

$$Pf(W) = \{ id_W + \alpha \mid \alpha \in \mathbf{p}.\mathrm{End}_{\mathbb{Z}_p}(W) \} \subset \mathrm{End}_{\mathbb{Z}_p}(W) \tag{5.2.20}$$

is a finitely generated uniformly powerful pro-p group, and that

$$\mathfrak{pf}(W) = \mathbf{p}.\mathrm{End}_{\mathbb{Z}_p}(W) \subset \mathrm{End}_{\mathbb{Z}_p}(W) \tag{5.2.21}$$

is a powerful $\mathbb{Z}_p$-Lie lattice, i.e., $\mathfrak{pf}(W) \in ob(\mathrm{Lie}^{u.p.f}_{\mathbb{Z}_p})$. We also set

$$Al(W) = \mathbb{Z}_p.id_W + \mathbf{p}.\mathrm{End}_{\mathbb{Z}_p}(W) \subset \mathrm{End}_{\mathbb{Z}_p}(W) \tag{5.2.22}$$

which is a compact $\mathbb{Z}_p$-algebra. Using $log\colon Pf(W) \to \mathfrak{pf}(W)$ and $exp\colon \mathfrak{pf}(W) \to Pf(W)$ given by $exp(x) = \sum_{k\in\mathbb{N}_0} \frac{1}{k!}x^k$, we deduce that

$$\mathrm{Log}(Pf(W)) = \mathfrak{pf}(W). \tag{5.2.23}$$

Let $P \in ob(\mathfrak{P}f^{t.f.})$ and let $M \in ob(\mathfrak{C}_p(P))$ be a finitely generated free $\mathbb{Z}_p$-module. Let $\phi_M\colon P \to GL_{\mathbb{Z}_p}(M)$ denote the associated representation. We say that the representation ϕ_M is *powerful*, if $im(\phi_M) \leqslant Pf(M)$. A powerful representation ϕ_M induces a representation of $\mathbb{Z}_p$-Lie algebras

$$\mathrm{Log}(\phi_M)\colon \mathrm{Log}(P) \longrightarrow \mathfrak{pf}(M) \tag{5.2.24}$$

and $\mathbb{Q}_p$-algebras

$$\hat{A}(\phi_M)\colon \hat{A}[\![P]\!] \longrightarrow \mathrm{End}_{\mathbb{Q}_p}(\mathbb{Q}_p \otimes_{\mathbb{Z}_p} M). \tag{5.2.25}$$

Let G be a compact p-adic analytic group, $V \in ob(\mathfrak{fd}_p(G))$ and let $\phi\colon G \to GL_{\mathbb{Q}_p}(V)$ denote the associated representation. Then V contains a $\phi(G)$-stable open $\mathbb{Z}_p$-lattice M (Prop. 3.8.1(2)), and $H = \phi^{-1}(Pf(M)) \lhd_o G$ is an open normal pro-p-subgroup of G. Let $P \lhd_o H$ be a uniformly powerful characteristic subgroup of H (Thm. 5.1.1). By construction $P \lhd_o G$ is a uniformly powerful open normal subgroup of G and, $\phi|_P\colon P \to Pf(M)$ is a powerful representation. Let

$$L_{M,P}(\phi) = \mathbb{Q}_p \otimes \mathrm{Log}(\phi|_P)\colon L(G) \to \mathbb{Q}_p \otimes_{\mathbb{Z}_p} \mathrm{End}_{\mathbb{Z}_p}(M) \to \mathrm{End}_{\mathbb{Q}_p}(V) \tag{5.2.26}$$

denote the associated representation for the $\mathbb{Q}_p$-Lie algebra $L(G)$. It is independent of the chosen uniformly powerful characteristic subgroup $P \lhd_o H$ (5.2.18), and of the chosen $\phi(G)$-stable $\mathbb{Z}_p$-lattice M. So we have proved ([17, Thm. V.2.4.9]):

Proposition 5.2.3. *Let G be a compact p-adic analytic group. Then every module $V \in ob(\mathfrak{fd}_p(G))$ is canonically a left $L(G)$-module. In particular, there is an exact functor*

$$L(\_)\colon \mathfrak{fd}_p(G) \longrightarrow {}_{L(G)}\mathrm{TMod}^G. \tag{5.2.27}$$

Proof. It suffices to show that the action of $L(G)$ on $V \in ob(\mathfrak{fd}_p(G))$ is G-equivariant.

Let $M \leqslant V$ be a G-stable $\mathbb{Z}_p$-lattice, $\phi\colon G \to \mathrm{End}_{\mathbb{Q}_p}(V)$ the associated representation and let $P \triangleleft_o G$ be uniformly powerful such that $\phi(P) \leqslant Pf(M)$. The action of P on $M \in ob(\mathfrak{C}_p(G))$ is G-equivariant (§2.3). Therefore the action of $\hat{A}[\![P]\!]$ on $V = \mathbb{Q}_p \otimes M$ is G-equivariant and thus also the action of $\mathrm{Log}(P)$ on M. This proves the claim.

For a finitely generated uniformly powerful pro-p group P we call $V \in ob(\mathfrak{fd}_p(P))$ *powerful* if it contains a P-stable open $\mathbb{Z}_p$-lattice M such that the associated representation $\phi_M\colon P \to GL_{\mathbb{Z}_p}(M)$ is powerful. The main goal of this subsection is to prove the following theorem which can be seen as an extended version of Theorem V. 2.4.10 in [17].

Theorem 5.2.4. (a) *Let P be a finitely generated uniformly powerful group and let $V \in ob(\mathfrak{fd}_p(P))$ be a powerful $\mathbb{Z}_p[\![P]\!]$-module. Then there is a canonical isomorphism*

$$\mathbb{H}^\bullet(P, V) \simeq H^\bullet(L(P), L(V)). \tag{5.2.28}$$

(b) *Let G be a compact p-adic analytic group and $W \in ob(\mathfrak{fd}_p(G))$. Then there is a canonical isomorphism*

$$\mathbb{H}^\bullet(G, W) \simeq H^\bullet(L(G), L(W))^G. \tag{5.2.29}$$

If $P \triangleleft_o G$ is an open normal uniformly powerful group for which the restriction of W to P is a powerful representation, then P acts trivially on $H^\bullet(L(G), L(W))$.

Proof.(a) Let $F^\bullet$ be the filtration on $\mathbb{Z}_p[\![P]\!]$ given by (5.2.4). The $\mathbb{F}_p[t]$-algebra $gr^\bullet(\mathbb{Z}_p[\![P]\!])$ is isomorphic to the universal enveloping algebra of $gr^\bullet(P)$. Since $gr^\bullet(P)$ is a finitely generated free $\mathbb{F}_p[t]$-module, $gr^\bullet(\mathbb{Z}_p[\![P]\!])$ has a finite special projective resolution $(\bar{P}_\bullet, \bar{d}_\bullet, \bar{\varepsilon}_\bullet, \bar{s}_\bullet)$ with $\bar{d}_\bullet$ a morphism of graded $gr^\bullet(\mathbb{Z}_p[\![P]\!])$-modules and $\bar{\varepsilon}_\bullet$, $\bar{s}_\bullet$ morphisms of graded $\mathbb{F}_p[t]$-modules, and all the $\bar{P}_k$ are free $gr^\bullet(\mathbb{Z}_p[\![P]\!])$-modules of finite rank (Rem. 2.5.2).

By Serre's Lemma (Lemma 2.5.1), this implies that $\mathbb{Z}_p[\![P]\!]$ has a finite special projective resolution $(P_\bullet, d_\bullet, \varepsilon_\bullet, s_\bullet)$, and all morphisms are morphism of filtered $\mathbb{Z}_p[\![P]\!]$-modules or $\mathbb{Z}_p$-modules, as appropriate.

All the filtered modules $(P_k, F^\bullet)$ are finitely generated filtered free, and the underlying filtration is a valuation (Lemma 2.5.1(iii)). Let $P_k^* = \mathbb{Q}_p \otimes P_k$ and $d_\bullet^*$, $\varepsilon_\bullet^*$, $s_\bullet^*$ denote the canonical extensions of the mappings $d_\bullet$, $\varepsilon_\bullet$, $s_\bullet$ to $P_\bullet^*$, and let $F^\bullet$ denote the extension of the filtration as in (5.2.12). Let Q_k denote the completion of P_k^* with respect to the filtration $F^\bullet$ and let $d_\bullet^*$, $\varepsilon_\bullet^*$, $s_\bullet^*$ denote the extension of the mappings to $Q_\bullet$. By construction $(Q_\bullet, d_\bullet^*, \varepsilon_\bullet^*, s_\bullet^*)$ is a finite special projective resolution of the $\mathbb{Q}_p$-algebra $\hat{A}[\![P]\!]$ (2.4.3), and all Q_k are free $\hat{A}[\![P]\!]$-modules of finite rank.

Note that $\mathrm{Log}(P) \leqslant \hat{A}[\![P]\!]$ is a filtered $\mathbb{Z}_p$-Lie algebra and the filtration (induced by the filtration on $\hat{A}[\![P]\!]$) coincides with the p-adic filtration (suitably re-indexed for $p = 2$). Hence $(\mathbf{Ko}_\bullet(\mathrm{Log}(P)), d_\bullet, \varepsilon_\bullet, s_\bullet)$ is a special projective resolution and all the mappings are mappings of filtered $\mathfrak{U}_{\mathbb{Z}_p}(\mathrm{Log}(P))$-modules or $\mathbb{Z}_p$-modules (Rem. 2.5.2). Since all filtrations are actually valuations (5.2.7), one can construct a special projective resolution $(Q_\bullet^\circ, d_\bullet^\circ, \varepsilon_\bullet^\circ, s_\bullet^\circ)$ of the $\mathbb{Q}_p$-algebra $\hat{A}[\![P]\!]$ as in the previous case

(Prop. 5.2.2(3)). Note that Q_k° coincides with the completion of $\mathbb{Q}_p \otimes_{\mathbb{Z}_p} \mathbf{Ko}_k(\mathrm{Log}(P))$, and hence is a free $\hat{A}[\![P]\!]$-module of finite rank.

Since P_k is a finitely generated free $\mathbb{Z}_p[\![P]\!]$-module, $(P_\bullet, d_\bullet, \varepsilon_\bullet, s_\bullet)$ is a special projective resolution in $\mathfrak{C}_p(P)$ and also ${}_{\mathbb{Z}_p[\![P]\!]}\mathrm{Mod}$. As $\mathbb{H}\mathrm{om}_{\mathbb{Z}_p[\![P]\!]}(P_k, M) = \mathrm{Hom}_{\mathbb{Z}_p[\![P]\!]}(P_k, M)$ for all $M \in ob(\mathfrak{fd}_p(P))$, there is a canonical isomorphism

$$\gamma\colon \mathbb{H}^\bullet(P, V) \longrightarrow H^\bullet(\mathbb{Z}_p[\![P]\!], V), \tag{5.2.30}$$

where $\mathbb{H}^\bullet(P, \_)$ stands for rational cohomology ((3.8.5.)), and $H^\bullet(\mathbb{Z}_p[\![P]\!], \_)$ coincides with the usual cohomology functor for an augmented $\mathbb{Z}_p$-algebra (2.3.7).

Since $\mathbb{Z}_p[\![P]\!] \leqslant \hat{A}[\![P]\!]$ is a $\mathbb{Z}_p$-subalgebra, there is a canonical restriction mapping

$$\begin{aligned}\mathrm{res}^\bullet_{gr}\colon H^\bullet(\hat{A}[\![P]\!], V) &= \mathrm{Ext}^\bullet_{\hat{A}[\![P]\!]}(\mathbb{Q}_p, V)\\ &\longrightarrow \mathrm{Ext}^\bullet_{\mathbb{Z}_p[\![P]\!]}(\mathbb{Z}_p, V) = H^\bullet(\mathbb{Z}_p[\![P]\!], V)\end{aligned} \tag{5.2.31}$$

defined for all $V \in ob({}_{\hat{A}[\![P]\!]}\mathrm{Mod})$. If $V \in ob(\mathfrak{fd}_p(P))$ is powerful, every morphism $\phi \in \mathrm{Hom}_{\mathbb{Z}_p[\![P]\!]}(P_k, V)$ is continuous (with respect to the filtration $F^\bullet$ on P_k), and so can be lifted to a morphism $\phi^* \in \mathrm{Hom}_{\hat{A}[\![P]\!]}(Q_k, V)$. This shows that $\mathrm{res}_{gr}\colon \mathrm{Hom}_{\hat{A}[\![P]\!]}(Q_k, V) \to \mathrm{Hom}_{\mathbb{Z}_p[\![P]\!]}(P_k, V)$ is an isomorphism, and so $\mathrm{res}^\bullet_{gr}$ is an isomorphism.

Similarly, we can construct the morphism

$$\mathrm{res}^\bullet_{lie}\colon H^\bullet(\hat{A}[\![P]\!], V) \longrightarrow H^\bullet(L(P), V) \tag{5.2.32}$$

for all $V \in ob({}_{\hat{A}[\![P]\!]}\mathrm{Mod})$. Basically, the same argument as in the previous case shows that for a powerful $\mathbb{Z}_p[\![P]\!]$-module $V \in ob(\mathfrak{fd}_p(P))$, $\mathrm{res}^\bullet_{lie}$ is an isomorphism. We have produced the following canonical isomorphisms

$$\mathbb{H}^\bullet(P, V) \xrightarrow{\gamma^\bullet} H^\bullet(\mathbb{Z}_p[\![P]\!], V) \xleftarrow{\mathrm{res}^\bullet_{gr}} H^\bullet(\hat{A}[\![P]\!], V) \xrightarrow{\mathrm{res}^\bullet_{lie}} H^\bullet(L(P), V). \tag{5.2.33}$$

This proves the claim.
(b) All morphisms in (5.2.33) are G-equivariant, so the claim follows from Theorem 4.2.6(b).

This theorem has some immediate and important consequences:

Corollary 5.2.5. *Let G be a p-torsion-free compact p-adic analytic group, let $D_p(G) \in ob(\mathfrak{C}_p(G))$ be its compact dualizing module and let $\phi_{D_p(G)}\colon G \to \mathbb{Z}_p^*$ denote the associated representation of G. Then*

$$\phi_{D_p(G)}(g) = det(Ad(g)), \tag{5.2.34}$$

where $Ad\colon G \to Aut(L(G))$ is defined in (5.2.19).

Proof. Let $dim(G) = n$. By considering the Koszul complex we conclude that for $W \in ob({}_{\mathfrak{U}(L(G))}\mathrm{Mod})$,

$$H^n(L(G), W) = Hom_{L(G)}(\Lambda_n(L(G)), W). \tag{5.2.35}$$

Note that $x \in L(G)$ acts (from the left) on $\Lambda_n(L(G))$ by multiplication with $tr(ad(x))$, where $ad(x)(y) = [x, y]$, $x, y \in L(G)$. Moreover, $tr(ad(x^g)) = tr(ad(x))$ for all $g \in G$. The natural (right) G-action on $\Lambda_n(L(G))$ is given by multiplication with $det(Ad(g))$, $g \in G$. The Koszul resolution is a G-equivariant resolution of $\mathfrak{U}(L(G))$, so for $W \in ob({}_{L(G)}\mathrm{TMod}^G)$, (5.2.35) is a G-morphism. For $W = \Lambda_n(L(G))$, $id \in \mathrm{Hom}_{L(G)}(\Lambda_n(L(G)), \Lambda_n(L(G)))$ is a fixed point of the G-action, so $H^n(L(G), \Lambda_n(L(G)))^G = \mathbb{Q}_p$.

If $M \in ob(\mathfrak{C}_p(G))$ and $M \simeq \mathbb{Z}_p$ as a $\mathbb{Z}_p$-module, with the action of G given by multiplication with $det(Ad(g^{-1}))$, $g \in G$, it is easy to check that $L(\mathbb{Q}_p \otimes M) \simeq \Lambda_n(L(G))$ and the isomorphism is an isomorphism in ${}_{L(G)}\mathrm{TMod}^G$. Hence $\mathbf{H}^n(G, M) \neq 0$ and the claim follows by Corollary 4.5.2.

Corollary 5.2.6. *Let G be a p-torsion-free compact p-adic analytic group with $dim(G) = n$. Then the following are equivalent:*

(i) *$L(G)$ is a unimodular $\mathbb{Q}_p$-Lie algebra, i.e., $H^n(L(G), \mathbb{Q}_p) \neq 0$.*

(ii) *G is a virtually orientable Poincaré duality group.*

(iii) *G has strict cohomological p-dimension $n + 1$.*

Proof. (i)⇔(ii): For $P \in ob(\mathfrak{P}f^{t.f.})$, $Ad^{\times} : P \to Aut(\mathrm{Log}(P))$ is a powerful representation and $exp(tr(log(Ad^{\times}(g) - 1))) = det(Ad^{\times}(g))$. Now, by Corollary 5.2.5, P is orientable if and only if $\mathbb{Q}_p \otimes \mathrm{Log}(P)$ is unimodular.
(ii)⇔(iii): This follows from [23, Prop. I.19].

We call a p-adic analytic group *semi-simple* if $L(G)$ is a semi-simple $\mathbb{Q}_p$-Lie algebra. For these groups Lazard's theorem and a well known vanishing result for cohomology of semi-simple Lie algebras in characteristic 0 imply the following result:

Corollary 5.2.7. *Let G be a semi-simple p-adic analytic group.*

(a) *Let $W \in ob(\mathfrak{fd}_p(G))$ be such that for every open subgroup $H \leqslant_o G$, $res^G_H(W)$ has no trivial composition factor (i.e., $\mathbb{H}\mathrm{om}_{\mathbb{Z}_p[\![H]\!]}(\mathbb{Q}_p, res^G_H(W)) = 0$). Then $\mathbb{H}^{\bullet}(G, W) = 0$.*

(b) *Let $M \in ob(\mathfrak{C}_p(G))$ be a finitely generated $\mathbb{Z}_p$-module such that $W = \mathbb{Q}_p \otimes_{\mathbb{Z}_p} M$ satisfies the hypothesis of* (a)*. Then $\mathbf{H}^k(G, M)$ is finite for all $k \in \mathbb{N}_0$.*

Proof. (a) The hypothesis ensures that $Hom_{L(G)}(\mathbb{Q}_p, W) = 0$. Since $L(G)$ is semisimple, $W = W_1 \oplus \cdots W_r$ is is a direct sum of simple $L(G)$-modules W_i by Weyl's theorem. Hence the claim follows from Theorem 5.2.4 and a well known vanishing result for the cohomology of semi-simple Lie algebras in characteristic 0 ([15, Prop. XVIII.3.1]).
(b) Let $T(M) \leqslant M$ denote the p-torsion part of M. Then $T(M) \in ob(\mathfrak{F}_p(G))$ and $\mathbb{Q}_p \otimes_{\mathbb{Z}_p} M/T(M) \simeq W$. The claim follows from the long exact sequence, part (a), (3.8.6) and (4.2.1).

References

[1] D. J. Benson, *Representations and Cohomology*, I. *Basic Representation Theory of Finite Groups and Associative Algebras*, Cambridge Studies in Advanced Mathematics, 30, Cambridge University Press, 1991.

[2] D. J. Benson, *Representations and Cohomology*, II. *Cohomology of Groups and Modules*, Cambridge Studies in Advanced Mathematics, 31, Cambridge University Press, 1991.

[3] R. Bieri, B. Eckmann, Groups with homological duality generalizing Poincaré duality, *Invent. Math.*, **20** (1973), 103–124.

[4] N. Bourbaki, *General Topology*, Chap. 1–4, Elements of Mathematics, Springer-Verlag, Berlin, Heidelberg, New York, 1989.

[5] K. S. Brown, *Cohomology of Groups*, Graduate Texts in Mathematics, 87, Springer-Verlag, New York, Heidelberg, Berlin, 1982.

[6] A. Brumer, Pseudocompact algebras, profinite groups and class formations, *J. Alg.*, **4** (1966), 442–470.

[7] D. A. Buchsbaum, Exact categories and duality, *Trans. Amer. Math. Soc.*, **80** (1955), 1–34.

[8] H. Cartan, S. Eilenberg, *Homological Algebra*, Princeton Math. Ser., 19, Princeton, 1956.

[9] P. Day, The May Spectral Sequence, Ph.D. Thesis, Wadham College, Oxford, 1994.

[10] J. D. Dixon, M. P. F. du Sautoy, A. Mann, D. Segal, *Analytic pro-p Groups*, 2nd edition, Cambridge University Press, Cambridge, 1999.

[11] A. Douady, Cohomologie des groupes compacts totalement discontinus, Sém. Bourbaki, 1959/60, exposé 189.

[12] L. Evens, *The Cohomology of Groups*, Oxford University Press, Oxford, 1991.

[13] P. Gabriel, Des catégories abéliennes, *Bull. Soc. Math. France*, **90** (1962), 323–448.

[14] S. I. Gelfand, Yu. I. Manin, *Methods of Homological Algebra*, Springer, Berlin, Heidelberg, New York, 1996.

[15] C. Kassel, *Quantum Groups*, Graduate texts in mathematics, 155, Springer-Verlag, New York, Berlin, Heidelberg, 1995.

[16] J. L. Koszul, Homologie et cohomologie des algèbres de Lie, *Bull. Soc. Math. France*, **78** (1950), 65–127.

[17] M. Lazard, Groupes analytiques p-adiques, *IHES Publ. Math.*, **26** (1965).

[18] S. Mac Lane, *Homology*, Grundlehren der mathematischen Wissenschaften, 114, Springer-Verlag, Berlin, Heidelberg, New York, 1975, 3rd corrected printing.

[19] J. P. May, The cohomology of restricted Lie algebras and Hopf algebras; applications to the Steenrod algebra, Ph.D. Thesis, Princeton University, 1964.

[20] J. P. May, The cohomology of restricted Lie algebras and Hopf algebras, *J. Alg.*, **3** (1966), 123–146.

[21] J. McCleary, *User's Guide to Spectral Sequences*, Mathematical Lecture Series, 12, Publish or Perish, Inc., Wilmington, Delaware (U.S.A.), 1985.

[22] D. Quillen, The spectrum of an equivariant cohomology ring: II, *Ann. of Math.*, **94** (1971), 573–602.

[23] J-P Serre, *Cohomologie Galoisienne*, Lecture Notes in Mathematics, 5, Springer-Verlag, Berlin, Heidelberg, New York, 1994, cinquième édition, révisée et complétée.

[24] J-P Serre, Sur la dimension cohomologique des groupes profinis, *Topology*, **3** (1965), 413–420.

[25] E. H. Spanier, *Algebraic Topology*, McGraw-Hill, New York, 1966.

[26] Th. Weigel, P-central groups and Poincaré duality, *Trans. Amer. Math. Soc.*, to appear.

[27] J. S. Wilson, *Profinite Groups*, Oxford University Press, Oxford, 1998.

Department of Mathematics
UMIST
PO Box 88
Manchester M60 1QD, UK
peter.symonds@umist.ac.uk

Mathematical Institute
24-29 St. Giles
Oxford OX1 3LB, UK
weigel@maths.ox.ac.uk

Appendix

Further Problems

Most of the following problems arise out of topics discussed in the preceding chapters. Many others appear throughout the book, especially in Chapter 1.

Characterisations of p-adic analytic groups

Many such are known; see [1], Interlude A. See also Problem 23 below. It is not known if the following necessary conditions are also sufficient.

1. If G is a pro-p group and all its closed subgroups are finitely generated, must G have finite rank?

2. If G is a pro-p group and all its closed non-open subgroups are of finite rank, must G have finite rank?

3. Does every finitely generated pro-p group which is not p-adic analytic have a section isomorphic to the (obvious) inverse limit of the groups $C_p \operatorname{wr} C_{p^n}$ (i.e., what we usually call $C_p \operatorname{wr} \mathbb{Z}_p$)?

(It does have sections isomorphic to each $C_p \operatorname{wr} C_{p^n}$ by a theorem of Shalev, see [1] Chapter 3, Exercise 4.)

Burnside-type problems

Zelmanov has proved that every periodic pro-p group is locally finite, and that finitely generated pro-p groups of finite exponent are finite (see Chapter 1, Section 2.1). The following well-known problem remains open:

4. Does every periodic pro-p group have finite exponent?

(The finitely generated ones do by Zelmanov's Theorem. On the other hand, a positive solution in general would imply that every periodic compact Hausdorff group has finite exponent, by results of Hewitt and Ross [7], Herfort [6] and Wilson [15].)

5. If a random element of a pro-p group G has finite order with probability 1, is G a torsion group?

Just infinite pro-p groups and related topics

6. Does there exist a finitely presented pro-p group which contains an isomorphic copy of every countably based pro-p group?

(Both the Nottingham group and branch pro-p groups have this property, see Chapters 10 and 4.)

7. Is the Nottingham group finitely presented? If so, what is the minimal number of relations?

(See Section 3 of Chapter 6. A positive answer to the first part would solve Problem 6.)

8. Does every just infinite pro-p branch group contain a non-abelian free pro-p subgroup?

(It is proved in Chapter 5 that it contains a non-abelian free abstract group. The analogous property of groups satisfying the Golod–Shafarevich condition is established in Chapter 7.)

9. Let G be a just-infinite pro-p group. Define $\dim G$ to be the supremum over all embeddings in W_p, the automorphism group of the p-ary rooted tree, of its Hausdorff dimension in W_p. Show that G is a branch group if and only if $\dim G > 0$.

(This is suggested by the situation discussed in Chapter 11, Section 3. In general, naturally occurring Galois representations into $GL_2(\mathbb{Z}_p)$ have open image. Since W_p is not finitely generated, we conjecture that the new kind of Galois representations have large image in the sense of Hausdorff dimension. Combining this with the evidence that its just-infinite quotients are branch groups leads one to this purely group-theoretic conjecture.)

10. Find the cohomological dimension of the subgroup

$$T(r) = \{t + \sum_{i \geq 1} a_i t^{1+ip^r} : a_i \in \mathbb{F}_p\}$$

of the Nottingham group.

(This group is discussed in Chapter 10 Section 9. It is the Galois group of a certain arithmetically profinite extension of a p-adic field, used to solve a problem of Coates and Greenberg.)

11. Is it true that every hereditarily just infinite pro-p group of finite width is either analytic over a profinite ring or else commensurable with one of the groups $T(r)$?

(This seems unlikely but no other examples are known; see Chapter 2 Section 7.)

12. Find necessary and sufficient group theoretic conditions for a pro-p group to be analytic over $\mathbb{F}_p[[t]]$, or over some pro-p integral domain, in the sense of [1], Chapter 13.

(For a non-trivial necessary condition see Chapter 1 Section 5.1 of the present volume. The problem is the $\mathbb{F}_p[[t]]$ analogue of Hilbert's Fifth Problem, solved in the p-adic case by Lazard; see [1] Interlude A.)

13. Is it true that every pro-p group which is analytic over $\mathbb{F}_p[[t]]$, or over some pro-p integral domain, is linear over the associated pro-p domain?

(This is true for $\mathbb{Z}_p$; see Section 7.3 of [1]. It forms the basis of Lubotzky's Linearity Criterion [1] Interlude B.)

Presentations

14. Let G be a pro-p group of cohomological dimension > 2 for which $H^2(G, \mathbb{Z}/p\mathbb{Z})$ is a vector space of dimension 1 (the latter condition is equivalent to saying that G admits a presentation with a single defining relator). Does G admit a presentation with a single defining relator of the form u^p?

(The problem was stated by Serre [14] and corrected by Gildenhuys [4].)

The next two questions arise in the context of Chapter 11.

15. Let S be a finite set of odd primes and P_S the Galois group over $\mathbb{Q}$ of the maximal pro-2 extension of $\mathbb{Q}$ unramified outside S. It is known that $d(P_S) = r(P_S) = |S|$. Thus if $|S| = 1$, P_S is finite and if $|S| > 3$, P_S is infinite. Is it true that if $|S| = 2$, P_S is finite, and that if $|S| = 3$, P_S is infinite?

16. The groups P_S above have the property that every open subgroup has finite abelianisation. Give an example of an infinite pro-2 group G with $4d(G) = r(G)$, with every open subgroup having finite abelianisation, and with no infinite quotients that are linear over $\mathbb{Z}_p$ or $\mathbb{F}_p[[T]]$.

17. Let F be a free pro-p group of finite rank n, and let φ be an automorphism of F of order p^∞ (i.e., φ generates a subgroup isomorphic to $\mathbb{Z}_p$ in $\mathrm{Aut}(F)$). Is the subgroup of fixed points $\mathrm{Fix}_F(\varphi)$ finitely generated? Is the rank of $\mathrm{Fix}_F(\varphi)$ at most n?

(See Chapter 3 Section 9 for related results.)

18. *Profinite Poincaré Conjecture.* Let $\widehat{S}_p$ be the pro-p completion of the fundamental group S of a closed orientable surface of genus $g > 1$. Is it true that, modulo the natural action of $\mathrm{Aut}\,\widehat{S}_p$, there is only one epimorphism $\widehat{S}_p \to \widehat{F}_p \times \widehat{F}_p$, where $\widehat{F}_p$ is the free pro-p group of rank g?

(It is easy to show that up to equivalence there is only one epimorphism from the free pro-p group of rank $2g$ to $\widehat{F}_p \times \widehat{F}_p$. In the abstract situation both questions are very difficult and equivalent or related to the Poincaré Conjecture and the Andrews-Curtis Conjecture; see [5].)

Probabilistic questions

See also Problem 5. For a natural number k, $Q(G, k)$ denotes the probability that a random k-tuple generates (topologically) an open subgroup in the profinite group G.

19. Characterise the pro-p groups G for which $Q(G, k) = 1$ for some k.

(It is shown in Chapter 8 Section 4 that non-abelian free pro-p groups do not have this property while p-adic analytic pro-p groups do, as do certain simple algebraic groups over $\mathbb{F}_p[[t]]$, see Chapter 1 Section 5.3.)

20. Which pro-p groups have the property that for some k, the closed normal subgroup generated by a random k-tuple is open with probability 1?

(See Chapter 8, and [11].)

Subgroup growth and enumeration

21. Characterise the pro-p groups with polynomial normal subgroup growth.

(The metabelian case is done in [13]. The pro-p groups with polynomial subgroup growth are precisely the p-adic analytic ones, see [1] Chapter 3.)

22. Let G be a pro-p group with polynomial normal subgroup growth. Is the normal zeta function of G a rational function in p^{-s}?

(The answer is yes if G is p-adic analytic, in which case the zeta function counting all subgroups is also rational, see Chapter 9. The answer is also yes for the Nottingham group, see Chapter 6.)

23. What is the supremum of the numbers c with the following property: if G is a pro-p group such that $a_n(G) \leq n^{c \log_p n}$ for all large n, then G is p-adic analytic?

(The answer lies between 1/8 and 1/2, see Chapter 8 Section 4.)

24. Determine the eventual period of the 'pruned trees' of finite p-groups of coclass r, described in Section 7 of Chapter 9.

(Theorem 15 of Chapter 9 asserts the periodicity but doesn't specify the period. Conjecture P of Newman and O'Brien [12] states that each family of 2-groups of coclass r is ultimately periodic with period dividing 2^{r-1}.)

25. Prove that for each triple (n, c, d) the number $f(n, p, c, d)$ of isomorphism types of groups of order p^n and class at most c on d generators is PORC (polynomial on residue classes).

(This would imply Higman's PORC conjecture [8]. The behaviour of this function is discussed in Chapter 9 Section 6.)

26. Let $f(n, p, d)$ be the number of groups of order p^n having d generators. Does there exist a number γ such that $f(n, p, d) = p^{\gamma n^2 + o(n^2)}$? If so, what is the value of γ?

(This would be a sharpening of Theorem 5 in Chapter 8, Section 2. See also [10].)

27. Prove that the local zeta functions $\zeta_{F,p}$ and $\zeta^{\triangleleft}_{F,p}$ are uniform in p when F is a finitely generated free-nilpotent group.

(See Chapter 9. This question is related to Question 25, see Chapter 9 Section 6.)

28. A necessary condition for a sequence of numbers a_{p^n} to satisfy a linear recurrence relation (or equivalently for the associated Poincaré series $\sum_n a_{p^n} T^n$ to be a rational function in T) is that a_{p^n} grow polynomially in p^n. The following are sequences that have this latter property:

(i) the number of d-generated subgroups of index p^n in a finitely generated pro-p group (see Theorem 7 of Chapter 8);

(ii) the number of normal subgroups of index p^n in a finitely generated pro-p group which are the normal closure of r elements (Proposition 8 of Chapter 8);

(iii) the number of groups of order p^n defined by r relations (Theorem 9 of Chapter 8);

(iv) the number of $\mathbb{F}_p[[t]]$-subalgebras (or ideals) of index p^n in a finite dimensional $\mathbb{F}_p[[t]]$-Lie algebra.

Determine whether each sequence satisfies a linear recurrence relation.

(Note that (i) and (ii) satisfy a linear recurrence when the pro-p group is p-adic analytic. The first case appears explicitly in [2] and the second follows by adapting the proof there; the method is explained in Chapter 9 Section 2. If the Lie algebra in (iv) is $L \otimes \mathbb{F}_p[[t]]$ where L is a Lie algebra defined over $\mathbb{Z}$ then the numbers in (iv) satisfy a recurrence relation for almost all primes p [3]. In (iii) the number of generators is bounded, by the Golod–Shafarevich inequality ([1], Interlude D).)

29. Calculate further explicit examples of zeta functions of groups and Lie algebras. In particular, calculate $\zeta_{L\otimes\mathbb{Z}_p}(s)$ where $L = L(X_l)$ denotes the $\mathbb{Z}$-span of a Chevalley basis corresponding to the simple Lie algebra over $\mathbb{C}$ with Dynkin diagram of type X_l.

(These are rational functions of p^{-s} but very little is known about their general form. The only such example so far calculated is $X_l = A_1$ i.e. $\mathfrak{sl}_2(\mathbb{Z})$, see [9] and [3].)

30. Calculate examples of the zeta functions $\zeta_{c,d,p}(s)$ counting p-groups of class c on d generators, generalizing the classical calculation

$$\zeta_{1,d,p}(s) = \zeta_p(s)\zeta_p(2s)\cdots\zeta_p(ds).$$

References

[1] J. D. Dixon, M. P. F. du Sautoy, A. Mann and D. Segal, *Analytic pro-p Groups, 2nd edition*, Cambridge Studies in Advanced Math. 61, Cambridge Univ. Press, Cambridge, 1999.

[2] M. P. F. du Sautoy, Finitely generated groups, p-adic analytic groups and Poincaré series, *Annals of Math*, 137 (1993), 639–670.

[3] M. P. F. du Sautoy, The zeta function of $\mathfrak{sl}_2(\mathbb{Z})$, *Forum Math.* 12 (2000), 197–221.

[4] D. Gildenhuys, On pro-p groups with a single defining relator, *Invent. Math.* 5 (1968), 357–366.

[5] R. I. Grigorchuk and P. F. Kurchanov, Some questions of group theory related to geometry, in *Algebra, VII: Encyclopaedia Math. Sci.* 58, pp. 167–240, Springer, Berlin, 1993.

[6] W.N. Herfort, Compact torsion groups of finite exponent, *Arch. Math.* 33 (1979/80), 404-410.

[7] E. Hewitt and K.A. Ross, *Abstract Harmonic Analysis II*, Springer-Verlag, 1970.

[8] G. Higman, Enumerating p-groups, II, *Proc. London Math. Soc.* 10 (1960), 566–582.

[9] I. Ilani, Zeta functions related to the group $SL_2(\mathbb{Z}_p)$, *Israel J. Math.*, (to appear).

[10] A. Mann, Some questions about p-groups, *J. Austral. Math. Soc.* 67 (1999), 356–379.

[11] A. Mann, Enumerating finite groups and their relations, II, *(submitted)*.

[12] M. F. Newman and E. O'Brien, Classifying 2-groups by coclass, *Trans. Amer. Math. Soc.*, 351 (1999), 131–169.

[13] D. Segal, On the growth of ideals and submodules, *J. London Math. Soc.* (2) 56 (1997), 245–263.

[14] J-P. Serre, Structure de certains pro-p groupes, *Séminaire Bourbaki* 252 (1962/63).

[15] J.S. Wilson, On the structure of compact torsion groups *Mh. Math.* 96 (1983), 57–66.

Index

Progress in Mathematics

Edited by:

Progress in Mathematics is a series of books intended for professional mathematicians and scientists, encompassing all areas of pure mathematics. This distinguished series, which began in 1979, includes authored monographs and edited collections of papers on important research developments as well as expositions of particular subject areas.

We encourage preparation of manuscripts in some form of TEX for delivery in camera-ready copy which leads to rapid publication, or in electronic form for interfacing with laser printers or typesetters.

Proposals should be sent directly to the editors or to: Birkhäuser Boston, 675 Massachusetts Avenue, Cambridge, MA 02139, USA.

120 JOSEPH/ MEURAT/MIGNON/PRUM/ RENTSCHLER (eds). First European Congress of Mathematics, July, 1992, Vol. II
121 JOSEPH/ MEURAT/MIGNON/PRUM/ RENTSCHLER (eds). First European Congress of Mathematics, July, 1992, Vol. III (Round Tables)
122 GUILLEMIN. Moment Maps and Combinatorial Invariants of T^n-spaces
123 BRYLINSKI/BRYLINSKI/GUILLEMIN/ KAC (eds). Lie Theory and Geometry: In Honor of Bertram Kostant
124 AEBISCHER/BORER/KALIN/LEUENBERGER/ REIMANN (eds). Symplectic Geometry
125 LUBOTZKY. Discrete Groups, Expanding Graphs and Invariant Measures
126 RIESEL. Prime Numbers and Computer Methods for Factorization
127 HORMANDER. Notions of Convexity
128 SCHMIDT. Dynamical Systems of Algebraic Origin
129 DIJGRAAF/FABER/VAN DER GEER (eds). The Moduli Space of Curves
130 DUISTERMAAT. Fourier Integral Operators
131 GINDIKIN/LEPOWSKY/WILSON (eds). Functional Analysis on the Eve of the 21st Century: In Honor of the Eightieth Birthday of I. M. Gelfand, Vol. 1
132 GINDIKIN/LEPOWSKY/WILSON (eds). Functional Analysis on the Eve of the 21st Century: In Honor of the Eightieth Birthday of I. M. Gelfand, Vol. 2
133 HOFER/TAUBES/WEINSTEIN/ZEHNDER (eds). The Floer Memorial Volume
134 CAMPILLO LOPEZ/NARVAEZ MACARRO (eds). Algebraic Geometry and Singularities
135 AMREIN/BOUTET DE MONVEL/GEORGESCU. C_0-Groups, Commutator Methods and Spectral Theory of N-Body Hamiltonians
136 BROTO/CASACUBERTA/MISLIN (eds). Algebraic Topology: New Trends in Localization and Periodicity
137 VIGNERAS. Représentations l-modulaires d'un groupe réductif p-adique avec $l = p$
138 BERNDT/DIAMOND/HILDEBRAND (eds). Analytic Number Theory, Vol. 1 In Honor of Heini Halberstam
139 BERNDT/DIAMOND/HILDEBRAND (eds). Analytic Number Theory, Vol. 2 In Honor of Heini Halberstam
140 KNAPP. Lie Groups Beyond an Introduction
141 CABANES. Finite Reductive Groups: Related Structures and Representations
142 MONK. Cardinal Invariants on Boolean Algebras
143 GONZALEZ-VEGA/RECIO (eds). Algorithms in Algebraic Geometry and Applications
144 BELLAÏCHE/RISLER (eds). Sub-Riemannian Geometry
146 JARDINE. Generalized Etale Cohomology
147 DIBIASE. Fatou Theorems
148 HUANG. Two-Dimensional Conformal Geometry and Vertex Operator Algebras
149 SOURIAU. Structure of Dynamical Systems: A Symplectic View of Physics
150 SHIOTA. Geometry of Subanalytic and Semialgebraic Sets
151 HUMMEL. Gromov's Compactness Theorem for Pseudo-holomorphic Curves
152 GROMOV. Metric Structures for Riemannian and Non-Riemannian Spaces. Sean Bates (translator)

153 BUESCU. Exotic Attractors: From Liapunov Stability to Riddled Basins
154 BÖTTCHER/KARLOVICH. Carleson Curves, Muckenhoupt Weights, and Toeplitz Operators
155 DRAGOMIR/ORNEA. Locally Conformal Kähler Geometry
156 GUIVARC'H/JI/TAYLOR. Compactifications of Symmetric Spaces
157 MURTY/MURTY. Non-vanishing of L-functions and Applications
158 TIRAO/VOGAN/WOLF (eds). Geometry and Representation Theory of Real and p-adic Groups
159 THANGAVELU. Harmonic Analysis on the Heisenberg Group
160 KASHIWARA/MATSUO/SAITO/SATAKE (eds). Topological Field Theory, Primitive Forms and Related Topics
161 SAGAN/STANLEY (eds). Mathematical Essays in Honor of Gian-Carlo Rota
162 ARNOLD/GREUEL/STEENBRINK. Singularities: The Brieskorn Anniversary Volume
163 BERNDT/SCHMIDT. Elements of the Representaiton Theory of the Jacobi Group
164 ROUSSARIE. Bifurcations of Planar Vector Fields and Hilbert's Sixteenth Problem
165 MIGLIORE. Introduction to Liaison Theory and Deficiency Modules
166 ELIAS/GIRAL/MIRO-ROIG/ZARZUELA (eds). Six Lectures on Commutative Algebra
167 FACCHINI. Module Theory
168 BALOG/KATONA/SZA'SZ/RECSKI (eds). European Congress of Mathematics, Budapest, July 22-26, 1996. Vol. I
169 BALOG/KATONA/SZA'SZ/RECSKI (eds). European Congress of Mathematics, Budapest, July 22-26, 1996. Vol. II
168/169 Sets Vols I, II
170 PATERSON. Groupoids, Inverse Semigroups, and their Operator Algebras
171 REZNIKOV/SCHAPPACHER (eds). Regulators in Analysis, Geometry and Number Theory
172 BRYLINSKI/BRYLINSKI/NISTOR/TSYGAN/XU (eds). Advances in Geometry
173 DRÄXLER/MICHLER/RINGEL (eds). Computational Methods for Representation of Groups and Algebras: Euroconference in Essen
174 GOERSS/JARDINE. Simplicial Homotopy Theory
175 BAÑUELOS/MOORE. Probabilistic Behavior of Harmonic Functions
176 BASS/LUBOTZKY. Tree Lattices
177 BIRKENHAKE/LANGE. Complex Tori
178 PUIG. On the Local Structure of Morita and Rickard Equivalence Between Brauer Blocks
179 MORALES RUIZ. Differential Galois Theory and Non-Integrability of Hamiltonian Systems
180 PATERNAIN. Geodesic Flows
181 HAUSER/LIPMAN/OORT/QUIRÓS (eds). Resolution of Singularities: In Tribute to Oscar Zariski
182 BILLEY/LAKSHMIBAI. Singular Loci of Schubert Varieties
183 KAPOVICH. Hyperbolic Manifolds and Discrete Groups
184 du SAUTOY/SEGAL/SHALEV (eds). New Horizons in pro-p Groups
185 FARAUT/KANEYUKI/KORÁNYI/LU/ROOS. Analysis and Geometry on Complex Homogeneous Domains
186 KLEINERT. Units in Skew Fields
187 BOST/LOESER/RAYNAUD. Courbes Semi-stables et Groupe Fondamental en Géométrie Algébrique
188 DOLBEAULT/IORDAN/HENKIN/SKODA/TRÉPREAU (eds). Complex Analysis and Geometry: International Conference in Honor of Pierre Lelong
189 ANDRÉ/BALDASSARI. DeRham Cohomology of Differential Modules on Algebraic Varieties
190 van den ESSEN. Polynomial Automorphisms and the Jacobian Conjecture.
191 KASHIWARA/MIWA (eds). Physical Combinatorics